Expression Profiling of Human Tumors

EXPRESSION PROFILING OF HUMAN TUMORS

Diagnostic and Research Applications

Edited by

Marc Ladanyi, MD
Memorial Sloan-Kettering Cancer Center, New York, NY

William L. Gerald, MD, PhD
Memorial Sloan-Kettering Cancer Center, New York, NY

Humana Press
Totowa, New Jersey

© 2003 Humana Press Inc.
999 Riverview Drive, Suite 208
Totowa, New Jersey 07512

www.humanapress.com

All rights reserved. No part of this book may be reproduced, stored in a retrieval system, or transmitted in any form or by any means, electronic, mechanical, photocopying, microfilming, recording, or otherwise without written permission from the Publisher.

The content and opinions expressed in this book are the sole work of the authors and editors, who have warranted due diligence in the creation and issuance of their work. The publisher, editors, and authors are not responsible for errors or omissions or for any consequences arising from the information or opinions presented in this book and make no warranty, express or implied, with respect to its contents.

This publication is printed on acid-free paper. ∞

ANSI Z39.48-1984 (American Standards Institute) Permanence of Paper for Printed Library Materials.

Production Editor: Tracy Catanese.
Cover based on a design by Marc Ladanyi. Top panel provided by William Gerald, middle panel by Paul Meltzer, bottom panel by Marc Ladanyi.

For additional copies, pricing for bulk purchases, and/or information about other Humana titles, contact Humana at the above address or at any of the following numbers: Tel: 973-256-1699; Fax: 973-256-8341; E-mail: humana@humanapr.com, or visit our Website: http://www.humanapress.com

Photocopy Authorization Policy:
Authorization to photocopy items for internal or personal use, or the internal or personal use of specific clients, is granted by Humana Press Inc., provided that the base fee of US $20.00 per copy is paid directly to the Copyright Clearance Center at 222 Rosewood Drive, Danvers, MA 01923. For those organizations that have been granted a photocopy license from the CCC, a separate system of payment has been arranged and is acceptable to Humana Press Inc. The fee code for users of the Transactional Reporting Service is: [1-58829-122-7/03 $20.00].

Printed in the United States of America. 10 9 8 7 6 5 4 3 2 1

Library of Congress Cataloging-in-Publication Data

Expression profiling of human tumors : diagnostic and research applications / edited by Marc Ladanyi and William L. Gerald.
    p. cm.
  Includes bibliographical references and index.
  ISBN 1-58829-122-7 (alk. paper); 1-59259-386-0 (e-book)
  1. Oncogenes. 2. Gene expression. 3. Cancer--Molecular aspects. I. Ladanyi, Marc. II. Gerald, William L.

RC268.42 .E976 2003
616.99'4042--dc21

2002032924

Preface

Advances in our knowledge of the molecular basis of cancer are at the heart of the present revolution in clinical oncology. The identification of tumor-specific molecular alterations has led to new means of diagnosis and classification, and the characterization of critical pathways regulating tumor growth is providing the potential for less toxic, more effective targeted therapy. Nonetheless, these advances had previously occurred at an agonizingly slow pace, i.e., one gene at a time. That investigative pace has now been dramatically altered by the completion of a draft of the entire human genome and the development of miniaturized high-throughput technology for genetic analysis. These extraordinary accomplishments now permit not only the monitoring of every gene sequence in a single experiment, but also a comprehensive analysis of the complex coordinated programs and pathways that contribute to the clinical phenotype of cancers. This rapid and comprehensive approach to the investigation of tumor biology has the potential to dramatically shape the future of clinical oncology.

Expression Profiling of Human Tumors: Diagnostic and Research Applications is intended to provide an introduction and overview to comprehensive gene expression profiling of human tumors, one of the most promising new high-throughput investigative approaches in molecular biology. The intent was to provide not only a primer for the technology and analytical methods, but also an early assessment of the state-of-the-art with respect to both successes and pitfalls. These successes are significant and include methods of more precise diagnosis, and identification of prognostic markers, therapeutic targets, and gene expression patterns that predict therapeutic response. Nonetheless, there are significant challenges to further success, such as procurement and processing of appropriate samples, improvement and validation of technical approaches, and refinement of analytical methods for the resulting complex datasets. We have attempted to provide a balance between the basic science aspects of this work and its application to the clinical setting, but we have focused on the analysis of human tissue samples as providing the most direct means of translating findings to clinical practice. There are many complex issues that need to be considered as this type of work goes forward, and we hope this text will serve as a starting point for future discoveries.

The emphasis here on gene expression profiling is not intended to suggest that this should be considered the ultimate view of the molecular biology of the cancer cell. On the contrary, we all look forward to the day when analysis at the protein level is as comprehensive and provides as much detail as the present attempts of global gene transcript measurements. Obviously, the closer we come to assessment of the actual function of each molecule, the more accurate our abilities to correlate those with the clinical phenotype. Proteomics holds the promise to better achieve that goal, but is still in its infancy, with even greater hurdles to overcome than we presently face with sequence-based expression analysis. We leave that topic for future publications.

We would like to express our deep appreciation to the many authors who have provided overviews of work in their fields. These individuals have contributed their time and effort to provide highly useful information for others (sometimes while being badgered by the Editors!). We would also like to thank Ms. Fabienne Volel and Ms. Shirley Tung for excellent assistance. Finally we thank our families for their patience and support.

Expression Profiling of Human Tumors: Diagnostic and Research Applications clearly depicts the rapid advances that are occurring in clinically important areas and that will no doubt increasingly impact clinical care. We sincerely hope that our book provides information useful to all basic or clinical investigators concerned with the molecular basis of cancer and the improvement of cancer care.

Marc Ladanyi, MD
William L. Gerald, MD, PhD

Contents

Preface .... *v*
Contributors .... *ix*

PART I. INTRODUCTION

1 Introduction: *Present and Potential Impact of Expression Profiling Studies of Human Tumors*
   Marc Ladanyi and William L. Gerald 3

PART II. TECHNICAL ASPECTS

2 cDNA Microarrays
   Paul S. Meltzer 11

3 Oligonucleotide Microarrays
   Marina Chicurel and Dennise Dalma-Weiszhausz 23

4 Serial Analysis of Gene Expression (SAGE) in Cancer Research
   C. Marcelo Aldaz 47

5 Tissue Arrays
   Cyrus V. Hedvat 61

6 Microarray Data Analysis:
   Cancer Genomics and Molecular Pattern Recognition
   Pablo Tamayo and Sridhar Ramaswamy 73

7 The Role of Tumor Banking and Related Informatics
   Stephen J. Qualman, Jay Bowen, Sandra Brewer-Swartz, and Mary France 103

PART III. APPLICATIONS

8 Characterization of Gene Expression Patterns for Classification of Breast Carcinomas
   Irene L. Andrulis, Nalan Gokgoz, and Shelley B. Bull 121

9   Microarray Analysis of Colorectal Cancer
    Daniel A. Notterman, Carrie J. Shawber, and Wei Liu *147*

10 Gene Expression Analysis of Prostate Carcinoma
    William L. Gerald *173*

11 Classification of Human Lung Carcinomas
    by mRNA Expression Profiling
    Arindam Bhattacharjee and Matthew Meyerson *199*

12 Molecular Profiling of Bladder Cancer Using High-Throughput
    DNA Microarrays
    Marta Sánchez-Carbayo and Carlos Cordon-Cardo *219*

13 Gene Expression Profiling of Renal Cell Carcinoma
    and its Clinical Implications
    Masayuki Takahashi and Bin Tean Teh *235*

14 Expression Profiling of Pancreatic Ductal Adenocarcinoma
    Christine A. Iacobuzio-Donahue and Ralph H. Hruban *257*

15 Gene Expression in Ovarian Carcinoma
    Garret M. Hampton *277*

16 Classification of Pediatric Tumors Using DNA Microarrays
    Javed Khan and Marc Ladanyi *295*

17 Transcriptomes of Soft Tissue Tumors:
    Pathologic and Clinical Implications
    Sabine C. Linn, Rob B. West, and Matt van de Rijn *305*

18 Gene Expression Profiling in Lymphoid Malignancies
    Wing C. Chan and Louis M. Staudt *329*

19 Gene Expression Profiling of Brain Tumors
    Meena K. Tanwar and Eric C. Holland *345*

20 Expression Profiling of Bone Tumors
    Deborah Schofield, Daniel Wai, and Timothy J. Triche *359*

Index *393*

Contributors

C. MARCELO ALDAZ, MD, PhD • *Science Park Research Division, Department of Carcinogenesis, M. D. Anderson Cancer Center, The University of Texas, Smithville, TX*
IRENE L. ANDRULIS, PhD • *Department of Molecular and Medical Genetics, Fred A. Litwin Centre for Cancer Genetics, Samuel Lunenfeld Research Institute of Mount Sinai Hospital, Toronto, ON, Canada*
ARINDAM BHATTACHARJEE, PhD • *Agilent Technologies, Andover, MA*
JAY BOWEN, MS • *Department of Laboratory Medicine, Columbus Children's Hospital, Columbus, OH*
SANDRA BREWER-SWARTZ, MEn • *Department of Laboratory Medicine, Columbus Children's Hospital, Columbus, OH*
SHELLEY B. BULL, PhD • *Samuel Lunenfeld Research Institute of Mount Sinai Hospital, and Department of Public Health Sciences, University of Toronto, Toronto, ON, Canada*
WING C. CHAN, MD • *Department of Pathology and Microbiology, University of Nebraska Medial Center, Nebraska Medical Center, Omaha, NE*
MARINA CHICUREL, PhD • *Affymetrix, Inc., Santa Clara CA*
CARLOS CORDON-CARDO, MD, PhD • *Division of Molecular Pathology, Memorial Sloan-Kettering Cancer Center, New York, NY*
DENNISE DALMA-WEISZHAUSZ, PhD • *Affymetrix, Inc., Santa Clara, CA*
MARY FRANCE, BS • *Department of Laboratory Medicine, Columbus Children's Hospital, Columbus, OH*
WILLIAM L. GERALD, MD, PhD • *Department of Pathology, Memorial Sloan-Kettering Cancer Center, New York, NY*
NALAN GOKGOZ, PhD • *Fred A. Litwin Centre for Cancer Genetics, Mount Sinai Hospital, Toronto, ON, Canada*
GARRET M. HAMPTON, PhD • *Genomics Institute of Novartis Research Foundation, San Diego, CA*
CYRUS V. HEDVAT, MD, PhD • *Department of Pathology, Memorial Sloan-Kettering Cancer Center, New York, NY*
ERIC C. HOLLAND, MD, PhD • *Departments of Neurosurgery, Neurology, and Cell Biology, Memorial Sloan-Kettering Cancer Center, New York, NY*
RALPH H. HRUBAN, MD • *Department of Pathology, The Johns Hopkins Hospital, Baltimore, MD*
CHRISTINE A. IACOBUZIO-DONAHUE, MD, PhD • *Department of Pathology, The Johns Hopkins Hospital, Baltimore, MD*

JAVED KHAN, MD • *Oncogenomics Section, Pediatric Oncology Branch, National Cancer Institute, Bethesda, MD*
MARC LADANYI, MD • *Department of Pathology, Memorial Sloan-Kettering Cancer Center, New York, NY*
SABINE C. LINN, MD, PhD • *Department of Medical Oncology, Antoni van Leeuwenhoek Hospital, Amsterdam, The Netherlands*
WEI LIU, MD, PhD • *Departments of Pediatrics and Molecular Genetics, Robert Wood Johnson Medical School, New Brunswick, NJ*
PAUL S. MELTZER, MD, PhD • *Section of Molecular Genetics, Cancer Genetics Branch, Human Genome Research Institute, Bethesda, MD*
MATTHEW MEYERSON, MD, PhD • *Department of Adult Oncology, Dana-Farber Cancer Institute, Harvard Medical School, Boston, MA*
DANIEL A. NOTTERMAN, MD • *Departments of Pediatrics and Molecular Genetics, Microbiology, and Immunology, Robert Wood Johnson Medical School, New Brunswick, NJ*
STEPHEN J. QUALMAN, MD • *Department of Laboratory Medicine, Columbus Children's Hospital, Columbus, OH*
SRIDHAR RAMASWAMY, MD • *Department of Adult Oncology, Dana-Farber Cancer Institute, Harvard Medical School, Boston, MA*
MARTA SÁNCHEZ-CARBAYO, PhD • *Division of Molecular Pathology, Memorial Sloan-Kettering Cancer Center, New York, NY*
DEBORAH SCHOFIELD, MD • *Clinical Associate Professor of Pathology and Pediatrics, Keck School of Medicine, University of Southern California, Los Angeles, CA*
CARRIE J. SHAWBER, PhD • *Weill Medical College, Cornell University, New York, NY*
LOUIS M. STAUDT, MD • *Metabolism Branch, Center for Cancer Research, National Cancer Institute, Bethesda, MD*
MASAYUKI TAKAHASHI, MD • *Laboratory of Cancer Genetics, Van Andel Research Institute, Grand Rapids, MI; Department of Urology, The University of Tokushima School of Medicine, Tokushima, Japan*
PABLO TAMAYO, PhD • *Cancer Genomics Group, Whitehead Institute, Massachusetts Institute of Technology Center for Genome Research Cambridge, MA*
MEENA K. TANWAR, PhD • *Departments of Neurosurgery, Neurology, and Cell Biology, Memorial Sloan-Kettering Cancer Center, New York, NY*
BIN TEAN TEH, MD, PhD • *Laboratory of Cancer Genetics, Van Andel Research Institute, Grand Rapids, MI*
TIMOTHY J. TRICHE, MD, PhD • *Professor of Pathology and Pediatrics, Keck School of Medicine, University of Southern California, Los Angeles, CA*
MATT VAN DE RIJN, MD, PhD • *Department of Pathology, Stanford University Medical Center, Stanford, CA*
DANIEL WAI, MS • *Keck School of Medicine, University of Southern California, Los Angeles, CA*
ROB B. WEST, MD, PhD • *Department of Pathology, Stanford University Medical Center, Stanford, CA*

Part I
Introduction

1
Introduction: Present and Potential Impact of Expression Profiling Studies of Human Tumors

Marc Ladanyi and William L. Gerald

Expression profiling refers to the process of measuring the expression of thousands of individual genes simultaneously in a given tissue sample. The resulting patterns of gene expression reflect the molecular basis of the sample phenotype and can be used for sample comparisons and classification. In broad terms, there are two main methods of expression profiling: hybridization-based and sequencing-based. In the former, RNA is extracted from the sample, converted to cDNA or cRNA, and hybridized to a DNA microarray. DNA microarrays are either nylon membranes, glass slides, or synthetic "chips," to which are attached nucleic acid probes as cDNA clones or cDNA clone-specific oligonucleotides corresponding to hundreds to tens of thousands of genes *(1,2)* (*see* Chapters 2 and 3). Sequencing-based approaches involve high-throughput sequencing of cDNA libraries generated from tumors (the Cancer Genome Anatomy Project [CGAP], Web site: http://cgap.nci.nih.gov/) or specialized techniques for efficient mass sequencing of short "tags" of each cDNA molecule derived from a tumor (serial analysis of gene expression [SAGE]) *(3,4)* (*see* Chapter 4). Because of its labor-intensive nature, sequencing-based expression profiling will remain purely an investigative technique, but microarray-based expression profiling is likely to be eventually applied in a clinical setting, at least in some form. The general concepts of tumor gene expression profiling and its potential impact on clinical oncology and pathology have been explored in detail in recent commentaries and reviews *(5–9)*. The aim of the present book is to provide an overview of the emerging "wisdom" in cDNA microarray technology, experimental design, and data analysis (*see* Chapter 6), as applied to the study of human tumors. We hope that it will serve both as an introduction to the field for beginners, as well as a "progress report" for those already active in the field. Ultimately, we hope this synthesis of work in the area will contribute to improved approaches for the comprehensive molecular analysis of human cancer.

As presented in detail in the latter part of this book (*see* Chapters 8–20), investigators in the field of expression profiling of human tumors initially performed studies in which tumors of different morphology or different primary sites were shown, perhaps not surprisingly, to have clearly distinguishable patterns of gene expression *(10–13)*. In the evolving jargon of this field, this type of diagnostic classification by analysis of expression profiles is sometimes called "class prediction." These proof-of-principle

From: *Expression Profiling of Human Tumors: Diagnostic and Research Applications*
Edited by: Marc Ladanyi and William L. Gerald © Humana Press Inc., Totowa, NJ

studies served to validate cDNA microarray technology, and researchers soon shifted their focus to identification of molecularly defined tumor entities (usually with associated clinical relevance) that were inapparent by conventional pathologic analysis ("class discovery"). Indeed, new unsuspected biological subsets have been thus detected among cutaneous melanomas *(14)*, breast carcinomas *(15)*, and pediatric acute lymphoblastic leukemias *(16)*. In other cases, there has been what could be termed "class rediscovery" or "class confirmation." Initial expression profiling experiments in breast cancers *(17)* led to the rediscovery of the previously described, but subsequently neglected, immunohistochemical distinction of basal vs luminal cell type breast carcinoma *(18)*. Studies of B lineage diffuse large cell lymphoma (DLCL) have confirmed the presence of a subset of germinal center-derived cases (associated with a favorable prognosis) as anticipated by older data on the prognostic significance in B cell DLCL of BCL2 expression or the follicular lymphoma-associated *BCL2* rearrangement due to the t(14;18) *(19–22)*. Likewise, the observation that sarcomas or acute leukemias with different specific translocations have specific expression profiles *(11,16,23,24)* confirms the validity of previously established morphologic–cytogenetic entities.

Pathologists have for many years been performing a simple form of expression profiling, namely immunohistochemistry. In problematic cases, multiple immunostains are usually performed as a panel, which provides information useful to refine a differential diagnosis. These panels are necessary, since there is no single native protein whose expression is 100% specific and 100% sensitive for a given diagnosis. Presented with the results of these immunostains, the pathologist routinely performs a mental computation of combinatorial probabilities that takes into account the published specificity and sensitivity of each antibody for a given diagnosis, along with the risk of a technical false positive or false negative result in the case under study, while estimating the overall likelihood of the same combination of positive and negative results occurring in diagnoses lower on the differential diagnostic list. It is well accepted that moving from single immunostains to panels of immunostains has improved final diagnostic accuracy. In that sense, moving from panels of several immunostains to high-throughput expression profiling of hundreds or thousands of markers, with appropriate software to interpret the data and provide a diagnostic prediction with a probability score, would seem to be a logical and inevitable next step in this progression. Indeed, some argue that full-scale clinical implementation of expression profiling is feasible and desirable in the near future *(9,12,25)*. Others consider a more incremental scenario more realistic, with the demand for microarray-derived markers and subclassifications satisfied in the near term by the transfer of key markers into standard clinical immunohistochemistry laboratories *(6,8)*.

There is a considerable gulf between the practical world of hospital-based molecular diagnostic laboratories and some of the predictions prompted by high-throughput genomics work. Hopeful statements, such as one made in 1999 that "doctors will be offering gene expression profiles to some patients in the next 3 yr" *(25)*, failed to consider the many issues in moving complex assays from research laboratories to clinical laboratories. Because of regulatory, billing, quality control, and test validation concerns, combined with limited resources, academic molecular diagnostic laboratories are extremely selective in their test menus. Furthermore, it is not yet clear that, considering the present cost of microarrays, large-scale expression profiling for "class prediction" is more cost-effective than established diagnostic approaches, i.e., histopathology, supplemented in selected cases by immunohistochemistry, cytogenetics, or specific molecular assays, although many believe that this may only be a matter of time.

Obviously, aside from diagnostic classification, expression profiling has the potential to identify biological and clinical subsets that cannot be reliably recognized by pathologic analysis. Indeed, its greatest promise is the potential to provide prognostically useful information. In studies addressing prognostic classification, there has been some variability in the results obtained so far by different groups addressing similar questions. This variability is, perhaps, more notable in prognostic studies than in class prediction studies, and it serves as a reminder that the field of expression profiling of human tumors is so young that for most tumor types, the issues of interlaboratory and interplatform reproducibility are only beginning to be addressed. For example, two large breast cancer expression profiling studies differed considerably in the results of unsupervised clustering: one identified three major subsets, i.e., estrogen receptor-positive luminal cell type, basal cell type, and *ERBB2*-amplified type *(17)*, while the other detected only two major subgroups according to estrogen receptor status and lymphocytic infiltration *(26)*. Moreover, in the latter study, well-established clinical markers, such as *ERBB2* and estrogen receptor were not found within a list of 70 genes linked to breast cancer prognosis by the microarray analysis. Expression profiling studies of B cell DLCL provide another example of variable results. The first study used unsupervised clustering to identify genes whose expression dichotomized 42 B lineage DLCL into cases derived from germinal center B cells or activated B cells, respectively, and demonstrated the prognostic significance of this distinction *(27)*. When a second group used supervised analysis to identify prognostic subsets among 58 B lineage DLCL, these two lymphoma subsets were not reproduced, and a completely different set of prognostic markers were identified *(28)*. Moreover, a reanalysis of their data using the differentially expressed genes identified in the first study failed to confirm the prognostic significance of the germinal center B cell vs activated B cell distinction. Technically, these differences may be due to different microarray platforms, different statistical analytic methods, or the variability introduced by relatively small patient numbers. Biologically, the differences may also reflect the likely "multidimensional" nature of prognostically significant genes or pathways, i.e., the presence of a matrix of several independent and additive prognostic categories for given cancer types. Larger, more systematic, and more standardized studies will be needed to sort out these variability issues. Such studies will benefit from careful tumor ascertainment and banking, as outlined in Chapter 7.

In the short term, however, the immediate clinical impact of tumor expression profiling is in the identification of new diagnostic and prognostic markers, which can be studied individually by more conventional and accessible techniques. Expression profiling studies are performed using cDNA microarrays representing tens of thousands of genes, and the clustering algorithms typically use the several thousand genes most differentially expressed among the tumors in question. However, the end result is often a very reduced subset of differentially expressed genes (e.g., 1–100), which can essentially be used to replicate the same clustering. In many cases, the differences in the expression of these genes can be detected robustly at the protein level by immunohistochemistry. The availability of the tissue microarray approach accelerates the validation of these new immunohistochemical assays, as described in Chapter 5. We are already seeing the first wave of such markers moving from expression profiling studies into clinical application. A second larger wave of markers will result from widespread efforts in academic and commercial laboratories to generate new antibodies for the products of the many differentially expressed genes without currently available antibody reagents. Thus, regardless of the issues that might delay the clinical implementation of microarray-based expression profiling of human tumors, these studies are already having an immediate clinical impact through the high-throughput identification of new markers of diagnosis and prognosis and the resulting influx of new immunohistochemical assays into clinical laboratories.

REFERENCES

1. Schena, M., Shalon, D., Davis, R. W., and Brown, P. O. (1995) Quantitative monitoring of gene expression patterns with a complementary DNA microarray. *Science* **270,** 467–470.
2. Lockhart, D. J., Dong, H., Byrne, M. C., et al. (1996) Expression monitoring by hybridization to high-density oligonucleotide arrays. *Nat. Biotechnol.* **14,** 1675–1680.
3. Velculescu, V. E., Zhang, L., Vogelstein, B., and Kinzler, K. W. (1995) Serial analysis of gene expression. *Science* **270,** 484–487.
4. Strausberg, R. L. (2001) The Cancer Genome Anatomy Project: new resources for reading the molecular signatures of cancer. *J. Pathol.* **195,** 31–40.
5. Ladanyi, M., Chan, W. C., Triche, T. J., and Gerald, W. L. (2001) Expression profiling of human tumors: the end of surgical pathology? *J. Mol. Diagn.* **3,** 92–97.
6. Lakhani, S. R. and Ashworth, A. (2001) Microarray and histopathological analysis of tumours: the future and the past? *Nat. Rev. Cancer* **1,** 151–157.
7. Bertucci, F., Houlgatte, R., Nguyen, C., Viens, P., Jordan, B. R., and Birnbaum, D. (2001) Gene expression profiling of cancer by use of DNA arrays: how far from the clinic? *Lancet Oncol.* **2,** 674–682.
8. Alizadeh, A. A., Ross, D. T., Perou, C. M., and van de Rijn, M. (2001) Towards a novel classification of human malignancies based on gene expression patterns. *J. Pathol.* **195,** 41–52.
9. Ramaswamy, S. and Golub, T. R. (2002) DNA microarrays in clinical oncology. *J. Clin. Oncol.* **20,** 1932–1941.
10. Golub, T. R., Slonim, D. K., Tamayo, P., et al. (1999) Molecular classification of cancer: class discovery and class prediction by gene expression monitoring. *Science* **286,** 531–537.
11. Khan, J., Wei, J., Ringnér, M., et al. (2001) Classification and diagnostic prediction of cancers using gene expression profiling and artificial neural networks. *Nat. Med.* **7,** 673–679.
12. Su, A. I., Welsh, J. B., Sapinoso, L. M., et al. (2001) Molecular classification of human carcinomas by use of gene expression signatures. *Cancer Res.* **61,** 7388–7393.
13. Ramaswamy, S., Tamayo, P., Rifkin, R., et al. (2001) Multiclass cancer diagnosis using tumor gene expression signatures. *Proc. Natl. Acad. Sci. USA* **98,** 15,149–15,154.
14. Bittner, M., Meltzer, P., Chen, Y., et al. (2000) Molecular classification of cutaneous malignant melanoma by gene expression profiling. *Nature* **406,** 536–540.
15. Sorlie, T., Perou, C. M., Tibshirani, R., et al. (2001) Gene expression patterns of breast carcinomas distinguish tumor subclasses with clinical implications. *Proc. Natl. Acad. Sci. USA* **98,** 10,869–10,874.
16. Yeoh, E. J., Ross, M. E., Shurtleff, S. A., et al. (2002) Classification, subtype discovery, and prediction of outcome in pediatric acute lymphoblastic leukemia by gene expression profiling. *Cancer Cell* **1,** 133–143.

17. Perou, C. M., Sorlie, T., Eisen, M. B., et al. (2000) Molecular portraits of human breast tumours. *Nature* **406,** 747–752.
18. Dairkee, S. H., Puett, L., and Hackett, A. J. (1988) Expression of basal and luminal epithelium-specific keratins in normal, benign, and malignant breast tissue. *J. Natl. Cancer Inst.* **80,** 691–695.
19. Yunis, J. J., Mayer, M. G., Arnesen, M. A., Aeppli, D. P., Oken, M. M., and Frizzera, G. (1989) Bcl-2 and other genomic alterations in the prognosis of large cell lymphoma. *N. Engl. J. Med.* **320,** 1047–1054.
20. Offit, K., Koduru, P. R. K., Hollis, R., et al. (1989) 18q21 rearrangement in diffuse large cell lymphoma: incidence and clinical significance. *Br. J. Haematol.* **72,** 178–186.
21. Gascoyne, R. D., Adomat, S. A., Krajewski, S., et al. (1997) Prognostic significance of Bcl-2 protein expression and Bcl-2 gene rearrangement in diffuse aggressive non-Hodgkin's lymphoma. *Blood* **90,** 244–251.
22. Huang, J. Z., Sanger, W. G., Greiner, T. C., et al. (2002) The t(14;18) defines a unique subset of diffuse large B-cell lymphoma with a germinal center B-cell gene expression profile. *Blood* **99,** 2285–2290.
23. Khan, J., Simon, R., Bittner, M., et al. (1998) Gene expression profiling of alveolar rhabdomyosarcoma with cDNA microarrays. *Cancer Res.* **58,** 5009–5013.
24. Armstrong, S. A., Staunton, J. E., Silverman, L. B., et al. (2002) MLL translocations specify a distinct gene expression profile that distinguishes a unique leukemia. *Nat. Genet.* **30,** 41–47.
25. Friend, S. H. (1999) How DNA microarrays and expression profiling will affect clinical practice. *BMJ* **319,** 1306–1307.
26. van't Veer, L. J., Dai, H., van de Vijver, M. J., et al. (2002) Gene expression profiling predicts clinical outcome of breast cancer. *Nature* **415,** 530–536.
27. Alizadeh, A. A., Eisen, M. B., Davis, R. E., et al. (2000) Distinct types of diffuse large B-cell lymphoma identified by gene expression profiling. *Nature* **403,** 503–511.
28. Shipp, M. A., Ross, K. N., Tamayo, P., et al. (2002) Diffuse large B-cell lymphoma outcome prediction by gene-expression profiling and supervised machine learning. *Nat. Med.* **8,** 68–74.

Part II
Technical Aspects

2
cDNA Microarrays

Paul S. Meltzer

INTRODUCTION

Cancer can be viewed as a disease of disturbed genome function. The phenomena of aberrant growth, differentiation, invasion, and metastasis are the phenotypic manifestations of an underlying genetic process. Ultimately, irrespective of whether this is the result of point mutation, translocation, deletion, gene amplification, or methylation, the malignant phenotype is mediated by a characteristic pattern of gene expression. Identifying the genes whose expression differs between normal tissues and tumors and among tumor types has long been a focus of cancer researchers. This endeavor was tremendously accelerated by the development of technologies for the parallel analysis of gene expression. This chapter will focus on one of these technologies, cDNA microarrays. The ability to measure the expression of tens of thousands of genes in a tumor specimen has revolutionized our ability to describe cancers. A rapidly burgeoning literature offers hope that this improvement will translate into improved diagnosis and prognosis, as well as accelerate the discovery of new therapeutic targets.

The pivotal concept enabling cDNA microarray technology is simple. Rather than maintaining libraries of cDNA clones as stocks of bacteria mixed in suspension, libraries can be stored as collections of individual clones arrayed in microtiter plates. This essential aspect of expressed sequence tag (EST) library sequencing projects provides a residual physical resource that can be used for other purposes (1). Libraries in this format can be screened for individual genes of interest by replacing the traditional colony lift with a filter prepared by transferring bacteria from the source plates to a hybridization membrane (2). Such filters can also be hybridized with labeled cDNA prepared from a cell source of interest (3,4). By quantitating the hybridization signal, an estimate of the expression of the gene corresponding to each cDNA can be obtained. The cDNA microarray, which has now found wide use in all fields of biomedical research, is the much refined descendant of this simple concept. The fundamental elements, which are necessary to carry out this analysis, are arrayed libraries of cDNA clones, a means for producing hybridizable arrays of these cDNAs, a system to detect hybridization signal, and a means to quantitate those signals link them to the individual cDNAs and compare these data across sample sets. The following discussion will briefly consider the individual elements of the system and the considerations in experimental design that are of particular relevance to cancer research.

From: *Expression Profiling of Human Tumors: Diagnostic and Research Applications*
Edited by: Marc Ladanyi and William L. Gerald © Humana Press Inc., Totowa, NJ

cDNA LIBRARIES

One of the great attractions of the cDNA microarray platform is its flexibility, and in principle, one can construct arrays from any cDNA library that the investigator might select. One could construct clone arrays that represent specific pathways or protein classes or even use nonsequenced libraries from a tissue of interest. In practice, most researchers use clones that have been culled from EST sequencing projects. There are now over 4.4 million sequences in the National Center for Biotechnology Information (NCBI) database of ESTs. Given that the number of genes in the human genome is two orders of magnitude lower, it is apparent that there is considerable redundancy in the EST sequence data. Using the EST sequence data and the mRNA sequences of known genes, bioinformatic tools have been used to cluster sequences into groups representing individual transcripts. The most widely used system, Unigene, is maintained by NCBI *(5,6)*. Each individual cluster is designated by an identifier that can be used to extract the set of sequences that constitute that cluster. Clone sets are selected to represent each Unigene cluster. These clones must then be physically retrieved from their source plates and rearrayed into sets for microarray fabrication. Ideally, each clone is sequence-verified at the time it is rearrayed to maintain a high standard for sequence authenticity in the final rearrayed library.

Each strategy for microarray production has intrinsic strengths and weaknesses. Ultimately, one would like to have the option of constructing microarrays that include a complete representation of the genome. To accomplish this, it would be necessary to retrieve a cDNA clone for each gene, a goal that is limited by a number of factors, including the still incomplete annotation of the human genome sequence. It is relatively easy to access the genes that have been encountered multiple times in the course of sequencing EST libraries. However, although over 800 libraries have contributed data to the Unigene database, some genes, which may be expressed only in specialized tissues or at developmental stages that have not been sampled, may not be represented at all. Genes that have been sequenced only a few times may be difficult to locate, depending on how effectively libraries have been archived. These considerations have not posed major limitations for expression profiling studies of human cancers, but the possibility that key genes may be missing from a given microarray is important to bear in mind when considering the results of any study. In addition, the Unigene clustering system undergoes periodic revision (builds) as new data becomes available, so clusters are not stable over time. There are also significantly more clusters (over 100,000) than the estimated number of human genes, and there is certainly both noise (due to artifactual cDNA clones) and redundancy (multiple clusters for the same gene) within Unigene. Over 36,000 clusters are represented by only a single sequence. These are difficult to access and may include clones that represent library artifacts or genes with very low expression.

Another limitation inherent to cDNA libraries is the problem of preserving sequence authenticity. In general, for microarray applications, libraries of rearrayed sequence-verified clones are used. However, in the manipulation of tens of thousands of bacterial stocks, it is inevitable that a residual level of error remains, and investigators must bear this caveat in mind. Despite all these difficulties, cDNAs have major attractions. They are readily available at a relatively low cost and can be manipulated with familiar techniques. Once clones are obtained, an unlimited supply of DNA for printing can be obtained by polymerase chain reaction (PCR), and the clones themselves are a convenient source of probes for follow-up studies. The cDNA technology lends itself to specialized projects potentially utilizing special purpose libraries constructed from material of interest to an investigator and potentially enriched for disease-specific genes, which might not be included in generic clone collections. Finally, of the various expression microarray technologies, only cDNA arrays lend themselves to the determination of gene copy number by comparative genomic hybridization, an analysis that adds a potentially important dimension to tumor profiling studies *(7,8)*.

How big does a cDNA microarray have to be to generate useful information for tumor profiling? It is quite clear that full genome-scale arrays are not necessary, as the world literature to date falls short of this level. Most investigators conclude that they would like to use the largest available array, because the analysis is a destructive process, and sample sets may be more limiting than the arrays themselves. However, although this issue has not been studied systematically, there seems to be a decline in useful information as genes are added to arrays. If one imagines a list of genes ranked as to their frequency of expression in tissues, the lower portion of this list will contain genes that are very infrequently expressed and, therefore, are less likely to be expressed in any given tissue of interest. This tends to counterbalance the tendency of cDNA clone sets to be limited to the 10,000–20,000 Unigene clusters representing the most commonly expressed genes.

CONSTRUCTING MICROARRAYS

Once a clone set has been selected, fabricating microarrays is quite straightforward. The technology is dependent on the use of a robotic device to deposit DNA (typically a PCR product) from each clone on a solid support, usually a glass microscope slide *(9)*. As an alternative to glass, microarrays can also be printed on nylon membranes for use in radioactive detection systems rather than the fluorescence-based detection used for glass microarrays. Detailed protocols for cDNA spotting are readily available. Robots for printing microarrays are produced by several manufacturers. The printing procedure is sufficiently simple that many institutions have established facilities for constructing microarrays, and expertise in array fabrication is now quite widespread. Commercial sources of spotted cDNAs are now available and represent an alternative to locally fabricated microarrays. It should be noted that once a spotting facility has been established, only minor modifications are necessary to spot alternative DNAs, such as synthetic oligonucleotides.

There is one important consideration that investigators who plan to use microarrays for tumor classification should bear in mind. It is extremely important that data sets, which are designed to provide this type of information, be generated in as homogeneous a fashion as possible. This maximizes the possibility of recognizing smaller differences in expression between sample subgroups and minimizes the number of false positives due to nonuniformities in technique. Of the many sources of this type of error, slide-to-slide variation is perhaps the most important. Generally, within a print batch, this error is relatively small and well compensated for by the use of a two-color hybridization scheme. However, when comparing batches of slides printed at different times, a large number of variables can interact to result in significant nonuniformities between batches. This can present a problem, which is relevant to project design. For example, if a given printing system can generate a batch of 100 slides, then no more than 100 specimens can be compared within a single batch. Switching to a second batch of slides for the next 100 specimens may yield data that can still be useful, but it will not be as satisfactory for recognizing the smaller differences between samples, which may be of the greatest interest to clinical investigators. This difficulty can potentially be overcome by improvements in printing technique, but these are most likely to take place within an industrial environment. This issue, at the very least, deserves consideration in the design of tumor profiling studies using cDNA microarrays.

MEASURING GENE EXPRESSION ON cDNA MICROARRAYS (FIG. 1)

In order to generate the primary expression data, a labeled representation of the sample mRNAs must be prepared for hybridization to the microarray. Each feature on the array is referred to as a "probe," and the mixture derived from the sample is the "target." Fluorescence detection has emerged as the most useful methodology when coupled with the use of glass microarrays (9). Fluorescence allows for simultaneous hybridization of an unknown and a reference sample, each labeled with distinct fluorochromes. This forgives, to a large extent, any imperfections in array fabrication and allows a very accurate and sensitive measurement of the unknown relative to the reference source. As an alternative, radiolabeled targets can be hybridized to nylon membrane microarrays. This presents some difficulties in image analysis, but is a viable alternative if access to glass arrays is not possible.

For hybridization, the mRNA from the sample is converted to a labeled derivative by reverse transcription to cDNA. A modified nucleotide is included in the cDNA synthesis reaction. A fluorochrome can be incorporated directly, coupled to a reactive group (as in the aminoallyl labeling strategy), or used in secondary detection. The dynamic range and signal intensity are two of the critical variables affecting labeling methods. Investigators using tissue samples prefer to minimize sample requirements. The direct incorporation of a fluorescent dye requires 20–100 μg of total RNA, while aminoallyl labeling requires 1–20 μg RNA. These are quantities that are easy to achieve with small tissue specimens. The use of smaller samples requires an amplification step. This can be accomplished by incorporating one or more cycles of in vitro transcription driven from a bacteriophage RNA polymerase promoter incorporated in the primer used for cDNA synthesis. Using this approach, useful data has been obtained from minute quantities of RNA (10,11). Investigators using amplification techniques should be aware that consistent labeling techniques should be used for a given project. Microdissection, with comparison of tumor and normal cells, is particularly attractive as an approach to directly identify genes that are cancer- rather than tissue-specific in their expression pattern (12,13).

For two-color hybridization, it is necessary to select a reference sample. In principle, the primary requirement of this material is a similar pattern of gene expression to the tumors for which it will be compared. If many genes, which are strongly expressed in the tumors, are expressed in the reference sample at near background levels, then the sample-to-reference ratio will be unreliable. This requirement for similar expression may be difficult to meet. One approach is to use a related cancer cell line or, as an alternative, a pool of cell lines. There are distinct advantages to using a pool. Specifi-

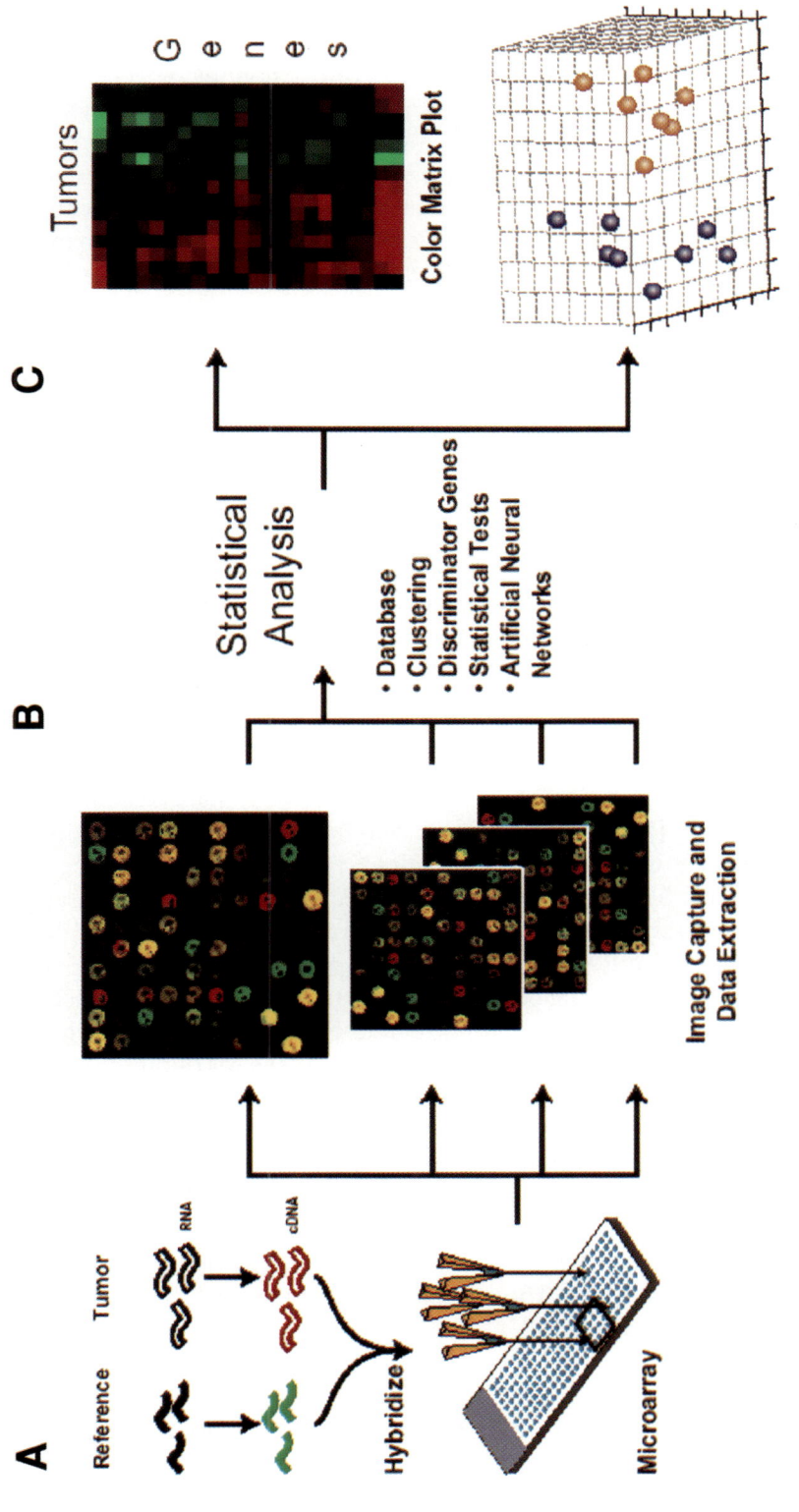

Fig. 1. cDNA microarray data flow. **(A)** Representations of the cellular mRNA pool are prepared by reverse transcription and hybridized to a robotically printed microarray. Each spot on the array represents a cDNA clone that can be assigned to a specific gene. Microarrays are typically hybridized with a mixture of tumor and reference samples, each labeled with a distinct fluorochrome. **(B)** After hybridization, the fluorescent images are captured in a scanner, and the quantitative hybridization signals are extracted for each array element using image analysis software. **(C)** Data from a series of samples is stored in a relational database, analyzed with statistical methods appropriate to the research question posed in the study, and displayed for inspection. Two of several available types of data display are used.

cally, each component of the pool will eliminate some low denominator genes, and variations between batches of cells are minimized across the pool. Ideally, a project is not started until sufficient reference RNA is available to complete the entire project. It is important to carry out test hybridizations to determine the suitability of a reference RNA before proceeding. Normal tissue truly representing the cancer progenitor cell is not generally available in sufficient quantity for use as a reference. The exception may be those situations where microdissected material will be amplified and where flanking normal cells might also be obtained for similar processing.

A recurring question in tumor processing is the influence of admixed normal stromal, endothelial, and inflammatory cells on the pattern of gene expression. Because it is significantly easier to generate data from whole tissues compared to microdissected cells, the vast majority of data in the literature has been obtained in this fashion. The signature of many important components of tumors, such as endothelial, smooth muscle, and inflammatory cells, can be recognized during data analysis, especially if suitable representatives of these cells are included in the database. One can argue that, since the biological properties of a tumor depend on the function of all the various cell types represented in the tumor tissue, information derived from the tissue as a whole actually adds value to the dataset. For example, it may be important to recognize subsets of tumors with higher content of inflammatory cells. It is important to note that, although subtraction *in silico* can provide a reasonable guide to the interpretation of expression patterns, one cannot formally prove that this result is correct without additional experimentation. In general, the difficulties of follow-up studies to verify conclusions drawn from *in silico* subtraction (*in situ* hybridization, immunohistochemistry, or reverse transcription PCR [RT-PCR]) must be weighed against the limitations imposed by cDNA amplification methods. In principle, analysis of microdissected malignant cells will provide a high degree of cell type specificity, but this comes at a considerable cost in terms of specimen processing, as well as carrying the risk of distorting the relative abundance of mRNAs in the amplified product.

After hybridization, a fluorescence image of the microarray is obtained with a scanning device, and the image file is processed with feature extraction software, which converts the raw image to numerical data corresponding to the level of fluorescence in each channel. Commercial instruments and software packages for this purpose perform well. Microarray users must become familiar with the properties of their scanner and use appropriate setting to maximize dynamic range and obtain consistent results between scans. Because it is impossible to use perfectly equivalent amounts of sample in each channel, it is necessary to normalize the sample and reference channels. Two strategies are in wide use, normalization by global intensity or by the use of a set of minimally variable housekeeping genes. The normalized processed data are then output as a spreadsheet for further analysis.

DATA ANALYSIS

Data from a series of tumor samples with expression levels for thousands of genes can present a challenge for analysis. A method for data storage and retrieval in a database is essential. This need not necessarily require the use of enterprise scale databases, but some form of data storage is required. For example, commonly available software, such as FileMaker Pro, can accommodate the needs of many projects.

Although expression profiling studies are sometimes contrasted with traditional hypothesis-driven research, in order to make sense out of microarray data, the researcher must still have a concept of the main questions which it is hoped that a given sample set might answer. Appropriate selection of analysis tools will depend on the questions to be addressed. Certain key questions pervade most cancer-related microarray research: Can two or more types of cancer be discriminated? What genes discriminate them most clearly? Are there genes that discriminate tumors from normal tissues? Are there subsets within tumors of the same apparent class? Are there correlations between expression profiles and other molecular or pathological properties of the tumor? Are there correlations between expression profiles and clinical variables, such as outcome and response to therapy? With what degree of confidence can it be said that these results are not due to chance alone? Do the genes, which arise from these analyses, fall into biologically recognizable pathways? Are these pathways relevant to the tumor phenotype or as potential targets for therapy?

These important questions and the need to develop the mathematical tools to address them have attracted the attention of computer scientists, engineers, and biostatisticians. Numerous computational approaches have been developed, and these will not be reviewed in detail here. However, certain important principles deserve emphasis. The questions listed above vary in difficulty. Some are very easy. For example, finding genes that distinguish colon cancer and glioblastoma will not be a great challenge, and numerous reports support the concept that different cancer types have distinct gene expression profiles *(14,15)*. On the other hand, finding genes that discriminate among stages of colon cancer might be significantly more challenging. There is no reason to be sure, *a priori*, that every question can be answered with confidence by gene expression profiling. For example, the chemosensitivity of a metastatic clone may not be predicted from the gene expression profile of the corresponding primary tumor. As the differences in gene expression narrow between groups of samples, which define clinically relevant groups, the analysis will be less and less forgiving of noise in the data, and the importance of the primary data quality increases. Similarly, larger numbers of samples in each group will be necessary to achieve statistically significant results. Once gene lists are developed, which appear to answer the question posed, they should be used to develop a formal rule-based classifier that might be applied to new samples *(16,17)*. Ideally then, experimental designs should also include a blinded test set, which can be used to validate the results obtained from a "training" set.

Because the number of genes in microarray data sets is always much larger than the number of samples, there will always be some number of genes that appear to differ significantly between groups based on chance alone. There is no method that can prove that this is not the case for any given gene, but probabilistic methods can provide an estimate of the probability that the results are due to chance fluctuations in the data. Alternative methods to address this issue include random permutation tests, leave-one-out analysis, and the introduction of gaussian noise into the data *(18–20)*. These methods help establish whether the data contain an overabundance of genes, which discriminate the samples compared to what would be expected at random. Additionally, it is important to use one of several available methods to rank the genes that discriminate among samples, in order to identify the genes that have the greatest impact on separating groups *(17,18)*. Even with apparently good results, there may be aspects of the data related to sample selection that will not be apparent until a confirmatory study is attempted. In the end, as in other types of clinical research, there is no substitute for a truly independent confirmatory study.

The methods used to analyze microarray data can be divided broadly into supervised and unsupervised approaches. Clinical correlative studies will utilize supervised methods that divide the samples into groups, for example, responders and nonresponders, according to a known variable, and then search for genes that differ between groups *(21–23)*. Alternatively, microarray datasets present the opportunity for class discovery. This entails the use of unsupervised techniques to search for properties of sample sets that emerge from the data analysis without utilizing the known classification data for the analysis. Unsupervised analysis provides the opportunity to discover unexpected complexities among samples sets. There are a number of excellent examples of the application of this approach to a variety of cancer types *(24–28)*.

VALIDATION OF MICROARRAY DATA

How reliable are microarray data? This question is somewhat laboratory-specific, depending on the precise methodology used. It is also dependent on where a given data point falls on the spectrum of gene abundance. Genes expressed at low levels will not be measured as accurately as more abundant transcripts. In general, when compared to conventional methods, microarray data from experienced laboratories are remarkably accurate *(29)*. Although validation by Northern blot or quantitative PCR methods may be required to confirm or extend important results, inherent data accuracy is usually not a major concern when a pattern of expression is reinforced by a large number of samples. It is somewhat problematic that there may not be an alternative technique that can be used to confirm the expression levels of dozens or hundreds of genes at the same level of accuracy as microarrays, especially when expression levels between sample groups vary by less than two-fold. In this case, the best validation will be obtained from microarray analysis of an additional sample set.

Confirmation at the protein level can also be difficult, since for most genes, a suitable antibody will not be available. Even in the case of genes for which good antibodies capable of staining tissue sections exist, the assay may not have the same dynamic range as hybridization-based methods. In-depth correlation of mRNA and protein expression levels for multiple genes will not be accomplished until accurate quantitative proteomic methods become available.

Tissue microarrays for *in situ* mRNA hybridization or immunohistochemistry provide the possibility of confirmatory studies on large numbers of samples *(30)*. Image analysis of mRNA *in situ* hybridization is remarkably quantitative and agrees well with cDNA microarray data *(31)*. This technology for analyzing a single gene in numerous samples nicely complements the ability of cDNA microarrays to analyze numerous genes in relatively small numbers of samples.

INTERPRETING GENE LISTS

Microarray analysis, whether supervised or unsupervised, ultimately generates lists of genes that discriminate among samples. Making sense of these gene lists presents a significant challenge. Gene names can be misleading, and the majority of genes are linked to little or no functional information. Currently, there are only limited tools that

29. Amundson, S. A., Bittner, M., Chen, Y., Trent, J., Meltzer, P., and Fornace, A. J., Jr. (1999) Fluorescent cDNA microarray hybridization reveals complexity and heterogeneity of cellular genotoxic stress responses. *Oncogene* **18,** 3666–3672.
30. Kononen, J., Bubendorf, L., Kallioniemi, A., et al. (1998) Tissue microarrays for high-throughput molecular profiling of tumor specimens. *Nat. Med.* **4,** 844–847.
31. Mousses, S., Bubendorf, L., Wagner, U., et al. (2002) Clinical validation of candidate genes associated with prostate cancer progression in the CWR22 model system using tissue microarrays. *Cancer Res.* **62,** 1256–1260.
32. Brazma, A., Hingamp, P., Quackenbush, J., et al. (2001) Minimum information about a microarray experiment (MIAME)-toward standards for microarray data. *Nat. Genet.* **29,** 365–371.

3
Oligonucleotide Microarrays

Marina Chicurel and Dennise Dalma-Weiszhausz

INTRODUCTION

Oligonucleotide Microarrays:
Tools for Decoding Cancer's Molecular Signatures

Oligonucleotide microarrays are fast earning a privileged position in the toolboxes of many cancer researchers. Driving their rise in prominence is a rapidly growing list of the technique's accomplishments, including the discovery of new tumor classes, the assignment of clinical samples to known tumor classes, the elucidation of molecular pathways underlying cancer behavior, the prediction of clinical outcomes, and the identification of potential targets for therapeutic intervention.

Oligonucleotide microarrays have uncovered molecular signatures and clues to the physiology of leukemia *(1)*, lymphoma *(2)*, melanoma *(3)*, as well as cancers of the breast *(4)*, lung *(5)*, prostate *(6)*, colon *(7)*, oral epithelium *(8)*, bladder *(9)*, ovaries *(10)*, and liver *(11,12)*. Recent highlights include the discovery of molecular signatures that help distinguish clinically relevant classes of breast tumors *(4)*, the identification of candidate genes involved in colon cancer progression *(7)*, and the elucidation of a molecular mechanism underlying metastasis in melanoma *(3)*. For example, oligonucleotide microarrays have contributed to several basic and clinical advances in the study of the most common malignant brain tumor in children, medulloblastoma. Oligonucleotide arrays allowed Pomeroy and coworkers to distinguish medulloblastomas from other brain tumors with very different prognoses *(13)*. They also provided clues as to the cellular origins of medulloblastomas, helping resolve a long-standing controversy regarding their classification. In addition, data from oligonucleotide arrays suggested the involvement of signaling pathways in metastatic medulloblastoma that could be inhibited by known drugs, including one recently approved by the Food and Drug Administration (FDA) for the treatment of a form of leukemia *(14)*. Perhaps most impressive, microarray-based profiling yielded the most significant predictor of medulloblastoma outcome currently available *(13)*.

This chapter uses recent examples to illustrate the power of monitoring the expression levels of thousands of genes at once to decode cancer's molecular signatures. It describes the design, manufacture, and use of oligonucleotide arrays and discusses the technology's strong points and limitations.

From: *Expression Profiling of Human Tumors: Diagnostic and Research Applications*
Edited by: Marc Ladanyi and William L. Gerald © Humana Press Inc., Totowa, NJ

WHY USE OLIGONUCLEOTIDE MICROARRAYS?

Introduction: The Challenge of Predicting Cancer's Next Move and the Power of Monitoring Many Genes at Once

Few diseases are as heterogeneous as cancer. Conventional histopathology has teased apart hundreds of different tumor classes. However, the difficulty of predicting cancer behavior betrays the existence of a much greater diversity. Tumors that look alike under the microscope can vary widely in their progression and their responsiveness to therapy. For instance, researchers have compiled a long list of breast tumor markers, including histologic criteria, lymph node metastases, the expression of steroid and growth factor receptors, estrogen-inducible genes, and mutations in the *TP53* gene, yet the ability to predict the disease's progression and its reaction to treatment is still poor. The handful of current distinctions of breast cancer fail to reflect the many biologically significant classes that have recently been predicted to exist *(15)*.

The fundamental problem is that cancer phenotypes result from alterations of many regulatory pathways and structural components, each including dozens of proteins. Single markers offer only a tiny window into the overall complexity *(16)*. Thus, finding better ways of classifying and subclassifying tumors to understand, predict, and modify their behaviors is a top priority in cancer research.

Technologies that can monitor many cellular components in parallel offer the possibility of capturing a more complete picture of a tumor and, potentially, predicting its behavior more reliably. Because of their key role as effectors of cell function, proteins would seem to be the ideal candidates for such larger scale monitoring. Indeed, much effort is being directed at improving and developing technologies to track proteins *(17,18)*. Current methods to monitor protein levels include Western blots, two-dimensional gels, chromatography, mass spectrometry, protein–fusion reporter constructs, and the characterization of polysomal RNA.

Protein-based approaches are generally more difficult to perform, less sensitive, and have a lower throughput than RNA-based methods *(17)*. Since changes in mRNA abundance usually reflect changes in protein levels, they can provide much information about a cell's physiological state. In addition, many of the mutations that affect tumor cells occur in signal transduction pathways that directly or indirectly regulate transcription factors. Based on data from several types of cancers, it is estimated that between 0.2–10% of all transcripts are differentially expressed between cancer and normal tissues *(7)*. Since a typical mammalian cell expresses 10,000–20,000 different transcripts, global analyses of gene expression patterns could yield as many as 2000 transcripts offering clues to the disease process and acting as potential diagnostic or prognostic markers.

The likelihood of finding distinguishing features is greatly enhanced by screening thousands of genes simultaneously, instead of single genes at a time. In addition, studies provide information on multiple fronts. A study designed to identify new tumor classes, for example, may also reveal clues about the basic biology of cancer, as well as suggest candidate genes for therapeutic intervention. In addition, the numerous results generated by large-scale gene expression experiments often include previously reported correlations that act as internal controls, providing some degree of built-in validation and replication *(17)*.

Another advantage of monitoring large-scale patterns of gene expression is that, by providing high resolution portraits of cancer, these technologies provide proxies for abnormalities in complete molecular pathways *(19)*. Single marker assays detect alterations at very specific nodes in a biochemical or signaling pathway and consequently, are unable to detect upstream or downstream disruptions, which may be functionally equivalent to alterations of the marker gene or protein. In contrast, monitoring many transcripts at once can provide patterns that reflect the functional status of entire pathways. The estrogen receptor (ER), for example, is often used as a prognostic marker for breast cancer. Almost 75% of breast cancers expressing ER will respond to tamoxifen treatment, whereas less than 5% of nonexpressing tumors respond. The standard method for assessing ER status is immunohistochemistry. Yet, by monitoring a single node in the ER pathway, alterations in other components of the pathway can be missed. Illustrating the potential of microarrays for bypassing this limitation, West and colleagues identified 100 genes whose expression correlated strongly with ER status, including genes involved in the estrogen pathway, as well as genes that encode proteins that synergize with ER *(4)*.

Many changes that distinguish cancers from each other and from normal tissues may be subtle or variable, making them individually unreliable. Yet, by monitoring many of these changes simultaneously, as a composite pattern, it is sometimes possible to identify more robust signatures. Pomeroy and coworkers, for example, found that the expression of a handful of genes encoding ribosomal proteins correlated with a poor outcome in medulloblastoma tumors. To achieve statistically significant predictions, however, they required additional genes whose expression correlated with a favorable outcome, including several genes characteristic of cerebellar differentiation *(13)*. A particularly striking illustration of the power of distributed patterns was provided by Alon et al. *(20)*, who showed that even quite subtle changes in gene expression, when taken together, can provide reliable information. They were able to distinguish tumor and normal colon tissue with expression profiling even after removing the 1500 genes, out of a total of 6500, that showed the most significant differences in expression *(20)*.

Oligonucleotide Microarrays: Specificity, Reproducibility, and Quantitation

There are many ways to monitor patterns of gene expression. A straightforward approach is to sequence cDNAs, or small parts of cDNAs, generated from the mRNAs expressed within a tissue. For example, the Cancer Genome Anatomy Project (CGAP), a program implemented by the National Cancer Institute to collect information on genes associated with cancer development has generated over one million human expressed sequence tag (EST) sequences from over 180 cDNA libraries *(21,22)*. Serial analysis of gene expression (SAGE) is another method that has provided much information on transcriptome patterns associated with cancer *(23)*. Like EST sequencing, SAGE relies on sequencing cDNAs, but streamlines the process by stringing together multiple sequence tags that uniquely identify individual transcripts.

The strategy offering the highest density of information output, however, is the use of microarrays of oligonucleotides or cDNAs. The basic concept is simple: labeled cDNA or cRNA targets derived from the mRNA of a tissue are hybridized to nucleic acid probes attached to a solid support. By monitoring the amount of label associated with each DNA location, it is possible to infer the abundance of each species represented on the array. Although hybridization has been used for decades to detect and quantify nucleic acids, the combination of the miniaturization of the technology and the large and growing amounts of sequence information, have enormously expanded the scale at which gene expression can be studied *(17,24)*.

Microarray technology combines at least five different components *(25,26)*: (*i*) a solid support; (*ii*) a device for either spotting the nucleic acid probes or for synthesizing them *in situ*; (*iii*) a fluidics system for exchanging solutions during hybridization and washing; (*iv*) a scanner to detect the labeled targets; and (*v*) computer programs to store, process, and mine the data (Fig. 2A). There are many variations to this basic theme (e.g., *27–29*), but the two most commonly used platforms are cDNA spotted arrays and *in situ* synthesized oligonucleotide arrays. cDNA arrays are made by robotically depositing DNA fragments, such as polymerase chain reaction (PCR) products derived from cDNA clones, onto coated glass slides (*see* Chapter 2). Oligonucleotide arrays, on the other hand, are manufactured by synthesizing oligonucleotides directly onto glass. Although oligonucleotide arrays can also be made by spotting, the term oligonucleotide microarray usually refers to *in situ* synthesized arrays.

Each technology has its strengths and weaknesses, and some researchers have suggested the two systems may be most effectively used in parallel *(30)*. Others have developed hybrid methods that attempt to capitalize on the strengths of both platforms *(31)*, although these methods suffer from their own limitations *(32,33)*.

Because cDNA arrays can be made in the laboratory, they are particularly useful in studies requiring very small batches of project-specific arrays. They are also well suited for studies in which the array design has to be modified frequently. In addition, because they rely on spotting rather than synthesis, cDNA arrays enable the study of genes that have not yet been sequenced.

cDNA arrays require managing banks of cDNA clones and are generally unable to provide the specificity, reproducibility, and standardized quantitation of transcript levels offered by oligonucleotide arrays (Table 1) *(34–37)*. In addition, oligonucleotide arrays yield reproducible results because their design and manufacture are highly stereotyped and consistent. This helps generate reliable results that are more easily compared between studies and are uniquely well suited for clinical applications. The high degree of control in design and manufacture also means less setup time for the end-user. Researchers do not have to manufacture or test the quality and reproducibility of the arrays, which translates into less setup time to get results. The scaleable manufacturing of oligonucleotide arrays allows for the production of a wide range of array sizes and facilitates the production of large sets of identical arrays. Oligonucleotide arrays are, therefore, particularly powerful for developing databases, because different samples from different sites can be readily compared. At the same time, the design and manufacturing of arrays is flexible, allowing for the creation of custom arrays, a process that has been streamlined by the development of on-line tools to search databases of previously manufactured probe sets in conjunction with biological databases. Even though it is now common to use a universal normalization control sample, cDNA arrays are often limited to providing ratios of RNA levels between an experimental condition and a control condition, whereas oligonucleotide arrays can provide measurements of transcript levels that allow for interexperimental comparisons.

Table 1
Performance Characteristics of GeneChip Microarrays

	Routine use	Current limit[a]
Starting material	5 μm total RNA	Single cell[b]
Detection specificity	1:100,000	$1:10^6$
Absolute quantitative accuracy	±2x	±10%
False positives	<2%	0%
Discrimination of related genes	70–90% identity	95% identity
Dynamic range (linear detection)	~500-fold	~10^4-fold
Probe pairs per gene/EST	11–20	4
Number of genes per array	~12,000	~40,000

[a]Experimental: under development at Affymetrix.
[b]Semiquantitative.

Oligonucleotide Microarrays and Tumor Profiling

A rapidly growing list of accomplishments attests to the powerful capabilities of oligonucleotide arrays. One of the earliest and clearest demonstrations of the potential of using these arrays for classifying tumors was reported by Golub and colleagues (1). By analyzing the expression patterns of 38 acute leukemia samples, the authors identified 1100 genes that are differentially expressed between acute myeloid leukemia (AML) and acute lymphoblastic leukemia (ALL). They then developed a class predictor based on the expression patterns of the 50 genes whose expression most closely correlated with the class distinction. The predictor proved 100% accurate, making strong predictions for 29 of 34 new samples. The results were particularly striking, because the procedures for collecting samples involved various preparation protocols and different sources of tissue, including both peripheral blood and bone marrow.

The authors also tested the feasibility of using microarrays to discover new tumor classes. Using an algorithm called a self-organizing map to automatically group the 38 leukemia samples based solely on their expression profiles, they independently generated classes that largely corresponded to AML, T-lineage ALL, and B-lineage ALL tumors. These results indicated that tumor classes could be discovered without previous biological knowledge.

Although the main contribution of this study lies in the proof-of-concept showing how microarray data can be used for class discovery and prediction, it also provided new clues about the physiology of leukemia. For example, AML correlated with the expression of high levels of the leptin receptor, a molecule previously found to have anti-apoptotic function in hematopoietic cells.

A wealth of recent studies using oligonucleotide arrays have similarly uncovered both expression-based classifiers and clues to cancer's physiology (e.g., 2,5–9,11–14,38). Pomeroy et al., for example, identified an expression signature that far outperforms current markers for predicting patient survival associated with medulloblastoma, while helping clarify the cancer's cellular origins (1). Experts argued whether medulloblastomas were members of a class of primitive neuroectodermal tumors (PNETs) arising from a common cell type in the subventricular germinal matrix, or whether they were distinct from PNETs, arising from progenitors of cerebellar granule cells. Based on expression patterns, Pomeroy and colleagues found that medulloblastomas and PNETs fell into distinct groups. They also discovered that two transcription factors, ZIC1 and neuronal stem cell leukemia (NSCL)-1, previously found to be specific to cerebellar granule cells, were among the genes that most highly correlated with the medulloblastoma group.

Other studies have identified genes involved in tumor progression. Clark and coworkers, for example, found 32 genes that correlated with metastasis in melanoma *(5)*. To test the biological significance of their array-based findings, the authors examined the effects of altering the expression of one of the genes, RhoC, a regulator of the actin cytoskeleton. Suggesting a causal role for RhoC in metastasis, cells that overexpressed the gene produced significantly more pulmonary metastatic nodules than control cells, while cells lacking functional RhoC suffered a dramatic reduction in their metastatic potential. Genes potentially involved in the transition from adenoma to carcinoma have also been identified through oligonucleotide microarrays, although not yet functionally tested *(9)*. Still other oligonucleotide array studies have revealed potential contributors to cancer initiation. For example, Alevizos et al. found that many of the genes that were down-regulated in oral cancer cells coded for enzymes in the xenobiotic pathway, which is a metabolic pathway that degrades toxic compounds including carcinogens *(10)*. Alterations in xenobiotic metabolism may thus contribute to an increased susceptibility to carcinogens, such as those present in tobacco and alcohol.

Oligonucleotide arrays have also illuminated the downstream players in several cancer-associated pathways, including those activated by BRCA1 *(39)*, EGR1 *(40)*, p53 *(41,42)*, and MYC *(43)*. Harkin et al., for example, established BRCA1-inducible cell lines and used oligonucleotide arrays to analyze gene expression following induction *(39)*. The authors found 23 genes and ESTs that increased their expression in response to BRCA1 induction.

Besides providing windows into the basic biology of cancer, microarray data are identifying candidate therapeutic targets. Based on their microarray studies, MacDonald et al. have suggested that inhibitors of platelet-derived growth factor receptor α (PDGFRA) and RAS proteins may be of therapeutic value against medulloblastoma *(2)*. The authors discovered that PDGFRA and downstream members of the RAS/mitogen-activated protein kinase signaling pathway were up-regulated in metastatic medulloblastoma tumors. As previously mentioned, several inhibitors of the pathway pinpointed in this study are currently available, including imatinib mesylate (Gleevec®, Novartis, Vienna, Austria), a drug recently approved by the FDA for the treatment of chronic myeloid leukemia.

Oligonucleotide arrays may also provide valuable information regarding drug response, potentially reducing the time-consuming and expensive process of testing drugs for safety in animal models, and helping future physicians tailor treatments to individual patients. Gerhold and coworkers, for example, recently demonstrated the feasibility of using oligonucleotide microarrays to measure expression profiles of genes involved in drug metabolism *(44)*.

DESIGNING OLIGONUCLEOTIDE MICROARRAYS

Many of the benefits of oligonucleotide arrays stem from the design of probes to optimize sensitivity, specificity, and reproducibility *(34–37)*. To illustrate the underlying principles, this section will focus on one of the most commonly used oligonucleotide microarrays, the GeneChip® array from Affymetrix (Santa Clara, CA, USA).

Unlike probes used in other arrays, *in situ*-synthesized probes can be designed based on a consistent set of rules to optimize hybridization. For example, palindromes can be avoided, reducing hairpin loops that interfere with intermolecular hybridization. The probes can also be optimized for hybridization under particular pH, salt, and temperature conditions by taking into account melting temperatures, and using empirical rules that correlate with desired hybridization behaviors.

Key to the design of GeneChip arrays is the match-mismatch probe strategy (Fig. 1B). For each probe designed to be perfectly complementary to a particular target sequence, a partner probe is designed that is identical except for a single base mismatch in its center. These probe pairs, called the perfect match probe (PM) and the mismatch probe (MM), allow the quantification and subtraction of signals caused by nonspecific cross-hybridization. The difference in hybridization signals between the partners, as well as their intensity ratios, serve as indicators of specific transcript abundance. Whereas evaluating the performance of features on cDNA microarrays relies on monitoring the hybridization of a few selected probes (e.g., *32*), PM/MM partners allow for global and specific evaluations of oligonucleotide probe performance.

Monitoring nonspecific signal in a probe-specific manner is particularly valuable for detecting and quantifying low abundance transcripts. Using arrays to perform a genome-wide analysis of gene expression in *Caernohabditis elegans*, for example, Hill and colleagues reported they could unambiguously detect control transcripts diluted as much as 1:300,000 *(45)*.

A huge advantage of selecting probes based on sequence is the possibility of uniquely identifying target transcripts. By using probes from regions of genes that greatly diverge between family members, GeneChip arrays are able to distinguish transcripts that are as much as 90% identical *(44)*. In addition, probes can be designed to distinguish between alternatively spliced transcripts, which is an important capability given estimates that at least 30% of all human genes are alternatively spliced.

In general, each target transcript is represented by between 11 and 20 different PM/MM partners *(46)*. This multiplicity is important for obtaining reliable results, because even when using a consistent set of rules for probe design, sequence-specific variations in hybridization are one of the most significant sources of array noise (Fig. 1B). Applying an analysis of variance (ANOVA) test to data from 21 oligonucleotide arrays containing more than 7,129 probe sets, Li and Wong reported that the variation due to probe effects was, in most cases, at least five times greater than that contributed by interarray variation *(47)*. Multiple probe pairs help deal with this variation, providing statistical measurements of probability and confidence for each queried transcript, resulting in improved tolerance of polymorphisms, hybridization inconsistencies, sequence similarities, and errors in sequence databases *(37)*. Multiple probe pairs also allow assessment of a sample's mRNA integrity; the lower the ratio between the hybridization signals of probes complementary to the 3' vs the 5' ends of a transcript, the less likely the sample has been degraded (Fig. 1B).

Wodicka et al. tested the ability of oligonucleotide arrays to reliably measure relative levels of transcripts. They hybridized the entire genome of the yeast *Saccharomyces cerevisiae* to expression arrays of the same organism *(48)*. Since most genes are

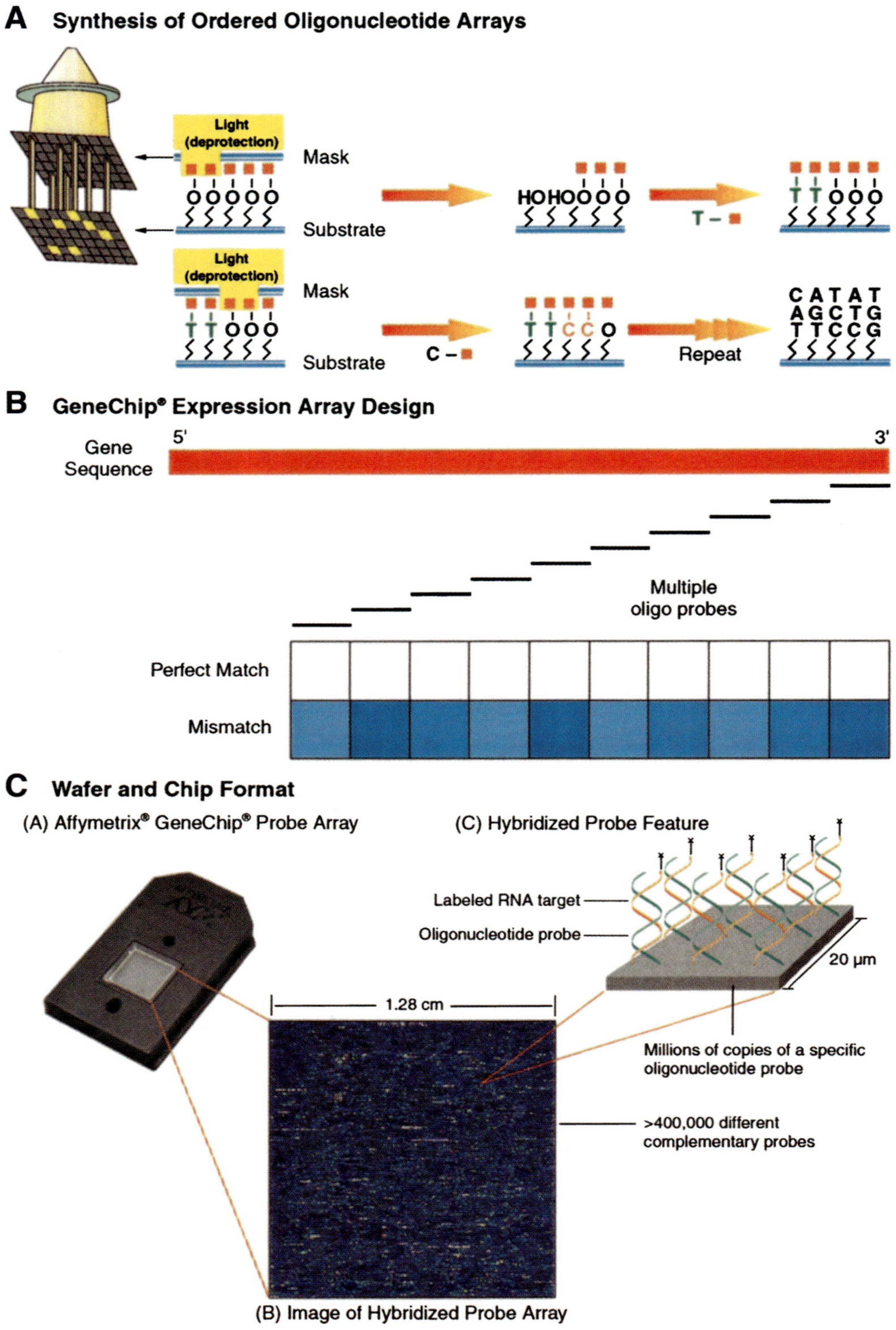

A Synthesis of Ordered Oligonucleotide Arrays

B GeneChip® Expression Array Design

C Wafer and Chip Format
(A) Affymetrix® GeneChip® Probe Array
(C) Hybridized Probe Feature
(B) Image of Hybridized Probe Array

Fig. 1. Schematic representation of the synthesis, design and wafer format of GeneChip oligonucleotide arrays. (**A**) Light-directed combinatorial chemistry. A light source is used to activate the surface of the wafer upon which the oligonucleotides will be synthesized.

present in the same copy number in the yeast genome, the authors expected to obtain very similar intensity signals across the entire array. Indeed, they found that the intensities of 80% of the hybridization signals were within two-fold of each other. Other studies have demonstrated that the results generated by oligonucleotide arrays are in close agreement with those obtained using Northern blot analysis *(46)*, SAGE *(49)*, and TaqMan® (Applied Biosystems, Foster City, CA, USA) *(44)*.

One of the major strengths of oligonucleotide arrays is their ability to yield standardized measurements of transcript abundance with a dynamic range of linear detection spanning at least 500-fold (Table 1) *(36,46,48)*. Several factors contribute to these quantitative abilities, including the use of PM/MM pairs, the inclusion of multiple probes per transcript (probe set), and the design of probes based on a consistent set of rules. The uniform conditions under which oligonucleotide probes are synthesized also contribute to the technique's quantitative reliability *(36,48)*. In contrast to the deposition of DNA onto spotted arrays, which can result in patchy distribution of the probe molecules and consequent variations in the signal detected, manufacturing probes *in situ* results in uniform coating of the array substrate.

MANUFACTURING OLIGONUCLEOTIDE ARRAYS

GeneChip arrays are manufactured using photolithographic methods and equipment adapted from the semiconductor industry (Fig. 1) *(34,37,50)*. The use of a solid-phase synthesis procedure circumvents the need for handling large numbers of clones or running PCR amplifications. The first step in the manufacturing process involves coating a 5-in square of glass with synthetic linkers modified with photolabile protecting groups. A chrome photolithographic mask with windows spanning between 18 and 20 µm² is then placed over the coated glass. The windows are distributed across the mask based on the first nucleotide in the sequence of each desired probe. When light is projected through the mask, the exposed linkers become deprotected and available for coupling to a nucleoside (Fig. 1A). The coated surface is then flushed with a solution containing either A, C, T, or G nucleosides, carrying removable protection groups. Uncoupled active sites are then capped by acylation so that, once again, the glass is covered by a lawn of protected molecules. A second mask is placed over the chip to allow the next round of deprotection and coupling. The process is repeated until the probes reach their full lengths, usually 25 nucleotides, which balances the needs for

---

(Figure 1 caption continued from previous page) The use of a photolithographic mask allows activation to be limited to specific sites, or features, such that nucleotide incorporation can be controlled to generate the desired sequences. Using solid-phase synthesis methods, nucleotides carrying a photolabile protecting group are sequentially added to the feature. In this example, a thymidine is added first. A different mask is used for the next synthetic step (cytosine). The process is repeated until the 25-nucleotide sequence is completed. Using strategies that streamline synthesis, the procedure can be accomplished in less than 25 steps. **(B)** Expression array design. Microarrays can be manufactured for tracking the expression levels of a multitude of genes. This design includes the PM sequence and the MM sequence. Each gene is represented by 11–16 oligonucleotide probes. **(C)** The glass wafer is sliced to create microarrays or "chips." The center image (B) shows a microarray's pattern of fluorescence emission. The picture on the left (A) depicts the packaging of the microarray. The picture on the right (C) shows a magnified array feature, in which millions of copies of a single oligonucleotide reside.

signal intensity and sequence specificity *(36)*. The distribution of probe partners is randomized to avoid potential biases from differences in hybridization conditions across the array.

Depending on the number of probes per array, between 50 and 400 chips can be generated from one 5-in square wafer. This parallel process significantly enhances reproducibility. In addition, the multiplicity of probes per gene and DNA molecules per probe location, each location harboring between 3 and 7 million DNA molecules, contributes to the array's robustness (Fig. 1C). Also, because the number of synthesis steps depends on the length of the probes and not the number of probes, array manufacture is scaleable *(37)*. Since each position in the sequence of an oligonucleotide can be occupied by one of four nucleotides, one might expect that, to synthesize a particular array of 25-mers, one would require 25 × 4, or 100, different masks. Yet, the number of masks can be significantly reduced by orchestrating the synthesis process, so that different probes are synthesized at different rates, and for some applications, single masks are used for more than one synthesis step. Harnessing the power of this approach, 5-in wafers with as many as 60 million 25-mer probes are being developed by Perlegen Sciences (Santa Clara, CA, USA).

The density of probe locations, or features, has also been rising steadily. Affymetrix has manufactured 1.28-cm^2 chips with over 400,000 features (Fig. 1C). The Human Genome U133 array set, sporting 18-µm^2 features and 11 probe partners per gene, is designed to contain over 1 million features representing approx 38,000 transcripts (44,000 probe sets) distributed across two standard-sized arrays. This very high density offers the possibility of monitoring global patterns of gene expression, while using fewer arrays and less labeled target. Also, since the underlying principles of design and manufacture remain constant, current arrays are backward compatible with older versions of scanners, software, and fluidics systems.

Further improvements in probe density are limited by two factors. In principle, hybridization efficiency can suffer if individual probe molecules are clustered too tightly *(37)*. Additionally, the resolving power of the current scanners (3 µm) that read the hybridization signals limits the degree to which feature size can be reduced *(37)*.

Although most oligonucleotide arrays in use today are manufactured following the steps outlined above, several potentially useful variations have been developed *(31,32,51,52)*. For example, instead of using expensive chrome masks with etched windows to specify each step of synthesis, Singh-Gasson developed a maskless synthesizer that uses virtual masks, i.e., computer-generated images that are projected onto a glass wafer using a digital micromirror array. The synthesizer has successfully produced oligonucleotide arrays with 76,000 features, each measuring 16 µm^2.

Another approach relies on technology developed originally for ink-jet printers *(31)*. Blanchard and colleagues developed a photolithographic procedure to generate arrays of hydrophilic islands surrounded by a highly hydrophobic background *(31)*. Water applied to the surface of the array forms separate droplets over the hydrophilic islands, which can be used as wells for performing DNA synthesis reactions. To deliver small amounts of synthesis reagents to each well, Blanchard et al. *(31)* built small ink-jet pumps, similar to those used in some ink-jet printers. Instead of delivering one of four different colored inks, each of the tiny pumps delivers one of the four DNA nucleosides as they are guided to the appropriate wells by an x-y stepping stage and a computer. Although the technique offers flexibility and low cost synthesis, ink-jet dispensers have some drawbacks. For example, the dispensers' glass heads are fragile and tend to clog. In addition, their operation can require larger volumes of fluid than other spotting devices (30).

Hughes et al. recently constructed a second-generation version of Blanchard's original synthesizer and tested its expression profiling capabilities (32). Using 60-mer oligonucleotides as probes, the team reported they could reliably detect transcripts present at a concentration equivalent to one copy per cell in a complex biological sample. However, the study only examined the hybridization behaviors of a small subset of sequences. In addition, the 60-mer probes failed to provide the hybridization specificity attainable with shorter probes. Whereas a single-base mismatch is sufficient to destabilize the hybridization of a 25-mer probe, 60-mers require, on average, five or more mismatches to reduce hybridization signals by more than 50%. Thus, the use of 60-mers results in decreased specificity due to cross-hybridization.

USING OLIGONUCLEOTIDE MICROARRAYS

Sample Preparation: Avoiding the Pitfalls

A major challenge in using either cDNA or oligonucleotide arrays for tumor profiling is navigating the many pitfalls associated with tissue selection and preparation. Low sample quality probably underlies many apparent inconsistencies across microarray studies and has seriously marred the clinical relevance of many reports.

One problem is that messenger RNA is a very labile molecule, which is susceptible to degradation by the RNases abundantly present in most tissues. Many of the traditional treatments used for preserving the morphology required for accurate tumor diagnosis are incompatible with the extraction of intact RNA. The crosslinking caused by formalin fixation, for example, severely compromises the quality of the RNA that can be extracted. On the other hand, protocols that help protect RNA, such as rapid freezing, can disrupt microscopic morphology (16).

To cope with these difficulties, some researchers are turning to alcohol-based fixation, which has the potential of preserving both morphology and RNA integrity (53,54). Even these methods fall short, however, if tissues are not processed quickly. As described by Perou and coworkers in their work with breast tumors, the prolonged handling of samples after surgical resection induces significant changes in gene expression patterns, including the induction of *c-fos* and *junB* (15). Key to obtaining high quality samples, then, is establishing effective collaborations between surgeons, pathologists, and researchers.

It is also important to assess RNA integrity once the samples have been collected. Stamey et al., for example, collected samples of prostatic tissue within 15 min of interruption of blood flow to the prostate and froze the tissue immediately, thereby providing excellent conditions for preserving RNA (8). Yet even under these conditions, 5 out of 22 samples showed signs of considerable degradation. After inspecting the RNA by agarose gel electrophoresis and spectrophotometry, the authors used a test microarray, the GeneChip Test3 array, to determine the ratio of 3' to 5' transcript levels of glyceraldehyde 3-phosphate dehydrogenase (GAPDH) and overall transcript levels for 24 housekeeping genes (55). Only 17 of the 22 samples were selected for subsequent experimental analysis based on having GAPDH ratios lower than three, and producing detectable signals for more than 40% of the housekeeping genes.

Tissue Heterogeneity: A Multitude of Complicating Factors

One of the strong points of expression profiling is its potential for providing robust molecular signatures. As previously mentioned, the signature provided by multiple markers can sometimes weather sample variability and heterogeneity much better than that provided by a single marker. Indeed, there is hope that gene expression data will allow the detection of a small number of cells diluted in complex mixtures. Martin and colleagues, for example, recently identified a set of 12 genes that allowed high sensitivity detection of disseminated tumor cells in the blood of 77% of patients with invasive breast cancer *(56)*.

However, heterogeneity still hinders the procurement of molecular information that is unambiguously associated with cancer and, consequently, can seriously limit its predictive value. To pinpoint mechanisms underlying tumor behavior, identify candidates for therapeutic intervention, and obtain predictors of tumor behavior with solid foundations in the disease process, experiments should be designed to account for and minimize heterogeneity within and between tissues, diversity in cell type composition, variability between individuals, and changes in tissues associated with disease progression.

In identifying expression patterns associated with disease progression, for example, it is useful to first distinguish tumors based on pre-established markers. Stamey et al. showed that the most robust histologic predictor of progression in radical prostatectomy samples was the amount of Gleason grade 4/5 tissue present in the largest tumor harbored by the prostate's peripheral zone *(57)*. The Gleason classification scheme is based on the architectural patterning of tumor glands, ranging from grade 1, corresponding to very well differentiated glands, which is similar in structure to those found in normal prostate, to grade 5, corresponding to very poorly differentiated tissue lacking individual separate gland units. Based on these findings, Stamey and coworkers focused on Gleason grade 4/5 tumors to search for expression patterns that correlated with disease progression *(8)*. In contrast, an independent study, which also examined expression profiles associated with prostate cancer, did not specifically select Gleason grade 4/5 tumors *(58)*. It is likely that at least some of the differences in results between the two studies can be attributed to the second study's failure to account for this kind of tumor heterogeneity.

Heterogeneity in the types of cells comprising tumors can also affect expression profiling. When Welsh et al., for example, screened the expression profiles of 27 ovarian tumors for genes whose expression correlated with malignancy, many of the genes they identified were characteristic of stromal tissue and infiltrating B cells *(12)*. Only after excluding 14 tumors, which expressed these genes at particularly high levels, did they obtain a list that was highly enriched for known or suspected markers of epithelial malignancy. Differences in smooth muscle content between tumors and normal tissues can also affect tumor profiling studies. Alon and colleagues found that colon tumors whose expression patterns clustered with normal tissues had higher muscle content than nonoutliers *(20)*. Similarly, outlying normal tissues that clustered with tumors had relatively low amounts of smooth muscle.

Even cancers that do not form solid tumors, such as lymphomas and leukemias, are burdened by the complications of tissue heterogeneity. Bone marrow aspirates, for example, contain many different cell types. Although procedures for isolating suspended cells, such as centrifugation through Ficoll® gradients (Amersham Pharmacia Biotech, Piscataway, NJ, USA), have been well established, the sorting process itself may induce changes in gene expression.

The selection of noncancerous controls also suffers from complications due to normal tissue heterogeneity. For example, the prostate can be anatomically divided into three zones that differ radically in their cancer-harboring potential. The peripheral zone gives rise to 80% of prostate cancers, the central zone appears to be cancer-resistant, and the transitional zone gives rise to benign prostatic hyperplasia and the remaining 20% of malignant cancers. Normal cells from the peripheral zone would appear to be, therefore, ideal controls for the study of most prostate cancers. Yet, in men over 50 yr old, the epithelium of the peripheral zone is frequently atrophied, which severely limits the amount of tissue available for collection. In addition, the available tissue often suffers from dysplasia, which gives rise to cancer and, therefore, cannot be considered a truly normal control. Even tissue that appears normal in the light microscope may be genotypically abnormal or exhibit a disrupted pattern of gene expression. Deng et al. observed genotypic abnormalities in normal tissue adjacent to breast carcinomas (59).

Other factors contributing to heterogeneity are the environmental and genetic background of the host. The treatment history of a patient can affect a tumor's expression pattern. Although Perou et al. showed that samples taken from the same breast tumor before and after treatment with doxorubicin were, in most cases, more similar to each other than either was to samples from other tumors, in 3 out of 20 cases, there was a significant change in tumor expression patterns that correlated with drug treatment (15). In addition, the genetic background of the host may help dictate a tumor's behavior by directly affecting the physiology of tumor cells or the behaviors of surrounding tissues that interact with the tumor.

In Vitro, In Situ, and In Silico Methods to Cope with Tissue Heterogeneity

A first step towards dealing with tissue heterogeneity is documenting tissue collection procedures carefully so that data can be more readily compared across studies. In addition, approaches that can help circumvent at least some of the problems caused by heterogeneity include the use of cell lines, performing *in silico* subtractions, and using microdissection tools (16).

Cells in culture have provided valuable information about cancer behavior and complemented tumor profiling studies. Many studies examining cancer-associated signaling pathways, for example, have made good use of cell lines carrying expression-inducible genes (39,41). However, cell lines often fall short of recapitulating the in vivo situation. Gene expression is greatly influenced by environmental conditions, including the presence of soluble factors, extracellular matrix molecules, and interactions with various cell types, all of which are conditions that are radically altered in cell culture. Indeed, the colon carcinoma cell lines included by Alon and coworkers expressed such unique patterns of gene expression that cluster analysis separated them into a group distinct from both tumor and normal in vivo tissues (20).

Cell lines can help deal with tissue heterogeneity in ways other than acting as simplified model systems for cancer cells. Using microarrays sporting 9703 cDNAs to obtain expression profiles of 60 cancer cell lines, Ross et al. discovered that the expression profiles of cancerous human breast tissue had recognizable counterparts in specific cell lines, thus providing a window into tumor cell type composition *(60)*. Tissue samples expressed genes associated with breast cell lines, such as the ER gene, as well as genes associated with stromal and immune cell lines.

Such information suggested the possibility of using *in silico* analyses to tease out the compositions of heterogeneous tumors. Indeed, by analyzing the expression profiles of 65 crudely enriched breast tumors and 17 cell lines of stromal, endothelial, mammary epithelial, and immune cellular origins, Perou et al. were able to identify at least eight groups of genes expressed by primary tumors that reflected the presence of multiple cell types *(15)*. Using algorithms to approximately gauge the relative abundance of the noncancerous cell subpopulations, they were then able to subtract these gene clusters from the data and generate a list enriched in tumor-specific transcripts.

The approach suffers from a few limitations, however. It is difficult to capture a truly complete expression profile of a complex mixture of cells, because low abundance transcripts tend to be swamped by abundant species. The time it takes for an RNA or DNA molecule to find and hybridize with its complementary partner is determined by the concentration of the partners and the diversity of sequences present in the mixture. Within a given amount of time, transcripts present at low levels in a highly diverse mixture are less likely to find their partners than abundant RNAs. This is potentially problematic, since low abundance transcripts often code for regulatory proteins, such as transcription factors, with key cellular functions.

Another complication is the emerging realization that noncancerous cells within a tumor can have gene expression signatures that are dramatically different from those of healthy tissues. A comparison of the expression patterns of endothelial cells derived from the blood vessels of malignant colorectal tissue and endothelial cells from normal vessels, for example, revealed 79 differentially expressed transcripts *(61)*. These results suggest that to perform accurate subtractions, it may be necessary to obtain expression profiles of noncancerous cells specifically associated with tumors.

Laser capture microdissection (LCM) is now making it much easier to obtain such small subpopulations from a variety of tissues *(62)*. More importantly, in some cases, it is offering a superior alternative to *in silico* subtraction by providing well-isolated samples of cancerous cells, which can be normalized to total DNA, total number of cells, or levels of control markers *(10,53,63–65)*. The study of small areas of tissue in particular, such as precancerous lesions, stands to benefit greatly from LCM.

LCM relies on a laser beam to transfer cells selected under a microscope onto a polymer film. To release the captured cells' RNA, the film is incubated with an extraction buffer. For example, by applying this technique to both normal and cancerous oral epithelial cells, Alevizos et al. recently identified 404 transcripts associated with oral cancer using oligonucleotide arrays *(10)*. The clustering of normal samples and tumor samples into two highly distinct groups indicated that LCM succeeded in isolating pure homogenous samples.

Although the difficulty of obtaining sufficient amounts of RNA for hybridization to microarrays has limited the adoption of this technology, several studies *(10,63,64,66)* have developed amplification procedures that circumvent this problem. Luzzi and colleagues recently used two rounds of linear amplification to perform expression profiling of ductal carcinoma using oligonucleotide arrays *(64)*. The authors used LCM-derived RNA to synthesize double-stranded cDNA coupled to a T7 promoter, but instead of producing labeled target directly from the cDNA as is usually done, they used the cDNA to generate unlabeled antisense RNA. A second round of amplification was then performed by using the amplified RNA as a template for synthesizing cDNA again. After second-strand synthesis, this amplified cDNA was used to produce labeled target. Based on expression levels of internal controls, the authors showed that their procedure was reproducible and yielded a sensitivity and precision comparable to standard methodologies that rely on 200–1000 times more RNA.

Despite these advances, LCM has its limitations. LCM requires specialized equipment and training. In addition, working with highly localized samples of tissue is not always desirable. Sampling a small subset of cells in a heterogeneous tumor harboring multiple malignant clones, which behave very differently from each other, for example, can provide a biased view and an oversimplification of the global picture *(67)*.

Target Preparation, Hybridization, and Signal Detection

Most target preparation protocols use total RNA or poly(A)$^+$ RNA as a template for synthesizing cDNA or cRNA while incorporating biotinylated or fluorescently labeled molecules (Fig. 2A) *(37)*. This step usually results in only a modest amplification of the starting material due to the ability of DNA and RNA polymerases to read templates multiple times. If a greater amplification is required, as described for LCM, additional synthetic steps can be inserted in the protocol. However, it is important that these amplifications maintain relative abundance levels to provide accurate results.

To monitor the subsequent hybridization step, labeled targets are spiked with labeled control transcripts. Each GeneChip array, for example, contains probe sets for several prokaryotic genes that serve as hybridization controls when complementary labeled RNAs are mixed with the experimental target. Typical hybridizations are performed at 45°C for 16 h. Washing of the arrays, and staining when biotin-tagged targets are used, can be done either manually or robotically. Using the GeneChip Fluidics Station 400, arrays can be processed in over 1 h.

Laser scanners are used to detect the hybridization signals. These devices have been steadily improving since the first generation of scanners was developed by adapting confocal microscopes. Current scanners have a 3-µm resolution, scan over 20 lines per second, detect as few as 400 phycoerythrin-labeled molecules in a 20 × 20 µm feature, and generate confocal images consisting of 25 million pixels in less than 10 min *(37)*.

The push to collect more data faster is fostering the development of, not only higher density arrays, but also methods to streamline array processing. The capacity of fluidics systems is being increased, and systems for automating the loading of arrays for scanning are being developed. Zarrinkar et al. recently increased the capacity for parallel processing by using "arrays of arrays," glass wafers that contain 49 individual oligonucleotide arrays *(68)*. During hybridization, the arrays and samples are kept isolated from one another by a silicone seal. Since washing and staining do not require array-specific handling, the hybridization chamber is subsequently converted into a flow cell for applying solutions over the entire set of arrays as a single unit. Using this technol-

A Expression Assay Format

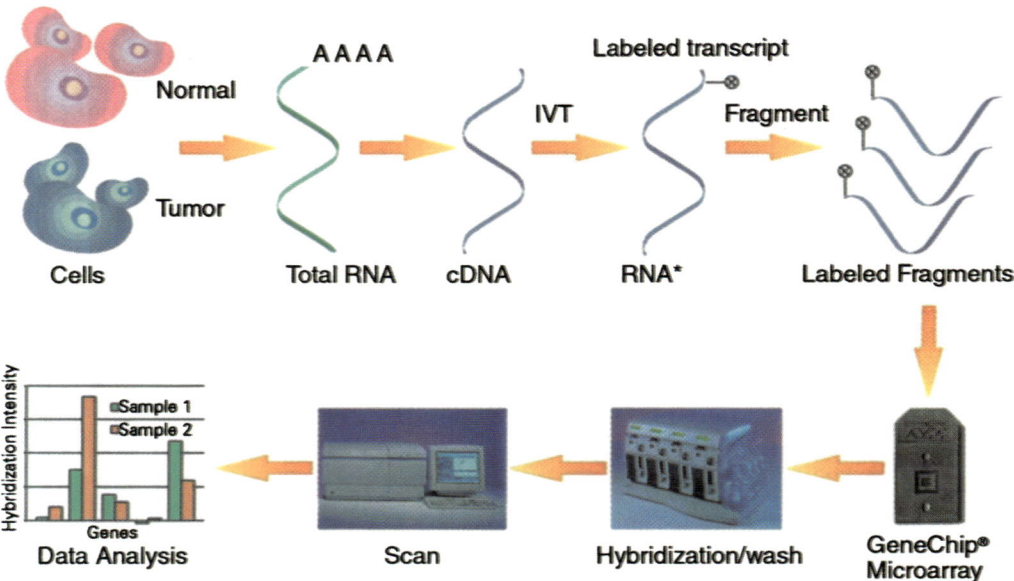

B Predicting the Tumor Class of an Unknown Sample

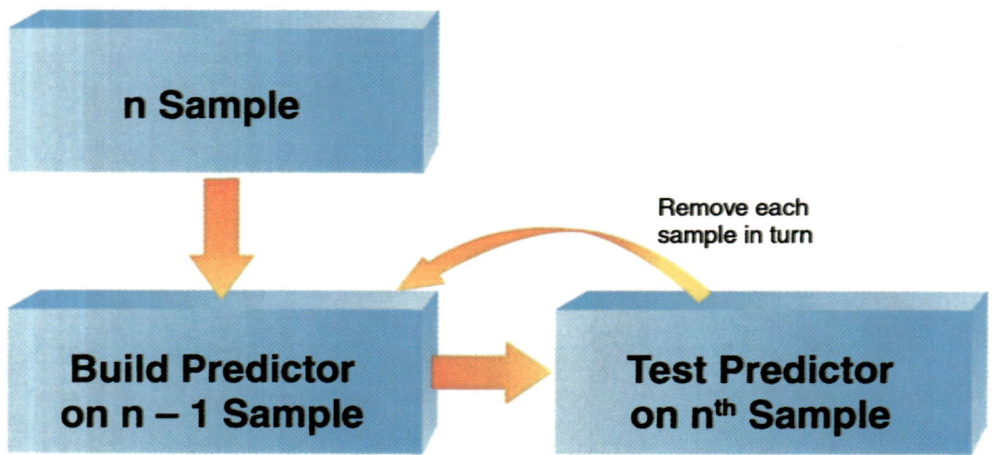

Fig. 2. Expression assay format and experimental design for tumor profiling. (**A**) Total poly(A) RNA from tumor, as well as normal cells are isolated. Double-stranded cDNA is synthesized, and labeled cRNA is obtained through in vitro transcription. The labeled cRNA is then hybridized to the probe arrays. The probe array is subsequently scanned, and fluorescence emission is prepared for data analysis. (**B**) The leave-one out approach is a commonly used strategy to build classifiers of tumor types and subtypes. For n samples, an n-1 predictor is built. This predictor is used to calculate the classification of the missing sample. Each of the samples is then removed, in turn, and classified based on the "overall predictor," which takes into account all other (n-1) samples.

ogy, Welsh and colleagues completed a tumor profiling study of ovarian cancer in a single experiment and in a fraction of the time required by conventional methods (12).

Data Analysis

Most of the time invested in tumor profiling studies, however, occurs during data analysis. The process often continues even after the analysis of a dataset has been published, as new statistical methods offer the possibility of extracting new or higher quality information.

The first step in dealing with the vast amounts of data generated by individual arrays (a single 12,000-gene GeneChip array produces 87 megabytes of information [33]) is to prepare the data for mining. Nonspecific signals are removed by subtracting MM signals from PM signals and calculating the ratio of their intensities. To assess the threshold above which differences in PM and MM values are meaningful, the background noise due to random variations in pixel intensity must be determined and removed. Finally, transcript levels are calculated by combining the data from probes representing the same transcript. Software packages to perform these calculations have been steadily improving. The latest package produced by Affymetrix, the Microarray Suite 5.0, for example, provides measures of statistical significance (p values) and confidence limits for the results of each probe set, and allows the user to adjust the balance between sensitivity and specificity. Initiatives to increase the openness of the microarray data analysis and management tools are currently underway.

Regardless of whether the goal of a study is to find new genes and pathways involved in cancer, discover new tumor classes, or develop predictors for known tumor classes, the key to subsequent data analysis is to reduce the number of experimental variables and to home in on a small subset of informative genes. A common first step is to set a statistical threshold to identify genes whose expression varies significantly between, but not within, two conditions of interest, such as normal vs tumor tissues. A student t-test, for example, can be applied to single out genes with differences in expression levels below a particular p value. The method used, however, should be compatible with the statistical distribution of the data.

A number of algorithms can then be used to group together tissue samples or genes with similar properties. Because of the large number of expression measurements, however, the significance of these correlations must be evaluated. Common approaches include setting aside samples for independently testing genes of interest (Fig. 2B) and applying permutation tests in which the data are scrambled or noise is introduced to determine how much the identified correlations differ from correlations that could arise randomly.

Statistical significance, however, is not always indicative of biological significance. Expression patterns may result from chance associations or from artifacts of sample preparation. Hughes et al. (69) have found that spurious correlations may emerge when profiling cells with unstable genomes. The authors discovered that yeast strains, which either lacked or harbored extra chromosomes, had altered transcript levels corresponding to the chromosomal alterations. The mRNA abundance of nearly every gene on trisomic chromosomes was increased, while that of monosomic chromosomes was decreased. Since many cancer cells are aneuploid, these correlations are likely to surface frequently in tumor profiling. Indeed, Perou and colleagues found that, due to an amplification of a region of chromosome 17, breast tumors overexpressing *Erb-B2* also overexpressed several neighboring genes *(15)*. If not accounted for, these correlations could be misleading when identifying mechanisms and potential therapeutic targets.

Microarray studies aimed at discovering new tumor classes or novel relationships between genes often use clustering analyses. These methods rely on unsupervised algorithms, in which the data are searched for patterns without imposing preconceived assumptions. They draw inferences from the expression data alone, without incorporating known biological, clinical, or demographic information. Self-organizing maps (SOMs) *(70)*, hierarchical algorithms *(71)*, and k-mean clustering algorithms *(72)* are examples of such techniques. Although some of these algorithms allow users to impose some constraints on the clusters generated *(70)*, the methods' main strength lies in providing systematic and unbiased analyses of expression data. As previously described, Golub et al. *(3)* first demonstrated the potential of this approach by applying a SOM to automatically group 38 leukemia samples based on the expression pattern of 6817 genes. The SOM-generated clusters closely paralleled known ALL and AML classes. Since this proof-of-principle study, many studies have revealed entirely new classes of tumors (e.g., *7,15,73,74*).

For some applications, it is useful to incorporate prior knowledge into the analyses. When Pomeroy and coworkers applied an SOM algorithm to a group of 60 medulloblastoma samples, for example, the tumors segregated into two well-defined groups that correlated with a *bona fide* biological parameter, i.e., the expression of ribosomal proteins, but the groups were not significantly correlated with clinical outcome *(1)*. To direct their search towards a clinically relevant classification scheme, the authors subsequently applied a k-nearest neighbors algorithm, which is a supervised learning algorithm. Supervised algorithms can be trained to search for expression patterns associated with particular attributes and then used to predict those attributes in new unknown samples. Supplying the algorithm with expression data that had been sorted based on patient survival, the authors identified a set of genes that predicted patient outcome extremely accurately. Eighty-five percent of the patients that were predicted to have good outcomes survived over a 5-yr period, compared to only 22% of those predicted to have poor outcomes. The expression-based predictor proved superior to previously identified prognostic factors, including disease distribution and the expression of the TrkC neurotrophin receptor. Other supervised algorithms that have yielded promising tumor classifiers include weighted voting algorithms *(3,4)*, the support vector machine method *(75)*, Bayesian models *(6)*, and artificial neural networks *(76)*.

However, many challenges remain. Comparisons between some of the methods have revealed strengths and weaknesses of each algorithm and provide some guidelines for selecting among the many available options. For example, nearest neighbor methods *(1)* are relatively intuitive and user-friendly, but provide less insight into mechanisms underlying class distinctions than classification trees *(77)*. Systematic comparisons of many methods, however, have not been performed, and since the development of classifier algorithms is an active area of research, new options are continuously emerging.

Much effort is currently being directed at developing more systematic methods for the initial selection of gene subsets, which should help optimize noise reduction and, consequently, improve classifiers' predictive abilities. In addition, statisticians are developing methods to more reliably detect variations in low abundance transcripts, which can be easily lost in the experimental noise. This is of particular importance in the study of cancer biology, since alterations of cellular pathway regulators, many of which are expressed at low levels, are frequently associated with the disease *(33)*.

Another ongoing challenge is the development of improved visualization methods to facilitate data examination and interpretation *(78)*. Eisen et al. developed a now widely used visualization method in which relationships among genes or tissues are represented by phylogeny-like trees *(71)*. More recently, Kim et al. used a three-dimensional approach to present their data *(79)*. Groups of related genes were represented as mountains, with the entire expression signature appearing as a mountain range. Distances correlated with gene similarity and mountain heights represented the density of genes in a similar location.

THE WAY AHEAD

Making the most of the wealth of tumor profiling data greatly depends on building databases to share and compare data across studies. Indeed, ongoing efforts to catalog expression data are already bearing fruit, but many challenges remain. As described by Aach and colleagues, a useful way of managing expression data systematically is to convert them into estimates of relative abundance (ERAs), which are measurements of the fractional abundance of an RNA in a single condition *(80)*. Although GeneChip arrays provide measurements of transcript levels that can be readily transformed into this format, other methods, such as cDNA arrays, do not. Additional factors interfere with establishing consistent standards, even when using ERAs. There is a lack of agreement as to how the multiple measurements of single transcripts generated by GeneChip experiments, for example, should be combined or selected for further analysis *(80)*. In addition, a lack of consistency in transcript nomenclature and incompatibilities in the software for acquiring, storing, managing, and analyzing results also hinder data comparison.

Like any new clinical tool, microarrays will have to undergo a rigorous appraisal of their sensitivity, specificity, and predictive value *(24)*. In addition, the full maturation of these tests will depend on the adoption of statistical methods that not only predict the probability of a tumor belonging to a particular category, such as ER positive or negative as in the case of breast tumors, but provide a formal assessment of the uncertainty associated with that determination *(6)*. Expression patterns that significantly stray from a canonical signature used as a classifier might reflect intra-tumor heterogeneity or a state of transition between two tumor classes. In either case, knowledge of this degree of uncertainty should help health professionals weigh the therapeutic alternatives.

Some investigators worry that the high costs of oligonucleotide microarrays will limit the speed with which they are introduced into clinical practice *(24)*. They predict that high costs will initially restrict their use to well-funded research units. However, several factors counter this concern. The clinical benefits provided by microarrays will almost certainly outweigh their costs. The ability to predict therapeutic outcomes, for example, should greatly reduce the number of failed treatments, and result in huge savings and dramatic improvements in the quality of patient care. The prices of oligonucleotide arrays have been steadily dropping, while the amount of information per chip has increased from approx 250 to nearly 20,000 transcripts.

The most critical factor for bringing microarray-based tests to maturity will probably be the large-scale validation of the many new and rapidly emerging discoveries. Well-designed studies that integrate clinical and histological data and are based on large numbers of samples are sorely needed. Indeed, several studies have already suggested the importance of sample number *(1,2,4,15,74,81)*. There is little doubt that DNA microarrays, and oligonucleotide arrays in particular, have a bright future as tools for tumor profiling. But as is true of any technology in its infancy, it will require responsible upbringing.

ACKNOWLEDGMENTS

M. C. and D. D.-W. contributed equally to this manuscript. We gratefully acknowledge Elizabeth Kerr and Thane Kreiner for critically reading the manuscript and Janet Warrington and Mamatha Mahadevappa for helpful discussions.

REFERENCES

1. Golub, T. R., Slonim, D. K., Tamayo, P., et al. (1999) Molecular classification of cancer: class discovery and class prediction by gene expression monitoring. *Science* **286,** 531–537.
2. Shipp, M., Ross, K., Tamayo, P., et al. (2002) Diffuse large B-cell lymphoma outcome prediction by gene expression profiling and supervised machine learning. *Nat. Med.* **8,** 68–74.
3. Clark, E. A., Golub, T. R., Lander, E. S., and Hynes, R. O. (2000) Genomic analysis of metastasis reveals an essential role for RhoC. *Nature* **406,** 532–535.
4. West, M., Blanchette, C., Dressman, H., et al. (2001) Predicting the clinical status of human breast cancer by using gene expression profiles. *Proc. Natl. Acad. Sci. USA* **98,** 11,462–11,467.
5. Bhattacharjee, A., Richards, W. G., Staunton, J., et al. (2001) Classification of human lung carcinomas by mRNA expression analysis. *Proc. Natl. Acad. Sci. USA* **98,** 13,790–13,795.
6. Stamey, T. A., Warrington, J. A., Caldwell, M. C., et al. (2001) Molecular genetic profiling of Gleason grade 4/5 prostate cancers versus benign prostatic hyperplasia. *J. Urol.* **166,** 2171–2177.
7. Notterman, D. A., Alon, U., Sierk, A. J., and Levine, A. J. (2001) Transcriptional gene expression profiles of colorectal adenoma, adenocarcinoma, and normal tissue examined by oligonucleotide arrays. *Cancer Res.* **61,** 3124–3130.
8. Alevizos, I., Mahadevappa, M., Zhang, X., et al. (2001) Oral cancer in vivo gene expression profiling assisted by laser capture microdissection and microarray analysis. *Oncogene* **20,** 6196–6204.
9. Thykjaer, T., Workman, C., Kruhoffer, M., et al. (2001) Identification of gene expression patterns in superficial and invasive human bladder cancer. *Cancer Res.* **61,** 2492–2499.
10. Welsh, J. B., Zarrinkar, P. P., Sapinoso, L. M., et al. (2001) Analysis of gene expression profiles in normal and neoplastic ovarian tissue samples identifies candidate molecular markers of epithelial ovarian cancer. *Proc. Natl. Acad. Sci. USA* **98,** 1176–1181.
11. Tackels-Horne, D., Goodman, M. D., Williams, A. J., et al. (2001) Identification of differentially expressed genes in hepatocellular carcinoma and metastatic liver tumors by oligonucleotide expression profiling. *Cancer* **92,** 395–405.
12. Graveel, C. R., Jatkoe, T., Madore, S. J., Holt, A. L., and Farnham, P. J. (2001) Expression profiling and identification of novel genes in hepatocellular carcinomas. *Oncogene* **20,** 2704–2712.

13. Pomeroy, S. L., Tamayo, P., Gaasenbeek, M., et al. (2001) Gene expression-based classification of outcome prediction of central nervous system embryonal tumors. *Nature* **415,** 436–442.
14. MacDonald, T. J., Brown, K. M., LaFleur, B., et al. (2001) Expression profiling of medulloblastoma: PDGFRA and the RAS/MAPK pathway as therapeutic targets for metastatic disease. *Nat. Genet.* **29,** 143–152.
15. Perou, C. M., Sorlie, T., Eisen, M. B., et al. (2000) Molecular portraits of human breast tumours. *Nature* **406,** 747–752.
16. Liotta, L. and Petricoin, E. (2000) Molecular profiling of human cancer. *Nat. Rev. Genet.* **1,** 48–56.
17. Lockhart, D. J. and Winzeler, E. A. (2000) Genomics, gene expression and DNA arrays. *Nature* **405,** 827–836.
18. Pandey, A. and Mann, M. (2000) Proteomics to study genes and genomes. *Nature* **405,** 837–846.
19. Golub, T. R. (2001) Genome-wide views of cancer. *N. Engl. J. Med.* **344,** 601–602.
20. Alon, U., Barkai, N., Notterman, D. A., et al. (1999) Broad patterns of gene expression revealed by clustering analysis of tumor and normal colon tissues probed by oligonucleotide arrays. *Proc. Natl. Acad. Sci. USA* **96,** 6745–6750.
21. Strausberg, R. L. and Klausner, R. D. (2001) Reading the molecular signatures of cancer, in *Microarrays and Cancer Research* (Warrington, J. A., Wong, D., and Todd, R., eds.), Biotechniques Press, Westborough, MA, pp. xi–xvi.
22. Schuler, G. D. (1997) Pieces of the puzzle: expressed sequence tags and the catalog of human genes. *J. Mol. Med.* **75,** 694–698.
23. Velculescu, V. E., Zhang, L., Vogelstein, B., and Kinzler, K. W. (1995) Serial analysis of gene expression. *Science* **270,** 484–487.
24. Aitman, T. J. (2001) DNA microarrays in medical practice. *BMJ* **323,** 611–615.
25. The Chipping Forecast. (1999) *Nat. Genet.* **21(Suppl),** 1.
26. Blohm, D. H. and Guiseppi-Elie, A. (2001) New developments in microarray technology. *Curr. Opin. Biotechnol.* **12,** 41–47.
27. Walt, D. R. (2000) Techview: molecular biology. Bead-based fiber-optic arrays. *Science* **287,** 451–452.
28. Sosnowski, R. G., Tu, E., Butler, W. F., O'Connell, J. P., and Heller, M. J. (1997) Rapid determination of single base mismatch mutations in DNA hybrids by direct electric field control. *Proc. Natl. Acad. Sci. USA* **94,** 1119–1123.
29. Edman, C. F., Raymond, D. E., Wu, D. J., et al. (1997) Electric field directed nucleic acid hybridization on microchips. *Nucleic Acids Res.* **25,** 4907–4914.
30. Rose, S. D. (2001) Spotted arrays: technology overview, in *Microarrays and Cancer Research* (Warrington, J. A., Wong, D., and Todd, R., eds.), Eaton Publishing, Westborough, MA, pp. 1–12.
31. Blanchard, A. P., Kaiser, R. J., and Hood, L. E. (1996) High-density oligonucleotide arrays. *Biosens. Bioelectron.* **6/7,** 687–690.
32. Hughes, T. R., Mao, M., Jones, A. R., et al. (2001) Expression profiling using microarrays fabricated by an ink-jet oligonucleotide synthesizer. *Nat. Biotechnol.* **19,** 342–347.
33. Triche, T. J., Schofield, D., and Buckley, J. (2001) DNA microarrays in pediatric cancer. *Cancer J.* **7,** 2–15.
34. Fodor, S. P., Read, J. L., Pirrung, M. C., Stryer, L., Lu, A. T., and Solas, D. (1991) Light-directed, spatially addressable parallel chemical synthesis. *Science* **251,** 767–773.
35. Lipshutz, R. J., Fodor, S. P., Gingeras, T. R., and Lockhart, D. J. (1999) High density synthetic oligonucleotide arrays. *Nat. Genet.* **21,** 20–24.
36. Lockhart, D. J., Dong, H., Byrne, M. C., et al. (1996) Expression monitoring by hybridization to high-density oligonucleotide arrays. *Nat. Biotechnol.* **14,** 1675–1680.

37. Warrington, J. A., Dee, S., and Trulson, M. (2000) Large-scale genomic analysis using Affymetrix GeneChip probe arrays, in *Microarray Biochip Technology* (Schena, M., ed.), Biotechniques Books, Natick, MA, pp. 119–148.
38. Hippo, Y., Yashiro, M., Ishii, M., et al. (2001) Differential gene expression profiles of scirrhous gastric cancer cells with high metastatic potential to peritoneum or lymph nodes. *Cancer Res.* **61,** 889–895.
39. Harkin, D. P., Bean, J. M., Miklos, D., et al. (1999) Induction of GADD45 and JNK/SAPK-dependent apoptosis following inducible expression of BRCA1. *Cell* **97,** 575–586.
40. Svaren, J., Ehrig, T., Abdulkadir, S. A., Ehrengruber, M. U., Watson, M. A., and Milbrandt, J. (2000) EGR1 target genes in prostate carcinoma cells identified by microarray analysis. *J. Biol. Chem.* **275,** 38,524–38,531.
41. Zhao, R., Gish, K., Murphy, M., et al. (2000) Analysis of p53-regulated gene expression patterns using oligonucleotide arrays. *Genes Dev.* **14,** 981–993.
42. Kannan, K., Amariglio, N., Rechavi, G., et al. (2001) DNA microarrays identification of primary and secondary target genes regulated by p53. *Oncogene* **20,** 2225–2234.
43. Coller, H. A., Grandori, C., Tamayo, P., et al. (2000) Expression analysis with oligonucleotide microarrays reveals that MYC regulates genes involved in growth, cell cycle, signaling, and adhesion. *Proc. Natl. Acad. Sci. USA* **97,** 3260–3265.
44. Gerhold, D., Lu, M., Xu, J., Austin, C., Caskey, C. T., and Rushmore, T. (2001) Monitoring expression of genes involved in drug metabolism and toxicology using DNA microarrays. *Physiol. Genomics* **5,** 161–170.
45. Hill, A. A., Hunter, C. P., Tsung, B. T., Tucker-Kellogg, G., and Brown, E. L. (2000) Genomic analysis of gene expression in *C. elegans*. *Science* **290,** 809–812.
46. de Saizieu, A., Certa, U., Warrington, J., Gray, C., Keck, W., and Mous, J. (1998) Bacterial transcript imaging by hybridization of total RNA to oligonucleotide arrays. *Nat. Biotechnol.* **16,** 45–48.
47. Li, C. and Wong, W. H. (2001) Model-based analysis of oligonucleotide arrays: expression index computation and outlier detection. *Proc. Natl. Acad. Sci. USA* **98,** 31–36.
48. Wodicka, L., Dong, H., Mittmann, M., Ho, M. H., and Lockhart, D. J. (1997) Genome-wide expression monitoring in Saccharomyces cerevisiae. *Nat. Biotechnol.* **15,** 1359–1367.
49. Ishii, M., Hashimoto, S., Tsutsumi, S., et al. (2000) Direct comparison of GeneChip and SAGE on the quantitative accuracy in transcript profiling analysis. *Genomics* **68,** 136–143.
50. Fodor, S. P., Rava, R. P., Huang, X. C., Pease, A. C., Holmes, C. P., and Adams, C. L. (1993) Multiplexed biochemical assays with biological chips. *Nature* **364,** 555–556.
51. Singh-Gasson, S., Green, R. D., Yue, Y., et al. (1999) Maskless fabrication of light-directed oligonucleotide microarrays using a digital micromirror array. *Nat. Biotechnol.* **17,** 974–978.
52. Beier, M. and Hoheisel, J. D. (2000) Production by quantitative photolithographic synthesis of individually quality checked DNA microarrays. *Nucleic Acids Res.* **28,** E11.
53. Sgroi, D. C., Teng, S., Robinson, G., LeVangie, R., Hudson, J. R., Jr., and Elkahloun, A. G. (1999) In vivo gene expression profile analysis of human breast cancer progression. *Cancer Res.* **59,** 5656–5661.
54. Goldsworthy, S. M., Stockton, P. S., Trempus, C. S., Foley, J. F., and Maronpot, R. R. (1999) Effects of fixation on RNA extraction and amplification from laser capture microdissected tissue. *Mol. Carcinog.* **25,** 86–91.
55. Warrington, J. A., Nair, A., Mahadevappa, M., and Tsyganskaya, M. (2000) Comparison of human adult and fetal expression and identification of 535 housekeeping/maintenance genes. *Physiol. Genomics* **2,** 143–147.
56. Martin, K. J., Graner, E., Li, Y., et al. (2001) High-sensitivity array analysis of gene expression for the early detection of disseminated breast tumor cells in peripheral blood. *Proc. Natl. Acad. Sci. USA* **98,** 2646–2651.

57. Stamey, T. A., McNeal, J. E., Yemoto, C. M., Sigal, B. M., and Johnstone, I. M. (1999) Biological determinants of cancer progression in men with prostate cancer. *JAMA* **281**, 1395–1400.
58. Luo, J., Duggan, D. J., Chen, Y., et al. (2001) Human prostate cancer and benign prostatic hyperplasia: molecular dissection by gene expression profiling. *Cancer Res.* **61**, 4683–4688.
59. Deng, G., Lu, Y., Zlotnikov, G., Thor, A. D., and Smith, H. S. (1996) Loss of heterozygosity in normal tissue adjacent to breast carcinomas. *Science* **274**, 2057–2059.
60. Ross, D. T., Scherf, U., Eisen, M. B., et al. (2000) Systematic variation in gene expression patterns in human cancer cell lines. *Nat. Genet.* **24**, 227–235.
61. St. Croix, B., Rago, C., Velculescu, V., et al. (2000) Genes expressed in human tumor endothelium. *Science* **289**, 1197–1202.
62. Emmert-Buck, M. R., Bonner, R. F., Smith, P. D., et al. (1996) Laser capture microdissection. *Science* **274**, 998–1001.
63. Leethanakul, C., Patel, V., Gillespie, J., et al. (2000) Distinct pattern of expression of differentiation and growth-related genes in squamous cell carcinomas of the head and neck revealed by the use of laser capture microdissection and cDNA arrays. *Oncogene* **19**, 3220–3224.
64. Luzzi, V., Holtschlag, V., and Watson, M. A. (2001) Expression profiling of ductal carcinoma in situ by laser capture microdissection and high-density oligonucleotide arrays. *Am. J. Pathol.* **158**, 2005–2010.
65. Leethanakul, C., Patel, V., Gillespie, J., et al. (2000) Gene expression profiles in squamous cell carcinomas of the oral cavity: use of laser capture microdissection for the construction and analysis of stage-specific cDNA libraries. *Oral Oncol.* **36**, 474–483.
66. Ohyama, H., Zhang, X., Kohno, Y., et al. (2000) Laser capture microdissection-generated target sample for high-density oligonucleotide array hybridization. *BioTechniques* **29**, 530–536.
67. Hampton, G. M. (2001) Monitoring gene expression in human cancer using synthetic oligonucleotide arrays, in *Microarrays and Cancer Research* (Warrington, J., Todd, R., and Wong, D., eds.), Eaton Publishing, Westborough, MA.
68. Zarrinkar, P. P., Mainquist, J. K., Zamora, M., et al. (2001) Arrays of arrays for high-throughput gene expression profiling. *Genome Res.* **11**, 1256–1261.
69. Hughes, T. R., Roberts, C. J., Dai, H., et al. (2000) Widespread aneuploidy revealed by DNA microarray expression profiling. *Nat. Genet.* **25**, 333–337.
70. Tamayo, P., Slonim, D., Mesirov, J., et al. (1999) Interpreting patterns of gene expression with self-organizing maps: methods and application to hematopoietic differentiation. *Proc. Natl. Acad. Sci. USA* **96**, 2907–2912.
71. Eisen, M. B., Spellman, P. T., Brown, P. O., and Botstein, D. (1998) Cluster analysis and display of genome-wide expression patterns. *Proc. Natl. Acad. Sci. USA* **95**, 14,863–14,868.
72. Tavazoie, S., Hughes, J. D., Campbell, M. J., Cho, R. J., and Church, G. M. (1999) Systematic determination of genetic network architecture. *Nat. Genet.* **22**, 281–285.
73. Bittner, M., Meltzer, P., Chen, Y., et al. (2000) Molecular classification of cutaneous malignant melanoma by gene expression profiling. *Nature* **406**, 536–540.
74. Sorlie, T., Perou, C. M., Tibshirani, R., et al. (2001) Gene expression patterns of breast carcinomas distinguish tumor subclasses with clinical implications. *Proc. Natl. Acad. Sci. USA* **98**, 10,869–10,874.
75. Brown, M. P., Grundy, W. N., Lin, D., et al. (2000) Knowledge-based analysis of microarray gene expression data by using support vector machines. *Proc. Natl. Acad. Sci. USA* **97**, 262–267.
76. Khan, J., Wei, J. S., Ringner, M., et al. (2001) Classification and diagnostic prediction of cancers using gene expression profiling and artificial neural networks. *Nat. Med.* **7**, 673–679.

77. Zhang, H., Yu, C. Y., Singer, B., and Xiong, M. (2001) Recursive partitioning for tumor classification with gene expression microarray data. *Proc. Natl. Acad. Sci. USA* **98,** 6730–6735.
78. Young, R. A. (2000) Biomedical discovery with DNA arrays. *Cell* **102,** 9–15.
79. Kim, S. K., Lund, J., Kiraly, M., et al. (2001) A gene expression map for Caenorhabditis elegans. *Science* **293,** 2087–2092.
80. Aach, J., Rindone, W., and Church, G. M. (2000) Systematic management and analysis of yeast gene expression data. *Genome Res.* **10,** 431–445.
81. Alizadeh, A. A., Eisen, M. B., Davis, R. E., et al. (2000) Distinct types of diffuse large B-cell lymphoma identified by gene expression profiling. *Nature* **403,** 503–511.

4
Serial Analysis of Gene Expression (SAGE) in Cancer Research

C. Marcelo Aldaz

INTRODUCTION

Due to considerable research investment in the Human Genome Project, we will have access to the complete sequence information on all the genes encoded by the human genome within the next few years. Although this information will be of great value, it will not be sufficient to provide full understanding of the complex pathophysiology of diseases such as cancer. A more daunting task that lies ahead in this "post-genome era," is to understand how the different genes are altered in the various cancers and how complex gene interactions produce particular outcomes.

Studies aimed at tackling that challenging next level of complexity are well underway. In fact, in the last few years we have witnessed a revolution in technical approaches to analyze the transcriptome of the cancer cell. Numerous techniques have been developed for the analysis of global gene expression changes, in which thousands of gene targets can be assayed simultaneously. One of the first efforts utilized for analyzing gene expression in a global fashion has been basically a "brute force" approach to sequence as many clones as possible from tissue-specific cDNA libraries; the cDNAs analyzed in this way are known as expressed sequence Tags or ESTs. These efforts were led by consortia such as that of MERCK/Washington University and, in particular, by the Cancer Genome Anatomy Project (CGAP) *(1–3)*. These data were made available to the public via the GenBank® dbEST database and via the Unigene sequence cluster database (http://www.ncbi.nlm.nih.gov/UniGene).

More recently, other approaches were developed for analyzing gene expression changes, as described in various chapters in this text. Several of these currently very popular approaches rely on the microarraying of cDNAs or oligonucleotides on solid matrices. The readout in these techniques relies on comparative hybridization with labeled probes (i.e., cDNAs). All of these techniques require expensive hardware for arraying, scanning, and analysis of the experiments. In parallel with such developments, a completely different technical approach for the analysis of global gene expression was described a few years ago by Velculescu, Kinzler, and Vogelstein *(4)*. Serial analysis of gene expression (SAGE) is an extremely powerful, efficient, and comprehensive approach for analyzing gene expression profiles. Since its inception,

From: *Expression Profiling of Human Tumors: Diagnostic and Research Applications*
Edited by: Marc Ladanyi and William L. Gerald © Humana Press Inc., Totowa, NJ

SAGE has become one of the leading functional genomics methodologies, and various groups in academia and industry have demonstrated the utility of SAGE for novel gene and pathway discovery, for biomarker identification, and for gene expression profiling of numerous disease and normal conditions in multiple species.

ADVANTAGES AND PRINCIPLES OF THE SAGE METHODOLOGY

SAGE provides a statistical description of the mRNA population present in a cell without prior selection of the genes to be studied, and this constitutes the major advantage of SAGE vs gene expression chip-based assays. In other words, microarray approaches are limited to the study of the genes represented in the chip (i.e., the genes or ESTs had to have been cloned or their sequence known *a priori*). Secondly, for the interpretation of results, the cDNA and oligonucleotide microarray approaches usually rely on the comparative expression of specific transcripts in a population relative to the expression of other transcripts (e.g., housekeeping control transcripts) in the same population *(5)*. Alternatively, normalization of expression can be performed by comparing the relative expression of transcripts from a specific sample to a master control sample (mixture of RNAs from multiple cell lines) *(6)*. Thirdly, and not less important, there are now multiple gene expression microarray platforms available that introduce a multitude of variables (starting by the content and annotation of each array), making comparisons of similar studies from various laboratories very difficult. In fact, very recently a detailed study was reported matching mRNA measurements from two different microarray techniques (cDNAs vs Affymetrix [Santa Clara, CA, USA]). The conclusion of the study was that, in general, corresponding measurements from the two platforms showed very poor correlation. The authors further suggested that gene or probe-specific factors influence measurements differently in the two platforms *(7)*. Thus, this constitutes a major hurdle for comparative studies across platforms and ultimately for reaching meaningful conclusions on the validity of specific observations on global gene expression patterns.

Major advantages of the SAGE method are: (*i*) that the information generated is digital in format; (*ii*) that the data obtained can be directly compared with data generated from any other laboratory or with data available in public databases; and (*iii*) the information generated is virtually "immortal," and it has the advantage of being constantly updated and subject to reinterpretation, since the more we learn on the identification of new transcripts, the more complete and accurate the SAGE datasets become.

SAGE IS BASED IN THREE MAIN PRINCIPLES

The first of the three main principles of SAGE is that a short sequence tag of 14 bp is sufficient to identify uniquely the 3' end of most possible transcripts, provided that the tag is obtained from a defined position within the transcript. To this end, in one of the first steps of SAGE, a 4-bp cutter enzyme is used (usually *Nla*III) to anchor the site of the tag as the last restriction site for such enzyme (CATG for *Nla*III) prior to the poly(A) tail of each mRNA (see method below and Fig. 1). Four-base pair cutter enzymes such as *Nla*III are able to digest the DNA, on average, every 256 bp. Second, the concatenation of tags and cloning of individual unique concatamers of ditags allows for the efficient sequencing of multiple transcript tags per clone (typically 30–50 tags/clone). Thus, this method is several-fold (30- to 50-fold) more efficient

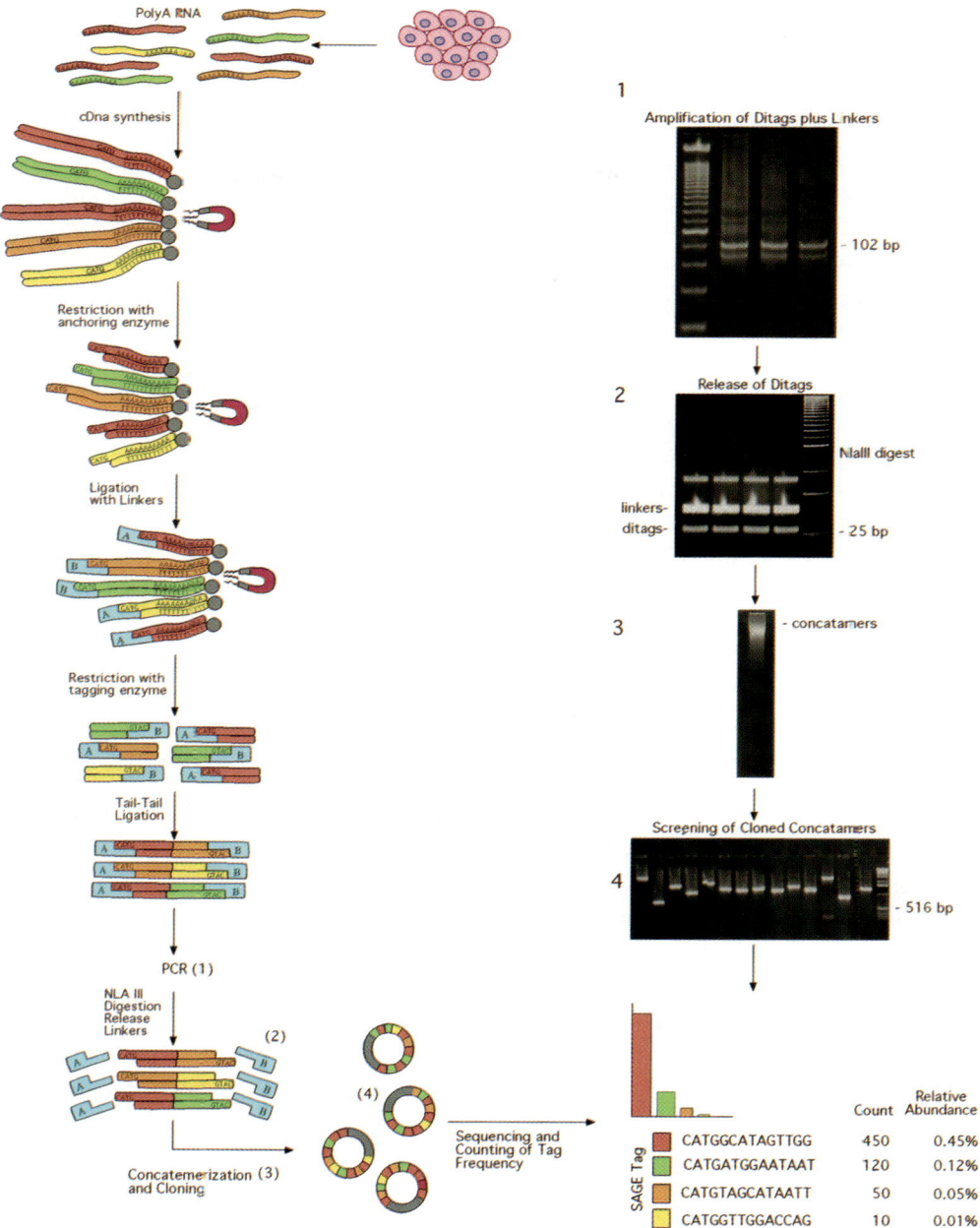

Fig. 1. Schematic depiction of the SAGE procedure. As template for cDNA synthesis, poly(A) RNA is isolated or directly captured from cell lysates using oligo(dt)-coated beads. The anchoring enzyme most frequently used is *Nla*III (a 4-bp cutter), which leaves the 3' of transcripts attached to the beads. The sticky CATG 5' overhangs are used to ligate specific linkers. These linkers have built in the recognition motif for the tagging enzyme, usually *Bsm*FI. This enzyme cuts the cDNA 14–15 bp 3' from the recognition motif, releasing the tags with the linkers. These tags are ligated tail to tail and amplified by PCR generating a 102-bp product (gel image 1). After purification, ditags are released from the linkers, giving rise to a product about 26 bp in size (gel image 2). These ditags are concatenated (gel image 3), cloned (gel image 4), and sequenced. The abundance of each tag in the cloned products is directly proportional to the abundance of the corresponding transcript in the original sample. The relative abundance of a transcript is calculated by dividing each specific tag count by the total number of tags sequenced.

than sequencing of ESTs, in which each clone represents a single transcript. Third, the expression level of a transcript is directly proportional to the number of times a specific tag is observed in the final count.

An additional key feature of the SAGE technique that makes it quite powerful, more than other differential gene expression techniques such as differential display, is that in the initial step, one obtains both quantitative information on the abundance of each mRNA and a partial sequence. This approach generates sequence information that allows not only the identification of all the transcripts being expressed in a normal or cancerous cell at any given time, but also provides quantitative information on the relative abundance of each of the transcripts. These data allow for the immediate analysis of the statistical significance of differences between samples.

One disadvantage of SAGE is that the methodology is not ideally suited for the comparison of multiple samples (e.g., hundreds of samples) in a relatively short time, in contrast with microarray approaches. However, this is only a limitation of the sequencing power of the laboratory performing the studies. Obviously, if the sequencing power is increased, the ability of analyzing multiple samples increases proportionally.

SUMMARY OF THE SAGE PROCEDURE

The SAGE procedure has previously been described in detail by Velculescu et al. *(4)*. The various steps of the SAGE methodology are schematically depicted in Fig. 1. Briefly, polyadenylated RNA is prepared, and double-stranded cDNA is synthesized. Alternatively, mRNAs can be directly captured from cell lysates (MicroSAGE procedure). Biotinylated-oligo(dT), or oligo(dT) bound to magnetic beads, is used as primer for first strand synthesis. The captured cDNAs are then digested with an "anchoring" restriction enzyme, usually *Nla*III, which leaves a 3' overhang. The 3' fragments are then isolated using the magnetic beads or streptavidin-coated beads. Two linkers, each containing the recognition sequence for a "tagging" restriction enzyme (type IIs restriction enzyme), usually *Bsm*FI, are ligated onto the *Nla*III overhangs. The tagging enzyme produces a staggered cut, offset by about 14–15 bp 3' from the recognition sequence. Subsequent digestion with *Bsm*FI and blunt end fill-in produces fragments of each cDNA molecule containing unique 14- to 15-bp sequences (including the *Nla*III sequence) that provide a "SAGE tag" specific to each expressed gene. The abundance of each tag in the population is proportional to the abundance of the corresponding mRNA in the original RNA population. These tags are then ligated tail-to-tail, amplified by polymerase chain reaction (PCR), and the linkers are released by digesting with *Nla*III again. The resulting ditags are purified, concatenated, and cloned. Each concatamer insert results in a randomly organized "series" of ditags of approx 20–24 bp, each flanked by the recognition sequence of the primary anchoring enzyme *Nla*III, i.e., the CATG sequence. Approximately 30–40 individual tags are produced per clone. In a typical experiment, approx 2000–3000 clones are sequenced to yield SAGE libraries of 60,000–100,000 transcript tags.

By sequencing the cloned tag concatamers and determining the frequency distribution of the total tag population (i.e., determining the abundance of each tag species), one obtains a statistical picture of the relative abundance of the different mRNAs expressed in the original cell population. Statistical analysis and comparison between different samples is performed following Zhang et al. *(8)*.

DATA ANALYSIS AND BIOINFORMATICS

SAGE tag information can be extracted from analyzing the sequencing files by various means or by using proprietary software developed by the K. Kinzler laboratory (John Hopkins University, School of Medicine, Baltimore MD, USA). In order to match tag identity, the experimental tag library information obtained is compared with virtual tag libraries generated in silico from nucleotide sequences obtained from all cDNA sequences reported and archived at the GenBank databases (National Center for Biotechnology Information [NCBI]). These comparisons with existing databases can be performed using the mentioned SAGE software, via resources developed by the NCBI (SAGEmap resources) or by using simple tag extraction procedures that can be developed using commercially available software tools (e.g., FileMaker Pro).

Upon analysis of the experimental SAGE library, one obtains quantitative information on three types of transcripts: (*i*) tags identifying known genes; (*ii*) tags identifying anonymous expressed sequences (i.e., EST clusters) obtained from the GenBank EST databases; and (*iii*) tags identifying transcripts with no matches in any of the available databases. This last feature is in itself a very important difference with other techniques for the study of global gene expression, since it allows one the potential identification and cloning of novel genes.

The SAGE software (Johns Hopkins University) has also various built-in statistical analysis tools that can be used to analyze datasets. The program is able to calculate the relative likelihood that a difference would be seen by chance for an individual tag or an entire project. For each p chance calculation, simulations can be performed assuming the null hypothesis, and the p chance value represents the fraction of simulations that yielded a difference equal or greater than the observed difference. This is a relative probability of obtaining the observed differences due to random variation. The general approach and rationale followed has been previously described *(8,9)*.

Further analysis of SAGE data, e.g., the comparison of multiple SAGE libraries, can be performed using other publicly available software tools such as the clustering programs developed at Stanford University *(10)* (Fig. 2).

PUBLICLY AVAILABLE SAGE RESOURCES

Investigators in collaboration with the NCBI have developed a very valuable centralized resource for the archiving and analysis of SAGE data (SAGEmap). This on-line resource has been developed and maintained by the National Cancer Institute's initiative named the CGAP *(1,11)*. The main goal of this effort is to facilitate the study of gene expression from normal tissues, cancer tissues, and model systems. This database can be found at (http://www.ncbi.nlm.nih.gov/SAGE). At the time of this writing, the SAGEmap database consisted of a total of over 5.5 million transcript tags representing more than 110 libraries from various tissues and cell lines, all of which is information that has been deposited by various laboratories from around the U.S.

It is essential for the correct matching of SAGEtag to genes that the true 3' end of transcripts be identified. Thus, the SAGEmap resource is linked to the Unigene NCBI database, which contains the unique clusters of complete cDNAs and the corresponding matching ESTs sequences to date that have been deposited in GenBank. The best possible SAGE tags for each Unigene cluster are identified, a link is provided, and

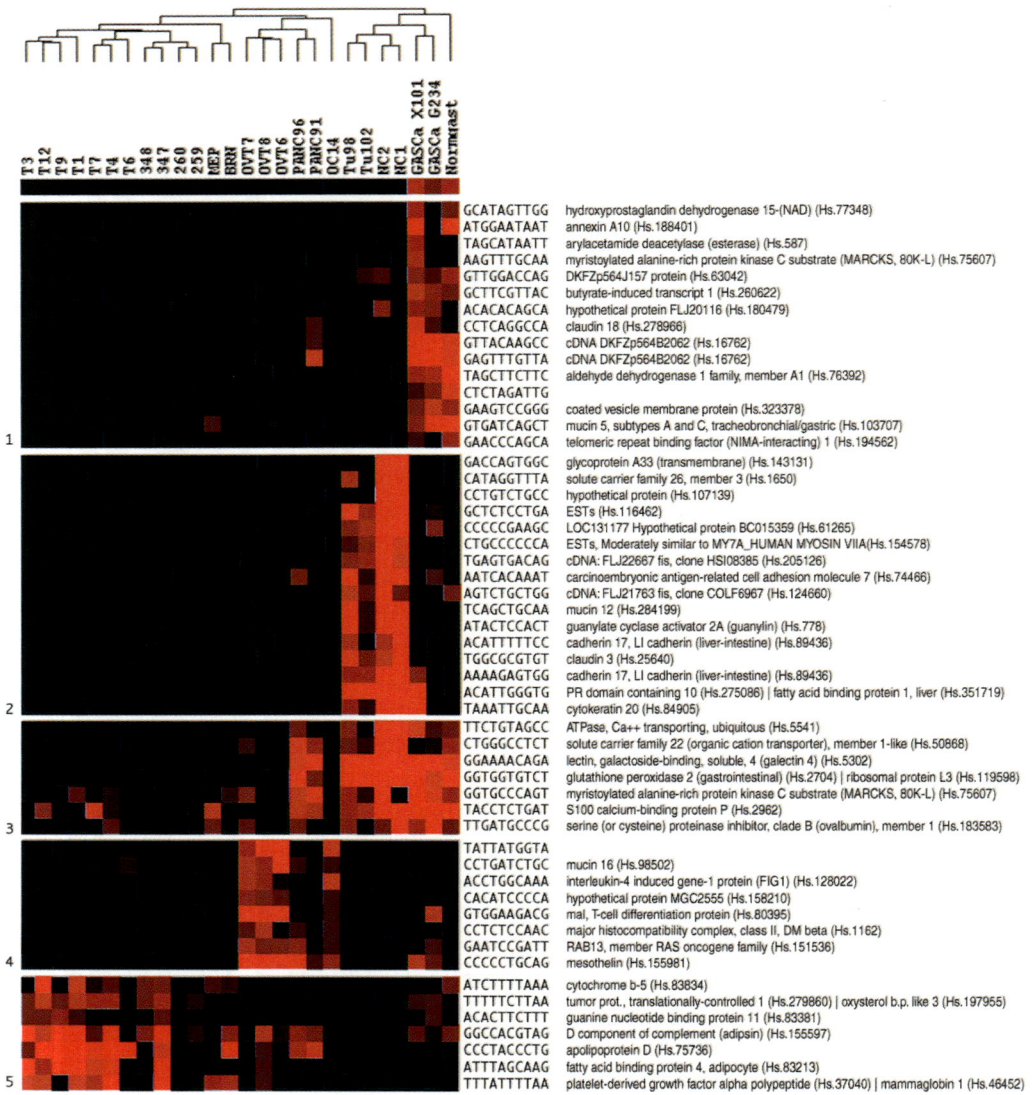

Fig. 2. Representative hierarchical clustering analysis of normal and tumor SAGE libraries. Breast carcinomas: T3, T12, T9, T1, T7, T4, T6, 347, (348 corresponding metastasis), 259, (260 corresponding metastasis); normal breast tissues: MEP and BRN. T1–T12 from the Aldaz laboratory and the other samples as previously reported by Porter et al. *(30)*. Ovarian carcinomas: OVT6, OVT7, OVT8, and OC14. Pancreatic carcinomas: Panc96 and Panc91. Colon tumors: Tu98 and Tu102; normal colon: NC1 and NC2. Gastric carcinoma xenografts: GASCa X101 and CASCa G234; normal gastric tissue: normagast. All the SAGE libraries analyzed with exception of T1–T12 were downloaded from (http://www.ncbi.nlm.nih.gov/SAGE). Libraries were normalized to 50,000 tags/library. The absolute abundance of each SAGE tag correlates with red color intensity, black with tag not present. Only some representative clusters are shown. Cluster 1, genes highly expressed in gastric tissue-derived samples, normal and tumors; cluster 2, tags highly abundant in colon samples; cluster 3, tags highly expressed in gastrointestinal tissues and pancreatic carcinomas; cluster 4, ovarian carcinoma gene cluster, note that mesothelin is also expressed in pancreatic tumors *(25)*; cluster 5, genes predominantly expressed in breast tumors.

virtual analysis on the relative expression of each of such tags can be performed by analyzing their relative expression (virtual Northern blot) in the various archived tumor and normal SAGE libraries.

Novel algorithms and tools are under development in various laboratories from around the world to characterize with greater certainty the 3' end untranslated regions (UTR) of genes, as well as to better define the complexity of alternatively spliced transcripts, all of these being of much importance in correct SAGE tag identification.

TECHNICAL ADVANCES IN THE USE OF SAGE

Various investigators reported improvements and adaptations of the SAGE methodology for multiple novel applications, in particular when dealing with small samples and hence low amounts of template RNA. Examples of this are the methods known as MicroSAGE, SAGE lite and the SAGE Adaptation to Downsized Extracts (SADE) *(12–15)*. SADE was used in studies aimed at characterizing mouse kidney and brain transcriptomes. In the kidney studies, it was shown that it is possible to generate expression libraries from specific microdissected nephron segments. The identification of molecular markers for specific brain areas was also described *(14)*. Other examples that allow the analysis of specific cell subpopulations using SAGE rely on the direct capture of mRNAs in cell extracts from purified cell populations obtained via various methods, such as immunopurification or microdissection *(15,16)*.

SAGE USES IN HUMAN GENOME MINING AND ANNOTATION

As previously mentioned, SAGE data is perpetual, and available databases become constantly updated in parallel with our constant improvement in understanding the complete human genome. A recent development of value in helping to better mine the human genome was the report of a modification of the SAGE technique that allows one to obtain longer SAGE tags. This will lead to a better and unique gene identification of the 3' ends of genes and, in turn, facilitates the isolation of novel genes. This improvement known as LongSAGE relies on the use of the enzyme *Mme*I as the tagging enzyme, which allows one to obtain 21-bp long tags instead of the 14-bp long tags obtained with the standard method *(17)*.

A significant effort is also underway on the construction of an accurate Human Transcriptome Map, which will be a Web-based application that can be used to generate expression profiles of any chromosomal region. This approach is based on the mapping of the individual SAGE tags to specific chromosomal regions *(18)*.

Representative Uses of SAGE in Cancer-Related Studies

In the following section, I will mention some representative cancer SAGE studies of the many published on several tumor types by various groups; this enumeration of studies does not attempt to be an exhaustive review on the topic.

Colon and Pancreatic Cancer

The first comprehensive analysis of global gene expression in human cancer was performed using SAGE *(8)*. The utility of this approach was demonstrated by comparing normal colon samples with colon and pancreatic tumor samples and cancer cell lines *(8)*. The data demonstrated the power of the SAGE technique and also gave a very good idea of the complexity of the problem. The investigators analyzed approx 300,000 transcripts representing approx 45,000 different genes, expressed at levels ranging from 1 to >5000 transcript copies/cell. They identified 289 transcripts that were expressed at significantly different levels between the colon tumors and normal colon tissue. Of these, 108 transcripts were expressed at higher levels in the colon cancers (an average increase of 13-fold). Interestingly, they did not find large differences when comparing primary tumors with tumor cell lines, indicating many of the changes in gene expression were retained when comparing primary tumors and cell lines. However, this must be interpreted with caution, because other SAGE studies have found significant differences in the profiles of bulk tumors and cell lines, similar to findings in cDNA microarray studies. Zhang et al. *(8)* also identified, for the first time, a series of transcripts with potential to serve as novel cancer biomarkers in the clinic. Investigators demonstrated that findings by SAGE are easily validated by other expression methodologies, such as Northern analysis or reverse transcription PCR (RT-PCR) or real-time quantitative RT-PCR.

It is precisely in the area of identification of novel tumor markers (for all tumor types) where SAGE has been demonstrated to be the most useful. For instance, tissue inhibitor of metalloproteinase type I (*TIMP1*) was identified as a potential marker for pancreatic cancer by SAGE, but serum levels of this protein were observed elevated only in a small fraction of patients with pancreatic cancer when compared with normal controls. However, when the serum detection of this marker was combined with that of additional serum markers, *CA19-9* and carcinoembryonic antigen, the detection of pancreatic cancer patients increased to 60% of 85 patients *(19)*. This demonstrated that by analyzing SAGE data, it was possible to identify cancer serum biomarkers, and by combining suboptimal markers, it was possible to considerably improve the specificity of cancer detection.

In recent studies, comparison of SAGE libraries from pancreatic adenocarcinomas was performed with that of normal tissues, and thus, additional markers were identified, and these included lipocalin, trefoil factor 2, and prostate stem cell antigen (*PSCA*). *PSCA* was confirmed overexpressed in 60% of primary pancreatic adenocarcinomas *(20)*.

St. Croix et al., using a modified SAGE procedure (MicroSAGE) and tumor endothelial cell purification steps, described studies that allowed the identification of differentially expressed genes in endothelial cells isolated from normal vs malignant colorectal tissues *(15)*. Several genes were identified as potential pan endothelial markers, including angiomodulin insulin-like growth factor binding protein *IGFBP7*), hevin, various collagens, Von Willebrand factor, osteonectin (or secreted protein acidic and rich in cysteine [*SPARC*]), *IGFBP4*, *CD146*, and others. A total of 79 genes were identified as differentially expressed between normal and tumor-derived endothelium. Representative genes from the tumor-specific group, triethylenemelamine (*TEM*) genes, were also tested in tumor samples from other tissues (lung, breast, brain, pancreas), and a similar pattern of expression to that observed in colon cancers was observed *(15)*.

Very recently, SAGE libraries were prepared from microdissected colon metastatic cells, aided by immunopurification of colon tumor epithelial cells. The tags purified in this manner were compared with SAGE libraries from normal and malignant, but nonmetastatic, colon epithelium. Numerous transcripts were identified as of interest; among them the transcript for *PRL-3* was found consistently overexpressed in metastatic lesions. *PRL-3* encodes a tyrosine phosphatase that is located at the cytoplasmic membrane. Furthermore, the investigators demonstrated that the *PRL-3* gene, located in chromosome 8q24.3, is amplified in a fraction of metastatic colon tumors *(16)*.

These last two studies illustrate the power of SAGE approaches to analyze the expression profiles of specific tumor subpopulations by combining microSAGE approaches and cell purifications steps. Both studies hold promise to identify novel targets for potential therapeutic intervention in tumor angiogenesis *(15)* and metastasis development *(16)*.

Central Nervous System and Childhood Tumors

SAGE was also used in the gene expression characterization of pediatric tumors, such as medulloblastoma and rhabdomyosarcoma. It was determined that the genes orthodenticle homolog 2 (*OTX2*), zinc finger protein 1 (*ZIC1*), and hairy homolog (*HES6*) were highly overexpressed in medulloblastoma tumors *(21)*. Interestingly, homologs of these genes encode proteins with known developmental functions in *Drosophila*. It was determined that *OTX1* and *OTX2*, in particular, appear to be specific nuclear markers for medulloblastoma and, thus, have potential diagnostic–prognostic value. *HES6*, a helix-loop-helix protein involved in differentiation and putative inhibitor of *HES1* transcriptional repressor activity, was found overexpressed in both tumor types (reported at the SAGE 2000 Conference by Michiels et al.).

In other studies, SAGE analysis of neuroblastoma cells identified the human homologue gene of the *Drosophila* Delta gene (Delta-like 1 [*DLK1*]) as unusually highly expressed. This finding, in turn, suggested involvement of the Delta-Notch pathway in neuroblast differentiation *(22)*.

Recently, SAGE analysis of primary glioblastoma cells led to the identification of the homolog of the melanoma-associated antigen gene family (*MAGE-E1*) as overexpressed in glioblastoma cells. *MAGE-E1* expression was only detected in brain and ovary, among normal tissues. Although the function of this gene is unknown, it holds potential to serve as a glioma marker *(23)*.

Other studies identified numerous candidate gene markers, which are overexpressed in glioblastomas, but not in normal tissues; examples of this are the genes Annexin A1 and *GPNMB (11,24)*. Similarly, the gene neuronatin was identified as being expressed in medulloblastomas, but not in normal cerebellum. Recently, a novel gene encoding a ring finger B-box coiled-coil protein named (*GOA*) was detected by SAGE as overexpressed in astrocytomas *(25)*.

Lung Cancer

SAGE was also used to analyze the transcriptome of non-small cell lung cancer comparing with normal lung tissues. One of the overexpressed genes in this tumor type was the *PGP9.5* transcript. This gene was detected in over 50% of primary tumors and cell lines, and advanced tumors were more likely to overexpress *PGP9.5 (26)*. *PGP9.5* is a ubiquitin hydrolase normally expressed in the neuroendocrine cells of the bronchial epithelium. A yeast two-hybrid screening approach was used to identify potential

PGP9.5 interacting proteins. Among the interacting proteins, the *RAN-BPM*, ubiquitin-conjugating enzyme (*UBC9*), and Jun activation domain binding protein 1 (*JAB1*) genes were identified. The *JAB1* was originally identified as co-activator of *c-Jun*, and recently it was shown to promote the degradation of p27^{kip1}. It was observed that JAB1 and *PCP9.5* co-localize in cell perinuclear and nucleolar regions. The investigators speculated that this complex may contribute to the degradation and hence inactivation of *p27^{kip1}* and that such effect may be of relevance in lung cancer progression (reported by Hibi et al. at the SAGE 2000 Conference).

Recently, hierarchical clustering was used to analyze SAGE data, comparing normal lung epithelial cells and non-small cell lung cancers. One hundred fifteen transcripts were identified that clearly distinguished both groups (i.e., normal and tumors), and furthermore, it was possible to differentiate non-small cell lung cancer histological subtypes. Adenocarcinomas were characterized by high level of expression of small airway-associated or immunologically related proteins, and the *p53* target genes *p21* (*CDKN1A*) and *14-3-3* were consistently underexpressed. Squamous cell carcinomas were characterized by overexpression of genes involved in detoxification or antioxidation. These observations were validated by real-time PCR analyses in larger numbers of samples, importantly indicating that an analysis of a limited number of SAGE libraries was sufficient to provide information significant for defining tumor-specific molecular signatures, which could then be extrapolated to a larger scale analysis *(27)*.

Ovarian, Breast, and Prostate Cancer

SAGE was also employed for the generation of gene expression profiles from ovarian primary tumors, ovarian cancer cell lines, and normal ovarian surface epithelial cells. Again these profiles were useful to identify differentially expressed genes. Many of the genes found up-regulated in ovarian cancer represent surface or secreted proteins such as claudin-3 and -4, mucin-1, Ep-Cam, and mesothelin *(28)*. Interestingly, the gene mesothelin was recently reported as also highly expressed in the vast majority of ductal pancreatic adenocarcinomas *(29)* (*see* also Fig. 2, cluster 4). The lipid homeostatic proteins *ApoE* and *ApoJ* were also found highly expressed in ovarian cancer. Interestingly, glutathione peroxidase 3 (*GPX3*) was observed as highly expressed mostly in ovarian clear cell tumors and has the potential to represent a specific marker for this subtype *(28)*.

In breast cancer studies, epithelial populations from normal breast epithelium and ductal carcinoma *in situ* (DCIS) lesions were obtained and analyzed with SAGE. Various chemokine and cytokine genes, such as *HIN1*, leukemia inhibitory factor (*LIF*), interleukin (*IL*)-8, and growth-related oncogene (*GRO*), were observed to have a decreased expression in DCIS when compared with normal tissue. Several transcripts were also identified as specific for DCIS; an example of this is psoriasin, an S100 binding protein *(30)*. It was also reported that promoter methylation is the main mechanism by which expression of the gene *HIN1* is silenced in many DCIS and invasive breast cancer lesions. This gene is a small cytokine with no homology to other genes, and it is likely to be involved in growth control *(31)*. The *p53* gene target *14-3-3* had also been shown previously by SAGE to exhibit lower expression in breast carcinoma cells in vitro due to hypermethylation when compared to normal epithelial cells *(32)*.

The effects of estrogen on gene expression in breast cancer estrogen-dependent cells were also investigated using SAGE. Charpentier et al. *(33)* demonstrated that a discrete number of genes were found to be up-regulated as the result of estradiol treatment. The resulting SAGE libraries from this study are available to the public as a searchable database at (http://sciencepark.mdanderson.org/ggeg). Among the up-regulated transcripts, five novel genes were identified and cloned *(E2IG1–5)*. *E2IG1* is a putative serine–threonine kinase with homology to small heat-shock proteins, and E2IG4 is a leucine-rich protein very likely secreted to the extracellular environment. The paracrine–autocrine factors stanniocalcin 2 (*STC2*) and inhibin β-B (a transforming growth factor [*TGF*] β-like factor) were also identified as highly up-regulated by the estrogen treatment. Interestingly, *STC2* and *E2IG1* were found exclusively overexpressed in estrogen receptor positive breast cancer cell lines and primary tumors and, thus, have the potential to serve as breast cancer biomarkers of use in the clinic *(33)*. The observation of *STC2* as an estrogen target and its association with estrogen receptor, first detected by SAGE studying MCF7 cells, was later confirmed by cDNA microarrays in analysis of primary breast carcinomas *(5,34)*.

Studies are also underway to generate a high-resolution transcriptome analysis of breast cancer using SAGE, breast cancer libraries are being generated at the 100,000 tag level per library, and over 1.8 million breast cancer SAGE tags have already been sequenced. Combining this effort with the information already available in public databases will allow us to obtain a comprehensive gene expression profile of breast cancer lesions *(35)*.

SAGE studies were also performed with prostate cancer samples; some specific transcripts were observed up-regulated in tumor epithelial cells, while others were found increased in tumor stroma *(36)*. In other studies, the transcript for *PMEPA1*, a gene mapping to chromosome 20q13, was detected by SAGE to be up-regulated by androgen treatment in the LNCaP prostate cancer cells and appears to be a direct target for transcriptional regulation by the synthetic androgen R1881 *(37)*.

SAGE IN PATHWAY DISSECTION AND ANIMAL MODELS OF HUMAN CANCER

SAGE was also used to better define and dissect specific molecular pathways, such as the adenomatous polyposis coli (*APC*)/β-catenin pathway. The *c-Myc* gene was identified as a downstream target of APC action, since it is down-regulated by the *TCF4*/β-catenin transcription complex *(38)*. In turn, also using SAGE, the *CDK4* gene was identified as a target up-regulated by *c-Myc (39)*. The peroxisome proliferator-activated receptor δ (*PPARδ*) was identified as a target of the *APC*/β-catenin/*TCF* pathway using SAGE *(40)*. In other studies, various targets in the *p53* gene pathway were also identified using this methodology, such as the various *p53*-induced genes (*PIGs*) in a model of *p53*-induced apoptosis *(41)*, and the *14-3-3*-ε protein was also first discovered as a *p53* target via SAGE analyses *(42)*. N-myc-transfected neuroblastoma cells were also analyzed with SAGE, and numerous downstream targets of interest were identified *(43)*. As previously mentioned, SAGE was also used to analyze cell pathways transcriptionally regulated by sex steroid hormones *(33,37)*.

Studies were also recently performed analyzing the transcriptional response of human tumor cells to hypoxic conditions. These studies led to the identification of at least 10 new hypoxia-regulated genes, all induced to a greater extent than vascular endothelial growth factor (*VEGF*), which is a known hypoxia-induced mitogen that promotes blood vessel growth *(44)*.

Recent SAGE studies were performed using a *p53* null mouse model of mammary epithelial in vivo preneoplastic progression. This led to the identification of several new and unsuspected targets directly or indirectly dysregulated by the absence of p53 in normal mammary epithelium in vivo. These studies also allowed us to analyze the dramatic physiologic effects of hormonal treatment in mammary gland differentiation *(45)* (database available at [http://sciencepark.mdanderson.org/ggeg]). In other studies using a mouse model of skin carcinogenesis, the gene expression profile of squamous cell carcinomas induced by UV-light has been compared with that of normal skin *(46)*.

In summary, since its inception, the use of SAGE has grown dramatically. The numerous publications using this methodology for a multitude of applications have validated the approach and demonstrated the power of this methodology for the analysis of global gene expression. As discussed, it was used in numerous cancer-related studies and has been particularly useful for the identification of novel tumor markers. One of the main advantages of the SAGE approach has also been its value as a powerful gene discovery tool. Rather than a competing methodology with other global gene expression approaches, SAGE is a complementary approach that in conjunction with other methodologies contributes to achieving a more comprehensive and quantitative picture of the transcriptome under study.

ACKNOWLEDGMENTS

SAGE studies from the Aldaz laboratory were supported in part by The Susan G. Komen Foundation and by National Cancer Institute (NCI) U19 grant CA84978. I am grateful to Dr. Michael MacLeod for critical reading of the manuscript, to Rebecca Deen for manuscript preparation, and to Joi Holcombe for artwork.

REFERENCES

1. Riggins, G. J. and Strausberg, R. L. (2001) Genome and genetic resources from the Cancer Genome Anatomy Project. *Hum. Mol. Genet.* **10,** 663–667.
2. Adams, M. D., Soares, M. B., Kerlavage, A. R., Fields, C., and Venter, J. C. (1993) Rapid cDNA sequencing (expressed sequence tags) from a directionally cloned human infant brain cDNA library. *Nat. Genet.* **4,** 373–380.
3. Williamson, A. R. (1999) The Merck Gene Index project. *Drug Discov. Today* **4,** 115–122.
4. Velculescu, E., Zhang, L., Volgelstein, B., and Kinzler, W. (1995) Serial analysis of gene expression. *Science* **270,** 484–487.
5. Gruvberger, S., Ringner, M., Chen, Y., et al. (2001) Estrogen receptor status in breast cancer is associated with remarkably distinct gene expression patterns. *Cancer Res.* **61,** 5979–5984.
6. Perou, C. M., Sorlie, T., Eisen, M. B., et al. (2000) Molecular portraits of human breast tumours. *Nature* **406,** 747–752.
7. Kuo, W. P., Jenssen, T. K., Butte, A. J., Ohno-Machado, L., and Kohane, I. S. (2002) Analysis of matched mRNA measurements from two different microarray technologies. *Bioinformatics* **18,** 405–412.
8. Zhang, L., Zhou, W., Velculescu, V., et al. (1997) Gene expression profiles in normal and cancer cells. *Science* **276,** 1268–1272.

9. Velculescu, E., Zhang, L., Zhou, W., et al. (1997) Characterization of the yeast transcriptome. *Cell* **88,** 243–251.
10. Eisen, M. B., Spellman, P. T., Brown, P. O., and Botstein, D. (1998) Cluster analysis and display of genome-wide expression patterns. *Proc. Natl. Acad. Sci. USA* **95,** 14,863–14,868.
11. Lal, A., Lash, A. E., Altschul, S. F., et al. (1999) A public database for gene expression in human cancers. *Cancer Res.* **59,** 5403–5407.
12. Datson, N. A., van der Perk-de Jong, J., van den Berg, M. P., de Kloet, E. R., and Vreugdenhil, E. (1999) MicroSAGE: a modified procedure for serial analysis of gene expression in limited amounts of tissue. *Nucleic Acids Res.* **27,** 1300–1307.
13. Peters, D. G., Kassam, A. B., Yonas, H., O'Hare, E. H., Ferrell, R. E., and Brufsky, A. M. (1999) Comprehensive transcript analysis in small quantities of mRNA by SAGE-lite. *Nucleic Acids Res.* **27,** e39.
14. Virlon, B., Cheval, L., Buhler, J. M., Billon, E., Doucet, A., and Elalouf, J. M. (1999) Serial microanalysis of renal transcriptomes. *Proc. Natl. Acad. Sci. USA* **96,** 15,286–15,291.
15. St. Croix, B., Rago, C., Velculescu, V., et al. (2000) Genes expressed in human tumor endothelium. *Science* **289,** 1197–1202.
16. Saha, S., Bardelli, A., Buckhaults, P., et al. (2001) A phosphatase associated with metastasis of colorectal cancer. *Science* **294,** 1343–1346.
17. Saha, S., Sparks, A. B., Rago, C., et al. (2002) Using the transcriptome to annotate the genome. *Nat. Biotechnol.* **20,** 508–512.
18. Caron, H., van Schaik, B., van der Mee, M., et al. (2001) The human transcriptome map: clustering of highly expressed genes in chromosomal domains. *Science* **291,** 1289–1292.
19. Zhou, W., Sokoll, L. J., Bruzek, D. J., et al. (1998) Identifying markers for pancreatic cancer by gene expression analysis. *Cancer Epidemiol. Biomarkers Prev.* **7,** 109–112.
20. Argani, P., Rosty, C., Reiter, R. E., et al. (2001) Discovery of new markers of cancer through serial analysis of gene expression: prostate stem cell antigen is overexpressed in pancreatic adenocarcinoma. *Cancer Res.* **61,** 4320–4324.
21. Michiels, E. M., Oussoren, E., Van Groenigen, M., et al. (1999) Genes differentially expressed in medulloblastoma and fetal brain. *Physiol. Genomics* **1,** 83–91.
22. van Limpt, V., Chan, A., Caron, H., et al. (2000) SAGE analysis of neuroblastoma reveals a high expression of the human homologue of the *Drosophila* Delta gene. *Med. Pediatr. Oncol.* **35,** 554–558.
23. Sasaki, M., Nakahira, K., Kawano, Y., et al. (2001) *MAGE-E1*, a new member of the melanoma-associated antigen gene family and its expression in human glioma. *Cancer Res.* **61,** 4809–4814.
24. Polyak, K. and Riggins, G. J. (2001) Gene discovery using the serial analysis of gene expression technique: implications for cancer research. *J. Clin. Oncol.* **19,** 2948–2958.
25. Vandeputte, D. A., Meije, C. B., van Dartel, M., et al. (2001) *GOA*, a novel gene encoding a ring finger B-box coiled-coil protein, is overexpressed in astrocytoma. *Biochem. Biophys. Res. Commun.* **286,** 574–579.
26. Hibi, K., Westra, W. H., Borges, M., Goodman, S., Sidransky, D., and Jen, J. (1999) *PGP9.5* as a candidate tumor marker for non-small-cell lung cancer. *Am. J. Pathol.* **155,** 711–715.
27. Nacht, M., Dracheva, T., Gao, Y., et al. (2001) Molecular characteristics of non-small cell lung cancer. *Proc. Natl. Acad. Sci. USA* **98,** 15,203–15,208.
28. Hough, C. D., Sherman-Baust, C. A., Pizer, E. S., et al. (2000) Large-scale serial analysis of gene expression reveals genes differentially expressed in ovarian cancer. *Cancer Res.* **60,** 6281–6287.
29. Argani, P., Iacobuzio-Donahue, C., Ryu, B., et al. (2001) Mesothelin is overexpressed in the vast majority of ductal adenocarcinomas of the pancreas: identification of a new pan-

creatic cancer marker by serial analysis of gene expression (SAGE). *Clin. Cancer Res.* **7,** 3862–3868.
30. Porter, D. A., Krop, I. E., Nasser, S., et al. (2001) A SAGE (serial analysis of gene expression) view of breast tumor progression. *Cancer Res.* **61,** 5697–5702.
31. Krop, I. E., Sgroi, D., Porter, D. A., et al. (2001) *HIN-1*, a putative cytokine highly expressed in normal but not cancerous mammary epithelial cells. *Proc. Natl. Acad. Sci. USA* **98,** 9796–9801.
32. Ferguson, A. T., Evron, E., Umbricht, C. B., et al. (2000) High frequency of hypermethylation at the 14-3-3 sigma locus leads to gene silencing in breast cancer. *Proc. Natl. Acad. Sci. USA* **97,** 6049–6054.
33. Charpentier, A. H., Bednarek, A. K., Daniel, R. L., et al. (2000) Effects of estrogen on global gene expression: identification of novel targets of estrogen action. *Cancer Res.* **60,** 5977–5983.
34. Bouras, T., Southey, M. C., Chang, A. C., et al. (2002) Stanniocalcin 2 is an estrogen-responsive gene coexpressed with the estrogen receptor in human breast cancer. *Cancer Res.* **62,** 1289–1295.
35. Aldaz, C. M., Hawkins, K., Drake, J., Laflin, K., Gaddis, S., and Sahin, A. (2002) High resolution transcriptome analysis of breast cancer using SAGE. *Proc. Am. Assoc. Cancer Res.* **43,** 2236.
36. Waghray, A., Schober, M., Feroze, F., Yao, F., Virgin, J., and Chen, Y. Q. (2001) Identification of differentially expressed genes by serial analysis of gene expression in human prostate cancer. *Cancer Res.* **61,** 4283–4286.
37. Xu, L. L., Shanmugam, N., Segawa, T., et al. (2000) A novel androgen-regulated gene, *PMEPA1*, located on chromosome 20q13 exhibits high level expression in prostate. *Genomics* **66,** 257–263.
38. He, T. C., Sparks, A. B., Rago, C., et al. (1998) Identification of *c-MYC* as a target of the APC pathway. *Science* **281,** 1509–1512.
39. Hermeking, H., Rago, C., Schuhmacher, M., et al. (2000) Identification of *CDK4* as a target of *c-MYC*. *Proc. Natl. Acad. Sci. USA* **97,** 2229–2234.
40. He, T. C., Chan, T. A., Vogelstein, B., and Kinzler, K. W. (1999) *PPARdelta* is an *APC*-regulated target of nonsteroidal anti-inflammatory drugs. *Cell* **99,** 335–345.
41. Polyak, K., Xia, Y., Zweier, J. L., Kinzler, K. W., and Vogelstein, B. (1997) A model for p53-induced apoptosis. *Nature* **389,** 300–305.
42. Hermeking, H., Lengauer, C., Polyak, K., et al. (1997) *14-3-3* sigma is a *p53*-regulated inhibitor of G2/M progression. *Mol. Cell* **1,** 3–11.
43. Boon, K., Caron, H. N., van Asperen, R., et al. (2001) *N-myc* enhances the expression of a large set of genes functioning in ribosome biogenesis and protein synthesis. *EMBO J.* **20,** 1383–1393.
44. Lal, A., Peters, H., St. Croix, B., et al. (2001) Transcriptional response to hypoxia in human tumors. *J. Natl. Cancer Inst.* **93,** 1337–1343.
45. Aldaz, C. M., Hu, Y., Daniel, R., Gaddis, S., Kittrell, F., and Medina, D. (2002) Serial analysis of gene expression in normal p53 null mammary epithelium. *Oncogene* **21,** 6366–6376.
46. Klein, R. D., Hawkins, K. A., Lubet, R., Fischer, S. M., and Aldaz, C. M. (2002) Comparative serial analysis of gene expression (SAGE) studies of normal mouse skin and UV-induced carcinomas. *Proc. Am. Assoc. Cancer Res.* **43,** 5676.

5
Tissue Arrays

Cyrus V. Hedvat

INTRODUCTION
Background and Practice

The development of high-throughput technologies, including DNA arrays and proteomics approaches, has led to a tremendous increase in data acquisition. Numerous groups have reported differentially expressed genes and proteins in a variety of normal and malignant tissues. In fact, many groups are using high-throughput techniques to define new tumor classifications. As important genes are identified by various high-throughput technologies, it is critical to correlate these studies with tissue expression to define the precise cell of origin of a particular transcript or protein. In many DNA array experiments, for example, nonpurified cells are used as a starting material. The RNA from cells to be analyzed is commonly admixed with that from a variety of other cells. In the case of tumor tissues, contaminating stromal cells, blood vessels (endothelial cells and smooth muscle), and other normal cells are typically present. It is difficult, therefore, to be certain that a differentially expressed gene is derived from the cell of interest rather than a "contaminating" cell. Some studies have validated findings using standard techniques based on archival formalin-fixed paraffin-embedded tissues,- including immunohistochemistry (IHC) and *in situ* hybridization (ISH). ISH and IHC are techniques that can localize the expression of a gene or protein to specific cells in tissues. These techniques can be performed with fixed tissues, while fresh or frozen tissue is required for DNA microarray studies. The amount of archival paraffin-embedded tissue far exceeds the tissue that is adequately preserved for RNA or protein extraction and, thus, can be used to expand the scope and significance of these studies. In the past, the use of these standard approaches to analyze the *in situ* expression of genes or proteins in tissues has been a slow and labor-intensive process, requiring the processing of numerous slides at a rate of one gene product per slide. Furthermore, a significant amount of tissue is required to perform many tests on a single specimen. Although automated stainers from various manufacturers can facilitate these techniques, they are not widely used and may not be of particular use to investigators who work with newly characterized reagents. The use of high density arrays composed of many tissue samples provides a highly efficient means to validate and extend molecular studies of human cancers.

From: *Expression Profiling of Human Tumors: Diagnostic and Research Applications*
Edited by: Marc Ladanyi and William L. Gerald © Humana Press Inc., Totowa, NJ

Development of Tissue Multiblocks

The benefit of analyzing a large number of tissues simultaneously is not a recent realization. Various laboratories have developed techniques, including a variety of multiblock preparations to analyze a large number of tissues on a single slide. Battifora developed a method to create a multispecimen tissue block formed from rods with a relatively small cross-sectional area *(1)*. While many specimens could be located in a compact area, it was difficult to track the identity of the various specimens. Battifora then reported an improved technique in which tissues were placed in groups in parallel grooves in a mold and embedded in a stacked manner. They were arranged so that a section of the block includes a spaced array of cross-sections of each of the embedded specimen strips *(2)*. While this method formed tissue samples into a grid pattern in which it was possible to track the identities of individual samples, the method was time-consuming and not suitable for assembling a single block from hundreds of core samples from individual blocks. Also, the original block was damaged by the procedure. This technique never gained widespread popularity, possibly due to its inherent technical difficulties. It did have the advantage that it could be applied readily to the study of fresh tissues. Other "homemade" techniques have been used to place multiple tissues in a single block with cork borers of various diameters and arranging the cores in molten paraffin. The number of cores, however, was still limiting, and it was difficult to form a perfect grid pattern. More recently developed techniques require specialized equipment, but result in hundreds of consistently arranged cores of tissue in a single block.

The following description of tissue microarray construction and use is not intended to be a step-by-step protocol (a manual is included with the instrument), rather it is intended to provide the reader with some specific details of block design and to address problems that can be prevented with careful planning. Several reviews on the subject have been written that provide additional details of the technique *(3,4)*.

TECHNIQUES

In 1998, investigators from the National Human Genome Research Institute (NHGRI) at the National Institutes of Health (NIH), in collaboration with the University of Tampere in Finland and the University of Basel in Switzerland, developed an instrument that allows hundreds of tissues to be placed in a single block, which they called a "tissue chip" *(5)* (Beecher Instruments, Silver Spring, MD, USA) (Fig. 1), which is also known as a tissue microarray (TMA). The construction of a TMA involves assembling a paraffin block containing hundreds of tissue cores (0.6 mm in diameter) derived from different "donor" blocks (Fig. 2). Since the cores are very small, minimal damage is done to the original block (Fig. 3). The technology is based on precision micrometers, which move a needle (0.6–2 mm in diameter) in the x- and y-axis. Micrometer drives are used to position the punch assembly with respect to the recipient block. The procedure from start to finish and the approximate required time to complete the process is shown in Table 1.

Block Construction

The majority of the work must be done prior to punching the first core and requires a significant amount of planning. The key to the process is the selection of the starting

Tissue Arrays

Fig. 1. The manual tissue arrayer. 0.6-mm punches are shown in detail in inset.

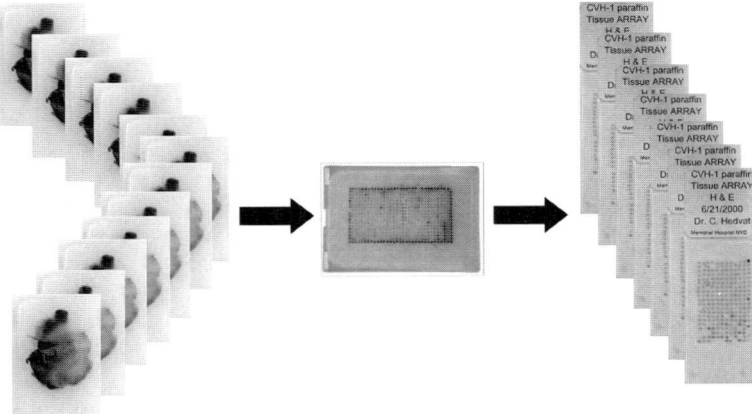

Fig. 2. Diagram of the arraying process. Cores are taken from multiple donor blocks, transferred to a single recipient block with the manual tissue arrayer, and sections can be cut.

material. Sources of material that can be used to prepare tissue arrays include formalin-fixed (or other fixative) paraffin-embedded tissue or paraffin-embedded cell blocks (prepared from tissue culture cell lines). In order to create a cell block of sufficient thickness to core for the array, at least 10^9 to 10^{10} cells are required. The cells are centrifuged in a 1.5-mL microfuge tube or a 15-mL plastic conical tube to pellet the cells, fixed in formalin or another suitable fixative, and embedded in paraffin.

The tissue block should contain sufficient tissue to ensure that the tissue will be represented in the majority of sections. Ideally, donor blocks should be at least 2 to 3 mm thick. Blocks that have been sectioned multiple times for other studies should be avoided if possible. Thinner blocks can be used, but the cores may be lost from some or all tissue array sections. Biopsy material with little original tissue should also be used

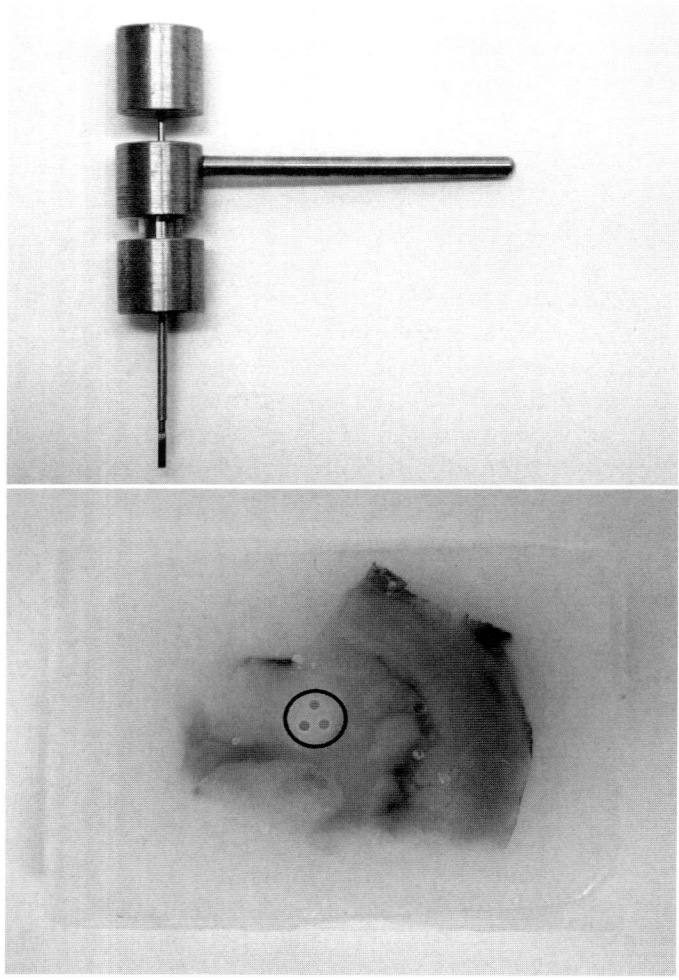

Fig. 3. A closeup view of an individual 0.6-mm punch with a tissue core is shown (upper panel). Minimal tissue is taken from a donor block, which is shown with three 0.6-mm cores removed.

Table 1
Tissue Array Preparation

Step	Time required
1. Select patients.	Variable
2. Collect blocks and slides.	Days to weeks
3. Review stained sections to select areas to core.	Days
4. Prepare tissue array.	2 to 3 days
5. Prepare sections.	Days

with caution, since removal of cores may prevent future clinical or research use of the block, and cores derived from these blocks are frequently lost in many of the TMA sections.

At least one practice array blocks should be constructed from noncritical tissue by the initial user to become familiar with the technique prior to working with valuable tissue. A video of the entire process can be viewed at (http://reresources.nci.nih.gov/tarp). First, all of the blocks and corresponding hematoxylin and eosin (H&E) stained sections, which are to be included in the array, are collected. An H&E stained section is reviewed, usually by a surgical pathologist, and the desired area is marked on the slide. The slide is then aligned with the tissue in the corresponding donor block to localize the area to be cored. In some cases, the area can be identified and targeted on the paraffin block itself. This eliminates any problems that might result from slide-block mismatch, which can result from tissue shrinkage during processing. A well-lit area should be used for the procedure, possibly with a spotlight on the recipient block to ensure precise transfer of tissue cores.

The instrument has two punches, which are thin-walled stainless steel tubes sharpened like a cork borer, one slightly larger than the other. A stainless steel stylet pushes the core of either paraffin or tissue out of the tube. The "recipient punch" is slightly smaller and is used to create a hole in a recipient paraffin block. The "donor punch" is larger and is used to obtain a core sample from a donor block of embedded tissue of interest. The inner diameter of the donor punch tube is designed to correspond to the outer diameter of the recipient punch tube. Thus, the sample snugly fits in the recipient block, and a precise array can be created.

An empty standard paraffin block (45 × 20 mm) is placed in the block holder and locked into place. The micrometers should be zeroed at an appropriate starting position (at least 5 mm from the edge of the block) prior to insertion of the first core. The first step is to create a hole in the recipient block with the acceptor punch. This punch removes a core of paraffin and creates a hole. The inner stylet should be elevated, such that the end of the punch is empty to ensure complete removal of paraffin from the recipient block, which prevents excessive paraffin buildup. A removable bridge is used to support the donor block over the recipient block to prevent damage during punching. A core is punched from the indicated area of the donor block (aligned with the H&E section) and transferred to the newly created hole in the recipient block. The needles are then moved along either the x- or y-axis to create a spacing of 0.1–0.3 mm between cores, and the next punch is transferred to the recipient block. The procedure is continued hundreds of times to form a TMA block. Correct placement of the donor and recipient punches (which differ only slightly in diameter) in the turret and consistently switching between donor and recipient punches is required for a uniform array with tightly fitting tissue cores that will not dislodge during sectioning. Incubating the block at 37°C for approx 15 min also helps prevent core loss or movement. The stylets should be cleaned intermittently to prevent paraffin buildup on the surface, which will hinder smooth operation, and may result in damage to donor blocks. They must also be inspected frequently for damage and replaced as necessary. Several replacement punches should be available for such a purpose. The movement of the turret is limited in the y-axis, requiring that the block be rotated 180° prior to being approximately two-thirds complete, if utilizing the entire recipient block area. Otherwise, the micrometer drive digital reading will advance, but the punch holder will not. A row can be skipped when rotating the block, since it is difficult to exactly align the core based on the distance from the previous row. The micrometer should also be zeroed at this point.

Using a core diameter of 0.6 and 0.1 mm edge–edge spacing, approx 600 individual cores (200 donor blocks, if using 3 cores/block) can be placed in a single block. Although punches are available in several diameters (0.6, 1.0, 1.5, and 2.0 mm), 0.6 mm is the most commonly used. This smaller diameter results in less damage to the donor block and allows more samples to be arrayed in the recipient block. Larger core sizes do allow more tissue representation, but may result in greater damage to the original block. Each core typically measures 3–5 mm in length. The spacing between adjacent cores is determined by the user, but should be at least 0.1 mm. Many groups use a spacing of 0.1 mm, while others use 0.2 or 0.3 mm. The main advantage to a smaller distance between adjacent cores is that a greater number can be accommodated in a single block. A greater distance between cores slightly limits the number of cores that can be placed in a single block. A representative TMA block and corresponding section are shown in Fig. 4A. Figure 4B shows the amount of tissue that is represented in a single core and the amount of spacing between cores.

Sections are cut from the TMA block with a microtome (4–6 µm thick), with standard handling or with a specialized tape-transfer system (Paraffin Tape-Transfer System; Instrumedics, Hackensack, NJ, USA). Sections should be cut by an experienced histotechnologist, since specific issues are encountered with this type of block. Representative results from a TMA section are shown in Fig. 4. Depending on the types of tissues arrayed, the block may be difficult to section with the standard method and may require use of the tape-transfer system for consistent high quality sections. Many groups prefer to use the tape-transfer system, which is based on slides coated with a UV-polymerized adhesive, resulting in a covalent attachment of the tissue to the glass slide. This type of attachment does not interfere with subsequent studies performed with the slides. The tissue treated in this manner will not be lost from the slide during subsequent harsh or prolonged treatments, such as ISH. The use of the adhesive-coated slides does, however, add to the cost of the procedure, and some of the morphology is not as sharp as those sectioned with a standard microtome. Although some practice is required to ensure technical expertise with the tape-transfer system, an experienced histotechnologist is not required to produce the sections. The microtome blade should be changed frequently when performing this procedure.

Approximately 100–200 sections can be cut from a single tissue array block. In an effort to maximize the number of sections obtained from a block, many sections may be cut at one time to minimize the "trimming" that occurs every time the block is reinserted in the microtome. Sections can be deparaffinized and then dipped in paraffin to prevent oxidation and loss of antigenicity in subsequent IHC experiments, if there is to be a prolonged interval between sectioning and staining *(6)*.

The orientation of the block is critical, since once sections are cut, it is difficult to determine the first row. Known tissue, morphologically distinct from that used in recipient block, such as normal tissue or tissue such as lung infiltrated with a dye, can be placed in specific nonsymmetric locations (such as in the first core position) to provide orientation points. Some groups divide the recipient block into quadrants, which can facilitate orientation.

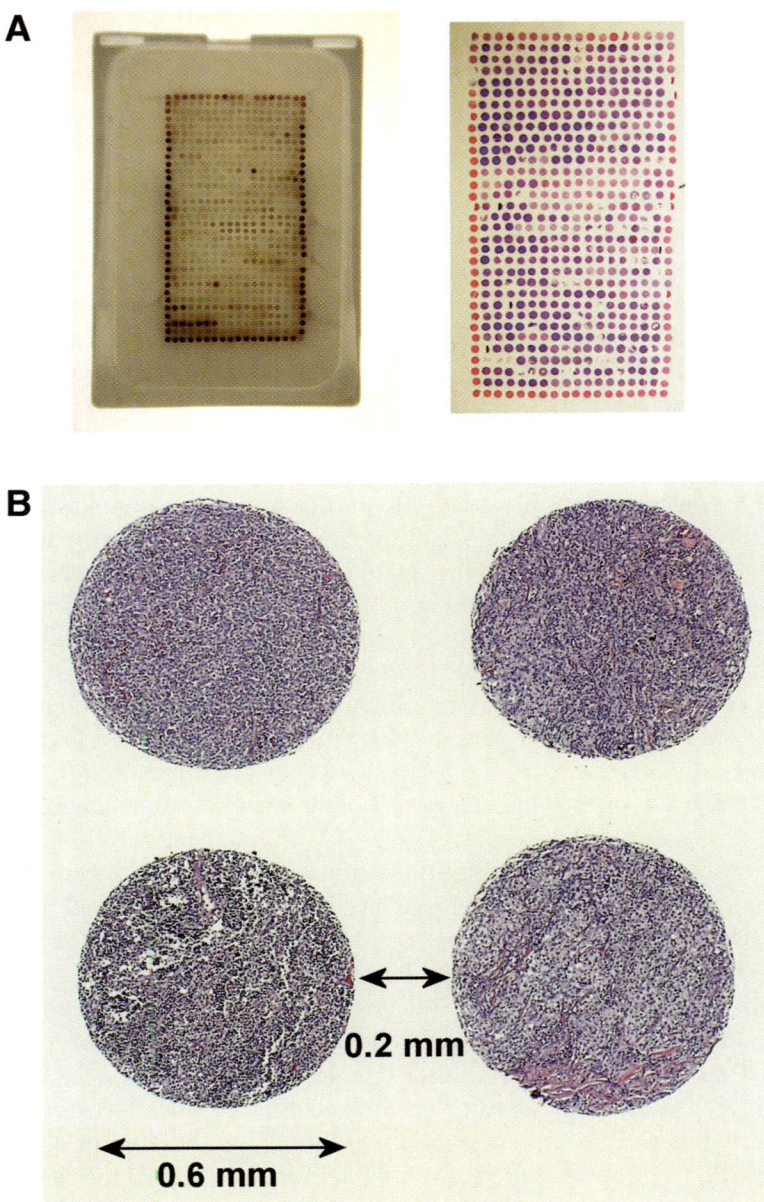

Fig. 4. (A) Representative tissue array block is shown with 0.6-mm cores and 0.2 mm spacing (left) and corresponding H&E stain section (right). **(B)** A representative photomicrograph of four adjacent 0.6-mm cores is taken from H&E stained lymphoma tissue array section (shown at 40×).

Types of Tissue Arrays

Potential users of these techniques include investigators from academic institutions and diagnostic laboratories to pharmaceutical and biotech companies. The type of block constructed will vary depending on the needs of the investigator. Different types of

TMA blocks can be constructed, including, but not limited to, tissue from multiple tumors types, tissue representing disease progression, tissue from a particular clinical study, or normal controls.

Multitumor TMAs are useful for the initial screening of many tumors for expression of a particular protein. For example, the expression of a newly described oncogene found in one tumor type can be analyzed in other tumor types. In one study, investigators constructed a TMA from 17 different tumor types (397 individual tumors) and 20 normal tissues and studied oncogene (CCND1, cMYC, and ERBB2) amplifications with fluorescence *in situ* hybridization (FISH) *(7)*. Amplification of cyclin D1 was found in several tumor types, including lung, breast, head and neck, and bladder carcinomas, and in melanoma. Similarly, TMAs from multiple normal tissues can be used to determine the normal range of expression of a protein. These studies need not be limited to human tissues. Normal or disease tissues from mice of different strains or transgenic or knock-out animals could be used. In order to study neoplastic progression, a TMA constructed from normal, hyperplastic, dysplastic–*in situ* disease, and various tumor grades can be used to examine the expression of a protein thought to be differentially expressed during this process. For some of these studies, it may be necessary to define more than one area to be sampled, and multiple cores from either areas of different grades or stages of neoplastic development must be sampled to adequately represent these areas on the resulting TMA.

One study constructed a TMA containing primary, recurrent, and metastatic prostate cancer and used FISH to detect gene amplifications *(8)*. Examination of large clinical cohorts can be facilitated by using TMAs, such as one study that examined Ki-67 expression in a group of prostate cancers *(9)*. Some studies have attempted to validate cDNA microarray data with tissue arrays *(10–12)*. In renal cell carcinoma, TMA analysis has been used to verify differential vimentin expression *(11)*. Insulin-like growth factor binding protein *(IGFBP2)* expression (discovered by cDNA microarray analysis of gliomas) was associated with progression and poor patient survival based on protein expression in TMA studies *(12)*. Since DNA microarray studies require high quality RNA and are, therefore, limited to using snap-frozen tissue, the number of specimens that can be analyzed is limited. Thus, the differential expression of a single gene is usually not statistically significant, although groups of genes can be described that, together, are considered adequate to define separate groups. TMAs can be used to increase the number of specimens studied so that the expression of a single protein, such as IGFBP2 in gliomas, can reach the level of statistical significance as an independent variable.

Frozen Tissue Arrays

Paraffin TMA methods are limited by the source of donor blocks. Tissues in paraffin-embedded blocks may be unsuitable for some studies. For example, fixation in formalin results in loss of immunoreactivity for some antigens *(13)*. For FISH studies, more consistent results are achieved with ethanol-fixed tissues *(5)*. Some of these limitations can be circumvented by the use of cryosections. Fixation of sections using this technique can be optimized for studies involving protein, DNA, or RNA. For IHC studies, the fixative and fixation time can be adjusted to maximize antigen retrieval. Tumor banks with fresh tissues that are rapidly frozen at <70°C are optimal as starting material for these methods. Use of frozen tissue allows adjacent tissue to be used for RNA extraction and DNA microarray analysis, providing a means of validation of results.

Two groups have reported the construction of frozen tissue multiblocks. The first method allows the construction of low density multiblocks from frozen tissue *(4)*. Currently, it is limited in the number of specimens the block can accommodate (48 3-mm cores), though the identity of the tissue can be determined by its position in the section.

A mold is created from optimal cutting temperature (OCT) with preformed cylindrical wells in a grid arrangement. The frozen tissue cores (3 mm in diameter) are taken from frozen donor tissue using a core needle and transferred to the mold to create the array. The use of adhesive-coated slides (Instrumedics) is recommended with both methods when cutting frozen tissue array sections, to a greater degree than with paraffin arrays, to ensure consistent high quality sections with maximal core representation.

The second method is a modified use of the manual tissue arrayer (Beecher Instruments) *(14)*. A recipient block is created from OCT, and donor cores are transferred from frozen tissue to the block with dry ice-chilled punches (0.6 or 1 mm in diameter). Using this technique, hundreds of cores can be accommodated in the recipient block. However, due to physical limitations of working with frozen tissue, a larger core size and greater spacing between adjacent core is necessary to produce a high quality frozen tissue array.

VALIDATION STUDIES

Since a very small amount of tissue is represented on a single slide, the question of adequate representation has been examined. In general, if a portion of the tumor is represented in an individual core, the resulting IHC staining pattern is reliable. There is no significant "edge effect" seen in individual stained cores. Figure 5 demonstrates that a TMA section produces readily interpretable results.

The data derived from TMA sections must be reliable in order to use the results in clinical correlation studies. Several studies have addressed the issue of concordance between results achieved with the TMA technique compared to standard whole tissue sections. Kononen et al. *(5)* used FISH to analyze six gene amplifications (ERBB2, MYC, CCND1, 17q23, 20q13, and MYBL2) and IHC to detect estrogen receptor and P53 protein expression in breast cancer tissue specimens and found that the frequencies agreed with published results. Other groups have found that redundancy is necessary for consistently reliable results.

One study examined estrogen receptor, progesterone receptor, and Her2/neu expression in breast cancer tissues *(6)*, comparing results from TMAs with whole sections. They concluded that two cores per case are sufficient to achieve a 95% concordance between TMAs and standard techniques. They argued that adding more cores per case does not significantly increase the accuracy of the technique, while it does limit the number of cases that can be studied on a single slide. Another study examined the expression of P53, Ki-67, and Rb markers, which must be quantitated rather than scored solely on their presence or absence *(15)*. In this study, three cores per case were significantly more accurate than two cores, resulting in concordance rates of up to 98%. Thus, greater accuracy is required for studies that are semiquantitative rather than requiring a simple categorical interpretation. In addition, the pattern of antigen expression in a

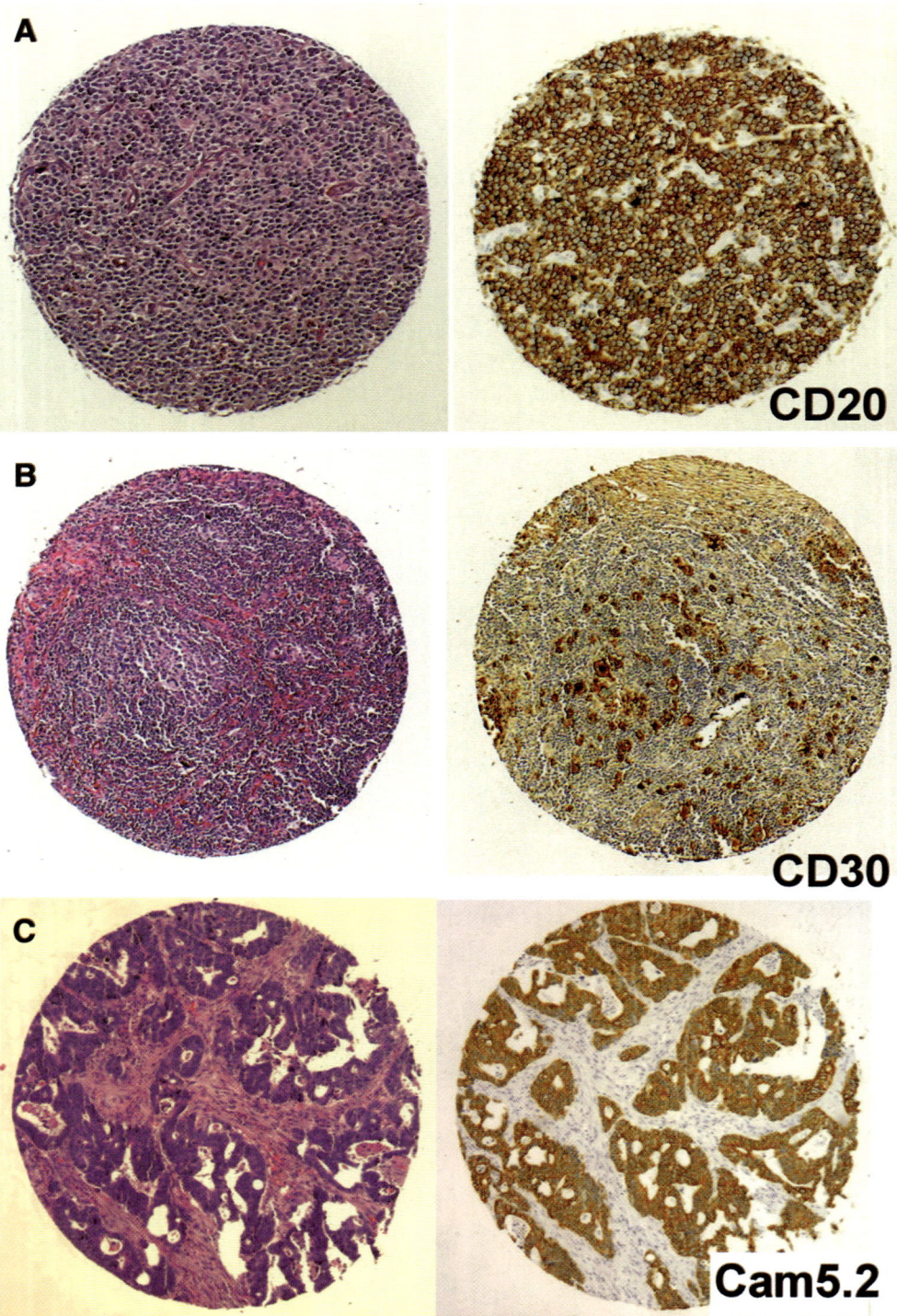

Fig. 5. (A) Diffuse large B cell lymphoma stained with H&E (left) and with anti-CD20 immunostain (right). **(B)** Hodgkin's lymphoma, mixed cellularity type stained with H&E (left) and anti-CD30 (right) demonstrating the presence of Reed-Sternberg cells. **(C)** Colonic adenocarcinoma stained with H&E (left) or anti-cytokeratin (cam5.2) (right), which highlights the epithelial cells. All photomicrographs are taken of 0.6 mm cores at 200×.

particular IHC experiment (whether expressed by all or only a minority of cells) will influence the reproducibility of results with the TMA method. Another study showed a greater discrepancy in expression of neuroendocrine markers in prostate cancer tissues than the other studies *(16)*. In this case, the expression of the proteins studied (chromogranin and synaptophysin) was focal rather than diffuse.

DATA ANALYSIS

Given the large amount of data on a single slide, automated analysis of the results would be extremely helpful to facilitate data acquisition. DNA arrays use an automated interpretation of the results based either on fluorescence intensity, amount of radioactivity, or phosphorimager analyses. In the case of DNA arrays, the precise placement of spots by a robot makes the interpretation relatively straightforward. Although it is difficult to make a perfectly aligned grid given the nature of paraffin, core placement with the newer device does result in relatively uniformly placed cores in a grid pattern. This arrangement makes it possible to automate the analysis of the slide. Unfortunately, automated analysis of either IHC or ISH data is not yet widely applicable except for some nuclear stains, such as Ki-67, which has been quantitated with some success (CAS 200; Bacus Labs). Some companies have developed automated slide scanners to be able to digitized entire sections (BLISS; Bacus Labs, Lombard, IL, and Interscope Technologies, Pittsburgh, PA). This technology has been applied most commonly in the field of telepathology, allowing the remote viewing of a slide for consultation purposes. The digitized image, however, can also be analyzed by a separate software package that can divide the image up into individual core images, which can then be analyzed. The images can then be input into a standard database so that accompanying clinical information can be linked to the images, and all stains from a single specimen can be grouped together. To date, the analysis of images is done in the standard way, by a pathologist who can score the result and input it into the database for export into statistical analysis software. Quantitation of the result is still performed in a subjective manner. These techniques do allow for archiving of images and construction of databases. In the future, these could be made available over the Internet, allowing interested individuals access to the primary data. A schema to combine these types of information has been proposed *(17)*.

FUTURE DIRECTIONS

The Tissue Array Research Program (TARP) of the National Cancer Institute (NCI) has been working to make tissue arrays available to the academic community. Since a single slide can contain all of the cases in a particular study, it is ideal for collaborative projects. Some limited attempts have been made to commercialize arrays. In addition, an automated tissue arrayer is now available (Beecher Instruments), although at a significantly higher cost. However, this instrument automates only array construction. The laborious task of case review and block selection and the designation of the area to be cored must still be performed manually. Clearly, as in the evolution of DNA technology, automation of as many of the steps as is possible is critical. TMAs will become a central technique to laboratories conducting translational research, since it provides an efficient method to confirm experimental results in tissues. As experimental studies with human tissues increase given the demand for greater clinical applicability of basic research, it will become necessary to conserve precious tissue resources.

Given the tremendous advances in digital imaging technology and increases in data storage capacity necessary to store large numbers of high quality images, improvements in automated image analysis should be achieved in the near future, which will enhance the utility of these techniques. Improvements in instrumentation should also make the process more efficient and reliable.

REFERENCES

1. Battifora, H. (1986) The multitumor (sausage) tissue block: novel method for immunohistochemical antibody testing. *Lab. Invest.* **55,** 244–248.
2. Battifora, H. and Mehta, P. (1990) The checkerboard tissue block. An improved multitissue control block. *Lab. Invest.* **63,** 722–724.
3. Bubendorf, L., Nocito, A., Moch, H., and Sauter, G. (2001) Tissue microarray (TMA) technology: miniaturized pathology archives for high-throughput in situ studies. *J. Pathol.* **195,** 72–79.
4. Hoos, A. and Cordon-Cardo, C. (2001) Tissue microarray profiling of cancer specimens and cell lines: opportunities and limitations. *Lab. Invest.* **81,** 1331–1338.
5. Kononen, J., Bubendorf, L., Kallioniemi, A., et al. (1998) Tissue microarrays for high-throughput molecular profiling of tumor specimens. *Nat. Med.* **4,** 844–847.
6. Camp, R. L., Charette, L. A., and Rimm, D. L. (2000) Validation of tissue microarray technology in breast carcinoma. *Lab. Invest.* **80,** 1943–1949.
7. Schraml, P., Kononen, J., Bubendorf, L., et al. (1999) Tissue microarrays for gene amplification surveys in many different tumor types. *Clin. Cancer Res.* **5,** 1966–1975.
8. Bubendorf, L., Kononen, J., Koivisto, P., et al. (1999) Survey of gene amplifications during prostate cancer progression by high-throughout fluorescence in situ hybridization on tissue microarrays. *Cancer Res.* **59,** 803–806.
9. Perrone, E. E., Theoharis, C., Mucci, N. R., et al. (2000) Tissue microarray assessment of prostate cancer tumor proliferation in African-American and white men. *J. Natl. Cancer Inst.* **92,** 937–939.
10. Hedenfalk, I., Duggan, D., Chen, Y., et al. (2001) Gene-expression profiles in hereditary breast cancer. *N. Engl. J. Med.* **344,** 539–548.
11. Moch, H., Schraml, P., Bubendorf, L., et al. (1999) High-throughput tissue microarray analysis to evaluate genes uncovered by cDNA microarray screening in renal cell carcinoma. *Am. J. Pathol.* **154,** 981–986.
12. Sallinen, S. L., Sallinen, P. K., Haapasalo, H. K., et al. (2000) Identification of differentially expressed genes in human gliomas by DNA microarray and tissue chip techniques. *Cancer Res.* **60,** 6617–6622.
13. Werner, M., Chott, A., Fabiano, A., and Battifora, H. (2000) Effect of formalin tissue fixation and processing on immunohistochemistry. *Am. J. Surg. Pathol.* **24,** 1016–1019.
14. Schoenberg Fejzo, M. and Slamon, D. J. (2001) Frozen tumor tissue microarray technology for analysis of tumor RNA, DNA, and proteins. *Am. J. Pathol.* **159,** 1645–1650.
15. Hoos, A., Urist, M. J., Stojadinovic, A., et al. (2001) Validation of tissue microarrays for immunohistochemical profiling of cancer specimens using the example of human fibroblastic tumors. *Am. J. Pathol.* **158,** 1245–1251.
16. Mucci, N. R., Akdas, G., Manely, S., and Rubin, M. A. (2000) Neuroendocrine expression in metastatic prostate cancer: evaluation of high throughput tissue microarrays to detect heterogeneous protein expression. *Hum. Pathol.* **31,** 406–414.
17. Manley, S., Mucci, N. R., De Marzo, A. M., and Rubin, M. A. (2001) Relational database structure to manage high-density tissue microarray data and images for pathology studies focusing on clinical outcome: the prostate specialized program of research excellence model. *Am. J. Pathol.* **159,** 837–843.

6
Microarray Data Analysis
Cancer Genomics and Molecular Pattern Recognition

Pablo Tamayo and Sridhar Ramaswamy

INTRODUCTION

Cancer Genomics

Cancer is a genetic malady, mostly resulting from acquired mutations and epigenetic changes that influence gene expression. Accordingly, a major focus in cancer research is identifying genetic markers that can be used for precise diagnosis or therapy. Over the last half-century, investigators have used reductionism to discover such markers through the study of simple genetic changes, like balanced chromosomal translocations. For example, fundamental insights into the nature of the *bcr-abl* gene translocation product resulted in the precise molecular classification of chronic myelogenous leukemia and recently led to the development of the molecularly targeted tyrosine kinase inhibitor STI571 (Gleevec; Novartis, East Hanover, NJ, USA) for the treatment of this disease. Ninety percent of human cancers, however, are epithelial in origin and display marked aneuploidy, multiple gene amplifications and deletions, and genetic instability, making resulting downstream effects difficult to study with traditional methods. Because this complexity probably explains the clinical diversity of histologically similar tumors, a comprehensive understanding of the genetic alterations present in all tumors is required.

The initial sequencing of the human genome, coupled with technologic advances, now make it possible to embrace the genetic complexity of common human cancers in a global fashion. Tools are currently available, or are being developed, for the identification of all changes that take place in cancer at the DNA, RNA, and protein levels. In particular, the use of DNA microarrays for the comprehensive analysis of RNA expression (expression profiling) in human tumor samples holds much promise (*see* review articles in ref. *1*).

A major challenge with this approach, however, remains the interpretation of complex and biologically "noisy" data in a way that yields new knowledge. We have, therefore, focused on developing first-generation approaches to gene expression data analysis that are suitable for this purpose. Without such analytic tools, DNA microarray data are useless. This chapter is meant to serve as an introduction to fundamental concepts and techniques that have been developed in gene expression data mining over the last 3 yr. It is not meant to be a comprehensive review of this rapidly expanding field,

From: *Expression Profiling of Human Tumors: Diagnostic and Research Applications*
Edited by: Marc Ladanyi and William L. Gerald © Humana Press Inc., Totowa, NJ

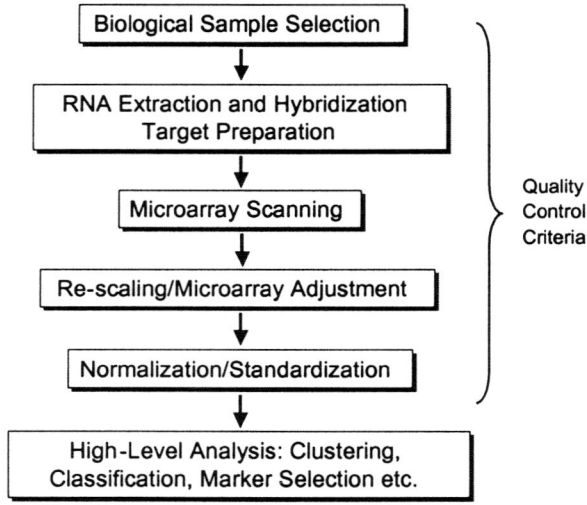

Fig. 1. Methodology for basic data analysis.

nor is it a step-by-step set of recipes. Most of the examples described come from our experience in cancer gene expression data analysis at the Whitehead/Massachusetts Institute of Technology (MIT) Center for Genome Research over the last 5 yr, but references to other works are also given when relevant to the discussion.

BASIC DATA ANALYSIS (FIG. 1)

Tumors are heterogeneous mixtures of different cell types, including malignant cells with varying degrees of differentiation, stromal elements, blood vessels, and inflammatory cells. Two tumors with similar clinical stages can vary markedly in grade and in relative proportions of different elements (e.g., prostatic adenocarcinoma). Tumors of different grades might potentially differ in gene expression, and different markers can be expressed either by malignant cells or by other cellular elements. Because this heterogeneity can complicate the interpretation of gene expression studies, sample selection is an important issue that must be kept in mind when analyzing tumor gene expression data.

Multiple sources of variation that must be understood in evaluating any microarray experiment include the following: (i) varying cellular composition among tumors; (ii) genetic heterogeneity within tumors due to selection and genomic instability; (iii) differences in sample preparation; (iv) nonspecific cross-hybridization of probes; and (v) differences between individual microarrays. In general, biologic variation is the major source of variation in gene expression experiments. Increasing the sample number can help in understanding the range of biologic variation in an experiment. Variation owing to technical factors can be addressed by replicating sample preparation or array hybridization. Although most high-throughput expression profiling centers have informal criteria for what constitutes bad data, however, there are no generally accepted guidelines (for approaches to microarray experimental design and the analysis of variation *see* refs. 2–6).

Basic data analysis consists of preparing datasets for higher level analysis, such as clustering or class prediction. This preprocessing of raw data can have profound effects on subsequent analysis and has to be done by considering the idiosyncrasies of the original gene expression technology platform (i.e., "chip type"). For example, cDNA microarrays generate gene ratio data between fluorescence intensities of experimental and control samples on a gene-by-gene basis. In contrast, oligonucleotide microarrays such as the GeneChip® (Affymetrix, Santa Clara, CA, USA) platform generate absolute expression values from a single sample. Each microarray platform generally has software packages that provide one file per sample, containing one gene per row. These sample files are usually combined into multisample files for further analysis. Our discussion of data analysis starts at this point.

RAW DATA QUALITY CONTROL

The quality of each microarray profile is generally assessed using measurements of overall microarray fluorescence intensity (e.g., mean, variance), the distribution of feature or spot intensities, and the proportion of total genes receiving significant signal. Any microarray that fails these quality control measures is generally excluded from downstream analysis. Replicate experiments for each sample can be used to focus on those gene measurements with the highest reproducibility *(7,8)*. With technologic improvements, however, raw data quality is presently quite good in experienced hands. Therefore, we currently emphasize the analysis of larger numbers of samples rather than studying fewer samples and more replicates.

SCALING

Raw gene expression data from multiple samples (chips) is generally scaled to compensate for global differences in chip intensities and microarray-to-microarray variation. This can be done using simple multiplicative factors to match overall mean intensities among microarrays. Other more sophisticated methods use model-based approaches to compensate for probe-specific biases *(9)*.

THRESHOLDING, FILTERING, AND NORMALIZATION

In some cases it may be desirable to threshold and ceiling the data, since very low and very high microarray fluorescence readings are less reliable and reproducible. As many clustering and classification algorithms work better with smaller number of genes, or are especially sensitive to noisy profiles, genes that show low or flat expression across multiple samples are usually filtered out of datasets. One of the simplest ways to do this is by using a variation filter, which tests for a minimum fold-change (max/min), and absolute variation (max − min) among samples and excludes genes not passing the corresponding thresholds. The precise parameters of variation filters are problem-, dataset-, and platform-dependent, and different thresholds and stringencies in the variation filter may be used depending on the particular analysis. After filtering, and before higher level data analysis, one may also consider normalizing each gene to a mean of 0 and variance of 1 across all samples. This strategy can be useful if one is interested in emphasizing relative rather than absolute differences in gene intensity.

HIGHER LEVEL DATA ANALYSIS: UNSUPERVISED AND SUPERVISED LEARNING

To date, the higher level computational analysis of gene expression data has centered on two approaches *(10)*. Unsupervised learning, or clustering, involves the aggregation of a diverse collection of data into clusters based on different features in a data set. For example, one could divide a group of people into clusters based on any combination of eye color, waist size, or height. Similarly, one can gather data about the various expressed genes in a collection of tumor samples and then cluster the samples as best as possible into groups based on the similarity of their aggregate expression profiles. Alternatively, one could cluster genes across all samples, to identify genes that share similar patterns of expression in varying biologic contexts. Such approaches have the advantage of being unbiased and allow for the identification of structure in a complex data set without making any *a priori* assumptions. However, because many different relationships are possible in a complex data set, the predominant structure uncovered by clustering may not necessarily reflect clinical or biologic distinctions of interest.

In contrast, supervised learning incorporates the knowledge of class label information to make distinctions of interest. A training data set is used to select those features that best make a distinction. These features are then applied to an independent test data set to validate the ability of selected features to make that distinction. For example, one could select a subset of expressed genes that are best able to distinguish between two cancer types and build a computational model that uses these selected genes to sort an independent unlabeled collection of those tumor types into the two groups of interest. However, supervised learning is dependent on accurate sample labels, which can be an issue given the limitations of histopathologic cancer diagnosis. Sometimes, results from unsupervised and supervised learning on a single data set can overlap, but this does not have to be the case.

An important issue with either analytic approach is that of statistical significance of observed correlations. A typical microarray experiment yields expression data for thousands of genes from a relatively small number of samples, and gene–class correlations, therefore, can be revealed by chance alone. This issue can be addressed by collecting more samples for each class studied, but this is often difficult with clinical cancer samples. Another approach is to perform exploratory data analysis on an initial data set and apply findings to an independent test set. Findings confirmed in this fashion are less likely a result of chance. Permutation testing, which involves randomly permuting class labels and determining gene–class correlations, has also been used to determine statistical significance *(10)*. Observed gene–class correlations that are stronger than those seen in permuted data are considered statistically significant.

UNSUPERVISED LEARNING: CLUSTERING

In unsupervised learning techniques, the structure in a data set is elucidated without using any *a priori* assumptions or knowledge as part of exploratory data analysis. The promise of these methods lies in their ability to provide a molecular grouping or taxonomy of samples or genes. One of the easiest ways to analyze data in this context is by using a clustering algorithm *(11–13)*. Objects of interest, usually genes or samples, are

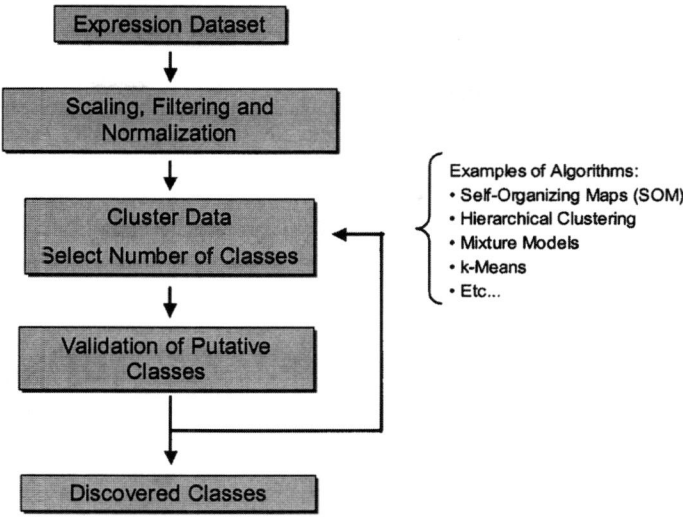

Fig. 2. General methodology for clustering gene expression data.

classified into groups according the how "close" they are to each other. This is accomplished by using a "distance," correlation, or "similarity" function in the clustering algorithm. For example, one can cluster a set of biological samples by their Euclidean distances by considering all gene expression values in a dataset:

$$\text{Distance (sample x, sample y)} = ([E^x_{\text{gene 1}} - E^y_{\text{gene 1}}]^2 + [E^x_{\text{gene 2}} - E^y_{\text{gene 2}}]^2 + \ldots)^{1/2}$$

Here, $E^x_{\text{gene 1}}$ is the expression value of gene 1 in the array corresponding to sample x. A clustering algorithm uses these distances to group samples or genes, and it returns an organization scheme to classify them (e.g., a set of clusters or a tree).

Unsupervised learning approaches, such as clustering, can be very useful when the underlying structure of the data is unknown; however, they have the disadvantage, if unguided, of sometimes producing results that may or may not be relevant to distinctions in the data that are biologically relevant. Clustering often rediscovers already known subclasses or differences if these distinctions are predominant (e.g., estrogen receptor positive vs negative breast cancers). However, this approach can also discover unanticipated relationships, and clustering methods have been used with relative success in a number of cancer classification problems. In practice, it is often challenging to interpret clusters that result from unsupervised learning in cancer datasets. A general methodology for clustering is shown in Fig. 2.

Some of the first work using this approach in analyzing gene expression involved time series data. Genes were grouped, or clustered, according to their behavior over time, first by eye *(14)* and then by an automated hierarchical technique *(15)*. Hierarchical clustering is an unsupervised learning method useful for dividing data into natural groups by organizing the data into a hierarchical tree structure (dendogram) based upon the degree of similarity between either samples or genes *(15)*. The lengths of branches in a dendogram reflect degree of relatedness. By examining dendogram branches, previously unanticipated relationships between samples and genes can be discovered in a gene expression dataset. Tamayo et al. (1999) *(16)* introduced the use of self-organiz-

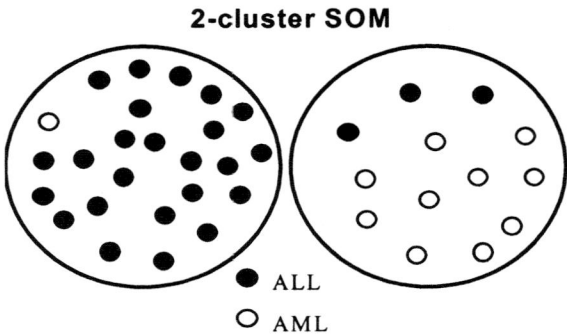

Fig. 3. Clustering of leukemia samples into two groups using a 2 × 1 SOM.

ing maps (SOMs) for unsupervised learning in the HL-60 model of leukemia differentiation and found that resulting gene clusters corresponded to pathways involved in the differentiation treatment of acute promyelocytic leukemia (APL). The SOM is a clustering algorithm, in which a grid of two-dimensional nodes (clusters) is iteratively adjusted to reflect the global structure in the expression dataset *(16)*. With the SOM, the geometry of the grid is randomly chosen (e.g., a 3 × 2 grid) and mapped to the k-dimensional gene expression space. The mapping is then iteratively adjusted to reflect the natural structure of the data. Resulting clusters are organized in a two-dimensional grid, where similar clusters lie near to each other and provide an automatic executive summary of the dataset.

Golub et al. *(10)* used a two-cluster SOM to automatically cluster an initial set of 38 leukemia samples into two classes based on the expression pattern of 6817 genes (Fig. 3). They then compared these SOM clusters to the known acute lymphoblastic leukemia vs acute myeloid leukemia (ALL/AML) distinction. As demonstrated, the two SOM clusters closely paralleled this morphological distinction with the first cluster containing mostly ALLs (24 out of 25 samples) and the second containing mostly AMLs (10 out of 13 samples). Thus, the clustering algorithm was effective, but not perfect at separating samples into biologically meaningful groups.

Golub et al. *(10)* also searched for further subclassifications of the leukemia samples by constructing a four-class (2 × 2) SOM (Fig. 4). The clustering algorithm was successful at separating the samples into more refined groups reflecting another important biological distinction: different ALL cell lineages (B and T cell).

Hierarchical clustering *(15)* was also applied to the same dataset (Fig. 5). Again, this clustering approach revealed three major leukemia subgroups, suggesting that robust gene expression differences between different tumor subtypes can be discovered using unsupervised learning.

Similar studies have recently been described for the subclassification of various tumor types including breast cancer *(17,18)*, lung cancer *(19)*, and melanoma *(20)*.

Clustering has yielded results that are interpretable in the context of *a priori* knowledge (i.e., known leukemia subclasses). However, in the absence of such knowledge, the biological interpretation of clustering results remains a challenge. Often, clustering results are not in themselves the desired results, but the starting point for further interpretation or experimentation. An area of active research, moreover, involves the statis-

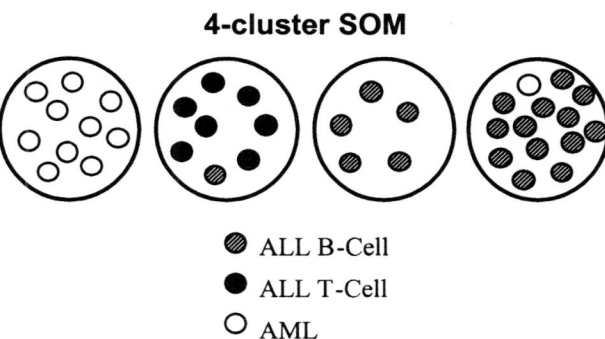

Fig. 4. Clustering of leukemia samples into four groups using a 4 × 1 SOM.

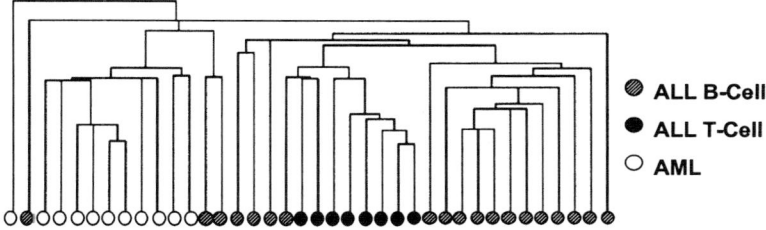

Fig. 5. Hierarchical clustering of leukemia samples based on the expression of the 330 most varying genes.

tical interpretation of clustering results. Often asked questions include, what constitutes a cluster and what is the statistical significance of a given clustering result? There are presently no good general answers for these important questions, although some groups have proposed the implementation of formal measures of clustering significance, such as the gap statistic *(21)*.

SUPERVISED LEARNING: PREDICTION

Supervised learning or class prediction methods represents another important paradigm in molecular classification and pattern recognition. The simplest analysis involves selecting the features (genes) most correlated with a phenotypic distinction of interest. These features or "marker genes" are biologically interesting in themselves, but they can also be used as the input of a classification algorithm that uses existing "labeled" samples to build a model to predict the labels for future samples. For example, marker genes in a cancer dataset can be fed into a computational classifier to distinguish cancer types on the basis of site and/or cell of origin or clinical outcome. This powerful approach, supervised machine learning or class prediction *(13,22)*, involves data collection, feature selection, model building, validation, and model testing on an independent dataset. Supervised learning classifiers can achieve highly accurate molecular classification if enough samples are available to "train" a classifier. In general, pairwise comparisons are less challenging than multiclass distinctions. In every case, the comparison of a supervised classifier has to be done against the best generally accepted clinical classification method, such as standard histopathology. In the next few sections, we will review in more detail the steps necessary to select and validate gene markers and to build classifiers.

SELECTING AND VALIDATING GENE MARKERS

Genes correlated with a binary class distinction, for example a morphological or clinical phenotype, can directly be identified and selected by using a "distance" metric, for example:

- Signal-to-noise ratio = $(\mu_A - \mu_B)/(\sigma_A + \sigma_B)$ [μ and σ are the means and standard deviation per class]
- t-test statistic = $(\mu_A - \mu_B)/(\sigma^2_A + \sigma^2_B)^{1/2}$ [μ and σ are the means and standard deviation per class]
- Pearson correlation coefficient

An example is shown in Fig. 6. The original dataset was created by joining 12 microarray datasets, 6 from normal kidney and 6 from renal cell carcinoma samples. Markers were selected by computing the signal-to-noise score. The mean and standard deviation of the expression values are computed in each class, and then the ratio of the difference of the means is divided by the sum of the standard deviations. For example, this calculation as applied to the profile of p53 shown below produces the following:

$$\text{Signal-to-noise ratio} = (\mu_{cancer} - \mu_{normal})/(\sigma_{cancer} + \sigma_{normal}) = 1.67$$

As can be seen in Fig. 7, this gene acts as a marker of the "cancer" phenotype by being expressed on average at higher level in cancer samples compared with normal ones. It is important to notice that the difference in absolute expression value may not always be large. In this example, p53 is a marker, but in general displays low values of expression.

This basic procedure of selecting *differentially* expressed genes is useful in two common analysis situations. The first is associated with selecting statistically significant markers for more detailed follow-up biological study (e.g., to identify genes that are differentially expressed in two different cancer types). Selected genes can then be subject to a literature search or to validation using other experimental assays (e.g., reverse transcription polymerase chain reaction [RT-PCR], immunohistochemistry, etc.). The second relates to the problem of *feature selection* or finding genes to feed into a supervised learning classifier. In this case, one is interested in selecting the subset of genes most likely to be useful in discriminating phenotypes of interest, either as single markers or in combination with others. This task is better viewed as a preprocessing step in a classification methodology. Gene selection is required, in part, because many supervised learning algorithms perform suboptimally with thousands of input variables and require some type of dimensionality reduction. A general methodology for supervised marker selection and classification is shown in Fig. 8. The training of classifiers will be discussed in detail in a subsequent section.

PERMUTATION TESTS

Once marker genes have been selected, one might want to decide how many of them to consider for further study. This is a difficult problem because typically there will be a gradual decrease in the score or correlations in such way that there is no well-defined

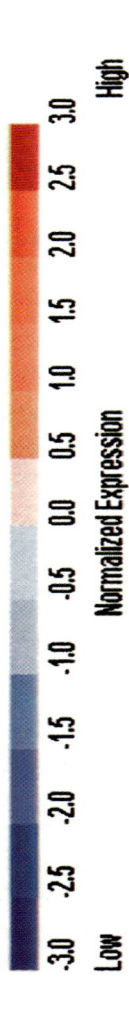

Fig. 6. Top 10 genes that differentiate normal kidney from renal carcinoma as selected from a microarray profiling experiment using the signal-to-noise (S2N) ratio score.

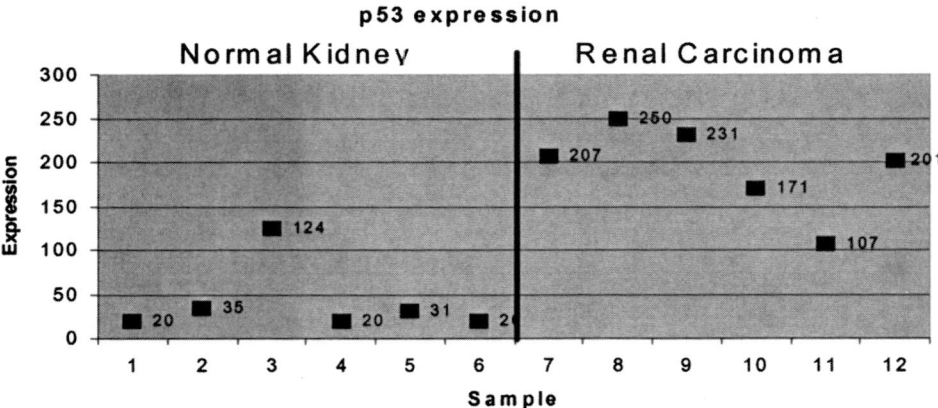

Fig. 7. P53 gene expression as a marker of the "cancer" phenotype in normal and neoplastic kidney samples.

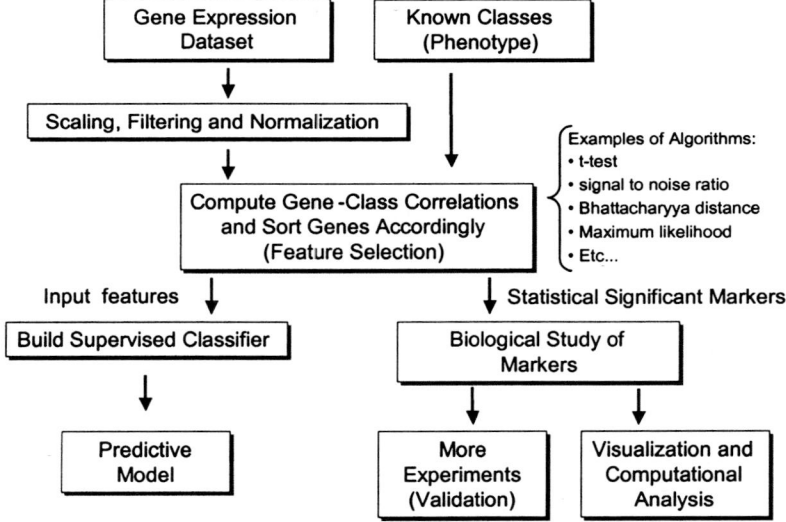

Fig. 8. Methodology for marker selection.

boundary between markers and nonmarkers. In most situations, the analysis will concentrate on the very top markers and exclude the rest. However, this problem can be addressed more formally by using permutation testing. This method *(10,23)* attempts to solve the marker selection problem by comparing the actual distribution of marker scores to a reference empirical distribution of scores obtained by permuting the phenotype class labels. The markers are viewed as close matches or "neighbors" of an ideal marker separating the classes. A histogram of scores for each of the ranked marker genes, corresponding to each permutation (neighborhood), is kept and the significance of an actual gene marker is obtained by finding the appropriate percentile in the histogram of the correspondingly ranked marker (i.e., the one with the same rank, e.g., best match, second best match, etc.). There are several advantages to performing a permuta-

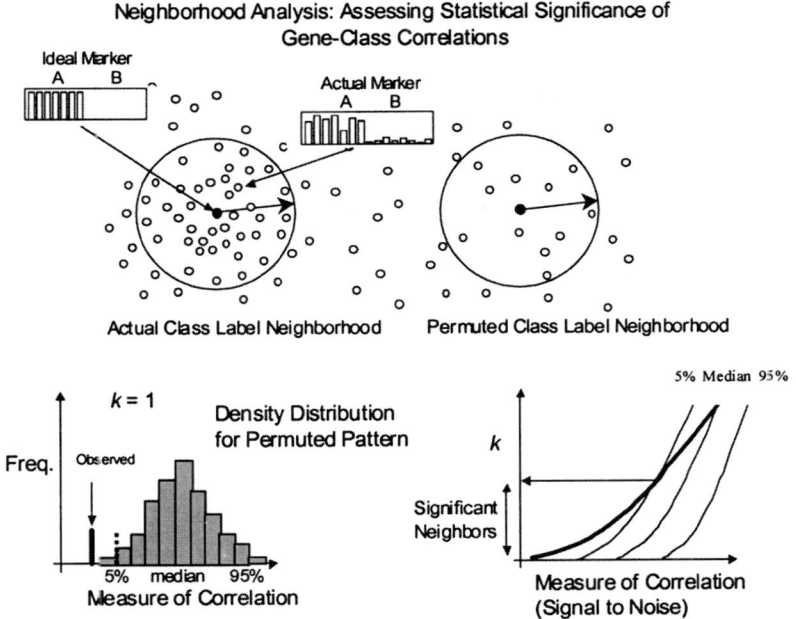

Fig. 9. Permutation test based assessment of significance for gene markers.

tion test: (*i*) the method does not assume a particular functional form for the distribution or correlation structure of genes; (*ii*) it is performed on the entire distribution of marker genes and, therefore, takes into account the gene-to-gene correlation structure; and (*iii*) it is a simple, intuitive approach that provides higher statistical power. In detail, the permutation test procedure for a given comparison of interest (e.g., markers high in class 0 and low in class 1) is as follows (Fig. 9):

- Generate signal-to-noise $(\mu_{class\ 0} - \mu_{class\ 1})/(\sigma_{class\ 0} + \sigma_{class\ 1})$ or other type of scores (*t*-test, Pearson, etc.) for all genes being considered using the actual class labels (phenotype) and sort them accordingly. The best match ($k = 1$) is the gene "closer" or more correlated to the phenotype using the signal-to-noise as a correlation function. In fact, one can imagine the reciprocal of the signal-to-noise as a "distance" between the "phenotype" and each gene as shown in Fig. 7.
- Generate 500 or more random permutations of the class labels (phenotype). For each case of randomized class labels, generate signal-to-noise scores and sort genes accordingly.
- Build a histogram of signal-to-noise scores for each value of k. For example, one for all the 500 top markers ($k = 1$), another one for the 500 second best ($k = 2$), etc. These histograms represent a reference distribution for the *k*th marker, and for a given value of k different genes contribute to it. Notice that the correlation structure of the data is preserved by this procedure. For each value of k, determine different percentiles (1%, 5%, 50%, etc.) of the corresponding histogram.
- Compare the actual signal-to-noise scores with the different significance levels obtained for the histograms of permuted class labels for each value of k. This test helps to assess the statistical significance of gene markers in terms of the distribution of class–gene scores using permuted labels.

For example, normal kidney vs renal carcinoma marker selection and permutation testing for each of the selected markers generates the following list shown in Table 1.

Table 1
Normal Kidney vs Renal Carcinoma Marker Selection and Permutation Testing

Number	Class	S2N Score	Perm 1%	Perm 5%	Median	Gene	Description
1	Normal	2.82	2.48	1.97	1.37	J03507	C7 Complement component 7.
2	Normal	2.21	1.89	1.69	1.22	HG3431-HT3616	Decorin, Alt. Splice 1.
3	Normal	2.08	1.83	1.56	1.12	Z30644	GB DEF = Chloride channel (putative).
4	Normal	2.07	1.76	1.47	1.07	J05257	DPEP1 Dipeptidase 1 (renal).
5	Normal	1.98	1.66	1.41	1.05	U27333	α-1,3 fucosyltransferase 6 (FCT3A).
6	Carcinoma	1.81	2.38	1.97	1.39	X56494	PKM2 Pyruvate kinase, muscle.
7	Carcinoma	1.78	1.99	1.74	1.21	X59798	CCND1 Cyclin D1.
8	Carcinoma	1.67	1.82	1.58	1.13	M22898	TP53 Tumor protein p53 (Li-Fraumeni syndrome).
9	Carcinoma	1.51	1.72	1.48	1.07	D50855	CASR Calcium-sensing receptor.
10	Carcinoma	1.47	1.66	1.43	1.04	HG662-HT662	Epstein-Barr virus small RNA-associated protein.

The class column represents the class for which the markers are high (low in the other class). The S2N score is the signal-to-noise of each marker. The Perm 1%, 5%, and 50% columns represent the percentiles in the histograms of signal-to-noise scores for permuted labels, for a given value of the rank order. These 10 markers shown all have signal-to-noise scores better than 5% of the random permutations ($p <= 0.05$).

Permutation tests assess the significance of gene markers in terms of class–gene correlations. If a group of genes fails to pass permutation testing, however, that by itself does not necessarily imply that it cannot be used to build an effective classifier *(24,25)*. In subtle phenotypes distinctions, for example, the top marker genes are often weak and may not show overwhelming statistical significance. This often results from a gene being expressed only in a subset of samples in a given class. However, such genes can still be effective when used in combination as input to a classifier. Examples of this phenomenon can be found in subsequent sections.

Other marker selection methods have been introduced in the literature. For example the SAM method of Tusher et al. *(26)* is similar to the one presented above, but includes a user-adjustable threshold to provide estimates of the false discovery rate. Dudoit et al. *(27)* have introduced a method based on step-down adjusted p values using Westfall and Young's *(28)* approach in the context of replicated cDNA experiments. Ideker et al. *(29)* used generalized likelihood tests to assess the statistical significance of differentially expressed genes in the context of two channel cDNA microarrays. Newton et al. *(30)* and Baldi and Long *(31)* used empirical Bayes hierarchical models to assess significance of differential expression. Lee et al. *(7)* combined the data from replicates to estimate posterior probabilities and identify differentially expressed genes. No systematic comparison of the error rates and statistical power of all these different methods have been published yet. Methods have also been proposed to combine both resampling and explicit control of the false discover rate *(32)*, such as the stepwise permutation-based procedures of Korn et al. *(33)*.

A logical extension of marker selection is *pattern discovery*, where one tries to find subpatterns, i.e., patterns not necessarily involving all of the samples, but that occur often and may represent groups of co-regulated or correlated genes. Califano et al. *(34)* introduced a pattern discovery algorithm (SPLASH) to expose more complex gene correlations. They extracted statistically significant subpatterns from expression array data using a geometric hashing algorithm. Although their statistical models were simplistic, their work represented one of the first analytic evaluations of subpattern significance in that context. Other attempts to elucidate complex gene–gene correlations or global correlation structure have used principal component analysis (PCA) *(20,35)*, singular value decomposition *(36)*, biclustering *(37)*, and plaid *(38)*. Hastie and associates introduced "gene shaving" as a global approach based on PCA to systematically expose coherent patterns of co-regulation in gene expression data *(39)*. All these methods are promising, but face the same challenge in terms of how to effectively separate biologically relevant signals from the noise.

CLASS PREDICTION

A general methodology for class prediction under the supervised learning paradigm is shown in Fig. 10. One starts by putting together the relevant samples into a single dataset, scaling and preprocessing the dataset, and by defining the target phenotype class based on morphology, tumor type, or treatment outcome clinical information. The dataset is split in train and test subsets if enough samples are available. If not enough samples are available, one can perform a leave-one out cross-validation, in which one sample is held, a predictor is trained on the remaining samples, the left out sample is classified by this predictor, and the process is repeated iteratively. Once a

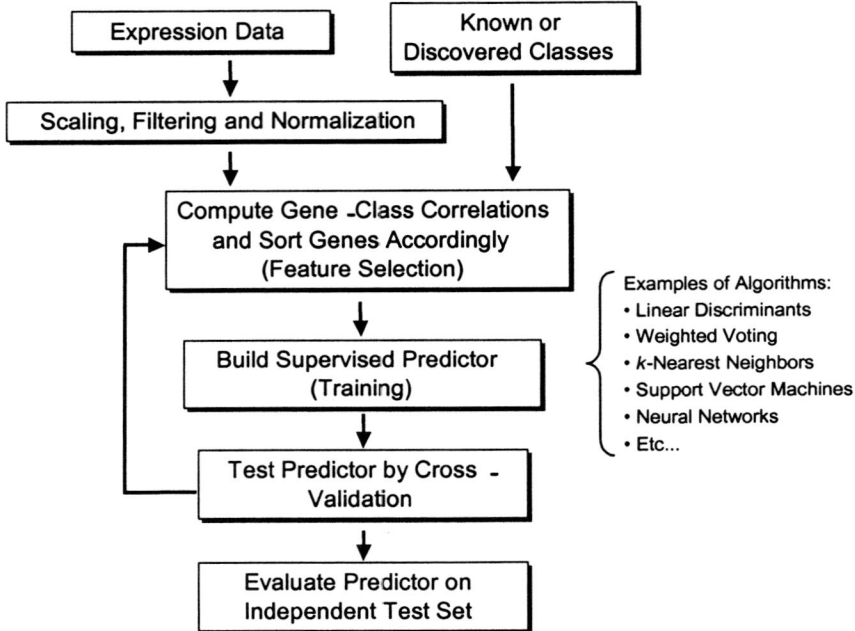

Fig. 10. Methodology for supervised learning.

proper training set has been defined, a marker selection methodology is applied. This step is, in general, useful and facilitates the training of most classification algorithms, although some classifiers, such as Naïve Bayes or support vector machines (SVM), can deal with thousands of variables effectively *(40,41)*. Feature selection is generally useful to facilitate subsequent validation of selected genes that are particularly informative in classification. Once markers have been selected, a classifier can be built using classification algorithms such as *(13,22,42)*:

- Linear or quadratic discriminants.
- *k*-nearest neighbors.
- Weighted voting.
- Naïve Bayes.
- Neural networks.
- SVM.
- Decision trees.

If the model has internal parameters that require tuning, this is typically done when training the predictor. In this way, several models are built using a different number of marker genes, and the final chosen model is the one that minimizes the total error in cross-validation. This model can then be validated on an independent test set. Detailed model-to-model performance comparisons require predictions with different instantiations of the train and test datasets and have to be made carefully as suggested by Salzberg *(43)*.

STATISTICAL SIGNIFICANCE OF A SUPERVISED CLASSIFIER

The statistical significance of a supervised classifier can be evaluated in several ways. One of the simplest is to compute a Fisher exact test of the classification confu-

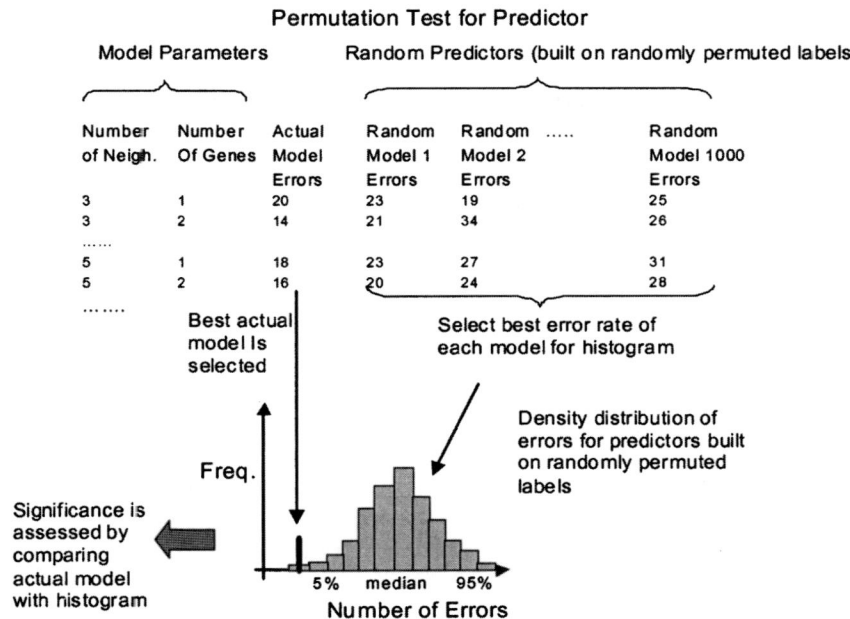

Fig. 11. Methodology to assess the statistical significance of a classifier.

sion matrix or use the proportional chance criterion to compare the observed with the expected classification accuracy for a random predictor *(24)*. When enough samples are available to produce independent train and test datasets, the proportional chance criterion is usually a sufficient measure of statistical significance *(24)*. A more sophisticated empirical approach, sometimes useful for weak classifiers or when there are not enough samples to create an independent test set and when cross-validation must be used, is the class label permutation *(44–46)*. The phenotype (sample) labels are randomly permuted 1000 or more times, and in each instance, predictive models are built and tested. Once this is done, one selects the best error rate for each of these 1000 random predictors and makes a histogram of these error rates. The error rate from the actual predictive model is then compared to this histogram to determine the statistical significance of this prediction (*see* Fig. 11).

Figure 12 shows the application of this permutation test for the k-nearest neighbor treatment outcome predictor in Pomeroy et al. *(35)*. This is a cross-validation model built on 60 medulloblastoma samples capable of distinguishing patients with "good" and "poor" prognosis on the basis of primary tumor gene expression profiles. An optimal model was defined using the following parameters:

Number of neighbors (k): 3, 5

Number of genes (ng): 1,2,3,4,5,6,7,8,9,10,15,25,50,100

Models were created using the actual treatment outcome labels and also for 1000 random permutation of those labels (keeping the gene expression data the same). The best predictive model used $k = 5$ and $ng = 8$, and correctly predicted 47 out of 60 cases as having "good" or "poor" prognosis. Random class label permutation

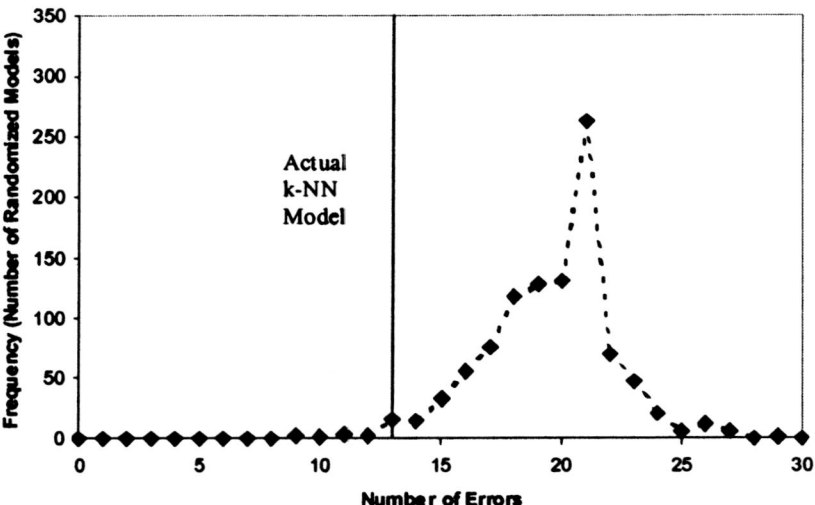

Fig. 12. Results of the permutation test for a k-nearest neighbor medulloblastoma treatment outcome predictor.

showed that there were 9 models with better performance (lower error rates) than the actual model. Based on this result, the statistical significance of this medulloblastoma outcome prediction study was $p = 0.009$ (9/1000).

PAIRWISE CLASSIFICATION: CLASSIFYING LEUKEMIA SUBTYPES

We next review microarray-based leukemia subclassification *(10,23)* as an example of a binary molecular classification problem. Acute leukemias arise from different precursor cells: lymphoid (ALL) and myeloid (AML). This distinction is critical for effective leukemia treatment planning and is currently done by assimilation of diverse information including morphological, cytogenetic, histochemical, and immunophenotypic analysis by an expert physician. Our initial analysis employed a set of 27 ALL and 13 AML samples. A permutation test of the gene markers revealed a striking excess density of genes correlated with the class distinction. We decided to employ a weighted voting classifier based on the top 50 genes. Sets of classifiers were first constructed in cross-validation experiments using the 40 leukemia samples. In one case, no prediction was made, because the confidence score fell below a predetermined threshold. For the remaining 39 cases, the prediction accuracy was 100%. While they initially chose 50 genes for the prediction algorithm, they also found that classifiers involving as few as 7 genes proved to be 100% accurate in the ALL/AML distinction. Interestingly, however, among the top 50 genes, no single gene yielded a perfect predictor. Correct classification, thus, requires multi-gene predictors. Other classification algorithms such as Naïve Bayes, k-nearest neighbors, and SVM produce similar results *(47)*.

The original weighted voting classifier was also tested on an independent collection of 34 AML and ALL samples. In three cases, the confidence score fell below the threshold for prediction, but the classifier made predictions in the remaining 30 cases, and

29 out of 30 were correct. The single error had the lowest confidence score of the samples, just barely passing the threshold. Overall, 69 of 70 samples were correctly classified either in cross-validation or using the independent test set (98.6%). Other algorithms also performed fairly well on the independent test set with the SVM model producing 100% accuracy.

The marker genes, shown in Fig. 13, are highly instructive. Some, including *CD22*, *CD11c*, *CD33*, and *CD79a*, encode cell surface proteins for which monoclonal antibodies have been previously demonstrated to be useful in distinguishing lymphoid from myeloid lineage cells. Others provide new markers of acute leukemia subtype. For example, the leptin receptor, originally identified as a cell surface receptor in adipocytes, but showed high relative expression in AML cells. The leptin receptor has been demonstrated to have anti-apoptotic function in hematopoietic cells. Some of the markers are typical markers of hematopoietic lineage, but others have biological function relevant to the cancer. For example, many of the genes encode proteins critical for S-phase cell cycle progression (Cyclin D3, Op18, and MCM3), chromatin remodeling (RbAp48), transcription (SNF2b and TFIIEβ), or cell adhesion (zyxin and integrin-α X), or are known oncogenes (*c-MYB*, *E2A*, *EWSR1*, and *HOXA9*).

PREDICTING TREATMENT OUTCOME: LYMPHOMA

Supervised learning classifiers are also well-suited to predict differential treatment outcome between histologically similar tumors. Here, we review the results of lymphoma treatment outcome prediction model of Shipp et al. *(48)*. Diffuse large B cell lymphomas (DLBCL) are the most common lymphoid neoplasm, and it accounts for up to 40% of adult (non-Hodgkin's) lymphomas. Using existing chemotherapeutic regimens only a subset of DLBCL patients is cured. Clinical prognostic models, such as the International Prognostic Index (IPI), are used to identify different DLBCL risks groups. The clinical factors used by the IPI (age, performance status, stage, number of extranodal sites, and serum lactate dehydrogenase [LDH]) are potentially surrogate markers for the true molecular heterogeneity of the disease and provide a useful but highly imperfect model for the identification of high-risk patients. Few molecular markers are, however, broadly useful for lymphoma risk stratification.

Our group studied 58 DLBCL patients uniformly treated with standard cyclophosphamide, doxorubicin, vincristine, and prednisolone (CHOP) chemotherapy, where long-term clinical follow-up was available *(48)*. These patients fell into two groups, including those with cured disease and those with fatal–refractory disease. They used supervised learning to determine differential treatment outcome on the basis of primary tumor gene expression profiles.

Top marker genes for the cured vs failure distinction were selected using the signal-to-noise ratio (Fig. 14). We developed a supervised classifier using a weighted voting algorithm *(23)* and used cross-validation testing to assess the performance of the classifier. Models containing between 8 and 16 genes yielded statistically significant predictions, with the highest accuracy obtained using 13 genes. This classifier separated the 58 patients into two groups according to the predicted class: predicted to be cured or predicted to have fatal–refractory disease based on the gene expression profiles of those 13 genes. A Kaplan-Meier plot of these results is shown in Fig. 15 ($p = 0.0013$ using a standard log-rank test).

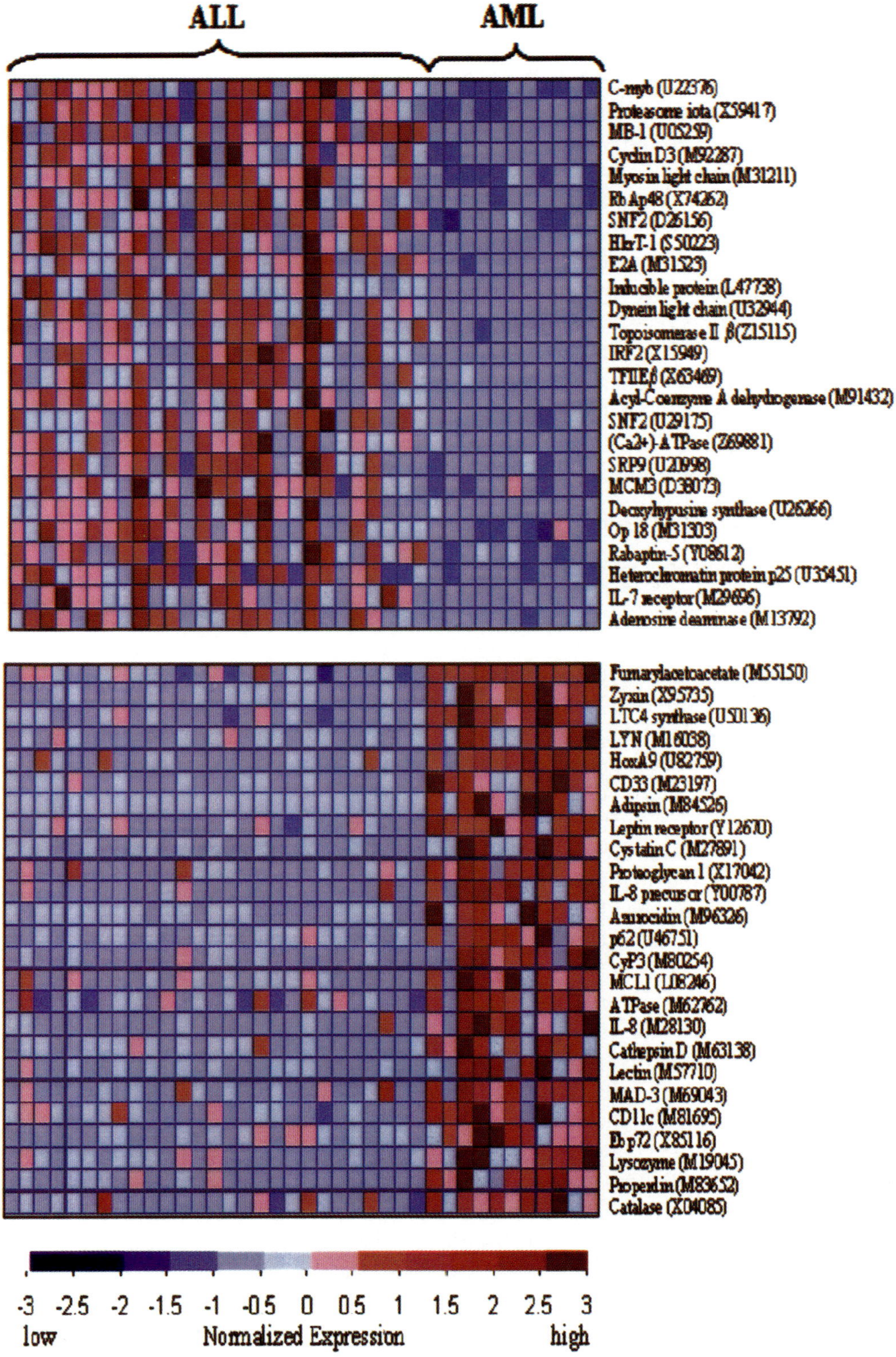

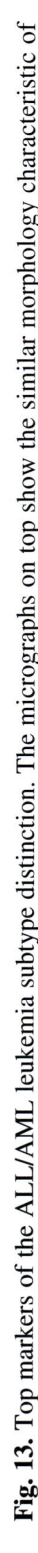

Fig. 13. Top markers of the ALL/AML leukemia subtype distinction. The micrographs on top show the similar morphology characteristic of these cells.

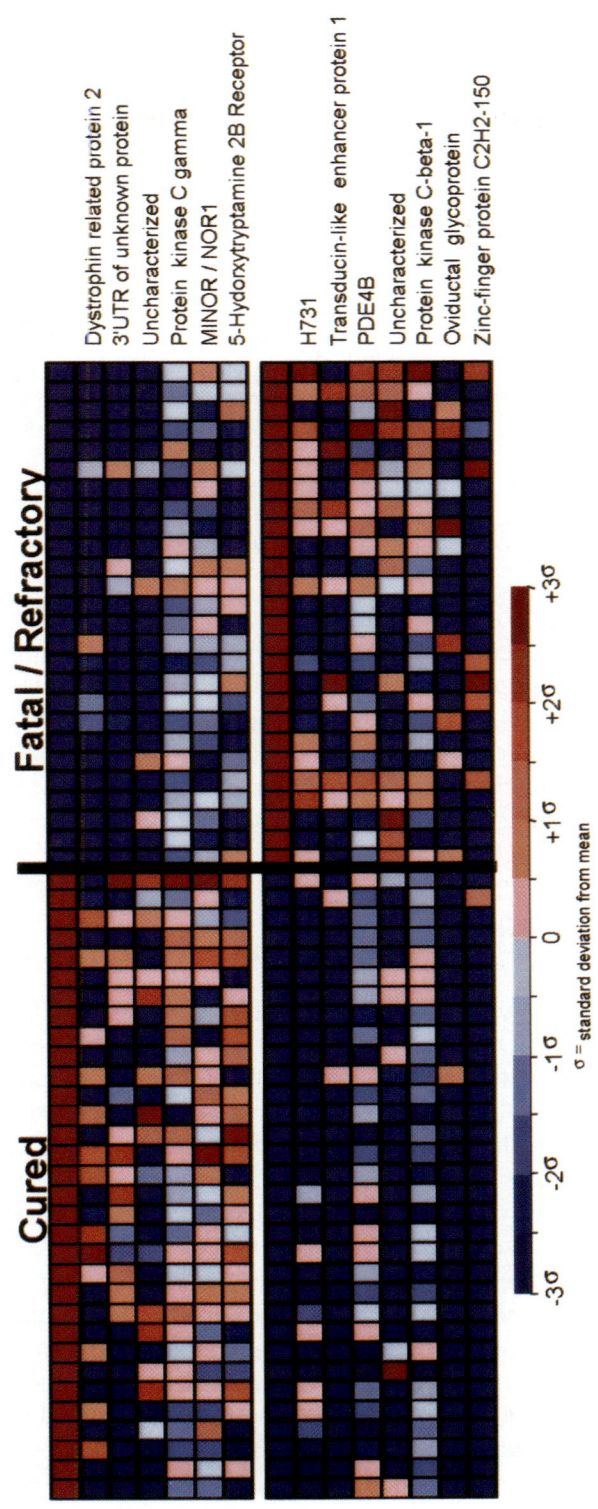

Fig. 14. Top markers of lymphoma treatment outcome.

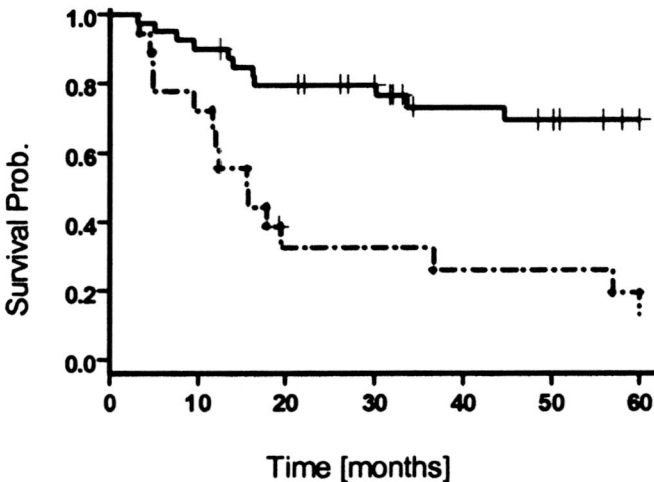

Fig. 15. Kaplan-Meier survival plot for the treatment outcome predicted groups.

Patients predicted by the classifier to be cured had dramatically improved long-term survival compared to those predicted to have fatal–refractory disease. The 5-yr overall survival (OS) is 70% vs 12%, with nominal log rank p value of 0.00004. As part of this study, we also built other supervised classification algorithms and obtained similar results. The fact that treatment outcome can be predicted solely based on gene expression patterns indicates the existence, at diagnosis, of a gene expression signature of outcome in DLBCL.

MULTICLASS CLASSIFICATION: CLASSIFYING MULTIPLE TUMOR TYPES

Multiclass classification problems are inherently more difficult than pairwise comparisons. In this section, we review our efforts to perform multiclass tumor classification (49,50). We explored the general feasibility of molecular cancer diagnosis of common human tumors solely on the basis of tumor gene expression profiles. We first created a gene expression database containing the expression profiles of 218 tumor samples representing 14 common human cancer classes and devised a multiclass classification method. Our analytical scheme is depicted in Fig. 16. First, the multiclass problem was divided into a series of 14 one-vs-all (OVA) pairwise comparisons. Each test sample was presented sequentially to these 14 pairwise classifiers, each of which either claimed or rejected that sample as belonging to a single class. This method resulted in 14 separate OVA classifications per sample, each with an associated confidence. Each test sample was then assigned to the class with the highest OVA classifier confidence. In mathematical terms: given m classes and m trained classifiers, a new sample takes the class of the classifier with the largest real valued output $class = arg\ max_{i=1...m} f_i$, where f_I is the real valued output of the ith classifier. A positive prediction strength corresponds to a test sample being assigned to a single class rather than to the ''all other'' class.

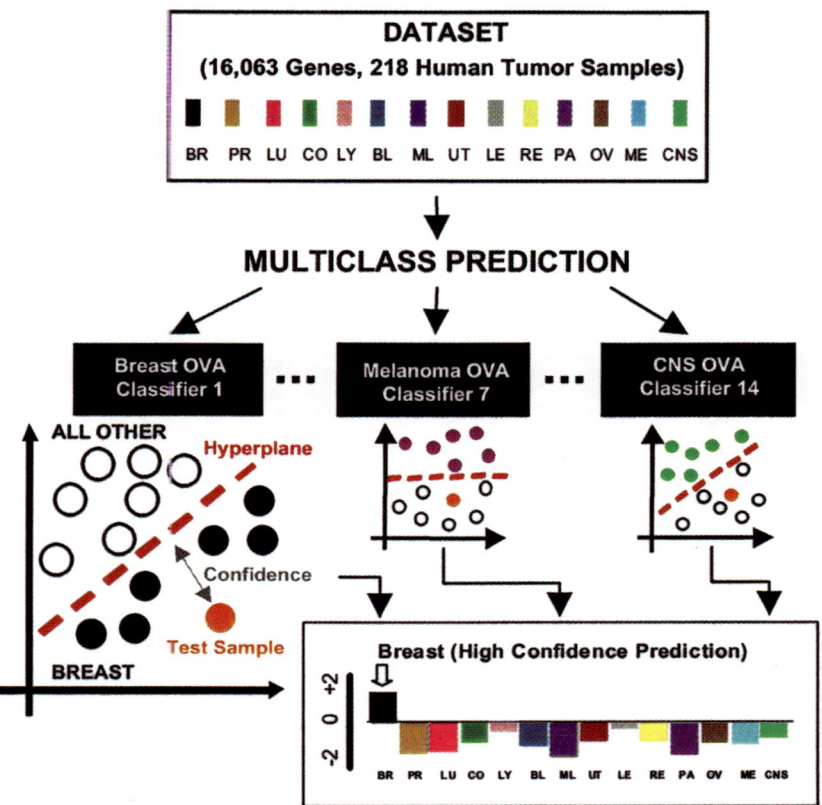

Fig. 16. Multiclass classification scheme. The multiclass cancer classification problem is divided into a series of 14 OVA problems, and each OVA problem is addressed by a different class-specific classifier (e.g., breast cancer vs not breast cancer). Each classifier uses the SVM algorithm to define a hyperplane that best separates training samples into two classes. In the example shown, a test sample is sequentially presented to each of 14 OVA classifiers and is predicted to be breast cancer, based on the breast OVA classifier having the highest confidence.

We then evaluated several classification algorithms for these OVA pairwise classifiers, including weighted voting, k-nearest neighbors, and SVM. Because the SVM algorithm consistently out-performed other algorithms, these results are described in detail. The SVM algorithm was used recently for pairwise gene expression-based classification *(41,47,51)* and has a strong theoretical foundation *(52,53)*. This algorithm considers all profiled genes to create descriptions of samples in this high-dimensional space, and then defines a hyperplane that best separates samples from two classes (Fig. 15). The position of an unknown sample relative to the hyperplane determines its membership in one or the other class (e.g., breast cancer vs not breast cancer). Fourteen separate SVM-based OVA classifiers classify each sample. The confidence of each OVA SVM prediction is based on the distance of the test sample to each hyperplane, with a value of 0 indicating that a sample falls directly on a hyperplane. The overall multiclass classifier assigns a sample to the class with the highest confidence among the 14 pairwise OVA analyses.

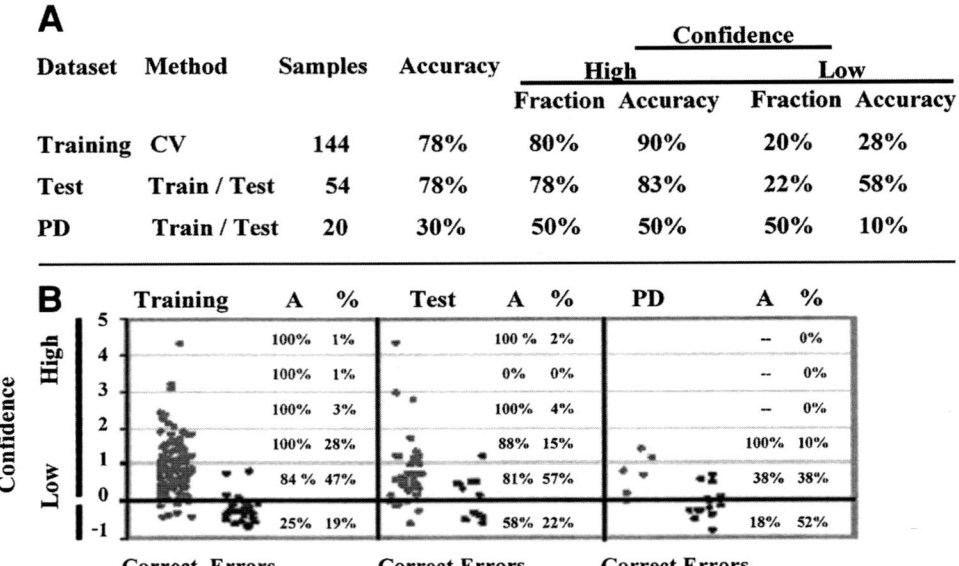

Fig. 17. Multiclass classification results. (a) Results of multiclass classification by using cross-validation on a training set (144 primary tumors) and independent testing with two test sets: Test (54 tumors; 46 primary and 8 metastatic) and PD (20 poorly differentiated tumors; 14 primary and 6 metastatic). (b) Scatter plot showing SVM/OVA classifier confidence as a function of correct calls (blue) or errors (red) for Training, Test, and PD samples. A, accuracy of prediction; %, percentage of total sample number.

The accuracy of this multiclass SVM-based classifier in cancer diagnosis was first evaluated by cross-validation in a set of 144 training samples. This method involves randomly withholding one of the 144 primary tumor samples, building a predictor based only on the remaining samples, then predicting the class of the withheld sample. The process is repeated for each sample, and the cumulative error rate is calculated. As shown in Fig. 17, the majority (80%) of the 144 calls were high confidence (defined as confidence >=0), and these had an accuracy of 90% using the patient's clinical diagnosis as the gold standard. The remaining 20% of the tumors had low confidence calls (confidence <0), and these predictions had an accuracy of 28%. Overall, the multiclass prediction corresponded to the correct assignment for 78% of the tumors. For half of the errors, the correct classification corresponded to the second- or third-most confident OVA prediction.

These results were confirmed by training the multiclass SVM classifier on the entire set of 144 samples and applying this classifier without further modification to an independent test set of 54 tumor samples. Overall prediction accuracy on this test set was 78%, a result similar to cross-validation accuracy and highly statistically significant when compared with class-proportional random prediction ($p < 10^{16}$). The majority of these 54 predictions (78%) were high confidence, with an accuracy of 83%, whereas low confidence calls were made on the remaining 22% of tumors with an accuracy of 58%. Again for one-half of the errors, the correct classification corresponded to the second- or third-best prediction. Of note, classification of 100 random splits of a combined training and test dataset gave similar results, confirming the stability of prediction for this collection of samples.

We next focused on the 28 samples that yielded low confidence predictions in cross-validation, as the multiclass predictor generally misclassifies these samples. We found that a large number (17 of 28) were moderately or poorly differentiated (high grade) carcinomas. It can be difficult to classify such tumors with traditional methods, because they often lack the characteristic morphological hallmarks of the organ from which they arise. It has been assumed that these tumors are nonetheless fundamentally molecularly similar to their better-differentiated counterparts, apart from a few differences that might account for their clinically aggressive nature. To directly test this hypothesis, the multiclass classifier was trained on the original 144-tumor dataset and then applied to an independent set of poorly differentiated tumors. Gene expression data were collected from 20 poorly differentiated adenocarcinomas (14 primary and 6 metastatic), representing five tumor types: breast, lung, colon, ovary, and uterus. The technical quality of this dataset was indistinguishable from the other samples in the study. However, these tumors could not be accurately classified according to their tissues of origin, compared with the high overall accuracy seen with lower grade tumors. Overall, only 6 out of 20 samples (30%) were correctly classified, which is statistically no better than what one would expect by chance alone ($p = 0.38$). Because the classifier relies on the expression of thousands of similarly weighted tissue-specific molecular markers to determine the class of a tumor, these findings indicate that poorly differentiated tumors do not simply lack a few key markers of differentiation, but rather have fundamentally distinct gene expression patterns.

DIMENSIONALITY REDUCTION AND PROJECTION: PRINCIPAL COMPONENTS ANALYSIS

Datasets with a large number of genes are in general difficult to visualize. Principal component analysis (PCA) is a dimensionality reduction method, which has been used to visualize complex gene expression datasets in two- and three-dimensional plots *(20,35,54,55)*. In this approach, one finds standardized linear combinations of variables, the "principal components," which are orthogonal and explain all of the variance in the original dataset. A typical method to obtain a simple projection (multidimensional scaling) of the dataset is to plot the top two or three principal components, which may account for a significant fraction of the variance, in a scatter plot. One can take this approach in a completely unsupervised manner, e.g., by using all genes that pass a data preprocessing step, or in a supervised way by projecting only the top marker genes of a phenotype of interest.

For example, principle component analysis can be applied to leukemia gene expression data. The initial set of genes is first subject to a variation filter resulting in a dataset with 612 genes that displayed the greatest variation across samples. In this case, the PCA method is used in an unsupervised way. Figure 18 shows a three-dimensional plot of these leukemia samples projected in the space of the top three principal components. This plot reveals the dominant structure of the dataset corresponding to the known morphological subclasses of leukemia, clearly separating ALL from the AML samples and separating the T cell ALL from B cell ALL samples.

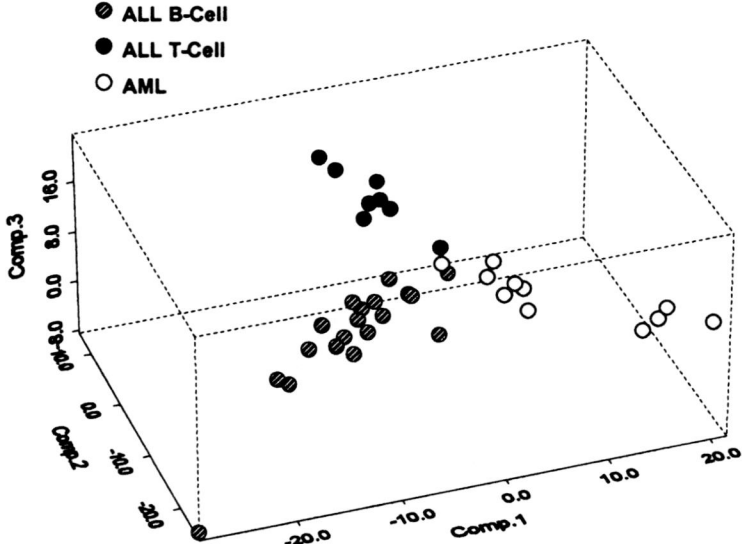

Fig. 18. Plot of the top three principal components for the 612 most highly varying genes in the leukemia subtypes dataset. The analysis is unsupervised and reveals the dominant structure of the dataset corresponding to the morphological subclasses.

CONCLUSIONS AND ANALYTICAL CHALLENGES IN MOLECULAR CLASSIFICATION

The analysis of cancer gene expression data is still in its infancy despite impressive recent progress. As expression profiling technologies mature, the identification of statistically significant patterns from relatively sparse and noisy data sets remains a major challenge. Although sophisticated data-mining techniques are already being used to analyze expression data, most of these techniques achieve robust performance with a large number of samples and a small number of variables *(56)*. However, gene expression data sets generally contain small numbers of samples, many profiled genes, and multiple sources of variation. Future advances will require adapting analytic and statistical techniques to this type of data. In addition, most published work has analyzed a relatively small number of samples, and most studies await independent confirmation.

A first generation of gene expression analysis methods has been used successfully in a variety of clustering and classification settings. For example, relatively successful models have been used to classify a variety of cancer types. Some examples include:

- Leukemias *(10,57,58)*.
- Lymphomas *(48,59–61)*.
- Ewing's sarcoma *(62)*.
- Brain cancer *(35,63,64)*.
- Breast cancer *(17,18,65)*.
- Lung cancer *(19,66,67)*.
- Prostate cancer *(68,69)*.
- Colon cancer *(70,71)*.
- Gastrointestinal tumors *(72)*.
- Ovarian cancer *(73–78)*.

- Melanoma *(20)*.
- Multiple tumors *(40,79,80)*.
- Soft tissue tumors *(81)*.

These studies have undoubtedly contributed to improve our understanding of cancer classification at the molecular level. However, in most cases, the complexity of the problem had to be simplified by treating genes as independent variables. While some studies expose co-regulation, they may not focus on the more complex patterns of interaction inherent in all biological processes and may further ignore the diversity of biological mechanisms within a phenotype. For example, in marker selection, one distinguishes between two phenotypes by determining which genes are up-regulated in one phenotype and down-regulated in the other. While this is a straightforward pattern to discover, we know it does not represent the true nature of genes' interactions. For example, it does not take into account distinct mechanisms that may yield the same biological state or subphenotypes and taxonomies that may be as yet unidentified. Even when clustering and classification methods are shown to be successful, it is often unclear exactly what the significant features or discovered patterns mean. Extracting more refined knowledge from the profiles and patterns is a serious scientific bottleneck.

Another important area relates to the integration of datasets generated in different laboratories using different profiling technologies. Many human cancer studies involve valuable or rare clinical specimens and are difficult to repeat. Ideally, one should be able to compare expression data sets obtained in any center, at any time, using any platform. However, this goal remains unrealized. Spotted array data is usually reported as ratios between experimental and control expression values and cannot be easily compared with oligonucleotide microarray data. Multiple expression profiling technologies require more sophisticated methods for data comparison and integration.

Despite initial sequencing of the human genome, we still have only a rudimentary knowledge of the physiologic roles of most genes. This represents a significant bottleneck in linking gene expression profiles to molecular mechanisms of transformation. There is a need for integrated databases, with complete annotation, comprehensive gene descriptions, and links to relevant genetic and proteomic information. In addition, as expression studies are performed in various species, integration of this information should prove as illuminating as interspecies gene sequence comparisons. Such databases will allow for an understanding of gene expression in the context of all other available biologic information. Although a number of commercial sources have started to create such databases, there is much room for improvement.

The challenges described above concern methodological and scientific issues. However, no computational approach is useful if it is not embodied in a set of software tools that scientists in the community can use. There are some academic codes available by Web download, but often they are not integrated and do not interoperate in a user-friendly environment. Available commercial codes are generally not current with the latest sophisticated techniques and often focus more on visualization of expression data than analysis and knowledge discovery. Since analysis of gene expression data remains a significant limitation in cancer genomics, the development of freely available and transparent analytic software continues to remain a major challenge.

ACKNOWLEDGMENTS

We are indebted to members of the Cancer Genomics Group, Whitehead/MIT Center for Genome Research, and the Golub Laboratory, Dana-Farber Cancer Institute/Harvard Medical School for many valuable and stimulating discussions.

REFERENCES

1. The Chipping Forecast. (1999) Special Supplement. *Nat. Genet.* **21,** 1.
2. Cheng, L. and Wong, W. H. (2001) Model-based analysis of oligonucleotide arrays: model validation, design issues and standard error application. *Genome Biol.* RESEARCH 0032.1–0032.11.
3. Tseng, G., Oh, M. K., Rohlin, L., Liao, J. C., and Wong, W. H. (2001) Issues in cDNA microarray analysis quality filtering, channel normalization, models of variation and assessment of gene effects. *Nucleic Acids Res.* **29,** 2549–2557.
4. Hunter, L., Taylor, R. C., Leach, S. M., and Simon, R. (2001) GEST: a gene expression search tool based on a novel Bayesian similarity metric. *Bioinformatics* **17,** 115S–122S.
5. Kerr, A. and Churchill, G. (2001) Experimental design for gene expression microarrays. *Biostatistics* **2,** 183–201.
6. Kerr, A. and Churchill, G. (2001) Statistical design and the analysis of gene expression microarrays. *Genet. Res.* **77,** 123–128.
7. Lee, M. T., Kuo, F., Whitmore, G. A., and Sklar, J. (2000) Importance of replication in microarray gene expression studies: statistical methods and evidence from repetitive cDNA hybridizations. *Proc. Natl. Acad. Sci. USA* **97,** 9834–9839.
8. Kerr, Afshari, Bennett, et al. Statistical analysis of a gene expression microarray experiment with replication. *Statistica Sinica*, in press.
9. Cheng, L. and Wong, W. H. (2001) Model-based analysis of oligonucleotide arrays: Expression index computation and outlier detection. *Proc. Natl. Acad. Sci. USA* **98,** 31–36.
10. Golub, T. R., Slonim, D. K., Tamayo, P., et al. (1999) Molecular classification of cancer: class discovery and class prediction by gene expression monitoring. *Science* **286,** 531–537.
11. Hartigan, J. (1975) *Clustering Algorithms.* John Wiley & Sons, New York.
12. Gordon, A. D. (1981) *Classification: Methods of the Exploratory Analysis of Multivariate Data.* Chapman and Hall, New York.
13. Duda, R. O., Hart, P. E., and Stork, D. G. (2000) *Pattern Classification, 2nd ed.* John Wiley & Sons, New York.
14. Cho, R. J., Campbell, M. J., Winzeler, E. A., et al. (1998) A genome-wide transcriptional analysis of the mitotic cell cycle. *Mol. Cell* **2,** 65–73.
15. Eisen, M., Spellman, P., Brown, P., and Botstein, D. (1998) Cluster analysis and display of genome-wide expression patterns. *Proc. Natl. Acad. Sci. USA* **95,** 14,863–14,868.
16. Tamayo, P., Slonim, D., Mesirov, J., et al. (1999) Interpreting gene expression with self-organizing maps: methods and application to hematopoietic differentiation. *Proc. Natl. Acad. Sci. USA* **96,** 2907–2912.
17. Perou, C. M., Jeffrey, S. S., van de Rijn, M., et al. (1999) Distinctive gene expression patterns in human mammary epithelial cells and breast cancers. *Proc. Natl. Acad. Sci. USA* **96,** 9212–9217.
18. Perou, C. M., Sorlie, T., Eisen, M. B., et al. (2000) Molecular portraits of human breast tumours. *Nature* **406,** 747–752.
19. Bhattacharjee, A., Richards, W. G., Staunton, J., et al. (2001) Classification of human lung carcinomas by mRNA expression profiling reveals distinct adenocarcinoma subclasses. *Proc. Natl. Acad. Sci. USA* **98,** 13,790–13,795.
20. Bittner, M., Meltzer, P., and Chen, Y. (2000) Molecular classification of cutaneous malignant melanoma by gene expression profiling. *Nature* **406,** 536–540.

21. Tibshirani, R., Walther, G., and Hastie, T. (2000) Estimating the number of clusters in a dataset via the Gap statistic." JRSSB.
22. Fukunaga, K. (1990) *Introduction to Statistical Pattern Recognition, 2nd ed.* Academic Press, New York.
23. Slonim, D. K., Tamayo, P., Mesirov, J. P., et al. (2000) Class prediction and discovery using gene expression data. Proceedings of the Fourth Annual International Conference on Computational Molecular Biology (RECOMB) New York, ACM Press, pp. 263–272.
24. Huberty, C. J. (1994) *Applied Discriminant Analysis.* John Wiley & Sons, New York.
25. Kearns, M. J. and Varzirani, U. V. (1997) *An Introduction to Computational Learning Theory.* MIT Press, Cambridge, MA.
26. Tusher, V. G., Tibshirani, R., and Chu, G. (2001) Significance analysis of microarrays applied to the ionizing radiation response. *Proc. Natl. Acad. Sci. USA* **98,** 5116–5121.
27. Dudoit, S., Yang, Y. H., Speed, T. P., and Callow, M. J. Statistical methods for identifying differentially expressed genes in replicated cDNA microarray experiments. *Statistica Sinica*, in press.
28. Westfall, P. H. and Young, S. S. (1983) *Resampling-Based Multiple Testing.* John Wiley & Sons, New York.
29. Ideker, T., Thorsson, V., Siegel, A. F., and Hood, L. E. (2000) Testing for differentially-expressed genes by maximum-likelihood analysis of microarray data. *J. Comput. Biol.* **7,** 805–817.
30. Newton, M. A., Kendziorski, C. M., Richmond, C. S., Blattner, F. R., and Tsui, K. W. (2001) On differential variability of expression ratios: Improving statistical inference about gene expression changes from microarray data. *J. Comput. Biol.* **8,** 37–52.
31. Baldi, P. and Long, A. D. (2001) A Bayesian framework for the analysis of microarray expression data: regularized t-test and statistical inferences of gene changes. *Bioinformatics* **17,** 509–519.
32. Yekuteli, D. and Benjamini, Y. (1999) Resampling based false discovery rate controlling procedure for dependent test statistics. *J. Stat. Plann. Inference* **82,** 171–190.
33. Korn, E., Troendle, J. T., McShane, L. M., et al. (2001, Aug.) Controlling the number of false discoveries. Application to high-dimensional genomic data. NCI–DCTD-003Technical report (http://linus.nci.nih.gov/~brb/TechReport.htm).
34. Califano, A., Stolovitzky, G., Tu, Y., et al. (1999) Analysis of gene expression microarrays for phenotype classification. Proceedings of the Eighth International Conference on Intelligent Systems for Molecular Biology, San Diego, pp. 75–85.
35. Pomeroy, S., Tamayo, P., Gaasenbeek, M., et al. (2002) Gene expression-based classification and outcome prediction of central nervous system embryonal tumors. *Nature* **415,** 436–442.
36. Alter, O., Brown, P. O., Botstein, D., et al. (2000) Singular value decomposition for genome-wide expression data processing and modeling. *Proc. Natl. Acad. Sci. USA* **97,** 10,101–10,106.
37. Cheng, Y. and Church, G. M. (2000) Biclustering of expression data. Proceedings of Intelligent Systems in Molecular Biology, August 19–23, 2000, La Jolla, CA.
38. Lazzeroni, L. and Owen, A. B. (2000) Plaid models for gene expression data. (Accessed on 2000 at http://www-stat.stanford.edu/~owen/reports/plaid.pdf).
39. Hastie, T., Tibshirani, R., Eisen, M. B., et al. (2000) "Gene shaving" as a method for identifying distinct sets of genes with similar expression patterns. *Genome Biol.* **1,** Research 0003.1–0003.21.
40. Ramaswamy, S., Tamayo, P., Rifkin, R., et al. (2001) Multiclass cancer diagnosis by using tumor gene expression signatures. *Proc. Natl. Acad. Sci. USA* **98,** 15,149–15,154.
41. Weston, J., Mukherjee, S., Chapelle, O., et al. (2001) Feature selection for SVMs, in *Advances in Neural Information Processing Systems 13.* (Solla, S. A., Leen, T. K., and Muller, K. M., eds.), MIT Press, Cambridge, MA.

42. Ripley, B. D. (1996) *Pattern Recognition and Neural Networks*. University Press, Cambridge.
43. Salzberg, S. (1999) On comparing classifiers: a critique on current research and methods. *Data Mining and Knowledge Discovery* **1,** 1–12.
44. Fisher, R. (1935) *The Design of Experiments, 3rd ed.* Oliver and Boyd, Ltd, London.
45. Lehman, E. C. (1986) *Testing Statistical Hypothesis, 2nd ed.* John Wiley & Sons, New York.
46. Good, P. (1994) *Permutation Tests: A Practical Guide to Resampling Methods for Testing Hypotheses.* Springer-Verlag, New York.
47. Mukherjee, S., Tamayo, P., and Slonim, D. (1999) Support vector machine classification of microarray data. CBCL Paper #182 Artificial Intelligence Lab. Memo #1676, Massachusetts Institute of Technology, Cambridge, MA.
48. Shipp, M., Ross, K., Tamayo, P., et al. (2002) Diffuse large B-cell lymphoma outcome prediction by gene expression profiling and supervised machine learning. *Nat. Med.* **8(1),** 68–74.
49. Ramaswamy, S., Osteen, R. T., and Shulman, L. N. (2001) in *Clinical Oncology 1st ed.* (Lenhard, R. E., Osteen, R. T., and Gansler, T., eds.) American Cancer Society, Atlanta, GA, pp. 711–719.
50. Yeang, C. H., Ramaswamy, S., Tamayo, P., et al. (2001) Molecular classification of multiple tumor types. *Bioinformatics* **17(Suppl 1),** S316–S322.
51. Brown, M. P., Grundy, W. N., Lin, D., et al. (2000) Knowledge-based analysis of microarray gene expression data by using support vector machines. *Proc. Natl. Acad. Sci. USA* **97,** 262–267.
52. Vapnik, V. N. (1998) *Statistical Learning Theory.* John Wiley & Sons, New York.
53. Evgeniou, T., Pontil, M., and Poggio, T. (2000) Regularization networks and support vector machines. *Adv. Comput. Math.* **13,** 1–50.
54. Mardia, K. V., Kent, J. T., and Bibby, J. M. (1979) *Multivariate Analysis.* Academic Press, London.
55. Yeung, K. Y. and Ruzzo, W. L. (2001) An empirical study on Principal Component Analysis for clustering gene expression data. (Accessed on 2001 at http://citeseer.nj.nec.com/yeung01empirical.html).
56. Friedman, J. H. (1994) An overview of computational learning and function approximation, in *Statistics to Neural Networks. Theory and Pattern Recognition Applications* (Cherkassy, V., Friedman, J., and Wechsler, H. W., eds.), Springer-Verlag, New York.
57. Yeoh, E. J., Ross, M. E., Shurtleff, S. A., et al. (2002) Classification, subtype discovery, and prediction of outcome in pediatric acute lymphoblastic leukemia by gene expression profiling. *Cancer Cell* **1(2),** 133–143.
58. Armstrong, S. A., Staunton, J. E., Silverman, L. B., et al. (2002) MLL translocations specify a distinct gene expression profile that distinguishes a unique leukemia. *Nat. Genet.* **30,** 41–47.
59. Alizadeh, A., Eisen, M., Davis, R. E., et al. (1999) The Lymphochip: a specialized cDNA microarray for the genomic-scale analysis of gene expression in normal and malignant lymphocytes. *Cold Spring Harb. Symp. Quant. Biol.* **64,** 71–78.
60. Alizadeh, A. A., Eisen, M. B., Davis, R. E., et al. (2000) Distinct types of diffuse large B-cell lymphoma identified by gene expression profiling. *Nature* **403,** 503–511.
61. Li, S., Ross, D. T., Kadin, M. E., Brown, P. O., and Waski, M. A. (2001) Comparative genome-scale analysis of gene expression profiles in T cell lymphoma cells during malignant progression using a complementary DNA microarray. *Am. J. Pathol.* **158(4),** 1231–1237.
62. Lessnick, S. L., Dacwag, C. S., and Golub, T. R. (2001) The Ewing's sarcoma oncoprotein EWS/FLI induces a p53-dependent growth arrest in primary human fibroblasts. *Cancer Cell* **1,** 393–401.

63. Sallinen, S. L., Sallinen, P. K., Haapasalo, H. K., et al. (2000) Identification of differentially expressed genes in human gliomas by DNA microarray and tissue chip techniques. *Cancer Res.* **60,** 6617–6622.
64. Ljubimova, J. Y., Lakhter, A. J., Loksh, A., et al. (2001) Black overexpression of {alpha}4 chain-containing laminins in human glial tumors identified by gene microarray analysis. *Cancer Res.* **61,** 5601–5610.
65. Sorlie, T., Perou, C. M., Tibshirani, R., et al. (2001) Gene expression patterns of breast carcinomas distinguish tumor subclasses with clinical implications. *Proc. Natl. Acad. Sci. USA* **98(19),** 10,869–10,874.
66. Garber, M. E., Troyanskaya, O. G., Schluens, K., et al. (2001) Diversity of gene expression in adenocarcinoma of the lung. *Proc. Natl. Acad. Sci. USA* **98(24),** 13,784–13,789.
67. Chen, J. J. W., Peck, K., Hong, T. M., et al. (2001) Global analysis of gene expression in invasion by a lung cancer model. *Cancer Res.* **61,** 5223–5230.
68. Singh Dinesh, Febbo, P. G., Ross, K., et al. (2002) Gene expression correlates of clinical prostate cancer behavior. *Cancer Cell* **1,** 203–209.
69. Welsh, J. B., Sapinoso, L. M., Kern, S., et al. (2001) Analysis of gene expression identifies candidate molecular markers and pharmacologic targets in prostate cancer. *Cancer Res.* **61,** 5974–5978.
70. Alon, U., Barkai, N., Notterman, D. A., et al. (1999) Broad patterns of gene expression revealed by clustering of tumor and normal colon tissues probed by oligonucleotide arrays. *Proc. Natl. Acad. Sci. USA* **96(12),** 6745–6750.
71. Mariadaso, J. M., Arango, D., and Corner, G. A. (2002) A gene expression profile that defines colon cell maturation in vitro. *Cancer Res.* **62,** 4791–4804.
72. Allander Susanne, V., Nupponen Nina, N., Ringner, M., et al. (2001) Gastrointestinal stromal tumors with KIT mutations exhibit a remarkably homogeneous gene expression profile. *Cancer Res.* **61,** 8624–8628.
73. Welsh, J. B., Zarrinkar, P. P., Sapinoso, L. M., et al. (2001) Analysis of gene expression in normal and neoplastic ovarian tissue samples identifies candidate molecular markers of epithelial ovarian cancer. *Proc. Natl. Acad. Sci. USA* **98,** 1176–1181.
74. Hough, C. D., Sherman-Baust, C. A., Pizer, E. S., et al. (2000) Large-scale serial analysis of gene expression reveals genes differentially expressed in ovarian cancer. *Cancer Res.* **60,** 6281–6287.
75. Hough, C. D., Cho, K. R., Zonderman, A. B., Schwartz, D. R., and Morin, P. J. (2001) Coordinately up-regulated genes in ovarian cancer. *Cancer Res.* **61,** 3869–3876.
76. Tonin, P. N., Hudson, T. J., Rodier, F., et al. (2001) Microarray analysis of gene expression mirrors the biology of an ovarian cancer model. *Oncogene* **20,** 6617–6626.
77. Bayani, J., Brentor, J. D., Pascale, F., et al. (2002) Parallel analysis of sporadic primary ovarian carcinomas by spectral karyotyping, comparative genomic hybridization, and expression microarrays. *Cancer Res.* **62,** 3466–3476.
78. Schwartz, D. R., Kardia, S. L. R., Shedden, K. A., et al. (2002) Gene expression in ovarian cancer reflects both morphology and biological behavior, distinguishing clear cell from other poor-prognosis ovarian carcinomas. *Cancer Res.* **62,** 4722–4729.
79. Lash, A. E., Tolstoshev, C. M., Wagner, L., et al. (2000) SAGEmap: a public gene expression resource. *Genome Res.* **10,** 1051–1060.
80. Su, A., Welsh, J. B., Sapinoso, L. M., et al. (2001) Molecular classification of human carcinomas using gene expression signatures. *Cancer Res.* **61,** 7388–7393.
81. Nielsen, T. O., West, R. B., Linn, S. C., et al. (2002) Molecular characterization of soft tissue tumours: a gene expression study. *Lancet* **359,** 1301–1307.

7
The Role of Tumor Banking and Related Informatics

Stephen J. Qualman, Jay Bowen, Sandra Brewer-Swartz, and Mary France

INTRODUCTION

The majority of this book deals with the diagnostic and research applications of molecular or tissue array technologies with regard to expression profiling of tumors. These efforts *(1)* will only be successful with the logical application of tumor banking and its associated informatics systems as the translational bridge linking new molecular information to its clinical significance. The design of tumor banks should be such that significant effort is devoted to obtaining data on clinical outcomes *(2)*, which permits investigators to know that such data are available for analysis as they pursue their molecular studies on bank-derived specimens. Tumor banking and its associated inventory informatics have been recognized for over a decade *(3,4)* as necessary tools to advance the science of molecular testing; however, it has only been in the last 2 to 3 yr that the linkage to clinical outcomes has been seen as crucial to achieving this goal *(3)*. It is estimated that by the year 2005 *(5)*, as much as 10% of clinical laboratory tests will be based on RNA or DNA analysis.

This chapter will deal with the broad concepts of tumor banking and informatics as practiced at the Biopathology Centers (BPC) located at the Children's Research Institute, Columbus, OH. The BPC provides banking and informatics services to a variety of cooperative groups, including the pediatric Cooperative Human Tissue Network (pCHTN), Children's Oncology Group (COG), Gynecologic Oncology Group (GOG), and Childhood Cancer Survivor Study (CCSS). The BPC deals with patient specimens that span the age range of newborns to the elderly; procuring tumor types that include the spectrum of pediatric solid tumors, selected adult gynecologic tumors (ovarian, cervical, uterine), and any second malignant (solid) neoplasm in childhood cancer survivors. In the last decade, the BPC has served nearly 85,000 tumor or related tissue specimens to over 300 different investigators. These specimens are linked to patient clinical outcome, because of the BPC's alliance with the aforementioned cooperative groups.

This chapter will describe the BPC's structure with regard to its physical repository and related tumor–tissue procurement activities, as well as its informatics system as it relates to data elements, security and data encryption, and specimen inventory. Finally,

From: *Expression Profiling of Human Tumors: Diagnostic and Research Applications*
Edited by: Marc Ladanyi and William L. Gerald © Humana Press Inc., Totowa, NJ

the metrics, which are used to assess both the BPC's quality control activities and the BPC's impact on advancing medical science, will be discussed. It is hoped that documentation of the BPC's experiences will serve as a paradigm for others who plan to procure, bank, and utilize human tumor tissues for research purposes, such as expression profiling.

BIOREPOSITORY STRUCTURE

Tissue Sources

Cooperative Group Affiliations

The types of specimens that are procured, processed, and stored by the BPC are dependent upon what is needed by the cooperative groups. By collecting every type of specimen associated with each group's protocols, the BPC can assure each group of uniformity in storage and quality control. While snap-frozen tumor tissue and paraffin blocks comprise the majority of preparations, the BPC stores everything from buccal cells to urine.

Cooperative group affiliations have been beneficial to the BPC, because they have allowed the BPC to be involved in protocol development, performance monitoring, and educating institutional members. The groups have benefited from utilizing the BPC banking and pathology services as well, because the BPC provides shipping containers, specimen procurement supplies, use of a courier account number, and customer service to the institutions. In addition, the groups know that specimens linked to their protocols are being managed by experienced staff who stay abreast of the ever changing shipping and patient protection regulations.

Local Institution(s)

The BPC is located on the Columbus Children's Hospital campus and maintains a relationship with the Pathology Department in order to access surgical and autopsy specimens processed by that department. It is the responsibility of the BPC to inform the surgical pathologists as to what cases and specimens are of interest to the bank researchers. Laminated wall charts are provided for posting in the surgical suite, the frozen section room, the histology laboratory, or the autopsy suite, which illustrate the types of tissue that can be procured according to tumor type (Fig. 1). These are also distributed to outside BPC procurement sites. Proof of consent to use the tissue for research must be obtained before the tissue is stored in the BPC repository.

Funding Sources

In addition to the pCHTN funding for tumor procurement, the BPC is supported by the GOG, COG and CCSS in order to provide these groups with banking services and pathology services. All funding is either direct or derivative funding from the National Institutes of Health (NIH). The baseline amount of funding required to establish a minimal tissue procurement service, independent of personnel costs, is estimated to be around $20,000 (2). This cost is also exclusive of space (4) and can be further broken down into initial equipment costs in freezing–storage apparatus, database software, and annual consumable costs (4).

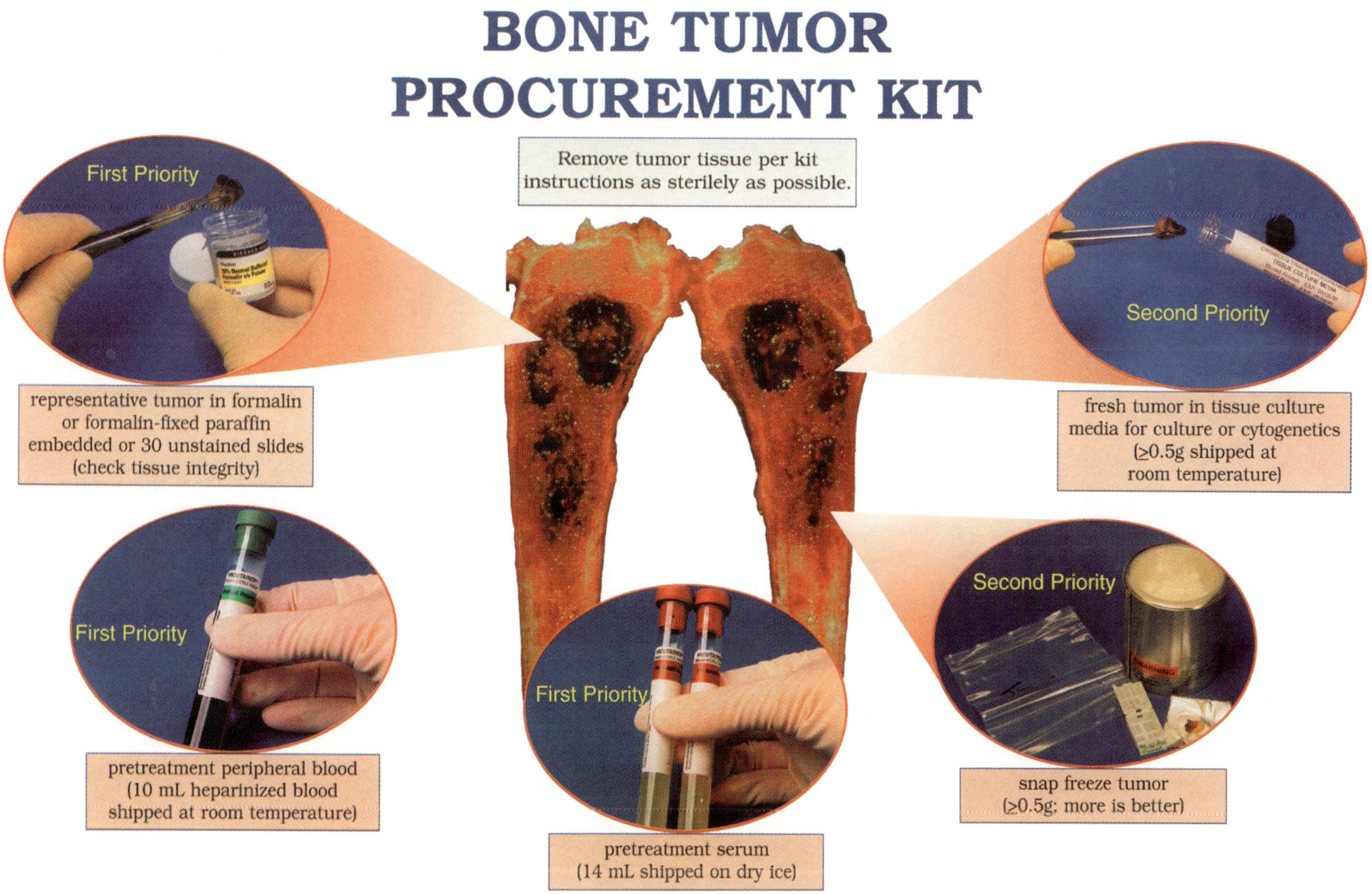

BONE TUMOR PROCUREMENT KIT

Remove tumor tissue per kit instructions as sterilely as possible.

Second Priority — fresh tumor in tissue culture media for culture or cytogenetics (≥0.5g shipped at room temperature)

Second Priority — snap freeze tumor (≥0.5g; more is better)

First Priority — pretreatment serum (14 mL shipped on dry ice)

First Priority — representative tumor in formalin or formalin-fixed paraffin embedded or 30 unstained slides (check tissue integrity)

First Priority — pretreatment peripheral blood (10 mL heparinized blood shipped at room temperature)

Fig. 1. Wall chart for posting in the surgical suite, the frozen section room, the histology laboratory, or the autopsy suite. Variations of this chart are created to illustrate specific types of tissue procurement, dependent on the tumor studied and the kit prepared for such purposes.

Space

The BPC presently occupies 2400 ft^2 of laboratory, storage, and office space. Approximately 640 ft^2 is used as laboratory space for histology and processing and distribution of specimens. With sufficient space, specimen processing and distribution can be done in the same laboratory.

Various types of storage space are needed by the BPC, including 300 ft^2 for liquid nitrogen vapor-phase freezers, –80°C freezers, and a cold room (4°C), while 150 ft^2 is needed for storage of slides, paraffin blocks, touch preparations, and procurement kits. A minimum of 150 ft^2 is estimated to be required for repository purposes *(2)*. Paperwork, which needs to be accessible to staff, is stored in filing systems within the office space of the BPC. Several rooms are utilized as office space for the BPC staff, totaling 1300 ft^2. Materials that are not frequently accessed may be stored off-site.

Equipment and Staff

Some of the pieces of equipment mentioned below are optional items, but were deemed as necessary due to the high activity level of the BPC. Laboratory-related staff consists of two histology technicians, two medical laboratory technicians, two research assistants, and one research coordinator. A minimum of two employees are probably required *(4)* to operate a biorepository, besides a pathologist to provide oversight; these include a bank coordinator, who maintains the tissue archive in all its aspects, and a histotechnician, to process tissues for distribution to investigators.

Processing and Distribution Laboratories

The specimens to be banked and/or served to bank researchers are received in the processing laboratory where they are weighed, labeled, and when necessary, further processed before being stored or served. The following equipment is routinely used by the processing and distribution laboratory staff: balances (precision to 0.01 g), centrifuge (Ficoll-Hypaque isolation of white blood cell pellets), chemical fume hood, laminar flow hood (sterile handling), and a refrigerator. To maintain tissue integrity, a heat sealer is used for sealing specimens in plastic bags prior to storage at –80°C and a Super Mutt Vacuum Sealer (Gramatech, Montebello, CA, USA) is used to package formalin-fixed tissues (paraffin blocks, scrolls, and slides) in mylar bags.

Storage Facility

Most tissues at the BPC are stored in liquid nitrogen vapor-phase freezers, because specimens kept at approximately –170°C are of higher integrity than those maintained at higher temperatures and show less desiccation *(2,6)*. Freezers (–80°C) are used for short-term tissue storage, and –20°C freezers are used for reagent storage. Manual defrost units are preferred, as the freeze–thaw cycles of automatic defrost units can degrade the quality of biologics and reagents.

Paraffin blocks, scrolls, and slides are best preserved in vacuum-sealed mylar bags with a commercial oxygen absorber at 4°C. Paraffin slides may also be dipped in molten paraffin prior to storage to preserve tissue antigenicity *(7)*. The BPC utilizes a cold room (walk-in cooler) (4°C) for storing its vacuum-sealed unstained slides and paraffin blocks.

Touch preparations, taken and fixed from fresh tumor specimens, were once standard submissions to the BPC, but are now best performed from segments of snap-frozen tissue stored at the bank. These can then be fixed and sent to the investigator with relatively good preservation of antigenicity and morphology.

Histology Laboratory

The histology technicians process fixed specimens and provide review materials to be used in diagnosing the tissues of the BPC. This includes processing formalin-fixed tissue into paraffin blocks and preparing an hematoxylin and eosin (H&E) section from each paraffin block for quality control review, as well as preparing fixed tissues for investigators (H&E sections, tissue scrolls for ploidy studies, immunoperoxidase slides, etc.). Frozen sections from optimal cutting temperature (OCT)-embedded frozen tissues are also routinely prepared for immunohistochemistry, *in situ* hybridizations, or nucleic acid extractions. The equipment located in the BPC histology laboratory includes a tissue microarray instrument, a microtome, a cryostat, a tissue processor to prepare tissue for embedding, an embedding center, and a slide-labeling machine (the latter labels all slides using the BPC's anonymous numbering system in order to protect patient confidentiality).

Specimen Procurement Kits and Storage

The BPC provides submitting institutions with specimen procurement kits to facilitate the procurement process and encourage specimen submission. The specimen procurement kits are reusable, insulated, multitemperature specimen containers (Insulated Shipping Containers, Phoenix, AZ, USA), which allow shipment of both frozen and ambient temperature specimens in the same package (*see* Fig. 2A). This kit maintains ambient and frozen temperatures for more than 48 h, despite the external environment through which it is transported (Fig. 2B).

The contents of kits are based on the particular group studies. Specimen procurement kits can include all the following kit supplies: OCT molds, 15-mL formalin jars in styrofoam mailers, foil and specimen baggies, 1.5-mL vials containing 1 mL of RNA*later*™ (Ambion, Austin, TX, USA) for tissue, 2-mL tubes for serum, 2-mL tubes for urine, 5-mL ethylenediaminetetraacetic acid (EDTA) tubes for anti-coagulated blood or bone marrow, 15-mL culture tubes, parafilm, charged glass slides in slide mailers, biohazard stickers, and dry ice labels.

Within the BPC, RNA is extracted from cryostat sections and tissues received in RNA*later*. Some nucleic acid preservation products may reduce the usefulness of specimens for proteomics, so this consideration should be taken into account in the design of tissue protocols. Tissues sent to investigators for RNA extraction by the BPC may utilize RNA*later* and obviate use of dry ice.

Long-term storage of specimens in vapor-phase liquid nitrogen freezers is done in small segments (e.g., 5 g or less) in plastic histology cassettes or mega-cassettes, with the specimens wrapped in aluminum foil to minimize desiccation. OCT-embedded specimens in cryomolds should likewise be wrapped in foil. Glass vials and pop-top plastic vials are not adequate storage containers for vapor-phase liquid nitrogen temperatures, as they may readily break or pop open *(2)*. Screw-cap cryovials work well for storage of serum or urine.

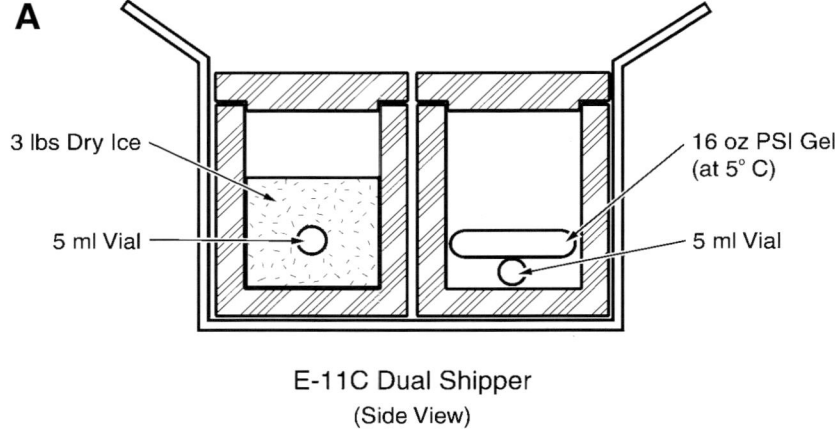

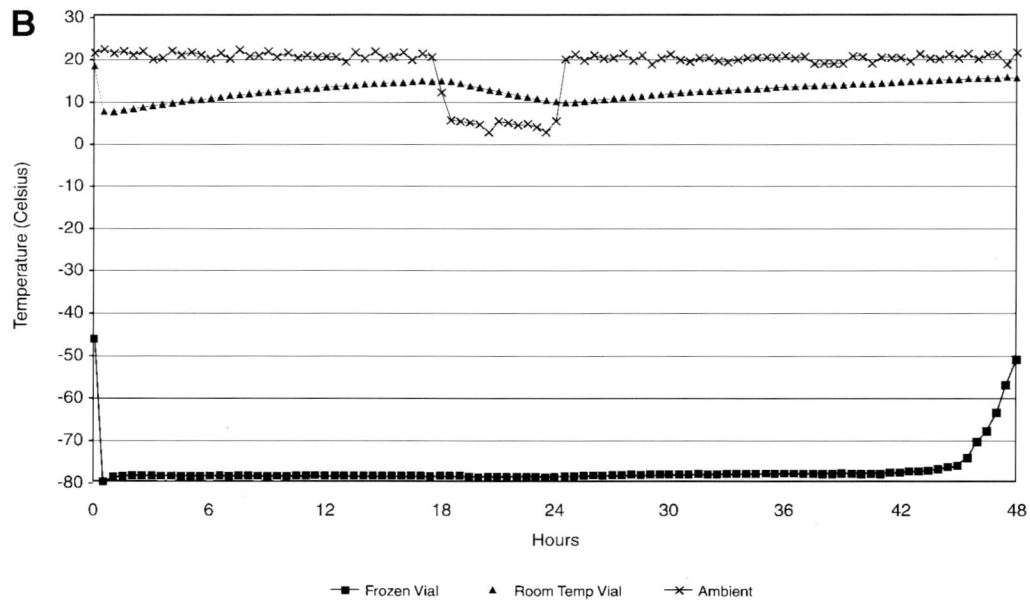

Fig. 2. (A) Insulated dual-chamber shipping container. One chamber contains dry ice for frozen specimens, while the other maintains specimens at room temperature using a gel pack. **(B)** Field testing of the kit shows dual maintenance of frozen and room temperature with associated specimens in the kit chambers (despite variation in outside ambient temperature) for a 48-h period.

Specimen Transport

Cooperative group-affiliated institutions must send specimens to the BPC through air-express couriers within 24 h of procurement, using the dual-chamber (Fig. 2) kits previously described. Shipments made via overnight carrier must be packaged to conform with International Air Transport Association (IATA) Section 1.5.0.2 and Code of Federal Regulations Section 172.7 (details available from the Web sites of various express courier companies and biohazardous shipping container companies). Researchers receiving specimens are billed for shipping charges and an additional specimen-processing fee associated with the types and numbers of specimens served.

DATA MANAGEMENT AND INFORMATICS

Data Management

The BPC staffs includes three informatics-related employees (computer engineer, programmer, and database liaison). The computer engineer is on staff during system design and redesign and is only cost-effective as a temporary position. The programmer is needed full-time, because systems are always being amended to meet the changing needs of the groups. The database liaison works with the staff in assuring that the systems are meeting the needs of the users (defining them for the programmer) and is also responsible for sharing data with cooperative groups. The majority of documents maintained by the BPC are registration documents, operative reports, and pathology reports. Copies of the associated pathology reports are sent along to researchers with submitted specimens, but are stripped of patient identifiers. To protect patient confidentiality, documents with patient identifiers are currently kept in lockable cabinets in locked rooms.

Informatics

Informatics can facilitate patient registration, specimen tracking, tissue cataloging, quality assurance, and specimen availability. The ability of databases to organize and present desired information can also aid in tracking informed consent and institutional compliance and be used to generate tissue bank inventory reports to match investigator requests with specimen availability.

Design Objectives for a Banking Inventory System

The components of a system must appear seamless to allow for efficient data entry, queries and report preparation, and must also allow for rapid deployment of new services. Consideration should also be made for the future, as systems will become increasingly diverse; supporting multiple architectures, platforms, and databases. Exchange of information between different databases at different institutions may also be or become a concern.

Ideally, an informatics system ensures that data is available over a long period of time, maintained in a standardized format, able to be disseminated to others as needed, and collected from collaborative sources and combined.

Informatics Programming Standards and Common Data Elements

Information Services departments may have database technology standards from which a database system can be based, but often there are none. Children's Research Institute Core Technologies (CRICT) are the standards used by Children's Research Institute in Columbus, OH, and were used as the starting point for the BPC's informatics system. These standards were then customized to meet the specific requirements of the project, resulting in a more specific set of standards called the BIOPATH II Database Architecture Guide *(8)*. These standards allowed 80% of all banking and pathology review systems to be part of the core data structure, leaving only 20% for customized programming.

The Cooperative Group Chairs, along with the National Cancer Institute, recognized the need for common data elements across data systems. As a result, the Intergroup Specimen Banking Committee was formed in 1999 to "identify existing standards, policies, and procedures, particularly those in common usage, which would address the needs of intergroup banking, rather than invent new ones. The goal was to deliver a set of recommendations that would require the least amount of work, change, and expense to implement, while providing sufficient guidance to the research community to facilitate useful banking efforts in support of correlative science, to avoid unnecessary conflict, and to ensure a sufficient standard of quality." (9).

Rather than establish standards of hardware, operating systems, or applications, the report instead set forth a standard for mandatory common data elements. These common elements provide for such things as uniform reporting, data exchange, and joint analysis.

Entities subject to these guidelines are medical institutions that submit specimens to one or multiple tissue banks–repositories, science review committees, investigators receiving specimens and responsible for submitting results, and statistical centers (9).

MINIMUM DATA REQUIREMENTS FOR BANKING (9)
- All samples and subsamples must have a unique identifier.
- The system must adhere to standard coding mechanism.
- The system must be able to maintain an up-to-date inventory by a unique identifier to include the original and current quantity of specimens and where they have been sent, when (if) they have been returned, and when they have been exhausted. The system must also be able to invalidate samples.
- The database containing information on sample locations (inventory) may reside at the bank and does not need to be accessible by systems outside the bank.
- The system must be able to flag (reserve) specimens for a particular study.
- The system must track the type of informed consent for research that is allowed.
- The system must track specimens and allow for withdrawal of consent.
- The system must be able to report and provide query results across groups to facilitate any combination of a bank per trial, a bank per disease site, or a bank per group per disease site. Projects may require samples accrued on more than one trial.
- The system must provide quality control–checks on imported data.

INFORMED CONSENT AND CONFIDENTIALITY

Tracking and cataloging informed consent is the key to banking informatics. Informed consent is the factor by which all tissues are qualified or disqualified for use by a potential researcher. The Intergroup Specimen Banking Committee recently endorsed the use of a three-item checkbox format for summarizing levels of informed consent (9). These levels allow for the patient to designate whether their tissue or case data may be used for: (*i*) cancer research; (*ii*) general medical research; and (*iii*) future patient contact for needed clinical follow-up.

LINKAGE AND DE-LINKAGE

Risks posed to subjects from research with their tissues are strongly related to the identifiability of individual sources of those tissues (10). Data records, which can be directly or indirectly associated with a person's name or other identifying information, are referred to as *linked data*. Such identifying elements are date of birth, treating institution, treating physician, medical record numbers, social security numbers, etc.

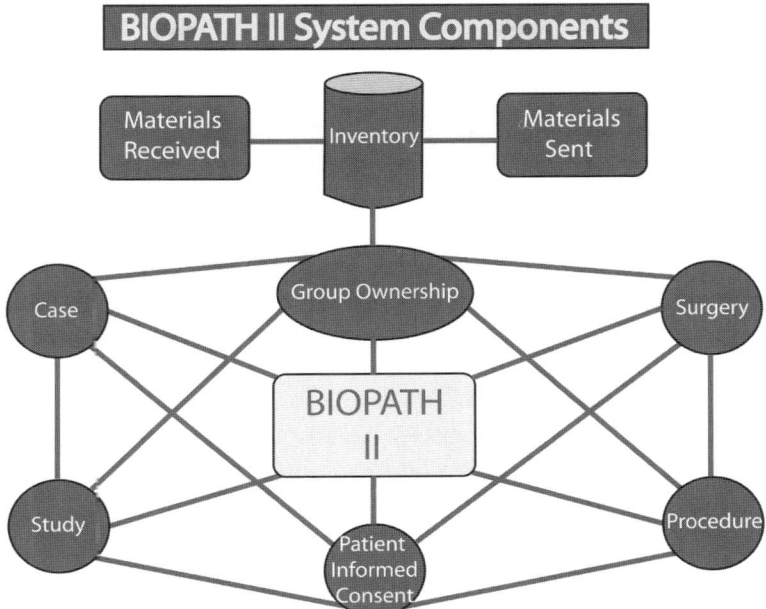

Fig. 3. Informatics design of BIOPATH II. A centralized inventory system is at the heart of the design, which can account for both cooperative group and patient-specific data, while maintaining a real-time inventory of specimen deposits, withdrawals, and residual tissues.

The BIOPATH II system uses specialized codes that, by default, "de-link" the specimen and clinical outcome data to maintain patient anonymity. If there is patient consent for the data to be linked, the data sets at the BPC and at a Cooperative Group Statistical Data Center are connected via a virtual private network (*see* section entitled Virtual Private Networks), and decoding can occur.

To permanently de-link data, any codes that will link or identify data or tissues with the donor must be removed, such that even the database manager can no longer trace a tissue or its related data back to the donor.

Centralized Inventory System

Inventory is at the heart of the tissue bank informatics design (Fig. 3). The inventory module is linked to cataloging, which provides for tagging and tracking of such things as informed consent, group ownership, and tissues reserved or specifically collected for a protocol or study. This ensures that specimens are tracked and distributed appropriately and in compliance with the patient's informed consent, as well as preventing the total depletion of a tissue sample.

The system design also streamlines tissue intake and serving of investigators. Materials are received and processed by a research assistant and entered directly into the system, while specimen shipments are processed through the same interface using the shipment module. The modules keep a precise record of materials as they are received and shipped, to produce a real-time inventory of what is available. When specimens are received and inventoried into the central system, the specimen is linked back to the specific cooperative group's case file. A full history of the specimen, from when it was received to when it was served, is maintained per cooperative group per case. Such history gives quick and easy access for reporting purposes.

Virtual Private Networks

A virtual private network (VPN) allows for usage of the Internet, normally a nonsecure medium, for transmission of confidential patient and specimen information using data encryption.

The use of virtual private networking has been fundamental in streamlining the patient enrollment and specimen registration process, reducing repetitive keying of data, and ensuring that information is synchronized between physically separate sites. This linking of databases can also be used to create virtual tissue banks from the inventory of different sites, thus allowing for the optimal matching of requests and distribution of tissues to reach a larger market.

The sensitive nature of medical information makes unauthorized disclosures and data alterations a concern. VPNs can be a cost-effective solution to provide secure point-to-point transactions utilizing encryption technologies. The primary requirements for implementation of a secure VPN are:

- Authentication; each end point checks the other and verifies that the transaction belongs to the secure point-to-point site before accepting the transaction or request.
- Strong data encryption to protect sensitive information.
- Transaction privacy (public–private key encryption).
- Encryption using a key unique to each information exchange session.
- Scalability; a VPN needs to be able to grow to accommodate an increase in connected sites.
- Ability to provide audit information.
- Immediate intrusion–attack detection and response; requires continuous evaluation of security policies and practices.

METRICS OF PERFORMANCE

Areas of Assessment

Creation of a physical biorepository, receipt of tissues, and development of an informatics system to monitor all aspects of specimen procurement and dispersal are merely a starting point for a successful tumor banking operation. Metrics or measures of the success of a banking operation are best assessed by looking at output, including what is served, who is served, and what is published. Metrics of what is served can include evaluations that are both morphologic and molecular in nature. This is in addition to written evaluations received from investigators served.

Morphologic Assessment

Central morphologic reviews are performed on the formalin-fixed tissues sent via the dual-chamber kits, which represent the mirror-image of samples of snap-frozen tumor and normal adjacent tissue also included in the kit. The histologic analyses are encoded on an informatics form for each histologic type of tumor examined and include such parameters as assessment of tumor diagnosis, percent tumor, percent stroma, and percent necrosis.

In a BPC sample of 7000 pediatric and gynecologic tumors examined over a 5-yr period, a diagnostic discrepancy rate of approx 10%, between central and institutional diagnoses, was identified. Although this discrepancy rate might at first appear to

Banking and Informatics

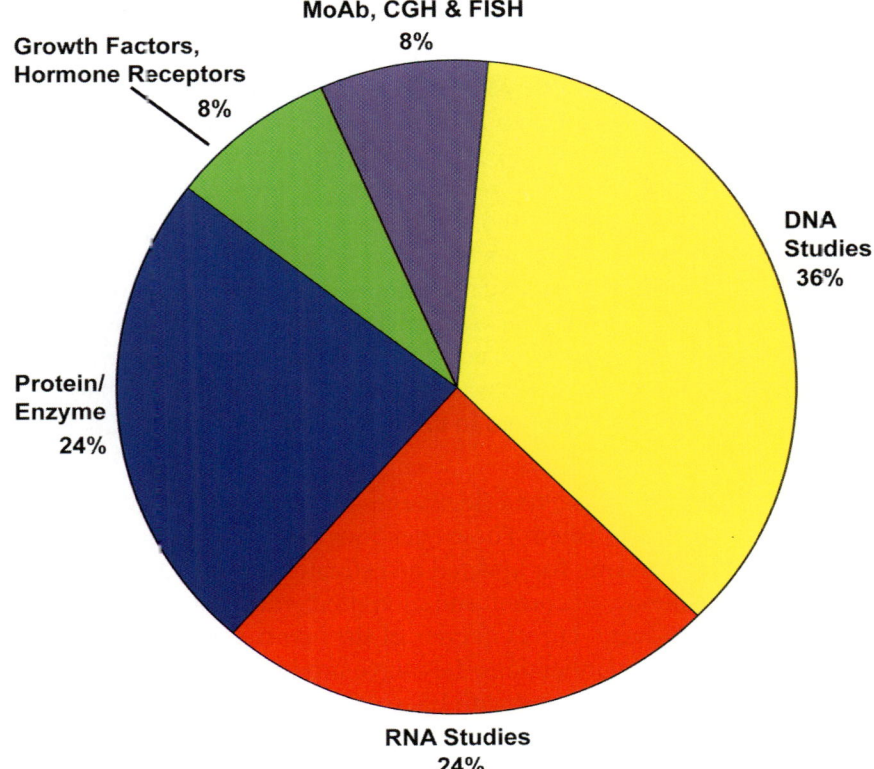

Fig. 4. Investigator uses of bank tissues as determined by direct query on feedback questionnaires. Sixty-percent (60%) of tissues are used for nucleic acid studies, nearly 25% for protein-based studies, and the rest for hybridizations (CGH and FISH) along with immunohistochemistry. MoAb, monoclonal antibody.

be quite high, these data in the main reflect sampling problems identified in tissue aliquots labeled by the institution as "normal" or "tumor," which were not reflective of the aliquot's content (e.g., tissue totally necrotic, normal tissue contaminated by tumor, etc.). Less than 2% of cases contained true discrepancies between central review and institutional diagnoses (most of which were minor, reflecting differing opinions as to tumor grade or subtype). The 10% discrepancy rate does emphasize the need for morphologic review of submitted tissues, so that extremely necrotic or contaminated tissue aliquots ("normal" with tumor) are not used by investigators. The central review diagnosis is the one reported to investigators.

Molecular Assessment

Confirmation of morphologic viability of tissues is only the first step in assessing the quality of tissues served to investigators. A survey of our pediatric and gynecologic investigators, as to uses of tissues for research (Fig. 4), revealed that the preponderance of specimens are used for various types of molecular studies (60% for RNA or DNA studies), with nearly one-third utilized for protein-based studies (including enzymes, growth factors, and hormone receptors). Studies involving hybridizations (fluorescence *in situ* hybridization [FISH], comparative genome hybridization [CGH]) or immunohistochemistry are in the minority (8% of total studies).

With such emphasis on molecular uses of BPC tissues, efforts to assess the molecular integrity of tissue became necessary. The BPC has looked at the molecular integrity of its gynecologic specimens by reverse transcription polymerase chain reaction (RT-PCR) and DNA and RNA electrophoresis performed on ovarian tissues *(11)*. RT-PCR assay for amplification of the mRNA gene product (177-bp product) of the *HPRT* housekeeping gene revealed adequate amplification in 70% of ovarian cases. RNA electrophoresis (followed by visual estimate of the 18s and 28s ribosomal RNA bands by ethidium bromide staining) also revealed RNA to be of good quality with minimal degradation in 70% of ovarian tissues. DNA electrophoresis showed the genomic DNA was of good quality in 100% of ovarian tissues tested. The challenge of adequate mRNA preservation is further discussed in the section entitled "Future Challenges."

Investigator Profiles and Publications

Another metric used to assess the effectiveness of a banking operation is defining who is served and what is published. These questions are answered during the feedback questionnaire process by asking for updated investigator biosketches and project summaries, along with a summary of current and submitted grant proposals (including those based on uses of BPC tissues). Direct queries of the investigators are also made on an annual basis regarding any publications they have produced based on BPC tissues, in part or total, and this list is compared with Internet listings of the same. All authors are asked to acknowledge BPC contributions within their papers.

These metric data can serve to validate a tumor bank's existence and successful function. One can also look more qualitatively at the contents of each scientific publication to define those papers that could be construed to be critical to the advancement of medical science, however, the details of this approach are beyond the scope of this chapter. Finally, it should be noted that the BPC's role is not to just serve the established investigator; a bank should also have the discretionary power to support the young investigator with exciting new scientific ideas.

FUTURE CHALLENGES

Tumor banking to advance medical genetic research has recently been identified as an international priority *(12)*. It has also changed the role of the pathologist from that of simply a tissue refiner to include a role as a data miner *(13)*. It is important in such an enterprise to establish a long-term commitment to substantial resources and secure funding *(12)*; however, the substantive challenges to the success of the enterprise in the future are really more based on tissue-, informatics-, or ethics-based questions.

Specific Challenges

The specific challenges which need to be addressed by any tumor banking operation in the future include: (*i*) doing more research with less tissue; (*ii*) addressing biohazards issues in all phases of the banking operation; and (*iii*) ensuring informed consent and maintenance of patient confidentiality.

More Research with Less Tissue

In their editorial on "Looking Forward in Pathology" *(1)*, NIH pathologists noted that "the emphasis is, and will continue to be, on getting more and more specific information from less and less material." The BPC's experience with tumor procurement has been that for most pediatric cancers, about 0.5 g of snap-frozen tumor is procured per case for research, while in adult gynecologic tumors, the average is more likely 1.0 g of tissue. In either case, neither amount is sufficient to service multiple investigators without careful forethought and planning.

While it is true that better education of surgeons, pathologists, and tumor procurement personnel may improve the yields of the procured tumor for research, ultimately, the amount of tissue made available for research, which would otherwise be discarded once diagnostic needs are met, is limited by institutional review boards (IRBs) *(12)*. This limitation of tissues procured is further magnified by the growth in use of core-needle biopsies and fine-needle aspirations as diagnostic procedures.

Given these limitations on amounts of tissue, preservation on-site of mRNA by timely snap-freezing of tissues in liquid nitrogen becomes paramount.

The ultimate role of the pathologist as a tumor banker in this setting is that of a tissue refiner. One of the chief complaints of investigators concerning tumor banks is the lack of sizeable specimens received for their research. It is incumbent upon the pathologists–bankers, as they review a research proposal, to educate the investigator both on the inherent limitations of tissue acquisition and on alternate research techniques that utilize less tissue. Fortunately, with the advent and increased usage of nucleic acid amplification techniques and FISH, such alternate techniques are more readily available; moreover, use of microarray technologies, combined with nucleic acid amplification, may allow for screening of thousands of genes *(1)*.

Biohazards in the Workplace

With the increasing numbers of individuals infected with hepatitis C or human immunodeficiency virus (HIV) in the U.S. *(14)*, it becomes incumbent upon any banking operation to operate under the mandate of "universal precautions," even though the BPC does not knowingly accept infected tissues as a stated policy. There are four steps in developing a biosafety program *(14)*, which include: (*i*) identifying governmental and accrediting agency requirements; (*ii*) identifying site-based risks and biosafety issues; (*iii*) developing written working guidelines to improve site-based biosafety; and (*iv*) implementing a training program.

The BPC maintains such a biosafety program in its banking operations and provides yearly universal precautions training to its procuring hospital personnel at cooperative group meetings. Training of the investigator's personnel in universal precautions is a written expectation of the BPC, which must be signed-off on by investigators, before they receive tissues *(6)*.

Biosafety hazards, as related to specimen containers and transport, have led to changes in the rules concerning transport of biologic specimens *(14)*. Information concerning such regulations is available through links at the Occupational Safety and Heath Administration Web site (http://www.osha.gov) *(14)*.

Informed Consent and Patient Confidentiality

In a recent editorial *(12)* on tumor banking, it was noted that "Ironically, it is the power of modern genetic analysis that creates the most difficult ethical dilemmas… concern that 'genetic research' may uncover information that is unwanted by the patient, has implications for family members, and could potentially lead to discrimination."

The keys to successfully addressing these dilemmas are to: (*i*) precisely define the meanings of a genetic test and genetic research *(15)*; (*ii*) define and implement model consent forms that address these definitions *(16)* and meet IRB expectations; and (*iii*) maintain the confidentiality of the flow of information from the institution procuring the specimens to the researcher testing them *(10)*. The tumor bank could effectively serve in the latter role by coding or de-linking tissues provided to researchers, managing updates of records, and handling requests for further follow-up data, while maintaining the fiduciary responsibility of protecting patient anonymity *(10)*.

With the evolution of final federal privacy rules concerning "Standards for Privacy of Individually Identifiable Health Information," tumor banks may need to take additional steps to successfully balance the demand for protection of patient privacy with the need to advance medical research. It is clear, from investigator surveys, that researchers are demanding more associated clinical data with their specimens. Improving the informed consent process to allow for acquisition of such data is the key to advancing such translational research.

REFERENCES

1. Abati, A. and Liotta, L. A. (1996) Looking forward in diagnostic pathology, the molecular superhighway. *Cancer* **78,** 1–3.
2. Grizzle, W. E. and Sexton, K. C. (1999) Development of a facility to supply human tissues to aid in medical research, in *Molecular Pathology of Early Cancer, Ch. 24.* (Srivastava, S., Henson, D. E., and Gazdar, A., eds.), IOS Press, Washington, D.C., pp. 371–383.
3. Naber, S. P., Smith, L. L., and Wolfe, H. J. (1992) Role of the frozen tissue bank in molecular pathology. *Diagn. Mol. Pathol.* **1,** 73–79.
4. Naber, S. P. (1996) Continuing role of a frozen-tissue bank in molecular pathology. *Diagn. Mol. Pathol.* **5,** 253–259.
5. Farkas, D. H., Kaul, K. L., Wiedbrauk, D. L., and Kiechle, F. L. (1996) Specimen collection and storage for diagnostic molecular pathology investigation. *Arch. Pathol. Lab. Med.* **120,** 591–596.
6. Grizzle, W. E., Aamodt, R., Clausen, K., LiVolsi, V., Pretlow, T. G., and Qualman, S. (1998) Providing human tissues for research: how to establish a program. *Arch. Pathol. Lab. Med.* **122,** 1065–1076.
7. Camp, R. L., Charette, L. A., and Rim, D. L. (2000) Validation of tissue microarray technology in breast carcinoma. *Lab. Invest.* **80,** 1943–1949.
8. Neville, R. (2000) Children's Research Institute core technologies database architecture guide. Internal document, Children's Hospital, Inc., Columbus, OH, May, 2000 (on file).
9. Report of the Intergroup Specimen Banking Committee, Cancer Therapy Evaluation Program, Division of Cancer Treatment and Diagnosis, National Cancer Institute, May 31, 1999 (on file).
10. Merz, J. F., Saukar, P., Taube, S. E., and LiVolsi, V. (1997) Use of human tissues in research: clarifying clinician and research roles and information flows. *J. Invest. Med.* **45,** 252–257.
11. Jewell, S. D., McCart, L. M., Williams, N., et al. (2002) Analysis of the molecular quality of human tissues: an experience from the Cooperative Human Tissue Network (CHTN). *Am. J. Clin. Pathol.* **118,** 733–741.
12. Balleine, R. L., Humphrey, K. E., and Clarke, C. L. (2001) Tumor banks: providing human tissue for cancer research. *Med. J. Aust.* **175,** 293–294.
13. Becich, M. J. (2000) The role of the pathologist as tissue refiner and data miner: the impact of functional genomics on the modern pathology laboratory and the critical roles of pathology informatics and bioinformatics. *Mol. Diagn.* **5,** 287–299.

14. Grizzle, W. E. and Fredenburgh, J. (2001) Avoiding biohazards in medical veterinary and research laboratories. *Biotech. Histochem.* **76,** 183–206.
15. Grizzle, W., Grody, W., Noll, W., et al. (1999) Recommended policies for cases of human tissue in research, education, and quality control. *Arch. Pathol. Lab. Med.* **123,** 296–300.
16. Taube, S. E., Barr, P., LiVolsi, V., and Pinn, V. W. (1998) Ensuring the availability of specimens for research. *Breast J.* **4,** 391–395.

Part III
Applications

8
Characterization of Gene Expression Patterns for Classification of Breast Carcinomas

Irene L. Andrulis, Nalan Gokgoz, and Shelley B. Bull

INTRODUCTION

Breast cancer is an important health problem that has proven to be a challenge for clinical and basic science research because of intrinsic tumor and cellular heterogeneity. In addition, the large number of genes potentially involved in controlling cell physiology complicates the accurate prediction of clinical behavior of breast carcinomas. The advent of gene expression microarray technology, which can be used for analyses of thousands of genes, provides a powerful tool to assist in determination of diagnosis, prognosis, and treatment.

Understanding the development of breast cancer from precursor lesions is critical for clinical treatment and prevention, however little is known of the molecular events involved in the progression to cancer. Once breast cancer has developed, the most important prognostic variable defining the natural history of breast cancer is the number of involved axillary lymph nodes (1,2). A number of predictive markers exist for axillary node positive breast cancer, and chemotherapy and hormonal therapy have been of benefit in reducing the risk of distant metastasis. As a group, women with axillary node negative (ANN) breast cancer have a good prognosis, however; approx 20% of individuals will experience a recurrence and die from systemic disease. There is no singular predictor of outcome for ANN breast cancer. Several clinical trials have indicated that some ANN patients may benefit from adjuvant chemotherapy or hormonal therapy (3–7). However, large numbers of women (including those in whom no recurrence will occur) must be treated in order to benefit those destined to relapse. It is, therefore, important to try to develop prognostic indicators for ANN patients who are more likely to experience a recurrence. Thus, currently available prognostic and predictive markers are not sufficient for the accurate determination of risk for many breast cancer patients. Better markers would obviously be of value in individualizing therapy.

Traditional predictors for patients with node positive breast cancer include tumor size, steroid hormone receptor status, menopausal status, and histologic grade (2,8–10). Adjuvant treatment for breast cancer includes hormonal therapy and chemotherapy. Most breast cancers express estrogen receptors (ER) at diagnosis, and ER status is used in the determination of adjuvant antiestrogen treatment, particularly with tamoxifen.

From: *Expression Profiling of Human Tumors: Diagnostic and Research Applications*
Edited by: Marc Ladanyi and William L. Gerald © Humana Press Inc., Totowa, NJ

Tamoxifen has been shown to be of benefit in the treatment and, more recently, prevention of breast cancer *(7,11)*. Unfortunately tamoxifen resistance can occur in individuals who were initially responsive to the drug. This is an area of active investigation, because of the biological and clinical importance, yet there remain many unanswered questions on the nature of the development of ER negative tumors and the mechanism of resistance to tamoxifen *(12)*.

In addition to traditional markers, other potential prognostic variables have been described, including cell proliferation *(13)*, DNA content *(14)*, neu/ERBB-2 amplification–overexpression *(15–24)*, epidermal growth factor receptor (EGFr) status *(25,26)*, nuclear grading *(27)*, p53 mutation *(28–31)*, cathepsin D *(32–35)*, p27 *(36,37)*, angiogenesis *(38–41)*, and amplification of 11q13 *(42–45)*.

A number of recent studies have reported the use of gene expression arrays to identify groups of co-expressed genes, to characterize genes by their expression profiles over a set of breast carcinoma samples, and to characterize molecular signatures of breast carcinomas. In this chapter, we describe how gene expression profiling is being used to classify specimens based on properties of the tumor, such as expression of ER and ERBB2, as well as *BRCA1* and *BRCA2* mutation status, to identify gene expression patterns related to nodal status and clinical outcome, and to predict therapeutic groups responsive to hormonal and chemotherapeutic agents. The ultimate objective of most of these studies is to assist in the development of treatment strategies based on tumor expression profiles.

SUMMARY OF ISSUES IN DESIGN AND ANALYSIS
Biospecimens and Patient Collections

To characterize gene expression patterns in human breast cancer, investigators have studied array profiles of breast epithelial cell cultures, breast cancer cell lines, and primary human breast tumors. Most papers describing array data using primary human tumor material include some histopathological characteristics of the tumors and/or clinical information on the patient. Aside from ER status of the specimens, which is generally indicated, studies vary widely in the amount of additional clinical information reported (*see* Table 1). There is also considerable variation in the source of patient tissue samples, including clinical trial participants, frozen tissue tumor banks, as well as unselected samples.

Human breast tumors are heterogeneous, and specimens available for molecular gene expression studies can be expected to contain contaminating normal material. In addition to the breast carcinoma cells, various amounts of stromal cells, adipose cells, endothelial cells and infiltrating lymphocytes will be present. Although microdissection methods can be used to enrich for cancer cells, and methods are being developed for reliable amplification of mRNA *(46)*, to date most expression profiling has been performed on breast carcinomas that have not been microdissected. As technology continues to improve, this is likely to change. Furthermore, it has been argued that, since the noncancer cells can also be identified by their unique expression signatures, analysis of the intact specimen may provide additional information related to cellular interactions in tumor progression *(47,48)*.

Various types of expression arrays have been used to investigate breast tumor specimens including cDNA arrays, oligonucleotide arrays, and membrane arrays (Table 1).

Table 1
Summary of Microarray Analyses in Breast Cancer

Group (reference)	Tumors	Tumor characteristics				Array characteristics
		ER status	LN status	Treatment	Outcome	Number of spots
Martin et al. (2000) *(55)*	18	+	+*	NR	+ (2 yr)	124[a]
Perou et al. (2000) *(50)*	42	+	+	+	NR	8102[b]
Bertucci et al. (2000) *(57)*	34	+	+	+	+	176[a]
Gruvberger et al. (2001) *(67)*	58	+	+	NR	NR	6728[b]
West et al. (2001) *(99)*	49	+	+	NR	NR	5000[c]
Sorlie et al. (2001) *(51)*	78[e]	+	+	+	+	8102[b]
van't Veer et al. (2002) *(52)*	117	+	+	NR	+	25,000[d]

*Stage.
[a]Membrane arrays.
[b]cDNA arrays.
[c]Oligonucleotide arrays.
[d]Ink-jet arrays.
[e]Includes 42 tumors in Perou et al. *(50)*.
ER, estrogen receptor; LN, lymph node; NR, not recorded.

For the cDNA and ink-jet spot arrays, a common reference control sample has been separately labeled and hybridized together with RNA from the cell or tissue sample of interest. Levels of gene expression in the test sample can be measured relative to the reference sample, thereby allowing for a consistent comparison between multiple experiments. It is desirable to choose a reference sample with diversity, so that the majority of the genes spotted on the array show some minimal level of fluorescence intensity *(48,49)*. Most studies of breast specimens have pooled mixtures of cell lines *(47,50,51)* or pooled cRNA derived from carcinomas *(52)*.

Approaches to Statistical Analysis and Assessment of Validity

Analytic strategy

As in other human cancer applications, the statistical analysis of gene expression data from microarray studies of breast carcinoma typically follows the process outlined in Table 2. The initial stage of image processing and quantitation, quality assessment, and data transformation and normalization to adjust for systematic between array differences is carried out for each array, with the methods dependent to a large extent on the specific array technology being used *(53)*. For cDNA spot arrays for example, we have observed systematic subarray effects that usually reflect the spatial position of the gene spot on the array slide. These effects can be identified by scatter plots and removed by location adjustments. To analyze a set of tumor samples together, scale adjustment and gene filtering across arrays is generally required. The latter is intended to eliminate genes that are not well measured or those that show no variation across the samples being analyzed.

Unsupervised classification methods, such as cluster analysis, are essentially exploratory in that they do not require any *a priori* information to group samples by

**Table 2
Statistical Analysis of Gene Expression Data**

Preprocessing of each array
- Image analysis.
- Quality assessment.
- Normalization.
- Diagnostic plots.

Selection of array sets and genes to be included in analysis
Unsupervised analysis methods
- Identification of clusters of samples with similar expression signatures.
- Identification of clusters of genes with similar expression profiles.

Supervised analysis methods
- Univariate single gene comparisons among groups of samples.
- Multivariate multiple gene comparisons among groups of samples.
- Prediction and validation of group membership for individual samples.

---

characteristics or to group genes by function. Rather, samples are grouped according to similarity in gene expression patterns, sometimes called signatures *(54)*, and genes are grouped according to similarity in gene expression profiles. The widely used hierarchical methods, for example, conduct clustering of samples and genes independently, although the results may depend on the sample composition and the genes under consideration. Then, once clusters of samples and/or genes have been identified, available information on gene function and sample characteristics is used in *post hoc* interpretation of the clusters. For example, tumor clusters may be compared with respect to patient outcomes, such as lymph node status or disease-free survival, or to externally derived tumor characteristics, such as ER or *p53* mutation status. Similarly, the significance of gene clusters may be interpreted using knowledge about particular genes within a cluster.

By comparison, supervised methods require clinical or molecular information about each of the samples, and generally focus on the identification of genes or combinations of genes selected so as to maximally distinguish among known sample characteristics. This approach often aims to develop a classifier, or prediction function, that uses a number of different gene expression measures from a particular sample to classify that sample into a predefined outcome group. The accuracy of the classifier is then assessed by comparing the predicted to actual group membership for each sample.

Statistical Validation of Microarray Technology

The validity of microarray technology is an important consideration in its application to human cancer studies that aim to improve the classification and treatment of cancer patients. These can include both internal and external measures as outlined in Table 3, as well as molecular validation, which is discussed further in the section entitled Molecular Validation of Expression Profiling Results.

Among the published reports, there have been a few examples of specific reproducibility studies of arrays, mostly in the earlier studies (*see* Table 4). In cDNA array applications, self–self comparisons involving replication of reference control vs control comparisons, in which no differences are expected, allows assessment of between

**Table 3
Validation Methods**

Internal validation
- Reproducibility studies, replication.
- Statistical.
  - Bootstrap for clustering.
  - Permutation for multiple testing.
  - Cross-validation for prediction.
  - Assessment of power by simulation.
- Molecular.
  - Confirmation by PCR, IHC.
  - Tissue arrays.

External validation
- Confirmation in independent samples.

array measurement error *(50)*. Sample vs reference control comparisons, with replications, also allow assessment of within and between sample variation *(47,55–58)*. In cDNA two-channel arrays with labeling by fluorescent dyes, dye swap repetitions are one form of replication that also can allow assessment of dye effects and separation of sample effects from array effects *(59,60)*.

Some authors *(53,61)* have recommended combining information from as many as three replicates per sample to decrease variability and increase measurement reliability. This is not always feasible with limited breast tumor material, but has been useful in studies of cell lines *(55,56)*. Only one of the published studies reports the use of dye-swap averaging in breast tissue samples *(52)*. In comparisons between groups of tumors, it is between patient variability that is critical, so the arraying of additional patient samples will increase overall precision and generalizability. For prediction of outcome for a single sample, however, measurement error is more critical, so averaging of replicate gene expression measures can increase the accuracy of prediction (*see* also the Discussion in Simon et al. *[49]*).

Other sources of evidence to support validity of the microarray technology as applied in a particular research setting include: (*i*) observations that multiple gene clones and sample replicates have similar gene expression; and (*ii*) observations that array measurements of previously reported candidate genes are associated with sample groups as expected.

Because of the large number of gene expression probes relative to the number of samples and the inherent random variability in the system, a number of statistical issues arise in microarray studies. In unsupervised classification, apparent clusters can be identified even in the absence of real structure. It is important to assess whether clusters are reproducible and are not due to chance variation. Statistical bootstrapping has been advocated as one approach to address this *(62,63)*. Similarly, permutation testing can address the problem of excess false positive findings produced by multiple testing of group differences for a large number of genes *(51,53,58,64)*. In supervised classification and prediction, the inherent overfitting can lead to overoptimistic estimates of the prediction accuracy, but this can be remedied by internal cross-validation in the so-called training set used to develop the classifier and by external validation in independent test samples.

Table 4
Aspects of Design and Analysis in Microarray Studies of Breast Carcinomas

Study (reference)	Microarray reproducibility studies and use of replication	Unsupervised classification of genes and tumors	Supervised gene selection and prediction of tumor groups
Perou et al. (1999) (68)	Self–self comparison of reference control vs reference control.	Hierarchical agglomerative (HA) cluster analysis: 1. Cell lines. 2. Cell lines and breast tissue.	
Sgroi et al. (1999) (55)	1. Variability estimated by duplicate hybridizations of three cell populations. 2. Duplicates averaged.		Genes ranked by expression differences between invasive or metastatic cells compared to normal cells.
Hilsenbeck et al. (1999) (91)	Repeat hybridizations using the same pool of RNA.		Principal components used to contrast three tumor type arrays and construct a prediction region to classify outlier genes.
Ross et al. (2000) (47)	1. Two of the cell lines, grown in three independent cultures. 2. Redundant array spotting of some genes.	HA clustering: 1. Cancer cell lines (equal to 60), repeated in two different gene sets. 2. Cell lines and breast tissue.	
Martin et al. (2000) (56)	1. Two hybridizations performed on different days with different replicate membranes. 2. Averaging of quadruplicate array spots. 3. Averaging of duplicates or triplicates of some cell lines.	Primary HA clustering: 1. Cell lines and breast tumor tissue samples, 124 genes. 2. Repeated for breast tumors, with 35 genes from four clinically relevant gene clusters.	Secondary Comparison of single gene expression and average gene cluster expression between tumor groups defined by ER status, tumor stage, grade, size, % S phase, and patient age.
Perou et al. (2000) (50)	Paired tumors, before and after chemotherapy. Used to select intrinsic gene set.	Primary HA clustering: 1. Cell lines and breast tissue. 2. Paired tumor samples.	Secondary Weighted voting supervised by two main tumor groups found in second cluster analysis, with internal cross-validation.

Study			
Bertucci et al. (2000) (57)	1. Comparison of duplicate spots. 2. Two independent hybridizations of same RNA. 3. One hybridization with the same probe on two independent arrays.	Primary HA clustering: 1. All samples (tumor and normal), all candidate genes. 2. Subset of tumors, subset of genes.	Secondary Supervised comparison of median expression for each gene by Mann-Whitney test: 1. Two subgroups of tumors identified in cluster analysis. 2. ER+ vs ER− tumors. 3. LN+ vs LN− tumors.
Alizadeh et al. (2001) (48)	Comparison of independent copies of gene clones on array.	HA clustering of microdissected breast tumor, cell lines, and breast tissue.	
Ross and Perou (2001) (83)	Independent propagations of the same cell line, obtained from different sources.	HA clustering of: 1. All cell lines, 1287 genes. 2. Cell lines and previously studied breast tissue samples, with intrinsic gene set of 476.	
Gruvberger et al. (2001) (67)	Not reported.	Secondary 1. HA clustering of: Tumors in training sample, genes selected by supervised weighted gene analysis. 2. Visualization by multidimensional scaling (MDS).	Primary Supervised by ER+ vs ER− tumors: 1. Artificial neural network, with internal cross-validation in training samples, external validation in test sample. 2. Repeated using different gene sets. 3. Weighted gene analysis.
West et al. (2001) (99)	Not reported.	Singular value decomposition (SVD) of top 100 genes in supervised single gene analysis, to create a smaller number of "super genes."	Primary Bayesian probit regression supervised by 1. ER+ vs ER−. 2. LN+ vs LN−. With internal cross-validation in training samples, external validation in test samples.

(continued)

Table 4 (*continued*)

Study (reference)	Microarray reproducibility studies and use of replication	Unsupervised classification of genes and tumors	Supervised gene selection and prediction of tumor groups
Sorlie et al. (2001) (*51*)	Not reported.	Primary HA clustering of: 1. Carcinomas and normal breast tissue, using intrinsic gene set. 2. Repeated in a subset of carcinomas. 3. Repeated in all samples, using gene set identified in supervised analysis. Secondary 1. HA clustering of genes identified as discriminators in supervised analysis. 2. Display by MDS.	Secondary 1. Univariate gene selection, supervised by patient survival, with permutation testing. 2. Comparison of overall and disease-free survival among five tumor groups identified in unsupervised clustering.
Hedenfalk et al. (2001) (*58*)	Estimation of experimental variance by hybridization of pairs of cDNA on different days.		Primary Supervised by tumor groups (*BRCA1*, *BRCA2*, sporadic no known family history): 1. Modified F-tests and *t*-tests to compare groups for each gene, with permutation to rank genes. 2. Prediction of tumor groups by a multi-gene compound covariate, with internal cross-validation. 3. Weighted gene analysis. 4. Mutual information scoring, with permutation.
Su et al. (2001) (*71*)	Not reported.	HA clustering of tumors to assess similarity of tumors within 11 cancer type classes.	Primary Supervised by cancer type: 1. Wilcoxon rank-sum test for each gene, comparing class with highest expression to others, to reduce number of genes.

van't Veer et al. (2002) (52)	1. Two hybridizations of each tumor with dye reversal, averaged for analysis.	HA clustering of all genes and all tumors (sporadic and *BRCA1*) to describe main tumor and dominant gene clusters.	2. Support vector machine to rank selected genes for class prediction, with internal cross-validation in training sample, and external validation in test sample. Supervised by tumor group: 1. Disease-free status at 5 yr (sporadic tumors only). 2. ER+ vs ER−. 3. *BRCA1* vs sporadic within ER− subgroup. Gene selection by single gene correlation of expression with group, internal cross-validation to rank genes for inclusion in the classifier. Tumors classified by correlation of their expression signature to the average signature of each tumor group, with internal cross-validation in training sample and external validation in a test sample.

Finally, it is desirable to have an assessment of the statistical power of a microarray data analysis. Although preliminary estimates of variation in gene expression within and between samples for a particular microarray system can provide relevant information for rough estimates of the number of arrays needed to detect single gene differences in gene expression between tumor subgroups *(49)*, *a priori* power determination for complex multi-gene methods is difficult in hypothesis-generating studies of a small sample size. Once analyses have been conducted, however, the ability of a particular study analysis to detect associations of interest can be assessed by statistical simulation studies based on the observed data. Van der Laan and Bryan offer one approach *(63)*.

Statistical Methods

In the published studies summarized in Table 4, unsupervised classification has mainly been based on clustering, supplemented by visual display using tree-like dendograms and multidimensional scaling plots. The cluster analyses have relied exclusively on hierarchical agglomerative methods *(65)*, and to date, other approaches such as k-means, self-organizing maps, and model-based clustering or simultaneous clustering of genes and samples have not been applied to breast carcinomas *(63,66)*. Similarly, the stability of the clusters identified has not been assessed using statistical validation methods, such as bootstrapping, although several authors have conducted sensitivity analyses by considering different sets of tumors and/or different sets of genes *(47,50,51,56,57)*.

More recently, there has been increasing application of supervised methods of analysis, corresponding to studies of tumor sample collections with richer clinical information. A broad range of univariate (one gene at a time) and multivariate (multiple genes together) techniques has been applied to identify genes that distinguish among predetermined groups of tumors, and to develop classifiers that can be used to predict the group membership of a tumor. These techniques include univariate two-group comparisons such as multiple *t*-tests, Wilcoxon and Mann-Whitney tests, and log-rank tests for survival outcomes, as well as the multivariate methods of weighted gene voting analysis, Bayesian binary regression, artificial neural nets, support vector machines, and survival models (*see* Table 4). In several cases, results at the univariate, single gene stage have been used to reduce the number of genes for consideration at the multivariate multi-gene stage. There has been good use of permutation testing and cross-validation to ensure internal validity and reduce false positive findings, as well as the use of external validation in independent samples.

For the most part, unsupervised classification of genes and tumors and supervised selection of genes have been used as complementary approaches in an informal fashion. For example, subgroups of tumors identified in cluster analyses have been used to define comparison groups for subsequent supervised analyses *(50,57)*. In other cases, supervised methods have been used to identify a subset of discriminating genes that were then clustered using unsupervised methods *(51,58,67)*.

RESULTS OF ARRAY STUDIES

Examination of Expression Profiles of Cultured Cells and Primary Tumors

Cultured Breast Epithelial Cells and Cancer Cell Lines

Perou et al. *(68)* performed in vitro experiments using cDNA microarrays of 5531 human clones to describe the expression patterns of hormone-manipulated cultured human mammary epithelial cells (HMEC) and 13 breast tumors. Clustering of co-expressed genes with consistent patterns of gene expression in breast tumors revealed two different gene expression parameters, expression levels of the genes in a "proliferation cluster" and an "interferon (IFN)-regulated cluster." In previous studies, *STAT1* and *STAT3* genes had been shown to have high levels of protein expression in primary breast tumors *(69)*. Perou et al. also demonstrated that a subset of breast cancer tumors had high levels of *STAT1* expression due to induction of IFN-regulated genes and subsequently showed high level of STAT3 protein expression in these tumors. In the histologically complex breast carcinomas, the authors also identified gene clusters for nonepithelial cells such as stromal cells and B lymphocytes that contributed to the gene expression patterns of the breast specimens.

In a more recent study, Ross et al. used cDNA microarrays to classify cancer cell lines according to their tissue of origin *(47)*. They performed molecular classification of 60 cancer cell lines (NCI60) derived from tumors of a variety of tissues and organs, using arrays of 9703 human cDNAs. They showed a consistent relationship between gene expression patterns and the tumors' tissue of origin. Based on the gene expression profiles, Ross et al. identified groups of genes they considered to represent epithelial, mesenchymal, stromal, and proliferation clusters. By comparing the gene expression signatures of two breast cancer specimens to a normal tissue specimen and to cultured breast cancer cell lines (including MCF7, T-47D, MDA-MB-231, BT-549, and Hs578T), they were able to distinguish between different cellular counterparts of breast tumors as reported by Perou et al. *(68)*. Expression of keratin 8 and keratin 19 in the ER positive (ER+) breast cancer cell lines T47D and MCF7 suggested that these cells had originated from luminal epithelial cells *(70)*. On the other hand, stromal-like cell lines Hs578T and BT549 had high levels of expression of collagen genes (*COL3A1*, *COL5A1*, *COL6A1*) and a smooth muscle cell marker (e.g., TAGLN), which were characteristic of stromal counterparts.

Su et al. *(71)* also classified human carcinomas by analyzing 100 primary carcinomas from 10 diverse tissues of origin including breast. Using expression arrays of 12,533 oligonucleotides, they identified highly restricted tumor-specific expression patterns and demonstrated the feasibility of predicting the tissue of origin of a carcinoma based on expression patterns.

Comparisons of Normal Breast Epithelial Cell and Breast Carcinoma Patterns

Several studies have compared expression profiles of normal breast epithelial cells and breast carcinomas *(68)*. Sgroi et al. *(55)* performed laser capture microdissection (LCM) on a single mastectomy specimen to isolate morphologically normal breast epithelial cells, invasive breast carcinoma cells, and metastatic (from a lymph node) cells and compared their expression patterns. Of 8084 cDNAs examined, 90 exhibited differences of twofold or greater between the normal and invasive or metastatic carcinoma cells. Among the differentially expressed genes, those exhibiting the greatest changes were apolipoprotein D *(72)*, SWI/SNF *(73)*, and heat-shock factor 1 (*HSF-1*) *(74)*.

Bertucci et al. *(57)* studied gene expression of eight normal tissue specimens from Clontech (Palo Alto, CA, USA) and 34 primary breast carcinomas using 176 gene arrays. Hierarchical clustering was performed on the tumors and genes, and despite the small sample size, they identified at least two subgroups of tumors with distinct clinical outcomes. They also compared gene expression between normal tissue and tumor specimens, between ER- and ER+ tumors, and between ANN tumors and tumors with 10 or more involved lymph nodes. The transcription factor GATA-binding protein 3 (*GATA3*), which has been previously shown to be correlated to ER gene expression in breast cancer *(75)*, showed high levels of expression in the ER+ tumor group. Other genes including the *MYB* proto-oncogene *(76)*, X-box binding protein 1, *p53 (77)*, and insulin-like growth factor (*IGF*) 2 *(78,79)* were also differentially expressed in ER+ compared to ER- tumors. They also found a correlation between ERBB2 expression and nodal status.

Tumor Classification Based on Expression Profiles

Expression Cluster Characteristics of Epithelial Cells

Perou et al. *(68)* examined tumor specimens from 42 individuals using an 8102 gene array and identified a cluster of genes with differential co-expression in a subgroup of tumors not recognized by the currently available histopathological methods. Using hierarchical clustering, they showed that tumors exhibited great variation in their pattern of gene expression. The molecular profiles not only identified similarities and differences among the tumors, but in many cases also pointed to a biological interpretation. One of the largest distinct clusters was a proliferation cluster, and the expression of the genes in this cluster varied widely between tumor samples and generally was well correlated with mitotic index. As expected, the genes encoding two widely used immunohistochemical markers of cell proliferation, Ki67 *(80,81)* and proliferating cell nuclear antigen (PCNA) *(82)*, were also in this cluster. The authors also showed that clinical measurement of ER protein levels in tumors correlated well with the variation of expression of ER on their array.

Tumor Classification Related to neu/ERBB-2 Expression

The normal human mammary gland contains two types of epithelial cells that can be distinguished by immunohistochemical staining. These are the basal–myoepithelial cells, which express keratins 5/6 and 17, and the luminal cells, which express keratins 8/18 *(70)*. To develop a system for classifying tumors on the basis of their gene expression pattern, Perou et al. *(50)* chose a subset of genes to use as the basis for cluster analysis. These 496 genes (termed the intrinsic gene subset) showed greater variation in expression pattern between different tumors than between paired samples from the same tumor. Using the intrinsic gene subset, they clustered the breast tumors into four main groups, which they described as ER+/luminal-like, basal-like, Erb-B2+, and normal breast-like, representing different features of breast tissue. The ER+ tumor group had high levels of expression of genes characterized as the luminal profile, including *GATA3 (75)*, and stained with antibodies against luminal cell keratins 8/18 *(70)*. In contrast, ER- tumors tended to be either basal-like, expressing keratins 5/6 and/or 17 *(70)*, or were in the group that had high expression of *ERBB*2 and related genes.

In a review by Ross and Perou *(83)*, gene expression signatures from 40 breast tissues and 16 breast tissue-derived cell lines were compared. First, they classified cell lines as basal–epithelial, luminal, and stromal-like–fibroblast subgroups. Then they integrated cell line analysis to breast tumor analysis and showed that cell lines, regardless of their presumed cell-type origin, clustered together in one branch separate from all of the tumors. However the cell lines were also similarly subdivided as the tumors into luminal/ER+, *ERBB2* normal breast, and basal-like subgroups. The most distinctive differences in gene expression between tumors and cell lines occurred in the proliferation cluster, indicating that most cell lines were growing much faster than in vivo tumor cells. These data might be used to identify which cell lines are best models for different breast subtypes and to define molecular fingerprints that distinguish the biology of different cell lines and tumor types.

Sorlie et al. *(51)* analyzed a total of 78 breast carcinomas, 3 fibroadenomas, and 4 normal breast tissue samples using the intrinsic gene subset *(50)* as the basis for tissue classification. Hierarchical clustering separated the tumors into two main branches. The first one contained three previously defined *(50)* subgroups (basal-like, ERBB2+, and normal breast-like) and, in the other branch, the luminal/ER+ group was divided into three subgroups (luminal subtypes A, B, and C). Tumors of the ERBB2 overexpressing subtype also exhibited high levels of expression of other genes in the amplicon, such as *GRB7 (84)*. In addition to keratins 5 and 17 *(70)*, the basal-like tumors also had high levels of laminin *(85)* and fatty acid binding protein 7 *(86)*. Luminal subtype A expressed high levels of ERα, as well as *GATA 3 (75)*, X-box binding protein 1, trefoil factor 3 *(87)*, and other genes.

Expression Patterns Related to p53 Mutation Status

Mutations in the *p53* gene are common in breast cancer and have been found to be of prognostic *(31)* and predictive significance in some studies *(88,89)*. Breast carcinomas with *p53* alterations are more likely to be ER- and exhibit ERBB2 overexpression *(88,89)*. It remains to be determined whether mutation in *p53* can be an independent marker of clinical behavior in this disease. Array data may be useful in exploring this question. Sorlie et al. *(51)* examined the correlation of *p53* status and tumor subclass in 69 tumors of their set, 30 of which had mutations in the *p53* gene. They found a difference in the distribution of *p53* mutations among subclasses. The ERBB2 positive and basal-like subclasses had *p53* mutations in 71 and 82% of tumors, respectively, whereas the luminal subtype A contained *p53* mutation only in 13% of the cases. Luminal subtype C shared features with the ERBB2 positive and basal-like subclasses, including p53 mutations in approx 80% of tumors.

We have performed array studies on a group of ANN breast cancers from a prospective cohort, which was designed to examine the prognostic importance of neu/ERBB2 amplification *(17)* and *p53* mutation (Bull et al., in preparation). We performed hierarchical cluster analysis of 81 potential candidate genes from the 19K arrays and found that the tumors clustered according to ER status and *p53* status (Fig. 1). We also compared mean and median log expression ratios to determine the association of several candidate genes with clinical properties of the tumors. We found that genes identified by other groups as being associated with ER positivity (such as *GATA3*) had higher levels in ER+ tumors in our set and that genes associated with ER negativity (e.g., P-cadherin) exhibited high levels in ER- specimens (Fig. 1). We also found GATA3 levels to be higher in the tumors without *p53* mutations.

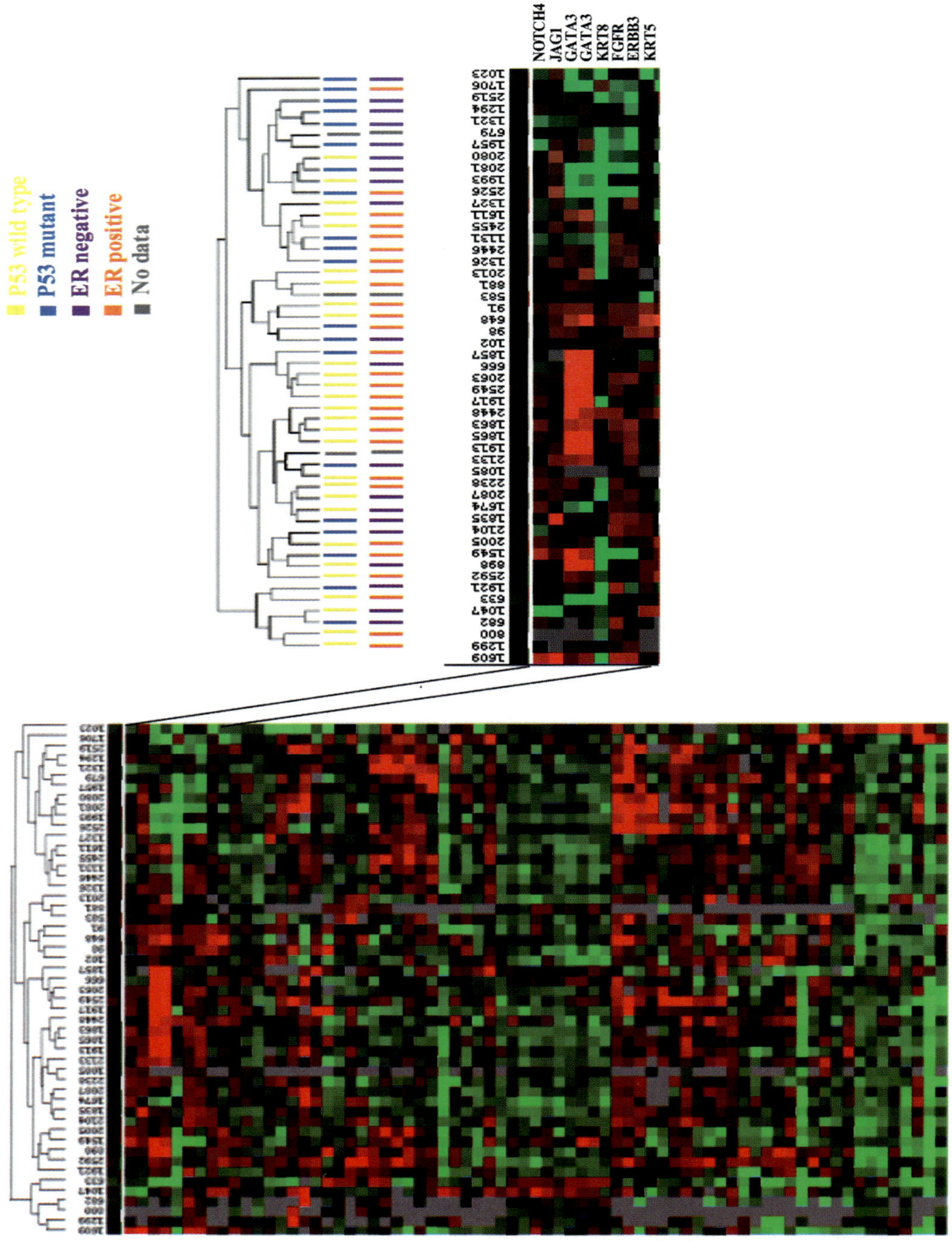

P53 wild type
P53 mutant
ER negative
ER positive
No data
NOTCH4
JAG1
GATA3
GATA3
KRT8
FGFR
ERBB3
KRT5

Fig. 1.

Expression Patterns Related to ER Status

Tamoxifen is a frequently prescribed drug for treatment and prevention of breast cancer *(7,11,90)*. ER+ tumors may initially respond to tamoxifen, but the development of tamoxifen resistance is an important clinical problem. A number of groups have compared microarray data of breast carcinomas based on ER status. Hilsenbeck et al. *(91)* compared gene expression arrays of estrogen-stimulated, tamoxifen-sensitive, and resistant MCF-7 cells grown in a nude mouse xenograft model. They used principal component analysis of the arrays to identify outlier genes that may be involved in tamoxifen resistance. An example of an outlier was *HSF-1 (74)*, which was found to be increased in tamoxifen-sensitive, relative to estrogen-stimulated, tumors.

Martin et al. *(56)* performed hybridization of RNAs from 18 tumors and 7 breast cancer cell lines to membrane arrays containing 124 genes. They identified two gene clusters associated with clinical ER status. One cluster included genes with higher expression in ER+ cells (e.g., p53 *[77]* and keratin 19 *[92]*), and the other contained genes (e.g., maspin) *(93)* with higher expression in ER-cells. Gruvberger et al. *(67)* specifically investigated gene expression patterns associated with this characteristic in 58 ANN breast carcinomas (28 ER+ and 30 ER-) using cDNA microarrays containing 6728 genes. This group used an artificial neural network to develop a classifier to predict the ER status of the tumors and ranked genes according to their contribution to the classifier. They chose the 100 most important genes to classify 47 training samples (23 ER+ and 24 ER-) and showed that all of the tumors were correctly classified. Some of the top genes in the classification were *ER1*, trefoil factor 3 *(87)*, *GATA3 (75)*, and cyclin D1 *(94)*, whose relative level of expression associated with ER positivity. In contrast, lipocalin2 *(95)*, P-cadherin *(96)*, ladinin1 *(97)*, fatty acid binding protein 7 *(86)*, and keratin 7 *(70)* were highly expressed in ER- tumors. The authors showed that the artificial neural network correctly predicted the ER status of 11 blinded breast cancer specimens in a test sample (five ER+ and six ER-). To find the molecular signatures for ER status, they also used weighted gene analysis *(98)*. One hundred thirteen genes showed significant variation between the two tumor types, and 50 of these genes also overlapped with the genes derived from the artificial neural network classification. Although some of the genes that were able to discriminate ER+ and ER- tumors were related to differences in estrogen responsiveness, the majority were not. They also used hierarchical clustering and multidimensional scaling methods to cluster tumors using the 113 genes selected in the supervised weighted gene analysis. This reproduced the ER+/- classification except for two ER+ tumors.

West et al. *(99)* also developed a classifier for breast cancers using Bayesian regression models. They used 38 arrays (18 ER+ and 20 ER-) as a training sample to select

---

Fig. 1. *(previous page)* Cluster analysis using candidate genes of microarray analysis of ANN breast cancer tumors on 19K cDNA arrays. 19K arrays were generated at the Microarray Facility, University Health Network, Ontario Cancer Institute (http://www.uhnres.utoronto.ca/services/microarray) and log base 2 ratios were normalized using subarray location and scale adjustment. Candidate gene expression levels were clustered using complete linkage clustering methods from Eisen's cluster software (ref. *65*). The clustered trees and images were displayed using Eisen's treeview software. Red squares denote positive expression, green squares denote negative expression, black squares denote zero expression, and grey squares denote missing values.

100 genes to include in a multivariate classifier, based on single gene comparisons of the two groups. They identified a number of genes in the estrogen pathway with higher expression, such as *ER*, *pS2* (trefoil factor 1) *(100)*, *GATA 3 (75)*, or lower expression, such as the serine proteinase inhibitor maspin *(93)*. They showed that their 100 gene classifier could correctly predict the ER status of an independent test sample of nine tumors. In addition, the authors used the same method for classification of tumors according to their lymph node status (25 node positive and 24 node negative). Although the gene expression patterns could classify the nodal status of the patients, more data was needed for sufficient predictive power for this clinical factor.

Expression Patterns Related to BRCA1 and BRCA2 Mutation Status

Hedenfalk et al. *(58)* studied the gene expression pattern of three different tumor types, tumors with mutations in the breast cancer susceptibility genes *BRCA1* and *BRCA2*, as well as sporadic breast tumors. Tumors with *BRCA1* mutations differ morphologically and immunohistochemically from tumors with *BRCA2* mutations and also from sporadic cases *(101)*. The authors analyzed the differential expression of these three tumor types using a panel of tumors from seven *BRCA1* mutation carriers, eight *BRCA2* mutation carriers, and seven sporadic cases. They identified 51 genes whose variation in expression best differentiated among these tumors. The authors were also able to show that the gene expression signatures of individual tumor samples could be used to accurately predict which gene mutations they carried. They identified nine genes to distinguish *BRCA1* mutation carriers from noncarriers and another set of 11 genes to distinguish tumors that possess *BRCA2* mutations from those that did not. Their multi-gene classifier correctly identified all of the *BRCA1* mutation-carrying tumors, however only five of eight tumors with *BRCA2* mutations and 13 out of 14 tumors without *BRCA2* mutations were correctly identified. Using three methods for supervised gene selection, the group also identified a common set of 176 genes that appear to distinguish *BRCA1* mutation positive tumors from *BRCA2* mutation positive tumors. These results show that the gene expression signatures of *BRCA1* mutation positive and *BRCA2* mutation positive tumors are distinct from each other and also distinct from sporadic tumors. Genes of two major biological processes, DNA repair and apoptosis, were found to be up-regulated in *BRCA1* tumors.

van't Veer et al. *(52)* performed microarray expression analysis that included tumors from 18 *BRCA1* mutation carriers and two *BRCA2* mutation carriers, together with 78 sporadic samples. In an unsupervised cluster analysis, 16 out of 18 *BRCA1* mutation carriers were classified into a group of tumors with predominantly ER- status or having higher lymphocytic infiltrate expression (see below). These results are consistent with studies indicating that most tumors with *BRCA1* mutations are ER- and show lymphocytic infiltration *(101)*. In addition, within the ER- tumors, a classification based on the supervised selection of 100 genes, which distinguished sporadic tumors from those with *BRCA1* mutation, was able to predict these two tumor groups.

Expression Patterns Related to Clinical Data

Expression Patterns Before and After Chemotherapy

One of the first studies to determine the expression profiles of a series of matched breast cancer specimens was performed by Perou et al. *(50)*. They analyzed 20 pairs of tumors taken before and after a 16-wk course of doxorubicin chemotherapy and found that independent samples from the same individual exhibited distinctive molecular portraits. Expression patterns for tumors from the same patient were more likely to be similar to each other than to those of any other patient.

Expression Patterns Related to Clinical Outcome

Sorlie et al. *(51)* extended the studies of Perou et al. *(50)* by increasing their sample size to 85 tissue specimens (including the 40 previously reported) from 84 individuals and including clinical outcome data. Survival analyses, which included 49 cases with locally advanced disease but no distant metastases, showed significantly different outcomes among the patients belonging to five subgroups of tumors identified in unsupervised cluster analysis. The basal-like and ERBB2 positive types were associated with the shortest survival times. Interestingly, there was a significant difference in outcome for patients in the luminal group, with the luminal C tumors having the worst outcome. Because the luminal C subgroup exhibits molecular signatures similar to those of the ERBB2 overexpressor and basal-like subtypes, overexpression of a common set of genes may be associated with poor outcome.

Expression Patterns Related to Disease-Free Survival in ANN Breast Cancer

In a recent study, van't Veer et al. *(52)* performed microarray expression analysis on tumors from 98 young (age at diagnosis <55 yr) breast cancer patients. They analyzed 34 tumors from ANN patients who developed metastases within 5 yr, 44 tumors from ANN patients who were disease-free after a period of at least 5 yr, 18 *BRCA1* mutation carriers, and 2 *BRCA2* mutation carriers.

Unsupervised two-dimensional cluster analysis of 98 breast tumors and 5000 genes was used to describe the main tumor and dominant gene clusters. There were two distinct clusters interpreted as representing good prognosis and poor prognosis tumors according to the disease-free survival status of the sporadic tumors in the cluster. The authors also investigated the association of these data with ER status of the patients. The majority of the ER- tumors clustered together in the poor prognosis branch of the tumor cluster. A gene cluster containing the ER gene and genes that are co-regulated with ER was found to have low expression in the poor prognosis tumor group while a second gene cluster containing genes that represent lymphocytic infiltration was found to have higher expression. Sixteen out of eighteen *BRCA1* carriers were also in the poor prognosis group together with ER- tumors and tumors with lymphocytic infiltration.

To identify gene expression patterns, which strongly predict outcome, van't Veer et al. *(52)* performed supervised classification using the 78 sporadic ANN patients and developed a classifier based on 70 genes to predict disease-free survival status at 5 yr. The prognosis classifier correctly predicted the actual outcome in 83% of the cases. Genes including cyclin E2 *(102)*, matrix metalloproteinases *(MMP)9 (103)* and *MMP1 (104)*, and others involved in cell cycle, invasion, metastasis, angiogenesis, and signal transduction were significantly up-regulated in tumors with poor prognosis signatures. Other genes, such as *ERBB2*, *ER*, and cyclin D1, that may have been expected to be associated with prognosis were not. However, it should be noted that the importance of expression of *ERBB2* as a prognostic marker is derived primarily from studies using immunohistochemistry and/or DNA copy number, not mRNA levels.

To validate the predictive power of these genes, the authors chose an additional independent set of tumors from 19 ANN breast cancer patients diagnosed at a young age. In the validation series, 12 individuals had developed metastatic disease within 5 yr, and 7 were disease-free. They investigated the same 70-gene prognosis classifier for the prediction of outcome and showed a comparable predictive power.

MOLECULAR VALIDATION OF EXPRESSION PROFILING RESULTS

Molecular and Immunohistochemical Techniques

Occasionally, investigators have used Northern blot analysis to validate the results obtained from gene expression arrays. For example, Bertucci et al. observed high levels of *GATA 3* expression in ER+ breast tumors on arrays and confirmed this by Northern blotting *(57)*. However, most genes are not expressed at the level of *GATA3* mRNA, and investigators do not have access to the amount of tumor RNA that is required for Northern blot analysis. Thus, other techniques, such as quantitative reverse transcription polymerase chain reaction (RT-PCR), real-time PCR, immunohistochemistry (IHC), and *in situ* hybridization are likely to be used more frequently for confirmation of array data *(47,48,50,52,55,68,71)*.

In some cases, the genes identified in microarray expression studies have been previously characterized genes for which antibodies were available *(47,48,50,52, 55,68,71)*. In cases of novel genes of potential importance, it may be possible to generate antibodies to be used for IHC studies. IHC analysis is the simplest method to further investigate the possible role of these genes in breast neoplasia. *In situ* hybridization can also be used as an adjuvant method to assess gene expression for selected genes of significant interest. Furthermore, IHC and *in situ* hybridization can be applied to tumor tissue arrays (*see* section entitled Tissue Arrays), to rapidly assess altered expression in larger panels of tumors and to examine the expression across a morphologic range from normal to invasive in individual cases.

Tissue Arrays

In addition, tumor arrays and IHC can be used to evaluate the potential importance of the novel molecular alterations in a large number of specimens. Tumor arrays represent a method in which hundreds of small samples from separate tumors are arrayed in a single paraffin block. This approach, as developed by Kallioniemi and colleagues, allows for the simultaneous analysis of genetic–protein markers in a single test, and is, thus, ideally suited to screening a large number of potential markers *(71,105,106)*. In a validation study, Camp et al. found that a minimum number of two tissue cores was necessary to be representative of expression of a marker in a tumor *(107)*. They also observed that cores from the center and tumor edge might be required for some markers.

Confirmation of Array Data with Other Techniques

Some studies have used a combination of techniques to validate the array data. For example, Sgroi et al. performed both RT-PCR and IHC *(55)*. Others have used IHC on breast tumor specimens, including a number of studies staining with antikeratin antibodies *(47,48,50)*. In at least one case, Western blot analysis was used to confirm gene expression *(91)*. Tissue arrays have advantages when combined with IHC and were utilized by Hedenfalk et al. to correlate expression array data with protein expression *(58)*.

THE FUTURE: UNRESOLVED QUESTIONS

The introduction of gene expression arrays to the rapidly moving field of molecular pathology has elicited excitement balanced with caution. The results to date are encouraging, because data are already available that show that expression profiling can be used to distinguish cell type-specific gene clusters (e.g., stromal, epithelial, mesenchymal, proliferation) and to classify breast tumors as basal-like, luminal-like, ERBB2 overexpressing, and normal breast-like. Furthermore, gene expression patterns related to ER status and *BRCA1* mutation status have been characterized, and profiles associated with good prognosis and poor prognosis groups of young ANN patients have been identified. However, because the development of the technology is recent, predicting the generalizability of the results to date awaits further studies. A number of studies include separate validation sets, but acknowledge that validation in additional cohorts will be necessary to confirm the data.

An unresolved question is whether ER- cells are derived from ER+ progenitors or whether both types are derived from normal populations that diverged during differentiation. Most mammary epithelial cells do not express ER, yet most primary breast carcinomas are ER+. Tamoxifen-resistant ER- cells may arise from estrogen-nonresponsive normal cells or may be derived from ER+ breast cancer cells during tumorigenesis. Further studies will be required to resolve these issues.

The work to date suggests that expression profiling will become an important research tool to aid in our understanding of tumor development and progression. The technology will provide a molecular complement to histology and IHC. Sensitive methods are being developed and refined, which will permit the examination of biological specimens containing a limited number of cells (such as premalignant breast lesions) and aid in the determination of the molecular events involved in the development and progression of breast neoplasia.

ACKNOWLEDGMENTS

We thank Lucie Collins for excellent technical assistance and Wenqing He for statistical assistance. We thank Drs. Jim Woodgett and Hilmi Ozcelik for helpful discussions. We gratefully acknowledge the support of the Canadian Breast Cancer Research Initiative, the National Cancer Institute of Canada (with funds from the Terry Fox Run) and the Network of Centres of Excellence in Mathematics (Canada).

REFERENCES

1. Fisher, B., Bauer, M., Wickerham, D. L., et al. (1983) Relation of number of positive axillary nodes to the prognosis of patients with primary breast cancer. An NSABP update. *Cancer* **52,** 1551–1557.
2. Carter, C. L., Allen, C., and Henson, D. E. (1989) Relation of tumor size, lymph node status, and survival in 24,740 breast cancer cases. *Cancer* **63,** 181–187.
3. Fisher, B., Redmond, C., Dimitrov, N. V., et al. (1989) A randomized clinical trial evaluating sequential methotrexate and fluorouracil in the treatment of patients with node-negative breast cancer who have estrogen-receptor-negative tumors. *N. Engl. J. Med.* **320,** 473–478.

4. Fisher, B., Costantino, J., Redmond, C., et al. (1989) A randomized clinical trial evaluating tamoxifen in the treatment of patients with node-negative breast cancer who have estrogen-receptor-positive tumors. *N. Engl. J. Med.* **320,** 479–484.
5. Mansour, E. G., Gray, R., Shatila, A. H., et al. (1989) Efficacy of adjuvant chemotherapy in high-risk node-negative breast cancer. An intergroup study. *N. Engl. J. Med.* **320,** 485–490.
6. The Ludwig Breast Cancer Study Group. (1989) Prolonged disease-free survival after one course of perioperative adjuvant chemotherapy for node-negative breast cancer. *N. Engl. J. Med.* **320,** 491–496.
7. Early Breast Cancer Trialists' Collaborative Group. (1992) Systemic treatment of early breast cancer by hormonal, cytotoxic, or immune therapy. 133 randomised trials involving 31,000 recurrences and 24,000 deaths among 75,000 women. *Lancet* **339,** 71–85.
8. McGuire, W. L., Clark, G. M., Dressler, L. G., and Owens, M. A. (1986) Role of steroid hormone receptors as prognostic factors in primary breast cancer. *NCI Monogr.* **1,** 19–23.
9. Henson, D. E., Ries, L., Freedman, L. S., and Carriaga, M. (1991) Relationship among outcome, stage of disease, and histologic grade for 22,616 cases of breast cancer. The basis for a prognostic index. *Cancer* **68,** 2142–2149.
10. Osborne, C. K. (1991) Receptors, in *Breast Diseases* (Harris, J. R., Hellman, S., Henderson, I. C., and Kinee, D. W., eds.), J.B. Lippincott, Philadelphia, pp. 301–325.
11. Fisher, B., Costantino, J. P., Wickerham, D. L., et al. (1998) Tamoxifen for prevention of breast cancer: report of the National Surgical Adjuvant Breast and Bowel Project P-1 Study. *J. Natl. Cancer Inst.* **90,** 1371–1388.
12. Sommer, S. and Fuqua, S. A. (2001) Estrogen receptor and breast cancer. *Semin. Cancer Biol.* **11,** 339–352.
13. Meyer, J. S. (1986) Cell kinetics in selection and stratification of patients for adjuvant therapy of breast carcinoma. *NCI Monogr.* **1,** 25–28.
14. Clark, G. M., Dressler, L. G., Owens, M. A., Pounds, G., Oldaker, T., and McGuire, W. L. (1989) Prediction of relapse or survival in patients with node-negative breast cancer by DNA flow cytometry. *N. Engl. J. Med.* **320,** 627–633.
15. Slamon, D. J., Clark, G. M., Wong, S. G., Levin, W. J., Ullrich, A., and McGuire, W. L. (1987) Human breast cancer: correlation of relapse and survival with amplification of the HER-2/neu oncogene. *Science* **235,** 177–182.
16. Slamon, D. J., Godolphin, W., Jones, L. A., et al. (1989) Studies of the HER-2/neu proto-oncogene in human breast and ovarian cancer. *Science* **244,** 707–712.
17. Andrulis, I. L., Bull, S. B., Blackstein, M. E., et al. (1998) neu/erbB-2 amplification identifies a poor-prognosis group of women with node-negative breast cancer. Toronto Breast Cancer Study Group. *J. Clin. Oncol.* **16,** 1340–1349.
18. van de Vijver, M. J., Peterse, J. L., Mooi, W. J., et al. (1988) Neu-protein overexpression in breast cancer. Association with comedo-type ductal carcinoma in situ and limited prognostic value in stage II breast cancer. *N. Engl. J. Med.* **319,** 1239–1245.
19. Paik, S., Hazan, R., Fisher, E. R., et al. (1990) Pathologic findings from the National Surgical Adjuvant Breast and Bowel Project: prognostic significance of erbB-2 protein overexpression in primary breast cancer. *J. Clin. Oncol.* **8,** 103–112.
20. Wright, C., Angus, B., Nicholson, S., et al. (1989) Expression of c-erbB-2 oncoprotein: a prognostic indicator in human breast cancer. *Cancer Res.* **49,** 2087–2090.
21. Allred, D. C., Clark, G. M., Tandon, A. K., et al. (1992) HER-2/neu in node-negative breast cancer: prognostic significance of overexpression influenced by the presence of in situ carcinoma. *J. Clin. Oncol.* **10,** 599–605.
22. Paterson, M. C., Dietrich, K. D., Danyluk, J., et al. (1991) Correlation between c-erbB-2 amplification and risk of recurrent disease in node-negative breast cancer. *Cancer Res.* **51,** 556–567.

23. Horak, E., Smith, K., Bromley, L., et al. (1991) Mutant p53, EGF receptor and c-erbB-2 expression in human breast cancer. *Oncogene* **6,** 2277–2284.
24. Gusterson, B. A., Gelber, R. D., Goldhirsch, A., et al. (1992) Prognostic importance of c-erbB-2 expression in breast cancer. International (Ludwig) Breast Cancer Study Group. *J. Clin. Oncol.* **10,** 1049–1056.
25. Sainsbury, J. R., Nicholson, S., Angus, B., Farndon, J. R., Malcolm, A. J., and Harris, A. L. (1988) Epidermal growth factor receptor status of histological sub-types of breast cancer. *Br. J. Cancer* **58,** 458–460.
26. Nicholson, S., Wright, C., Sainsbury, J. R., et al. (1990) Epidermal growth factor receptor (EGFr) as a marker for poor prognosis in node-negative breast cancer patients: neu and tamoxifen failure. *J. Steroid Biochem. Mol. Biol.* **37,** 811–814.
27. Fisher, E. R., Redmond, C., Fisher, B., and Bass G. (1990) Pathologic findings from the National Surgical Adjuvant Breast and Bowel Projects (NSABP). Prognostic discriminants for 8-year survival for node-negative invasive breast cancer patients. *Cancer* **65,** 2121–2128.
28. Thor, A. D., Moore, D. H. II, Edgerton, S. M., et al. (1992) Accumulation of p53 tumor suppressor gene protein: an independent marker of prognosis in breast cancers. *J. Natl. Cancer Inst.* **84,** 845–855.
29. Allred, D. C., Clark, G. M., Elledge, R., et al. (1993) Association of p53 protein expression with tumor cell proliferation rate and clinical outcome in node-negative breast cancer. *J. Natl. Cancer Inst.* **85,** 200–206.
30. Sommer, S. S., Cunningham, J., McGovern, R. M., et al. (1992) Pattern of p53 gene mutations in breast cancers of women of the midwestern United States. *J. Natl. Cancer Inst.* **84,** 246–252.
31. Hartmann, A., Blaszyk, H., Kovach, J. S., and Sommer, S. S. (1997) The molecular epidemiology of p53 gene mutations in human breast cancer. *Trends Genet.* **13,** 27–33.
32. Spyratos, F., Maudelonde, T., Brouillet, J. P., et al. (1989) Cathepsin D: an independent prognostic factor for metastasis of breast cancer. *Lancet* **2,** 1115–1118.
33. Tandon, A. K., Clark, G. M., Chamness, G. C., Chirgwin, J. M., and McGuire, W. L. (1990) Cathepsin D and prognosis in breast cancer. *N. Engl. J. Med.* **322,** 297–302.
34. Ravdin, P. M., Tandon, A. K., Allred, D. C., et al. (1994) Cathepsin D by western blotting and immunohistochemistry: failure to confirm correlations with prognosis in node-negative breast cancer. *J. Clin. Oncol.* **12,** 467–474.
35. Ferrandina, G., Scambia, G., Bardelli, F., Benedetti, P. P., Mancuso, S., and Messori, A. (1997) Relationship between cathepsin-D content and disease-free survival in node-negative breast cancer patients: a meta-analysis. *Br. J. Cancer* **76,** 661–666.
36. Catzavelos, C., Bhattacharya, N., Ung, Y. C., et al. (1997) Decreased levels of the cell-cycle inhibitor p27Kip1 protein: prognostic implications in primary breast cancer. *Nat. Med.* **3,** 227–230.
37. Porter, P. L., Malone, K. E., Heagerty, P. J., et al. (1997) Expression of cell-cycle regulators p27Kip1 and cyclin E, alone and in combination, correlate with survival in young breast cancer patients. *Nat. Med.* **3,** 222–225.
38. Weidner, N., Semple, J. P., Welch, W. R., and Folkman, J. (1991) Tumor angiogenesis and metastasis—correlation in invasive breast carcinoma. *N. Engl. J. Med.* **324,** 1–8.
39. Heimann, R., Ferguson, D., Powers, C., Recant, W. M., Weichselbaum, R. R., and Hellman, S. (1996) Angiogenesis as a predictor of long-term survival for patients with node-negative breast cancer. *J. Natl. Cancer Inst.* **88,** 1764–1769.
40. Toi, M., Inada, K., Suzuki, H., and Tominaga, T. (1995) Tumor angiogenesis in breast cancer: its importance as a prognostic indicator and the association with vascular endothelial growth factor expression. *Breast Cancer Res. Treat.* **36,** 193–204.
41. Obermair, A., Kucera, E., Mayerhofer, K., et al. (1997) Vascular endothelial growth factor (VEGF) in human breast cancer: correlation with disease-free survival. *Int. J. Cancer* **74,** 455–458.

42. Schuuring, E., Verhoeven, E., Mooi, W. J., and Michalides, R. J. (1992) Identification and cloning of two overexpressed genes, U21B31/PRAD1 and EMS1, within the amplified chromosome 11q13 region in human carcinomas. *Oncogene* **7**, 355–361.
43. Lammie, G. A., Fantl, V., Smith, R., et al. (1991) D11S287, a putative oncogene on chromosome 11q13, is amplified and expressed in squamous cell and mammary carcinomas and linked to BCL-1. *Oncogene* **6**, 439–444.
44. Tsuda, H., Hirohashi, S., Shimosato, Y., et al. (1989) Correlation between long-term survival in breast cancer patients and amplification of two putative oncogene-coamplification units: hst-1/int-2 and c-erbB-2/ear-1. *Cancer Res.* **49**, 3104–3108.
45. Lammie, G. A. and Peters, G. (1991) Chromosome 11q13 abnormalities in human cancer. *Cancer Cells* **3**, 413–420.
46. Wang, E., Miller, L. D., Ohnmacht, G. A., Liu, E. T., and Marincola, F. M. (2000) High-fidelity mRNA amplification for gene profiling. *Nat. Biotechnol.* **18**, 457–459.
47. Ross, D. T., Scherf, U., Eisen, M. B., et al. (2000) Systematic variation in gene expression patterns in human cancer cell lines. *Nat. Genet.* **24**, 227–235.
48. Alizadeh, A. A., Ross, D. T., Perou, C. M., and van de Rijn, M. (2001) Towards a novel classification of human malignancies based on gene expression patterns. *J. Pathol.* **195**, 41–52.
49. Simon, R., Radmacher, M. D., and Dobbin, K. (2001) Design of studies using DNA microarrays. National Cancer Institute Biometric Research Branch, Technical Report 004.
50. Perou, C. M., Sorlie, T., Eisen, M. B., et al. (2000) Molecular portraits of human breast tumours. *Nature* **406**, 747–752.
51. Sorlie, T., Perou, C. M., Tibshirani, R., et al. (2001) Gene expression patterns of breast carcinomas distinguish tumor subclasses with clinical implications. *Proc. Natl. Acad. Sci. USA* **98**, 10,869–10,874.
52. van't Veer, L. J., Dai, H., van de Vijver, M. J., et al. (2002) Gene expression profiling predicts clinical outcome of breast cancer. *Nature* **415**, 530–536.
53. Dudoit, S., Yang, Y. H., Callow, M. J., and Speed, T. P. (2000) Statistical methods for identifying differentially expressed genes in replicated cDNA microarray experiments. University of California, Berkeley, Department of Statistics, Technical Report 578.
54. Wu, T. D. (2001) Analysing gene expression data from DNA microarrays to identify candidate genes. *J. Pathol.* **195**, 53–65.
55. Sgroi, D. C., Teng, S., Robinson, G., LeVangie, R., Hudson, J. R., Jr., and Elkahloun, A. G. (1999) In vivo gene expression profile analysis of human breast cancer progression. *Cancer Res.* **59**, 5656–5661.
56. Martin, K. J., Kritzman, B. M., Price, L. M., et al. (2000) Linking gene expression patterns to therapeutic groups in breast cancer. *Cancer Res.* **60**, 2232–2238.
57. Bertucci, F., Houlgatte, R., Benziane, A., et al. (2000) Gene expression profiling of primary breast carcinomas using arrays of candidate genes. *Hum. Mol. Genet.* **9**, 2981–2991.
58. Hedenfalk, I., Duggan, D., Chen, Y., et al. (2001) Gene-expression profiles in hereditary breast cancer. *N. Engl. J. Med.* **344**, 539–548.
59. Kerr, M. K. and Churchill, G. A. (2001) Statistical design and the analysis of gene expression microarray data. *Genet. Res.* **77**, 123–128.
60. Kerr, M. K., Martin, M., and Churchill, G. A. (2000) Analysis of variance for gene expression microarray data. *J. Comput. Biol.* **7**, 819–837.
61. Lee, M. L., Kuo, F. C., Whitmore, G. A., and Sklar, J. (2000) Importance of replication in microarray gene expression studies: statistical methods and evidence from repetitive cDNA hybridizations. *Proc. Natl. Acad. Sci. USA* **97**, 9834–9839.
62. McShane, L. M., Radmacher, M. D., Freidlin, B., Yu, R., Li, M. C., and Simon, R. (2001) Methods for assessing reproducibility of clustering patterns observed in analyses of microarray data. National cancer Institute Biometric Research Branch, Technical Report 002.

63. van der Laan, M. and Bryan, J. (2001) Gene expression analysis with the parametric bootstrap. *Biostatistics* **2,** 1–18.
64. Tusher, V. G., Tibshirani, R., and Chu, G. (2001) Significance analysis of microarrays applied to the ionizing radiation response. *Proc. Natl. Acad. Sci. USA* **98,** 5116–5121.
65. Eisen, M. B., Spellman, P. T., Brown, P. O., and Botstein, D. (1998) Cluster analysis and display of genome-wide expression patterns. *Proc. Natl. Acad. Sci. USA* **95,** 14,863–14,868.
66. Hastie, T., Tibshirani, R., Eisen, M. B., et al. (2000) 'Gene shaving' as a method for identifying distinct sets of genes with similar expression patterns. *Genome Biol.* **1,** RESEARCH0003
67. Gruvberger, S., Ringner, M., Chen, Y., et al. (2001) Estrogen receptor status in breast cancer is associated with remarkably distinct gene expression patterns. *Cancer Res.* **61,** 5979–5984.
68. Perou, C. M., Jeffrey, S. S., van de, R. M., et al. (1999) Distinctive gene expression patterns in human mammary epithelial cells and breast cancers. *Proc. Natl. Acad. Sci. USA* **96,** 9212–9217.
69. Garcia, R. and Jove, R. (1998) Activation of STAT transcription factors in oncogenic tyrosine kinase signaling. *J. Biomed. Sci.* **5,** 79–85.
70. Heatley, M., Maxwell, P., Whiteside, C., and Toner, P. (1995) Cytokeratin intermediate filament expression in benign and malignant breast disease. *J. Clin. Pathol.* **48,** 26–32.
71. Su, A. I., Welsh, J. B., Sapinoso, L. M., et al. (2001) Molecular classification of human carcinomas by use of gene expression signatures. *Cancer Res.* **61,** 7388–7393.
72. Diez-Itza, I., Vizoso, F., Merino, A. M., et al. (1994) Expression and prognostic significance of apolipoprotein D in breast cancer. *Am. J. Pathol.* **144,** 310–320.
73. Chiba, H., Muramatsu, M., Nomoto, A., and Kato, H. (1994) Two human homologues of Saccharomyces cerevisiae SWI2/SNF2 and Drosophila brahma are transcriptional coactivators cooperating with the estrogen receptor and the retinoic acid receptor. *Nucleic Acids Res.* **22,** 1815–1820.
74. Yang, X., Dale, E. C., Diaz, J., and Shyamala, G. (1995) Estrogen dependent expression of heat shock transcription factor: implications for uterine synthesis of heat shock proteins. *J. Steroid Biochem. Mol. Biol.* **52,** 415–419.
75. Hoch, R. V., Thompson, D. A., Baker, R. J., and Weigel, R. J. (1999) GATA-3 is expressed in association with estrogen receptor in breast cancer. *Int. J. Cancer* **84,** 122–128.
76. Guerin, M., Sheng, Z. M., Andrieu, N., and Riou, G. (1990) Strong association between c-myb and oestrogen-receptor expression in human breast cancer. *Oncogene* **5,** 131–135.
77. Fernando, S. S., Wu, X., and Perera, L. S. (2000) p53 Overexpression and steroid hormone receptor status in endometrial carcinoma. *Int. J. Surg. Pathol.* **8,** 213–222.
78. Ellis, M. J., Jenkins, S., Hanfelt, J., et al. (1998) Insulin-like growth factors in human breast cancer. *Breast Cancer Res. Treat.* **52,** 175–184.
79. Peyrat, J. P., Bonneterre, J., Beuscart, R., Djiane, J., and Demaille, A. (1988) Insulin-like growth factor 1 receptors in human breast cancer and their relation to estradiol and progesterone receptors. *Cancer Res.* **48,** 6429–6433.
80. Schluter, C., Duchrow, M., Wohlenberg, C., et al. (1993) The cell proliferation-associated antigen of antibody Ki-67: a very large, ubiquitous nuclear protein with numerous repeated elements, representing a new kind of cell cycle-maintaining proteins. *J. Cell Biol.* **123,** 513–522.
81. Goodson, W. H. III, Moore, D. H., Ljung, B. M., et al. (2000) The prognostic value of proliferation indices: a study with in vivo bromodeoxyuridine and Ki-67. *Breast Cancer Res. Treat.* **59,** 113–123.
82. Pantel, K., Schlimok, G., Braun, S., et al. (1993) Differential expression of proliferation-associated molecules in individual micrometastatic carcinoma cells. *J. Natl. Cancer Inst.* **85,** 1419–1424.

83. Ross, D. T. and Perou, C. M. (2001) A comparison of gene expression signatures from breast tumors and breast tissue derived cell lines. *Dis. Markers* **17,** 99–109.
84. Stein, D., Wu, J., Fuqua, S. A., et al. (1994) The SH2 domain protein GRB-7 is co-amplified, overexpressed and in a tight complex with HER2 in breast cancer. *EMBO J.* **13,** 1331–1340.
85. Woodward, T. L., Lu, H., and Haslam, S. Z. (2000) Laminin inhibits estrogen action in human breast cancer cells. *Endocrinology* **141,** 2814–2821.
86. Das, R., Hammamieh, R., Neill, R., Melhem, M., and Jett, M. (2001) Expression pattern of fatty acid-binding proteins in human normal and cancer prostate cells and tissues. *Clin. Cancer Res.* **7,** 1706–1715.
87. May, F. E. and Westley, B. R. (1997) Trefoil proteins: their role in normal and malignant cells. *J. Pathol.* **183,** 4–7.
88. Berns, E. M., Foekens, J. A., Vossen, R., et al. (2000) Complete sequencing of TP53 predicts poor response to systemic therapy of advanced breast cancer. *Cancer Res.* **60,** 2155–2162.
89. Geisler, S., Lonning, P. E., Aas, T., et al. (2001) Influence of TP53 gene alterations and c-erbB-2 expression on the response to treatment with doxorubicin in locally advanced breast cancer. *Cancer Res.* **61,** 2505–2512.
90. Pritchard, K. I. (2001) Breast cancer prevention with selective estrogen receptor modulators: a perspective. *Ann. NY Acad. Sci.* **949,** 89–98.
91. Hilsenbeck, S. G., Friedrichs, W. E., Schiff, R., et al. (1999) Statistical analysis of array expression data as applied to the problem of tamoxifen resistance. *J. Natl. Cancer Inst.* **91,** 453–459.
92. Spencer, V. A., Coutts, A. S., Samuel, S. K., Murphy, L. C., and Davie, J. R. (1998) Estrogen regulates the association of intermediate filament proteins with nuclear DNA in human breast cancer cells. *J. Biol. Chem.* **273,** 29,093–29,097.
93. Zou, Z., Anisowicz, A., Hendrix, M. J., et al. (1994) Maspin, a serpin with tumor-suppressing activity in human mammary epithelial cells. *Science* **263,** 526–529.
94. Courjal, F., Louason, G., Speiser, P., Katsaros, D., Zeillinger, R., and Theillet, C. (1996) Cyclin gene amplification and overexpression in breast and ovarian cancers: evidence for the selection of cyclin D1 in breast and cyclin E in ovarian tumors. *Int. J. Cancer* **69,** 247–253.
95. Stoesz, S. P., Friedl, A., Haag, J. D., Lindstrom, M. J., Clark, G. M., and Gould, M. N. (1998) Heterogeneous expression of the lipocalin NGAL in primary breast cancers. *Int. J. Cancer* **79,** 565–572.
96. Palacios, J., Benito, N., Pizarro, A., et al. (1995) Anomalous expression of P-cadherin in breast carcinoma. Correlation with E-cadherin expression and pathological features. *Am. J. Pathol.* **146,** 605–612.
97. Marinkovich, M. P., Taylor, T. B., Keene, D. R., Burgeson, R. E., and Zone, J. J. (1996) LAD-1, the linear IgA bullous dermatosis autoantigen, is a novel 120-kDa anchoring filament protein synthesized by epidermal cells. *J. Invest. Dermatol.* **106(4),** 734–738.
98. Golub, T. R., Slonim, D. K., Tamayo, P., et al. (1999) Molecular classification of cancer: class discovery and class prediction by gene expression monitoring. *Science* **286,** 531–537.
99. West, M., Blanchette, C., Dressman, H., et al. (2001) Predicting the clinical status of human breast cancer by using gene expression profiles. *Proc. Natl. Acad. Sci. USA* **98,** 11,462–11,467.
100. Berry, M., Nunez, A. M., and Chambon, P. (1989) Estrogen-responsive element of the human pS2 gene is an imperfectly palindromic sequence. *Proc. Natl. Acad. Sci. USA* **86,** 1218–1222.
101. Lakhani, S. R., Jacquemier, J., Sloane, J. P., et al. (1998) Multifactorial analysis of differences between sporadic breast cancers and cancers involving BRCA1 and BRCA2 mutations. *J. Natl. Cancer Inst.* **90,** 1138–1145.

102. Geng, Y., Yu, Q., Whoriskey, W., et al. (2001) Expression of cyclins E1 and E2 during mouse development and in neoplasia. *Proc. Natl. Acad. Sci. USA* **98,** 13,138–13,143.
103. Scorilas, A., Karameris, A., Arnogiannaki, N., et al. (2001) Overexpression of matrix-metalloproteinase-9 in human breast cancer: a potential favourable indicator in node-negative patients. *Br. J. Cancer* **84,** 1488–1496.
104. McCarthy, K., Maguire, T., McGreal, G., McDermott, E., O'Higgins, N., and Duffy, M. J. (1999) High levels of tissue inhibitor of metalloproteinase-1 predict poor outcome in patients with breast cancer. *Int. J. Cancer* **84,** 44–48.
105. Kononen, J., Bubendorf, L., Kallioniemi, A., et al. (1998) Tissue microarrays for high-throughput molecular profiling of tumor specimens. *Nat. Med.* **4,** 844–847.
106. Nocito, A., Kononen, J., Kallioniemi, O. P., and Sauter, G. (2001) Tissue microarrays (TMAs) for high-throughput molecular pathology research. *Int. J. Cancer* **94,** 1–5.
107. Camp, R. L., Charette, L. A., and Rimm, D. L. (2000) Validation of tissue microarray technology in breast carcinoma. *Lab. Invest.* **80,** 1943–1949.

9
Microarray Analysis of Colorectal Cancer

Daniel A. Notterman, Carrie J. Shawber, and Wei Liu

INTRODUCTION

The microarray analysis of tumors that originate in solid organs, such as the lung, bowel, breast, ovary, and pancreas, offers special challenges and opportunities. The challenges lie in the complexity and heterogeneity of the normal and abnormal tissues of which these tumors are composed. The opportunities derive from the generally poor response to therapy displayed by these lesions and the early evidence that global expression analysis may provide the substrate for a new predictive taxonomy of cancer. Colorectal cancer has been extensively profiled at the DNA, histopathological, and clinical levels, and several groups have now added results from arrayed-based surveys of gene expression. Even so, it is not yet possible to offer a comprehensive correlation of these disparate sources of information. Doing so will be the work of the next generation of microarray science and bioinformatics.

Cancer of the colon and rectum is the second or third most common cause of cancer in adults in the U.S. (1,2). Clinical research has focused on prevention, early detection, and optimal selection of patients for adjunctive therapy. Aggressive preventive efforts may have been successful in reducing the number of new cases. According to The American Cancer Society, about 107,300 new cases of colon cancer (50,000 men and 57,300 women) and 41,000 new cases of rectal cancer (22,600 men and 18,400 women) will be diagnosed in 2002. Colon cancer is expected to cause about 48,100 deaths (23,100 men and 25,000 women) during 2002, while approx 8500 people (4700 men and 3800 women) will die from rectal cancer during 2002 (3).

The past decade has provided investigators with a richly annotated framework with which to correlate changes at the DNA level with the progression of this disease. There is yet very little broadly collected information regarding the sequential and cumulative changes in mRNA expression that occur during cancer initiation and progression in the colon. This is an important limitation, because abnormalities in gene expression are characteristic of neoplastic tissue (4). Traditional methods of analysis impose a practical limit on the number of candidate genes whose expression can be conveniently and simultaneously studied. Highly parallel technologies exploiting sample hybridization to oligonucleotide or cDNA arrays permit the expression levels of tens of thousands of genes to be monitored simultaneously and rapidly (5). As the task proceeds of comparing

From: *Expression Profiling of Human Tumors: Diagnostic and Research Applications*
Edited by: Marc Ladanyi and William L. Gerald © Humana Press Inc., Totowa, NJ

alterations in DNA and RNA with the clinical phenotype, it will be possible to advance the theoretical and experimental understanding of colon tumor formation in two complementary ways. They are directly, through the identification of individual genes, particularly those that become abnormal very early (during the transition from dysplastic to neoplastic tissue) and very late (during metastasis), and indirectly, by correlating distributed patterns of gene expression with the underlying genotype and the clinical phenotype. This is the work of building a new, molecular taxonomy of colorectal cancer, in which specific associations of mRNA expression pattern and DNA mutation draw the boundaries around classes of tumors with different clinical manifestations. This has already proved successful in early studies with different model systems involving B cell lymphoma, malignant melanoma, breast cancer, and lymphoblastic leukemia *(6–9)*.

In this chapter, we provide a summary of the normal microscopic anatomy of the colon, a basic description of the clinical manifestations of colorectal cancer, and summarize work using oligonucleotide arrays to monitor gene expression in colorectal neoplasms.

THE MICROSCOPIC ANATOMY OF THE COLON

The colon comprises four distinct layers, from inside to outside: the mucosa, submucosa, muscularis externa, and serosa. The colonic mucosa is lined by an absorptive and mucus-secreting columnar epithelium and contains tubular glands, the crypts of Lieberkühn. The epithelial stem cells, from which the malignant cells of colorectal cancer are thought to develop *(10,11)*, are found in the neck of the crypt. Epithelial cell division occurs in the lower one-third of the crypt, and new cells migrate from the crypt to the surface, replacing those that undergo apoptosis and slough into the lumen of the colon.

Just below the epithelial cell layer of the mucosa is the lamina propria, a cellular region that contains lymphocytes and related cells. A layer of smooth muscle, the muscularis mucosa, demarcates the mucosa from the submucosa. The submucosa contains connective and adipose tissue, blood vessels, and lymphatic channels. A second layer of smooth muscle, the muscularis externa, bounds it externally. The serosa is the outermost surface of the colon.

An obvious consequence of the fact that samples of colon tissue contain several tissue planes and cell types, is that expression profiles generated with undissected colon tissue (normal or malignant) contain patterns that represent each of these components. Fluctuation in the contribution of each tissue and cell type to the mass of the sample is characteristic of undissected bulk samples. This is due to both individual biological variation and to differences in surgical, pathological, and sampling technique. Random sample-to-sample variation introduces noise into the resulting expression pattern. Systematic differences between the tissue compositions of samples (e.g., between normal and neoplastic samples) will introduce bias into the expression profiles. For example, in our initial experience with gene expression profiles of colon cancer, we found that, compared with bulk samples of colon cancer, bulk samples of normal colon tissue usually displayed a greater expression of genes that resulted from smooth muscle and connective tissue *(12)*. This was attributed to inclusion of more of the intestinal wall in the normal samples than in the malignant samples when the tissue was resected.

To a certain extent, it is possible to compensate for bias of this type during data analysis, but it may be preferable to minimize it by fastidious microdissection. At present, efforts to microdissect solid tumors in general, and colorectal cancer in particular, emphasize purification and isolation of the malignant (usually epithelial) cells (e.g., see 13). However, by excluding nonmalignant cells from the analysis, the dialogue between transformed epithelium and supporting vascular and stromal tissue will be largely overlooked (14,15). Comprehensive expression monitoring of solid tumors requires breadth (all individual tissue components to be tested following microdissection and purification) as well as depth (the monitoring technique encompasses all relevant mRNAs). However, it must be acknowledged that, for some categories of solid tumor, apparently meaningful class discovery has been accomplished using undissected samples (16,17). Indeed, it may be the case that microdissection and cell type purification is important for some organs (e.g., colon), but not for others (e.g., breast). This difference could be predicated on the histological organization of the hollow organs compared with the solid glandular organs, such as breast or pancreas. At present, there is not a sufficiently large experience to understand whether microdissection is necessary (or desirable), and if so, under which circumstances.

CLINICAL MANIFESTATIONS OF COLORECTAL CANCER

A major goal of global expression analysis is to provide information that supports an enriched system of classification, either alone or in conjunction with clinical and genetic data. To place this effort in perspective, we sketch the clinical behavior of colon cancer and outline the major clinical–pathological classification systems.

Adenocarcinoma arises in the epithelial cells of the colon. The tumor consists of a mass of abnormal glands (hence the name, adenocarcinoma) that invades and, ultimately, may penetrate the deeper contiguous structures. While the malignant component of the carcinoma consists of epithelial cells, the tumor is also composed of vascular and connective elements, which are necessary to support and nourish the tumor.

Colorectal cancer principally affects those older than 40 yr of age, although it occurs occasionally in adolescents (18). Ninety percent of tumors are found in people older than 50 yr. The incidence rate varies about 20-fold in different parts of the world, with the highest in the West and the lowest in India (19). Migration from a low to a high incidence region is associated with an increase in disease risk. This suggests that the environment (probably the diet) can influence the incidence of colorectal cancer (20), although the occurrence of cancer predisposition syndromes (accounting for about 5% of all cases of colon cancer) such as familial adenomatous polyposis (FAP) and hereditary nonpolyposis colon cancer (HNPCC) clearly implicate a genetic role. In addition to the well-characterized genes associated with familial colon cancer syndromes, it is likely that there are a large number of susceptibility loci, each of which has a very small effect on an individual's predisposition to develop colon cancer. Taken together, these susceptibility genes define the probability that a specific individual, exposed to a particular environment, will develop colorectal cancer, or having developed cancer, will respond in a particular way to therapy.

The first morphologically distinct lesion is the congeries of hyperplastic and dysplastic glands, which are termed "aberrant crypt foci" (10,21). In aberrant foci, proliferative cells are no longer confined to the deeper crypt, but may extend to the surface. The tissue begins to accumulate in a mass, which eventually becomes an adenoma. Although adenomas are benign, and most do not progress to cancer during a person's lifetime, virtually all colon adenocarcinomas originate in adenoma *(22,23)*. Once formed, adenomas may continue to grow, accumulating more of the histological stigmata of the mature cancer. Large adenomas may contain foci of early cancer. Ultimately, the basement membrane beneath these cancer cells is breached, and an invasive adenocarcinoma arises *(10)*. The malignancy grows centrifugally, expanding into the lumen of the colon and invading the contiguous bowel wall. Once the muscularis mucosa is penetrated, and with increasing probability as the tumor grows larger, it penetrates local blood vessels and lymphatics. This sets the stage for regional and distant metastasis, to lymph nodes, the lungs, liver, brain, and other sites *(24)*.

The prognosis of colorectal neoplasia closely parallels the clinical stage. A widely used method for staging colorectal cancer, termed the tumor node metastasis (TNM) system, stratifies tumors by several parameters: (*i*) the depth to which the primary tumor invades the bowel wall (T1–T4); (*ii*) the absence or presence of metastases to regional lymph nodes (N0–N3); and (*iii*) the absence or presence of distinct metastasis (M0–M1) *(25,26)*. For example, a tumor that has invaded through the muscularis propria to the serosa (but not beyond) and is not associated with regional lymph node or distant metastasis would be staged T3, N0, M0. An earlier staging system (the Dukes system and its modifications) also classifies the tumor based on depth of colon wall penetration and the presence or absence of metastasis. In the current modification of the Dukes classification system, stage A tumors are confined to the colon wall, B tumors penetrate the wall, C tumors are associated with nodal spread, and D tumors are complicated by distant metastasis.

The chance of cure decreases with invasion and the presence of regional nodal or distant metastasis. Cure for patients with early (T1–T2, Dukes A and B) lesions and neither nodal nor distant metastasis is excellent *(27)*. Long-term survival decreases sharply with the presence of lymph node disease (approx 50% for a T3, N1, M0 tumor), and it is unusual in the presence of distant metastatic spread (e.g., T3, N1, M1) *(24)*.

In early stages (i.e., Dukes A and early B stage), treatment is surgical, and complete resection of the tumor is associated with cure. Various regimens of adjuvant chemotherapy are often employed for deeper stage B tumors and in the presence of nodal metastasis (stage C). It is likely that at least half of the stage B and C patients who do receive adjuvant chemotherapy would do just as well without this highly toxic form of management. Unfortunately, the clinical phenotype is not sufficiently precise to permit prospective discrimination of moderate disease patients who will or will not benefit from postsurgical chemotherapy. One of the most persuasive arguments for genome-wide expression profiling of colorectal cancer is that it might provide the opportunity to prospectively identify those patients who require and would benefit from adjuvant chemotherapy *(24)*.

MOLECULAR GENETICS OF COLON CANCER
Overview of Molecular Events

Colorectal cancer is a well-defined clinical model for studying the molecular events of tumor development and progression. There is a linear progression, during which the neoplasm develops from hyperplastic epithelium in aberrant crypt foci through adenoma to carcinoma and metastasis *(28)*. This sequence reflects the accumulation of specific genetic alterations. It is thought that five to seven genes must be altered sequentially in order for cancer to develop *(10)*. Two separate sequences of molecular pathogenesis have been postulated for colorectal cancer, based upon observations in families with inherited predisposition to colorectal cancer (Fig. 1). The first model is derived from observations initially made in patients with FAP. Family members with this condition (which accounts for 1–2% of colorectal cancer) inherit an inactivating mutation in the adenomatous polyposis coli *(APC)* gene. Inactivation of the second allele (occurring in about 1 of 10^6 colorectal epithelial stem cells) results in the formation of an adenoma *(20)*. In approx 85% of sporadic forms of colon cancer, a similar pathogenetic sequence is suspected, although sequential inactivation of both alleles of *APC* occurs as a somatic mutational event *(29–37)*. This sequence is characterized by tumors that display marked chromosomal instability (CIN), but in which the replication error repair pathway is intact (RER-), at least during the early phase of tumorigenesis. The second, more recently described sequence is observed in individuals with a second type of cancer predisposition, HNPCC. This inherited syndrome is caused by a germline mutation in one of several genes involved in the replication error repair pathway. A somatic mutation of this type also initiates approx 15% of sporadic colon cancer cases *(36,38)*. Reflecting the defect in replication surveillance, these RER+ tumors are also characterized by microsatellite instability (MIN).

As indicated in Fig. 1, it is likely that the initial molecular event in RER- tumors is the inactivation of both alleles of *APC* (or the constitutive activation of β-catenin), followed by the mutational activation of K-*ras*, and mutational inactivation or loss of the *p53* gene. In addition to these three well-characterized DNA alterations, RER- tumors exhibit CIN as evidenced by loss of heterozygosity (LOH) or gene amplification at many loci, as well as promoter hypermethylation resulting in silencing of the $p16^{INK4A}$ and $p19^{ARF}$ genes *(39,40)*. On the other hand, RER+ tumors have inactivated the replication repair genes, *hMSH2, hMLH1, hMSH3, hMSH6, hPMS1,* or *hPMS2* through mutations and/or by silencing promoter methylation *(38,41)*. Disruption of this DNA repair pathway results in mononucleotide repeat slippage in the transforming growth factor β receptor (TGFβR)-II, *Bax* and others genes *(38,42–45)*. In contrast to the RER- tumors, RER+ tumors maintain chromosomal integrity and are diploid, but do contain thousands of slippage mutations and exhibit MIN *(40)*.

RER- and RER+ tumors have different clinical phenotypes. RER+/MIN tumors are found predominantly in the proximal colon, they are bulky and poorly differentiated, progress rapidly from adenoma to invasive cancer, and have a somewhat more favorable prognosis. On the other hand, RER-/CIN tumors occur more commonly in the distal colon or rectum, progress slowly and inefficiently to invasive cancer, and have a less favorable prognosis following surgical resection.

Many of the abnormalities that disturb the equilibrium between cell proliferation and death do so by abrogating the normal function of one or more signaling pathways. Genetic abnormalities result in heritable changes in DNA sequence (e.g., mutations, insertions and deletions, gene amplifications, LOH). Epigenetic changes produce heritable changes in gene expression that occur without changes to the DNA sequence (e.g., methylation, chromatin remodeling) *(46–48)*. Dysregulation of gene expression

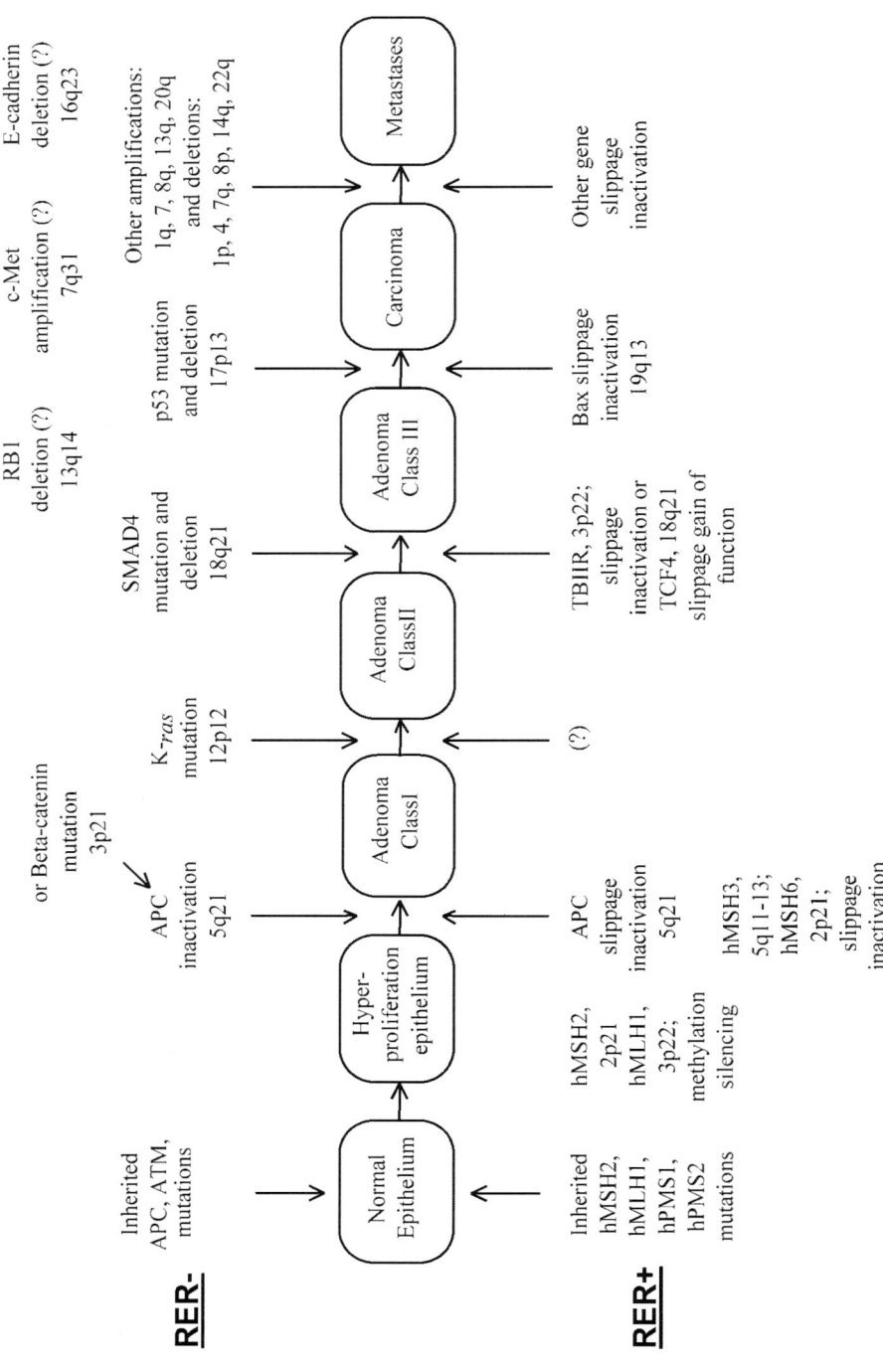

Fig. 1. Stages in the development of colon cancer. The two different model pathogenetic sequences (RER-/CIN and RER+/MIN) are shown. Abbreviations are indicated in the text. From ref. *24.*

can arise indirectly as cell autonomous events, secondary to upstream alterations, such as activation of K-*ras* or deletion of *SMAD4 (36,49,50)*. Environmental factors may trigger cellular responses that activate or inactivate signaling pathways; this is likely to be of key importance in colon cancer. Apart from genetic and epigenetic alterations, gene products (proteins) may be activated or inactivated through posttranslational events. This is of great importance in initiating and maintaining the transformed phenotype. While posttranslational modifications are only indirectly amenable to DNA microarray technology, they are being addressed by advances in high-throughput proteomics (e.g., *51*).

EXPRESSION PROFILES OF COLORECTAL TUMORS

As anticipated by the early experience in global expression monitoring of cancer *(6,12)*, many of the genetic and pathological changes that occur during neoplastic transformation should be reflected by altered patterns of gene expression. Thus, it is not surprising that cancer investigators have moved rapidly to exploit array technology or that these early studies have been quite productive. As noted earlier, colon cancer is a particularly good model system for expression monitoring, since many of the basic genetic alterations have been described and tied to specific stages in the development of the cancer. In addition to embellishing and expanding the catalogue of cancer pathway genes, by developing genome-wide profiles of colorectal cancer and its precursor lesions, expression monitoring may make it possible to learn if there are specific, cumulative, and robust changes in mRNA expression during the transition from normal through hyperplastic, dysplastic, adenomatous, malignant, and metastatic epithelium. It will be important to learn if changes in gene expression bear a predictable correlation with changes at the DNA and (eventually) protein level and whether there are specific reproducible transcriptional patterns that predict the response to therapy, the occurrence of metastasis, and ultimate survival.

Serial analysis of gene expression (SAGE) was one of the earliest large-scale efforts in expression analysis of colorectal cancer, and it was quite successful in identifying a catalogue of genes that are dysregulated in bulk samples of cancer and in cancer cell lines *(52,53)*. Although SAGE has the advantage of minimizing the ascertainment bias that is inherent in the use of microarrays with a predefined transcript set, the process is labor- and time-intensive. So far, the number of actual tissue samples reported in the literature is limited, e.g., *see* a SAGE database located at (http://www.ncbi.nlm.nih.gov/SAGE) and (http://www.sagenet.org/Cancer/Cancer.htm). Despite these issues, SAGE enabled a comprehensive comparison between normal and malignant colorectal tissue, and in limited respects, information gleaned from SAGE and microarray studies of colorectal cancer is comparable (*vide infra*).

To monitor mRNA expression in many different cancer samples and in precursor adenomas, our group first employed the GeneChip® (Affymetrix, Santa Clara, CA, USA) to measure the mRNA expression of approx 3200 full-length human cDNAs and 3400 expressed sequence tags (ESTs) in 40 tumor samples and 20 paired samples of normal tissue *(12)*. An important feature of this work is that bulk tissue samples were employed, and there was no effort to assure that the tissue composition of the normal and malignant samples was comparable. Extent of disease was estimated from surgical pathology reports (using a modification of the Duke staging system, see above), and a pathologist determined the percentage of the bulk tumor composed of malignant epithelial cells. In an analysis of the 40 malignant tumors and 20 normal samples, cluster analysis with an efficient deterministic annealing algorithm correctly discriminated the tumor from the normal samples with an error rate of <10% *(12)*. It was apparent that spatial clusters in the visual matrix often denoted functionally related genes. This is probably a consequence of co-regulation. For example, most of the genes and ESTs encoding ribosomal proteins cluster together, and their expression intensity is higher in tumors than in normal samples, which is consistent with previous observations *(54)*. A large cluster of genes was up-regulated in the normal compared with the tumor samples. This cluster was composed principally of smooth muscle genes, implying that some or all of the ability to partition normal from tumor samples might be due to differences in tissue composition rather than to an intrinsic property of the malignant epithelial cells. Indeed, normal tissue samples consistently displayed greater expression of identified smooth muscle genes.

Pairing 18 tumor samples with normal tissue from the same patients provided more information about gene dysregulation in colorectal cancer *(55)*. In developing this catalogue of dysregulated genes, a *t*-test was used to assign a *p* value to the average difference in expression between tumor and normal samples ($p < 0.001$ designated a significant change). To further cull the list of dysregulated genes, a fourfold change in average expression between tumor and normal was also required. These filters are arbitrary, and an advance in subsequent reports has been the use of various statistical approaches to place confidence intervals around clusters and changes in individual gene expression. A comparison of the average gene expression in normal and cancer tissue is shown in Fig. 2. Points lying above or below the upper and lower boundaries represent samples in which expression in tumors was either fourfold higher or fourfold lower, on average, than the corresponding normal sample. Genes for which expression differences achieved statistical significance are associated with an open circle ($p < 0.001$).

Tables 1 and 2 list the transcripts displaying a fourfold or greater increase or decrease in expression level, which was also significant at the $p < 0.001$ level. Nineteen transcripts (0.48 % of those detected at an intensity of 10 or greater) displayed 4- to 10.5-fold or greater expression in the tumors than the paired normal tissue ($p < 0.001$), while 88 transcripts (2.2%) displayed 4- to 38-fold lower expression in the cancer than the paired normal tissue ($p < 0.001$). Some of these genes seemed to reflect the excess smooth muscle and connective tissue in the normal samples and are omitted from the table. The complete data set, together with the associated *p* values, is contained in a file at (http://microarray.princeton.edu/oncology).

As previously reported, metalloproteinases were significantly more highly expressed in colonic neoplasia than in normal tissue (e.g., L23808, human metalloproteinase *[56]*, X05231, collagenase *[57]*). Other gene products were noted that are either linked to other forms of neoplasia or to the regulation of the cell cycle: GROα *(58–60)* 100 Kd co-activator *(61)*, ckshs2 *(62)*, CDC25B (an M-phase tyrosine phosphatase) *(63)*, and transcription factor IIIA (GTF3A) *(64,65)*.

As is the case in expression studies of breast cancer *(8,66)*, there is a large group of transcripts whose expression can be related to altered levels of metabolism (rather than cancer initiation, *per se*). In addition to the ribosomal protein genes mentioned earlier,

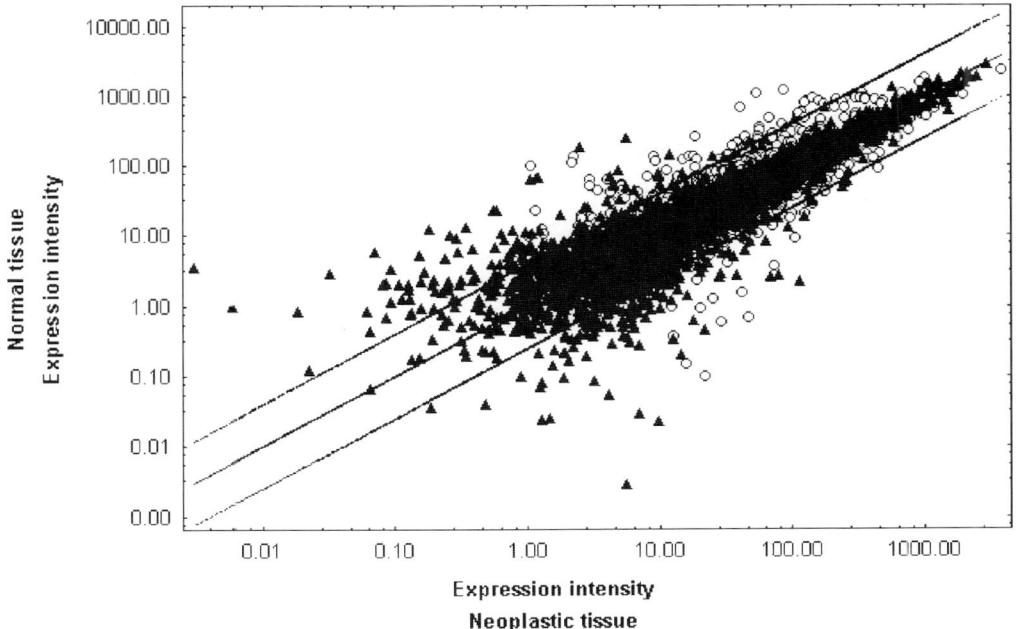

Fig. 2. Expression intensity in normal compared with tumor samples. The upper and lower boundaries represent a fourfold difference in the average of the expression of each gene between carcinoma and normal tissue. Open circle (O), genes for which average expression in carcinoma was significantly higher or lower than it was in the matched normal sample ($p < 0.001$). From ref. 55.

these include S-adenosylhomocysteine hydrolase *(67,68)*, pyrroline 5-carboxylate reductase *(69)*, and L-iditol-2 dehydrogenase *(70)*.

Table 2 omits some other gene products having proven association with colon cancer, because the *p* value associated with the expression difference was greater than 0.001 (ranging from 0.001–0.003). These included matrilysin (a matrix metalloproteinase) *(71,72)*, matrix metalloproteinase 12 (MMP12) *(73)*, osteopontin *(74)*, and TGFβ-induced gene product (BIGH3) *(75)*. Agrawal et al. recently confirmed that osteopontin is a lead marker in colon cancer progression *(76)*. Other transcripts are not included, because their increase in the tumors was less than the arbitrary cut-off of a fourfold change (nm23, c-myc). Some transcripts in Table 1 have already been shown to be down-regulated in cancer. Of particular interest, these are guanylin, which is known to be a product of colonic epithelial cells *(77,78)*, colon-mucosa-associated mRNA (down-regulated in adenocarcinoma [DRA]) *(79)*, tetranectin *(80,81)*, hevin *(82,83)*, and biliary glycoprotein (BGP1) *(84,85)*.

The forgoing indicates that it is possible to glean meaningful information from this analysis even though the tissue composition of the normal and neoplastic samples may differ. This is important, because it suggests that the computational techniques used to partition expression data sets may actually be quite robust, even in a context of substantial noise and bias. Such robustness is fortunate, since the medical community is eager to harness array technology to make clinical decisions and predictions. Even if

Comparisons with Adenomas

We wished to learn if some of the changes in mRNA expression found in carcinomas already exist in this precursor lesion *(20,21)*. To examine this, four adenomas (each paired with normal colon tissue from the same patient) were analyzed with the GeneChip. About 2400 unique accession numbers were in common between the somewhat different versions of the array used for these experiments *(55)*. Approximately 385 of this set were the product of genes whose expression changed significantly in the cancers (at $p < 0.01$). There was a close correlation ($r = 0.5$; $p < 0.0001$) between adenoma-associated changes in transcript expression and cancer-associated changes in transcript expression level. Forty-six transcripts were significantly dysregulated in both the adenoma and the cancer samples ($p < 0.01$) and are listed in Table 3. It is possible that the genes listed in Table 3 play a role at a relatively early stage of carcinogenesis. Of course, it is also possible that many of these abnormalities (as well as those in Tables 1 and 3) are merely bystanders to the neoplastic process, rather than reflecting a tumor-initiating mutation. Comparison between changes at the DNA, RNA, and clinical levels is needed to resolve this issue.

To further probe differences between normal tissue, adenomas, and carcinomas, cluster analysis was performed using both the adenomas and cancers, together with the paired normal samples. A hierarchical clustering algorithm was used for this purpose (Cluster 2.02, and the resulting expression map was visualized with Treeview 1.45, both shareware programs available at http://rana.stanford.edu/software) employing an average linkage rule. Before the program was used, expression levels less than or equal to zero were deleted, the remaining values were log-transformed, and both vectors of the data matrix were centered about the mean and normalized, in that order.

The resulting visual representation of the data with its associate dendrogram is shown in Fig. 3. The principal finding is that the phylogenetic tree is constructed so that the three tissue types are separated in a manner that respects the conventional pathological classification of this tumor. The carcinomas and their benign precursors, the adenomas, are placed on a different trunk than are the paired normal tissues, and these neoplasms are also separated from one another, occupying adjacent branches of the same trunk. Several clusters of genes appear to drive this partition. The three most obvious clusters are: Cluster 1 (Fig. 3, right upper panel) represented by a group of gene products that were more intensely expressed in adenoma than in either adenocarcinoma or normal tissue. This group contained several transcription factors, of which some have been implicated as oncogenes (XBP-1, SSRP1, ETS-2, SOX9), ribosomal proteins (S29 and S9), an inducer of apoptosis *(NBK)*, and a splicing factor (SRp30c). Many of these products are also listed in Table 1, as one would expect. Cluster 2 (Fig. 3, right lower panel) comprised a larger cohort of genes that were more highly expressed in adenocarcinoma than in adenoma or in normal tissue. This cluster contains many of the gene products already identified (Tables 1 and 2) as being more highly expressed in colorectal neoplasia than in normal tissue (e.g., Ckshs2, melanoma growth stimulatory activity (MGSA), matrilysin, and diverse products related to proliferation and metabolic rate).

Table 1
Transcripts More Highly Expressed in Adenocarcinoma Than in Paired Normal Tissue (24,55)

Accession no.	Description	Tumor/normal
X54489	HUMAN GENE FOR MELANOMA GROWTH STIMULATORY ACTIVITY.	10.5
U22055	HUMAN 100 KDA COACTIVATOR MRNA, COMPLETE CDS.	7.3
D14657	Human mRNA for KIAA0101 gene, complete cds.	6.5
M61832	Human S-adenosylhomocysteine hydrolase (AHCY) mRNA, complete cds.	6.0
M77836	Human pyrroline 5-carboxylate reductase mRNA, complete cds.	5.3
D21262	Human mRNA for KIAA0035 gene, partial cds (? nucleolar phosphoprotein).	5.2
M36821	Human cytokine (GRO-γ) mRNA, complete cds.	5.1
L23808	HUMAN METALLOPROTEINASE (HME) MRNA, COMPLETE CDS.	5.1
R08183	Similar to bovin hs 10 kDa protein 1(chaperonin 10)(HSPE1)(NM_002157).	4.8
L29254	Human (clone D21-1) L-iditol-2 dehydrogenase gene, exon 9 and complete cds.	4.7
H50438	M-PHASE INDUCER PHOSPHATASE 2 (*Homo sapiens*).	4.7
U33286	Human chromosome segregation gene homolog CAS mRNA, complete cds.	4.7
X54942	H. SAPIENS CKSHS2 MRNA FOR CKS1 PROTEIN HOMOLOGUE.	4.4
R32511	*Homo sapiens* cDNA clone 135395 3' (RNA POL II subunit).	4.3
T87871	*Homo sapiens* cDNA clone 115765 3' (myoblast cell surface antigen 24.1 DS).	4.2
X05231	Human mRNA for collagenase (identical to metalloproteinase 1).	4.2
R36977	Similar to *Homo sapiens* general transcription factor IIIA (GTF3A) (mRNA).	4.1
U17899	Human chloride channel regulatory protein mRNA, complete cds.	4.0
X54942	H. SAPIENS CKSHS2 MRNA FOR CKS1 PROTEIN HOMOLOGUE.	4.0

Intensity values <10 are adjusted to 10. Only transcripts with a fourfold difference or greater ($p < 0.001$) in expression intensity between tumor and normal are included. Transcripts shown in capital letters were confirmed by RT-PCR. Gene descriptions have been edited.

Table 2
Transcripts More Highly Expressed in Paired Normal Tissue Than in Adenocarcinoma (24,55)

Accession no.	Description	Normal/tumor
M83670	Human carbonic anhydrase IV mRNA, complete cds.	37.9
M97496	H. SAPIENS MRNA FOR GUAANYLIN, COMPLETE CDS.	20.2
X64559	H. SAPIENS MRNA FOR TETRANECTIN.	13.8
T54547	H. sapiens cDNA similar to M84526 COMPLEMENT FACTOR D PRECURSOR.	12.0
M95936	Human protein-serine/threonine (AKT2) mRNA, complete cds.	11.4
T55200	H. sapiens cDNA similar to gb:M10942_cds1 Human metallothionein-Ie gene.	8.4
T46924	H. sapiens cDNA similar to gb:U11863 AMILORIDE-SENS AMINE OXIDASE.	8.2
L11708	Human 17 β-hydroxysteroid dehydrogenase type 2 mRNA, complete cds.	8.1
T46933	H. sapiens cDNA clone 70843 3' (11-β dehydrogenase).	7.6
H54425	H. sapiens cDNA similar to gb:M10942_cds1 Human metallothionein-Ie gene.	7.4
M26393	Human short chain acyl-CoA dehydrogenase mRNA, complete cds.	7.2
M82962	Human N-benzoyl-L-tyrosyl-p-amino-benzoic acid hydrolase α-subunit mRNA.	7.1
J03037	Human carbonic anhydrase II mRNA, complete cds.	6.5
T72257	H. sapiens cDNA similar to gb:L07765 LIVER CARBOXYLESTERASE.	6.3
M84526	Human adipsin/complement factor D mRNA, complete cds.	5.9
T76971	H. sapiens cDNA similar to gb:X64177 H. sapiens mRNA for metallothionein.	5.7
H77597	H. sapiens cDNA similar to gb:X64177 H. sapiens mRNA for metallothionein.	5.6
T67986	H. sapiens cDNA clone 82030 3' similar to gb:X14723 CLUSTERIN PRECURSOR.	5.6
R99208	H. sapiens cDNA clone 200586 3' similar to gb:X76717 H. sapiens MT-11 mRNA.	5.6
U03749	Human chromogranin A (CHGA) gene, exon 8 and complete cds.	5.5
R93176	Soares 1NFLS H. sapiens cDNA similar to gb:M33987 CARB. ANHYDRASE I. #	5.3
L02785	H. SAPIENS COLON MUCOSA-ASSOCIATED (DRA), COMPLETE CDS. #	5.3
R94967	H. sapiens cDNA similar to gb:L11924 HEPATOCYTE GROWTH FACTOR.	5.2
J03037	Human carbonic anhydrase II mRNA, complete cds.	5.2
M74509	Human endogenous retrovirus type C oncovirus sequence.	5.2
L11708	Human 17 β-hydroxysteroid dehydrogenase type 2 mRNA, complete cds.	5.2
X77777	H. sapiens intestinal VIP receptor related protein mRNA.	5.1
R69552	H. sapiens cDNA clone 155302 3' (glutamate).	5.1

R50730	H. sapiens cDNA similar to gb:Z19585 THROMBOSPONDIN 4 PRECURSOR.	5.0
H43887	H. sapiens cDNA similar to gb:M84526 COMPLEMENT FACTOR D PREC.	4.7
U17077	Human BENE mRNA, partial cds.	4.5
U25138	Human MaxiK potassium channel β-subunit mRNA, complete cds.	4.5
X86693	H. SAPIENS MRNA FOR HEVIN LIKE PROTEIN.	4.5
H57136	H. sapiens cDNA similar to SP:A40533 A40533 CAMP-DEP PROTEIN KINASE.	4.5
X73502	H. Sapiens mRNA for cytokeratin 20. #	4.5
J03037	Human carbonic anhydrase II mRNA, complete cds.	4.4
R70806	H. sapiens cDNA similar to gb:X62535 DIACYLGLYCEROL KINASE.	4.4
T51913	H. sapiens cDNA similar to gb:S45630 α-CRYSTALLIN B CHAIN.	4.3
T50678	H. sapiens cDNA contains TAR1 repetitive element (α-tryptase).	4.3
Z50753	H. sapiens mRNA for GCAP-II/uroguanylin precursor #.	4.3
M58286	H. sapiens tumor necrosis factor receptor mRNA, complete cds.	4.3
U08854	Human UDP glucuronosyltransferase precursor (UGT2B15) mRNA, complete cds.	4.3
X52679	Human ASM-2 mRNA for sphingomyelin phosphodiesterase (EC 3.1.4.12).	4.2
T71025	H. sapiens cDNA similar to gb:J03910_rna1 Human.	4.1
M12272	H. sapiens alcohol dehydrogenase class I γ-subunit (ADH3) mRNA.	4.1
M26683	Human interferon-γ treatment inducible mRNA.	4.0
D90313	HUMAN MRNA FOR BILIARY GLYCOPROTEIN, BGP. #	4.0

Intensity values <10 are adjusted to 10. Only transcripts with a fourfold difference or greater ($p < 0.001$) in expression intensity between tumor and normal are included. Transcripts (41) representing smooth muscle or collagen are not shown. Transcripts shown in capital letters were confirmed by RT-PCR. Gene descriptions have been edited. #, also identified in the SAGE database.

Table 3
Forty-Six Transcripts Significantly Dysregulated in Both Adenoma and Cancer Samples ($p < 0.01$, no threshold) (24)

Accession no.	Description	Adenoma expression ratio	Cancer expression ratio
M61832	Human S-adenosylhomocysteine hydrolase (AHCY) mRNA, complete cds.	2.497	5.143
X54942	H. sapiens ckshs2 mRNA for Cks1 protein homologue.	2.687	4.069
Z46629	Homo sapiens SOX9 mRNA.	1.958	2.909
U18291	Human CDC16Hs mRNA, complete cds.	3.316	2.538
U04313	Human maspin mRNA, complete cds.	8.612	2.200
L09604	Homo sapiens differentiation-dependent A4 protein mRNA, complete cds.	1.568	1.937
M34458	Human lamin B mRNA, complete cds.	1.417	1.800
X68314	H. sapiens mRNA for glutathione peroxidase-GI.	6.452	1.776
Y00971	Human mRNA for phosphoribosyl pyrophosphate synthetase subunit II (EC 2.7.6.1).	1.473	1.763
Z23064	H. sapiens mRNA gene for hnRNP G protein.	1.867	1.613
Z49099	H. sapiens mRNA for spermine synthase.	1.741	1.600
X55715	Human Hums3 mRNA for 40S ribosomal protein s3.	2.079	1.585
M64716	Human ribosomal protein S25 mRNA, complete cds.	2.363	1.490
M86737	Human high mobility group box (SSRP1) mRNA, complete cds.	7.819	1.358
D10522	Human mRNA for 80K-L protein, complete cds.	0.345	0.742
X77366	H. sapiens HBZ17 mRNA.	0.390	0.667
X59841	Human PBX3 mRNA.	0.626	0.667
M63138	Human cathepsin D (catD) gene, exons 7, 8, and 9.	0.371	0.664
X64364	H. sapiens mRNA for M6 antigen.	0.403	0.649
M23254	Human Ca^{2+}-activated neutral protease large subunit (CANP) mRNA, complete cds.	0.575	0.603
L27943	Homo sapiens cytidine deaminase (CDA) mRNA, complete cds.	0.618	0.597
M92843	H. sapiens zinc finger transcriptional regulator mRNA, complete cds.	0.449	0.587
D14662	Human mRNA for KIAA0106 gene, complete cds.	0.389	0.526
X15880	Human mRNA for collagen VI α-1 C-terminal globular domain.	0.153	0.470
M80899	Human novel protein AHNAK mRNA, partial sequence.	0.707	0.459
X16354	Human mRNA for transmembrane carcinoembryonic antigen BGPa (formerly TM1-CEA).	0.157	0.416
L07648	Human MXI1 mRNA, complete cds.	0.340	0.405

Z24727	*H. sapiens* tropomyosin isoform mRNA, complete CDS.	0.268	0.405
X74295	*H. sapiens* mRNA for α-7B integrin.	0.444	0.402
X15882	Human mRNA for collagen VI α-2 C-terminal globular domain.	0.080	0.399
U27460	Human uridine diphosphoglucose pyrophosphorylase mRNA, complete cds.	0.379	0.393
L05144	*Homo sapiens* (clone λ-hPEC-3) phosphoenolpyruvate carboxykinase (PCK1) mRNA, complete cds.	0.187	0.282
J02851	Human 20-kDa myosin light chain (MLC-2) mRNA, complete cds.	0.067	0.270
L41351	*Homo sapiens* prostasin mRNA, complete cds.	0.332	0.263
M76378	Human cysteine-rich protein (CRP) gene, exons 5 and 6.	0.081	0.263
M95787	Human 22 kDa smooth muscle protein (SM22) mRNA, complete cds.	0.004	0.253
U28249	Human 11 kDa protein mRNA, complete cds.	0.103	0.247
U17077	Human BENE mRNA, partial cds.	0.049	0.200
L02785	*Homo sapiens* colon mucosa-associated (DRA) mRNA, complete cds.	0.025	0.197
M83186	Human cytochrome c oxidase subunit VIIa (COX7A) muscle isoform mRNA, complete cds.	0.138	0.169
M82962	Human N-benzoyl-L-tyrosyl-p-amino-benzoic acid hydrolase alpha subunit (PPH-α) mRNA, complete cds.	0.284	0.156
L11708	Human 17 β-hydroxysteroid dehydrogenase type 2 mRNA, complete cds.	0.151	0.123
X53416	Human mRNA for actin-binding protein (filamin) (ABP-280).	0.033	0.123
X54162	Human mRNA for a 64 kDa autoantigen expressed in thyroid and extra-ocular muscle.	0.197	0.121
X64559	*H. sapiens* mRNA for tetranectin.	0.885	0.092
M97496	*Homo sapiens* guanylin mRNA, complete cds.	0.009	0.058

The table is sorted in descending order by cancer expression ratio.

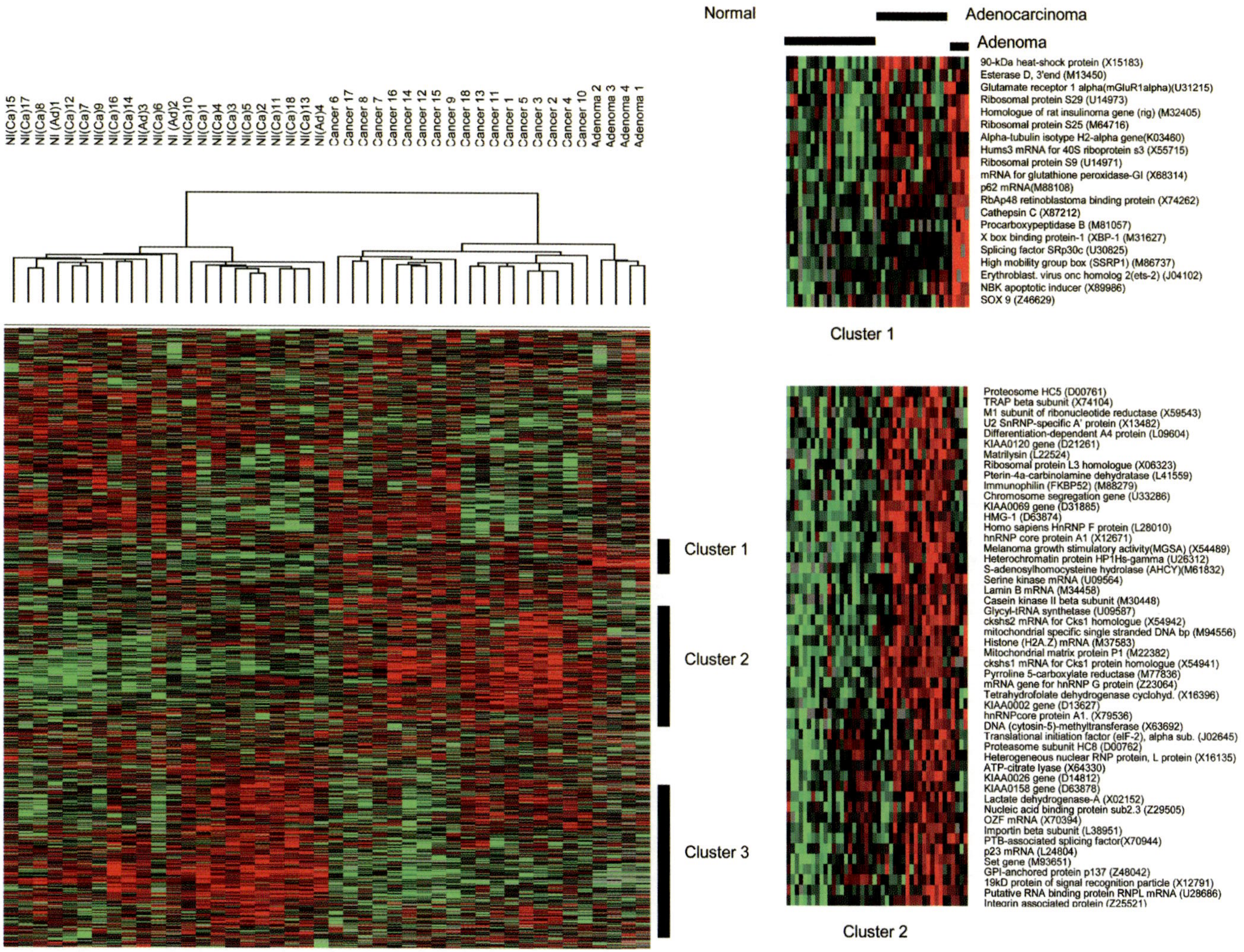
Nl(Ca)15
Nl(Ca)17
Nl(Ca)8
Nl (Ad)1
Nl(Ca)12
Nl(Ca)7
Nl(Ca)9
Nl(Ca)16
Nl(Ca)14
Nl(Ad)3
Nl(Ca)6
Nl (Ad)2
Nl(Ca)10
Nl(Ca)1
Nl(Ca)4
Nl(Ca)3
Nl(Ca)5
Nl(Ca)2
Nl(Ca)11
Nl(Ca)18
Nl(Ca)13
Nl(Ad)4
Cancer 6
Cancer 17
Cancer 8
Cancer 7
Cancer 16
Cancer 14
Cancer 12
Cancer 15
Cancer 9
Cancer 18
Cancer 13
Cancer 11
Cancer 1
Cancer 5
Cancer 3
Cancer 2
Cancer 4
Cancer 10
Adenoma 2
Adenoma 3
Adenoma 4
Adenoma 1
Cluster 1
Cluster 2
Cluster 3
Normal
Adenocarcinoma
Adenoma
90-kDa heat-shock protein (X15183)
Esterase D, 3'end (M13450)
Glutamate receptor 1 alpha(mGluR1alpha)(U31215)
Ribosomal protein S29 (U14973)
Homologue of rat insulinoma gene (rig) (M32405)
Ribosomal protein S25 (M64716)
Alpha-tubulin isotype H2-alpha gene(K03460)
Hums3 mRNA for 40S riboprotein s3 (X55715)
Ribosomal protein S9 (U14971)
mRNA for glutathione peroxidase-GI (X68314)
p62 mRNA(M88108)
RbAp48 retinoblastoma binding protein (X74262)
Cathepsin C (X87212)
Procarboxypeptidase B (M81057)
X box binding protein-1 (XBP-1 (M31627)
Splicing factor SRp30c (U30825)
High mobility group box (SSRP1) (M86737)
Erythroblast. virus onc homolog 2(ets-2) (J04102)
NBK apoptotic inducer (X89986)
SOX 9 (Z46629)
Cluster 1
Proteosome HC5 (D00761)
TRAP beta subunit (X74104)
M1 subunit of ribonucleotide reductase (X59543)
U2 SnRNP-specific A' protein (X13482)
Differentiation-dependent A4 protein (L09604)
KIAA0120 gene (D21261)
Matrilysin (L22524)
Ribosomal protein L3 homologue (X06323)
Pterin-4a-carbinolamine dehydratase (L41559)
Immunophilin (FKBP52) (M88279)
Chromosome segregation gene (U33286)
KIAA0069 gene (D31885)
HMG-1 (D63874)
Homo sapiens HnRNP F protein (L28010)
hnRNP core protein A1 (X12671)
Melanoma growth stimulatory activity(MGSA) (X54489)
Heterochromatin protein HP1Hs-gamma (U26312)
S-adenosylhomocysteine hydrolase (AHCY)(M61832)
Serine kinase mRNA (U09564)
Lamin B mRNA (M34458)
Casein kinase II beta subunit (M30448)
Glycyl-tRNA synthetase (U09587)
ckshs2 mRNA for Cks1 homologue (X54942)
mitochondrial specific single stranded DNA bp (M94556)
Histone (H2A.Z) mRNA (M37583)
Mitochondrial matrix protein P1 (M22382)
ckshs1 mRNA for Cks1 protein homologue (X54941)
Pyrroline 5-carboxylate reductase (M77836)
mRNA gene for hnRNP G protein (Z23064)
Tetrahydrofolate dehydrogenase cyclohyd. (X16396)
KIAA0002 gene (D13627)
hnRNPcore protein A1. (X79536)
DNA (cytosin-5)-methyltransferase (X63692)
Translational initiation factor (eIF-2), alpha sub. (J02645)
Proteasome subunit HC8 (D00762)
Heterogeneous nuclear RNP protein, L protein (X16135)
ATP-citrate lyase (X64330)
KIAA0026 gene (D14812)
KIAA0158 gene (D63878)
Lactate dehydrogenase-A (X02152)
Nucleic acid binding protein sub2.3 (Z29505)
OZF mRNA (X70394)
Importin beta subunit (L38951)
PTB-associated splicing factor(X70944)
p23 mRNA (L24804)
Set gene (M93651)
GPI-anchored protein p137 (Z48042)
19kD protein of signal recognition particle (X12791)
Putative RNA binding protein RNPL mRNA (U28686)
Integrin associated protein (Z25521)
Cluster 2

Fig. 3.

The SAGE database also contains a great deal of information regarding differences in gene expression between bulk samples of normal colon and colon cancer. Using libraries derived from two bulk normal colon tissues (SAGE NC1 and NC2) and two bulk colon tumors (Tu98 and Tu102) available at (http://www.ncbi.nlm.nih.gov/SAGE/sagexpsetup.cgi), the 100 SAGE tags most likely to be differentially expressed between bulk normal and tumor samples were identified with xProfiler, a tool provided at SAGEmap. Of these, one (M77349, TGFβ-induced gene product, BIGH3) was identified in the Affymetrix data set as overexpressed in colon cancer, and six were identified as more highly expressed in normal colon tissue (guanylin, uroguanylin, carbonic anhydrase I, and biliary glycoprotein). There are important differences between the SAGE and Affymetrix results, perhaps due to sample heterogeneity, the limited number of tissue samples analyzed for the SAGE databases, and the gene assignment differences in mapping oligonucleotide probes or SAGE tags. Clearly, it is important to develop a better understanding of the similarities and differences produced by alternative approaches to large-scale expression monitoring.

In a reanalysis of our colon cancer data, Getz et al. *(86,87)* employed a coupled two-way hierarchical clustering process (CTWC). The approach is an iterative one, in which the structure that emerges during an earlier partition of the data set is subsequently employed to develop relatively small stable clusters, in which a limited number of genes are used to represent each cluster. This approach results in the identification of submatrices of the total expression matrix, such that it may be simpler to identify biologically or clinically meaningful partitions of the samples than were obscured when examining the entire original data set. One advantage of this approach is that it reduces the noise that is introduced by the large number of operationally irrelevant genes. This may represent a useful approach to reduce the noise associated with sample and epithelial cell heterogeneity (*vide infra*). Interestingly, using CTWC, Getz et al. *(86)* showed that there is a high correlation between a subcluster containing epithelial genes (e.g., mucin) and a subcluster containing proliferative genes (e.g., ribosomal proteins). This strong correlation is limited to the tumor samples; it is not discerned in the normal samples. This suggests that the malignant epithelium in the cancer samples has a high rate of proliferation, while the epithelium in the normal samples does not, an idea which resonates with the anticipated biology. During the preparation of the mRNA samples for labeling and analysis, we employed two different methods *(12)*. For initial samples, mRNA was isolated, and for later samples, total RNA was isolated. CTWC correctly partitioned the early and late samples, uncovering this alteration in the protocol, and

---

Fig. 3. *(previous page)* (Left panel) Cluster map and phylogenetic tree resulting from a two-way pairwise average-linkage cluster analysis. The carcinomas and their benign precursors, the adenomas, are placed on an entirely different trunk than are the paired normal tissues. The adenomas and the carcinomas are also separated from one another, occupying adjacent branches of the same trunk. (Right upper panel, cluster 1) A cluster of gene products that are more intensely expressed in adenoma than in normal tissue or carcinoma. This grouping contained approx 8–10 genes, the mRNA expression of which appeared to be much higher in the adenoma samples than in associated normal tissue or the carcinomas. (Right lower panel, cluster 2) A cluster of gene products that are more highly expressed in carcinoma than in adenoma or normal tissue. From ref. *55*.

identified the 27 genes that drive this separation (*see* Fig. 8A in the supplementary information for ref. *87*). This suggests that CTWC, by defining smaller submatrices of the data set, may enhance the sensitivity and accuracy of sample classification.

In a subsequent refinement of our work, Kitahara et al. *(13)* reported a survey of RNA expression in samples of colon tissue taken from surgical specimens of adenocarcinoma. The samples were subjected to laser-capture microscopy, and normal and malignant cells were separately isolated (approx 10^4 cells/sample). Following T7 linear amplification, the aRNA was hybridized to cDNA microarrays (9216 cDNAs, of which 4220 represented genes with known function). Genes with differential expression were selected based on a permutation test, and selected changes were confirmed with reverse transcription polymerase chain reaction (RT-PCR) on amplified and nonamplified samples of RNA. Genes that demonstrated a significant change in more than half (i.e., at least five of eight samples) were considered as "commonly" up- or down-regulated. Using this criterion, 44 genes (0.48%) were up-regulated, and 191 (0.99%) were down-regulated. Remarkably, the percentage of down-regulated genes (0.48%) was identical to that which we observed, while the percentage of up-regulated genes was slightly lower than we reported (2.2%) *(55)*. Our value is probably a modest overestimate, due to the greater inclusion of smooth muscle and connective tissue in the normal bulk samples rather than the tumor. Several of the gene products identified as abnormally expressed were previously implicated in colorectal cancer, and others were known oncogenes. Unfortunately, the authors did not make their data available in a comprehensive, electronic format, so it has not been possible to apply other statistical approaches to their data set, nor to compare their results with what others have reported. It would be important to do so, because, as yet, the use of linear amplification prior to array analyses has not been systematically validated following laser microdissection.

More recently, Yu-Min Lin et al. *(88)* analyzed the gene expression profiles of 9 adenomas and 11 adenocarcinomas. The samples were microdissected and amplified following the procedure of Kitahara et al. *(13)*. A cDNA array, which embodied approx 23,000 cDNAs, was used for the analysis and was performed in duplicate. Using a similar two-way clustering algorithm operating on a subset of 771 genes (selected for reproducibility and detection in more than 80% of cases), they were able to partition adenoma from carcinoma, confirming a previous result *(55)*. Using a twofold variation as denoting up- or down-regulation, 51 genes were up-regulated in adenoma and carcinoma, as compared with normal epithelium, and 376 were down-regulated. In addition, a group of genes were differentially regulated between adenomas and carcinomas; interestingly, as they observe, nearly half of these genes play a role in "bioenergetics," including the proteins involved in the response to hypoxia. A comparison of Lin's results with the SAGE database (*vide supra*) and the data of Notterman et al. *(55)* revealed some overlap. The methods used to produce the expression profiles in these studies varied widely (cDNA microarray vs SAGE vs Affymetrix GeneChip; bulk tissue vs laser capture microscopy; T7 amplification of mRNA vs nonamplified mRNA). Therefore, it is not surprising that the resulting data sets are not congruent. Indeed, it is gratifying that both of the array experiments were able to partition adenoma and carcinoma on the basis of distributed gene expression patterns *(55,88)*.

These experiments with colon cancer and adenoma indicate the great potential of microarray analysis of solid tumors. They make it possible to enumerate a catalogue of genes that are dysregulated in both cancer and the precursor adenoma. Many of these abnormalities are consistent with the literature, and others may signify cancer pathway genes that are novel or have not previously been associated with colorectal cancer. Because it is possible to use an unsupervised clustering algorithm to reorganize the data in a way that respects the existing taxonomy (class prediction), we are encouraged to think that it may be feasible, with the addition of appropriate clinical and demographic information, to discover new, clinically relevant classes of colon cancer. As described elsewhere in this volume, this has been accomplished, in a limited way, in a variety of tumor types, including B cell lymphoma, melanoma, breast cancer, and lymphoblastic leukemia (6,7,9,17).

Differences in experimental and analytic techniques mean that comparisons across different colon expression data sets are not yet robust or even practical. This greatly encumbers an effort to generalize experimental results. One hopes that in the near future, scientists will adopt a common experimental approach and a uniform data management structure so that it will be possible to link the data sets from diverse microarray expression experiments (even involving different tumor types) originating in different laboratories (e.g., see MicroArray and Gene Expression Markup Language [MAGE-ML] described at http://xml.coverpages.org/mageML.html). This will facilitate comparisons and meta-analysis, in which genes common to the molecular pathology of diverse tumors can be evaluated.

It is worth emphasizing that without detailed clinical and genetic information, it is not possible to learn how particular expression patterns correlate with genotype and disease phenotype or to develop a molecular taxonomy in which newly discovered expression patterns serve as an analytic bridge between the underlying genetic substrate and the resulting clinical behavior of the tumor. As mentioned earlier, in the case of some tumor types, this effort has already been productive; unfortunately, this sort of correlation has lagged in the case of colorectal cancer.

There are concerns that are more fundamental. Analysis of neoplasia is often predicated upon a comparison of abnormal with normal tissue. We have already described the problem of sample heterogeneity—normal tissue is composed of hundreds of different cell types, each with its own expression phenotype. However, even if it were possible to purify samples to 100% normal epithelium, it remains that colon epithelial cells are not monolithic with respect to their state of differentiation (11). Cells from the neck of the crypt will have a different expression phenotype than those at the surface or different regions of the crypt. Each of these cells and their associated mRNA and protein profile is "normal." Therefore, following microdissection, the signal observed in the typical microarray experiment will be a composite of the signals from the different epithelial cell types. To an unknown extent, discrepancies between studies may be a function of variation in the specific blend of epithelial cells across individual preparations. Furthermore, it is possible that a comparison between normal cells at specific developmental stages (crypt neck vs crypt mouth and surface) and specific tumor samples would provide insight into the molecular trajectory followed by individual tumors. In this context, the field needs much more work in defining what is meant by normal.

There is an analogous problem in the interpretation of data from of tumor specimens. Samples of colon adenocarcinoma are heterogenous with respect to many genetic markers, even across individual samples *(89)*. It can be expected, therefore, that there will be tumor regional differences in expression phenotype; thus, the signal reported for a particular sample is a weighted composite of the sample's regional signals. Indeed, this source of heterogeneity will be enhanced by epithelial cell purification techniques. Furthermore, it is likely that certain tumor regions have a selective advantage over others and will be dominant with respect to determining the fate of the tumor (and the patient). The expression signature related to this region will be the most relevant for driving a molecular predictive model of tumor progression, but it will be obscured by the signals from the other regions of the tumor.

It is also worth noting that the normal tissue used in most microarray experiments is derived from the resected segment of colon. In fact, epithelial cells from this tissue may already display early preneoplastic characteristics. Failure to characterize carefully the genetic and histologic characteristics of these cells risks underestimating the difference between normal and abnormal.

CONCLUSION

This review of the major efforts in array analysis of colorectal cancer points to the great promise of this work, but indicates that an approach grounded exclusively on measurement of mRNA expression has three major limitations.

First, many of the observed expression changes reflect secondary differences in metabolic and proliferation rates and the tissue environment. Some of these changes in RNA quantity clearly play a role in fostering the growth of the tumor, but others do not. It is important to distinguish those expression changes that produce neoplastic growth from those changes that are secondary to this process of tumor initiation.

Second, there remain important concerns about sample heterogeneity (e.g., variation in smooth muscle composition due to sampling differences), cellular heterogeneity (e.g., "what is a normal epithelial cell?"), subtle abnormalities in normal tissue adjacent to neoplastic tissue, and regional differences in the genetic background and expression phenotype of individual tumors.

Third, some RNA changes are true experimental artifacts, imposed by the necessary period of tissue ischemia between separation of the tumor from its blood supply and immersion in liquid nitrogen *(90)* or by differences in experimental technique *(87)*.

In addressing these concerns, it seems reasonable that changes in RNA quantity that play a causal role in tumor initiation will involve genes that are also altered at the DNA level (or are just downstream to genes altered at the DNA level). It is likely that the best approach to managing issues of sample and cellular heterogeneity is to anchor an analysis of expression phenotype to the underlying genetic events and resulting clinical phenotype. Thus, for expression monitoring to reach its true potential in the service of cancer gene discovery, gene expression studies need to be coupled with a parallel effort to identify alterations at the DNA level and to tie both to the underlying clinical phenotype.

ACKNOWLEDGMENT

This work was supported in part by a grant from the New Jersey Commission on Science and Technology (grant no. 99-2042-007-12 and the National Cancer Institute (grant no. P01-CA65930-05). Samples were processed in the laboratory of Arnold J. Levine.

REFERENCES

1. Parker, S. L., Tong, T., Bolden, S., and Wingo, P. A. (1996) Cancer statistics, 1996. *CA Cancer J. Clin.* **46,** 5–27.
2. Boring, C. C., Squires, T. S., Tong, T., and Montgomery, S. (1994) Cancer statistics, 1994. *CA Cancer J. Clin.* **44,** 7–26.
3. American Cancer Society. (2002) Estimated New Cancer Cases and Deaths by Sex for All Sites, United States, 2002. Vol. 2000.
4. Zhang, L., Zhou, W., Velculescu, V. E., et al. (1997) Gene expression profiles in normal and cancer cells. *Science* **276,** 1268–1272.
5. Brown, P. O. and Botstein, D. (1999) Exploring the new world of the genome with DNA microarrays. *Nat. Genet.* **21,** 33–37.
6. Alizadeh, A. A., Eisen, M. B., Davis, R. E., et al. (2000) Distinct types of diffuse large B-cell lymphoma identified by gene expression profiling. *Nature* **403,** 503–511.
7. Bittner, M., Meltzer, P., Chen, Y., et al. (2000) Molecular classification of cutaneous malignant melanoma by gene expression profiling. *Nature* **406,** 536–540.
8. Perou, C. M., Sorlie, T., Eisen, M. B., et al. (2000) Molecular portraits of human breast tumours. *Nature* **406,** 747–752.
9. Yeoh, E. J., Ross, M. E., Shurtleff, S. A., et al. (2002) Classification, subtype discovery, and prediction of outcome in pediatric acute lymphoblastic leukemia by gene expression profiling. *Cancer Cell* **1,** 133–143.
10. Cummings, O. W. (2000) Pathology of the adenoma-carcinoma sequence: from aberrant crypt focus to invasive carcinoma. *Semin. Gastrointest. Dis.* **11,** 229–237.
11. Wright, N. A. (2000) Epithelial stem cell repertoire in the gut: clues to the origin of cell lineages, proliferative units and cancer. *Int. J. Exp. Pathol.* **81,** 117–143.
12. Alon, U., Barkai, N., Notterman, D. A., et al. (1999) Broad patterns of gene expression revealed by clustering analysis of tumor and normal colon tissues probed by oligonucleotide arrays. *Proc. Natl. Acad. Sci. USA* **96,** 6745–6750.
13. Kitahara, O., Furukawa, Y., Tanaka, T., et al. (2001) Alterations of gene expression during colorectal carcinogenesis revealed by cDNA microarrays after laser-capture microdissection of tumor tissues and normal epithelia. *Cancer Res.* **61,** 3544–3549.
14. Hanahan, D. and Weinberg, R. A. (2000) The hallmarks of cancer. *Cell* **100,** 57–70.
15. St. Croix, B., Rago, C., Velculescu, V., et al. (2000) Genes expressed in human tumor endothelium. *Science* **289,** 1197–1202.
16. Sorlie, T., Perou, C. M., Tibshirani, R., et al. (2001) Gene expression patterns of breast carcinomas distinguish tumor subclasses with clinical implications. *Proc. Natl. Acad. Sci. USA* **98,** 10,869–10,874.
17. van't Veer, L. J., Dai, H., van de Vijver, M. J., et al. (2002) Gene expression profiling predicts clinical outcome of breast cancer. *Nature* **415,** 530–536.
18. Datta, R. V., LaQuaglia, M. P., and Paty, P. B. (2000) Genetic and phenotypic correlates of colorectal cancer in young patients. *N. Engl. J. Med.* **342,** 137–138.
19. Potter, J. D. (1999) Colorectal cancer: molecules and populations. *J. Natl. Cancer Inst.* **91,** 916–932.
20. Kinzler, K. W. and Vogelstein, B. (1996) Lessons from hereditary colorectal cancer. *Cell* **87,** 159–170.
21. Takayama, T., Katsuki, S., Takahashi, Y., et al. (1998) Aberrant crypt foci of the colon as precursors of adenoma and cancer. *N. Engl. J. Med.* **339,** 1277–1284.
22. Winawer, S. J., Zauber, A. G., Ho, M. N., et al. (1993) Prevention of colorectal cancer by colonoscopic polypectomy. *N. Engl. J. Med.* **329,** 1977–1981.
23. Winawer, S. J., Zauber, A. G., Gerdes, H., et al. (1996) Risk of colorectal cancer in the families of patients with adenomatous polyps. National Polyp Study Workgroup. *N. Engl. J. Med.* **334,** 82–87.

24. Notterman, D., Shawber, C. A., and Levine, A. J. (2002) Tumor biology and microarray analysis of solid tumors:colorectal cancer as a model system, in *Microarrays and Cancer Research* (Warrington, J., Todd, C., and Wong, D., eds.), Eaton Publishing, Westboro.
25. American Joint Committee on Cancer. (1992) *Manual for Staging of Cancer*, in (Beahrs, O. H., ed.), Lippincott, Philadelphia.
26. American Joint Committee on Cancer. (1988) *Manual for Staging of Cancer*. JB Lippincott, Philadelphia.
27. Eisenberg, B., Decosse, J. J., Harford, F., and Michalek, J. (1982) Carcinoma of the colon and rectum: the natural history reviewed in 1704 patients. *Cancer* **49,** 1131–1134.
28. Vogelstein, B., Fearon, E. R., Hamilton, S. R., et al. (1988) Genetic alterations during colorectal-tumor development. *N. Engl. J. Med.* **319,** 525–532.
29. Joslyn, G., Carlson, M., Thliveris, A., et al. (1991) Identification of deletion mutations and three new genes at the familial polyposis locus. *Cell* **66,** 601–613.
30. Groden, J., Thliveris, A., Samowitz, W., et al. (1991) Identification and characterization of the familial adenomatous polyposis coli gene. *Cell* **66,** 589–600.
31. Kinzler, K. W., Nilbert, M. C., Su, L. K., et al. (1991) Identification of FAP locus genes from chromosome 5q21. *Science* **253,** 661–665.
32. Nishisho, I., Nakamura, Y., Miyoshi, Y., et al. (1991) Mutations of chromosome 5q21 genes in FAP and colorectal cancer patients. *Science* **253,** 665–669.
33. Miyoshi, Y., Nagase, H., Ando, H., et al. (1992) Somatic mutations of the APC gene in colorectal tumors: mutation cluster region in the APC gene. *Hum. Mol. Genet.* **1,** 229–233.
34. Powell, S. M., Zilz, N., Beazer-Barclay, Y., et al. (1992) APC mutations occur early during colorectal tumorigenesis. *Nature* **359,** 235–237.
35. Jen, J., Powell, S. M., Papadopoulos, N., et al. (1994) Molecular determinants of dysplasia in colorectal lesions. *Cancer Res.* **54,** 5523–5526.
36. Kinzler, K. W. and Vogelstein, B. (1998) Landscaping the cancer terrain. *Science* **280,** 1036–1037.
37. Smith, K. J., Levy, D. B., Maupin, P., Pollard, T. D., Vogelstein, B., and Kinzler, K. W. (1994) Wild-type but not mutant APC associates with the microtubule cytoskeleton. *Cancer Res.* **54,** 3672–3675.
38. Perucho, M. (1999) Correspondence re: C.R. Boland et al., A National Cancer Institute workshop on microsatellite instability for cancer detection and familial predisposition: development of international criteria for the determination of microsatellite instability in colorectal cancer. *Cancer Res.* **58,** 5248–5257, 1998. *Cancer Res.* **59,** 249–256.
39. Lengauer, C., Kinzler, K. W., and Vogelstein, B. (1997) DNA methylation and genetic instability in colorectal cancer cells. *Proc. Natl. Acad. Sci. USA* **94,** 2545–2550.
40. Breivik, J. and Gaudernack, G. (1999) Genomic instability, DNA methylation, and natural selection in colorectal carcinogenesis. *Semin. Cancer Biol.* **9,** 245–254.
41. Heyer, J., Yang, K., Lipkin, M., Edelmann, W., and Kucherlapati, R. (1999) Mouse models for colorectal cancer. *Oncogene* **18,** 5325–5333.
42. Parsons, R., Myeroff, L. L., Liu, B., et al. (1995) Microsatellite instability and mutations of the transforming growth factor beta type II receptor gene in colorectal cancer. *Cancer Res.* **55,** 5548–5550.
43. Markowitz, S., Wang, J., Myeroff, L., et al. (1995) Inactivation of the type II TGF-beta receptor in colon cancer cells with microsatellite instability. *Science* **268,** 1336–1338.
44. Rampino, N., Yamamoto, H., Ionov, Y., et al. (1997) Somatic frameshift mutations in the BAX gene in colon cancers of the microsatellite mutator phenotype. *Science* **275,** 967–969.
45. Yamamoto, H., Perez-Piteira, J., Yoshida, T., et al. (1999) Gastric cancers of the microsatellite mutator phenotype display characteristic genetic and clinical features. *Gastroenterology* **116,** 1348–1357.
46. Cavalli, G. and Paro, R. (1999) Epigenetic inheritance of active chromatin after removal of the main transactivator. *Science* **286,** 955–958.

47. Wolffe, A. P. and Matzke, M. A. (1999) Epigenetics: regulation through repression. *Science* **286**, 481–486.
48. McBurney, M. W. (1999) Gene silencing in the development of cancer. *Exp. Cell Res.* **248**, 25–29.
49. Arends, J. W. (2000) Molecular interactions in the vogelstein model of colorectal carcinoma. *J. Pathol.* **190**, 412–416.
50. Fearon, E. R. and Vogelstein, B. (1990) A genetic model of colorectal tumorigenesis. *Cell* **61**, 759–767.
51. Sreekumar, A., Nyati, M. K., Varambally, S., et al. (2001) Profiling of cancer cells using protein microarrays: discovery of novel radiation-regulated proteins. *Cancer Res.* **61**, 7585–7593.
52. Zhang, L., Zhou, W., Velculescu, V. E., et al. (1997) Gene expression profiles in normal and cancer cells. *Science* **276**, 1268–1272.
53. Velculescu, V. E., Zhang, L., Vogelstein, B., and Kinzler, K. W. (1995) Serial analysis of gene expression. *Science* **270**, 484–487.
54. Pogue-Geile, K., Geiser, J. R., Shu, M., et al. (1991) Ribosomal protein genes are overexpressed in colorectal cancer: isolation of a cDNA clone encoding the human S3 ribosomal protein. *Mol. Cell. Biol.* **11**, 3842–3849.
55. Notterman, D. A., Alon, U., Sierk, A. J., and Levine, A. J. (2001) Transcriptional gene expression profiles of colorectal adenoma, adenocarcinoma, and normal tissue examined by oligonucleotide arrays. *Cancer Res.* **61**, 3124.
56. Shapiro, S. D., Kobayashi, D. K., and Ley, T. J. (1993) Cloning and characterization of a unique elastolytic metalloproteinase produced by human alveolar macrophages. *J. Biol. Chem.* **268**, 23,824–23,829.
57. Whitham, S. E., Murphy, G., Angel, P., et al. (1986) Comparison of human stromelysin and collagenase by cloning and sequence analysis. *Biochem. J.* **240**, 913–916.
58. Owen, J. D., Strieter, R., Burdick, M., et al. (1997) Enhanced tumor-forming capacity for immortalized melanocytes expressing melanoma growth stimulatory activity/growth-regulated cytokine beta and gamma proteins. *Int. J. Cancer* **73**, 94–103.
59. Richmond, A., Balentien, E., Thomas, H. G., et al. (1988) Molecular characterization and chromosomal mapping of melanoma growth stimulatory activity, a growth factor structurally related to beta- thromboglobulin. *EMBO J.* **7**, 2025–2033.
60. Haghnegahdar, H., Du, J., Wang, D., et al. (2000) The tumorigenic and angiogenic effects of MGSA/GRO proteins in melanoma. *J. Leukoc Biol.* **67**, 53–62.
61. Tong, X., Drapkin, R., Yalamanchili, R., Mosialos, G., and Kieff, E. (1995) The Epstein-Barr virus nuclear protein 2 acidic domain forms a complex with a novel cellular coactivator that can interact with TFIIE. *Mol. Cell Biol.* **15**, 4735–4744.
62. Richardson, H. E., Stueland, C. S., Thomas, J., Russell, P., and Reed, S. I. (1990) Human cDNAs encoding homologs of the small p34Cdc28/Cdc2-associated protein of *Saccharomyces cerevisiae* and *Schizosaccharomyces pombe*. *Genes Dev.* **4**, 1332–1344.
63. Nagata, A., Igarashi, M., Jinno, S., Suto, K., and Okayama, H. (1991) An additional homolog of the fission yeast cdc25+ gene occurs in humans and is highly expressed in some cancer cells. *Nat. New Biol.* **3**, 959–968.
64. Drew, P. D., Nagle, J. W., Canning, R. D., Ozato, K., Biddison, W. E., and Becker, K. G. (1995) Cloning and expression analysis of a human cDNA homologous to *Xenopus* TFIIIA. *Gene* **159**, 215–218.
65. Arakawa, H., Nagase, H., Hayashi, N., et al. (1995) Molecular cloning, characterization, and chromosomal mapping of a novel human gene (GTF3A) that is highly homologous to Xenopus transcription factor IIIA. *Cytogenet. Cell Genet.* **70**, 235–238.
66. Perou, C. M., Jeffrey, S. S., van de Rijn, M., et al. (1999) Distinctive gene expression patterns in human mammary epithelial cells and breast cancers. *Proc. Natl. Acad. Sci. USA* **96**, 9212–9217.

67. Coulter-Karis, D. E. and Hershfield, M. S. (1989) Sequence of full length cDNA for human S-adenosylhomocysteine hydrolase. *Ann. Hum. Genet.* **53,** 169–175.
68. Cools, M. and De Clercq, E. (1990) Influence of S-adenosylhomocysteine hydrolase inhibitors on S-adenosylhomocysteine and S-adenosylmethionine pool levels in L929 cells. *Biochem. Pharmacol.* **40,** 2259–2264.
69. Herzfeld, A., Legg, M. A., and Greengard, O. (1978) Human colon tumors: enzymic and histological characteristics. *Cancer* **42,** 1280–1283.
70. Iwata, T., Popescu, N. C., Zimonjic, D. B., et al. (1995) Structural organization of the human sorbitol dehydrogenase gene (SORD). *Genomics* **26,** 55–62.
71. Momiyama, N., Koshikawa, N., Ishikawa, T., et al. (1998) Inhibitory effect of matrilysin antisense oligonucleotides on human colon cancer cell invasion in vitro. *Mol. Carcinog.* **22,** 57–63.
72. Itoh, F., Yamamoto, H., Hinoda, Y., and Imai, K. (1996) Enhanced secretion and activation of matrilysin during malignant conversion of human colorectal epithelium and its relationship with invasive potential of colon cancer cells. *Cancer* **77,** 1717–1721.
73. Murray, G. I., Duncan, M. E., O'Neil, P., Melvin, W. T., and Fothergill, J. E. (1996) Matrix metalloproteinase-1 is associated with poor prognosis in colorectal cancer. *Nat. Med.* **2,** 461–462.
74. Brown, L. F., Papadopoulos-Sergiou, A., Berse, B., et al. (1994) Osteopontin expression and distribution in human carcinomas. *Am. J. Pathol.* **145,** 610–623.
75. Skonier, J., Neubauer, M., Madisen, L., Bennett, K., Plowman, G. D., and Purchio, A. F. (1992) cDNA cloning and sequence analysis of beta ig-h3, a novel gene induced in a human adenocarcinoma cell line after treatment with transforming growth factor-beta. *DNA Cell Biol.* **11,** 511–522.
76. Agrawal, D., Chen, T., Irby, R., et al. (2002) Osteopontin identified as lead marker of colon cancer progression, using pooled sample expression profiling. *J. Natl. Cancer Inst.* **94,** 513–521.
77. Wiegand, R. C., Kato, J., Huang, M. D., Fok, K. F., Kachur, J. F., and Currie, M. G. (1992) Human guanylin: cDNA isolation, structure, and activity. *FEBS Lett.* **311,** 150–154.
78. Cohen, M. B., Hawkins, J. A., and Witte, D. P. (1998) Guanylin mRNA expression in human intestine and colorectal adenocarcinoma. *Lab. Invest.* **78,** 101–108.
79. Schweinfest, C. W., Henderson, K. W., Suster, S., Kondoh, N., and Papas, T. S. (1993) Identification of a colon mucosa gene that is down-regulated in colon adenomas and adenocarcinomas. *Proc. Natl. Acad. Sci. USA* **90,** 4166–4170.
80. Hogdall, C. K., Christiansen, M., Norgaard-Pedersen, B., Bentzen, S. M., Kronborg, O., and Clemmensen, I. (1995) Plasma tetranectin and colorectal cancer. *Eur. J. Cancer* **31A,** 888–894.
81. Hogdall, C. K. (1998) Human tetranectin: methodological and clinical studies. *APMIS Suppl* **86,** 1–31.
82. Bendik, I., Schraml, P., and Ludwig, C. U. (1998) Characterization of MAST9/Hevin, a SPARC-like protein, that is down-regulated in non-small cell lung cancer. *Cancer Res.* **58,** 626–629.
83. Nelson, P. S., Plymate, S. R., Wang, K., et al. (1998) Hevin, an antiadhesive extracellular matrix protein, is down-regulated in metastatic prostate adenocarcinoma. *Cancer Res.* **58,** 232–236.
84. Tanaka, K., Hinoda, Y., Takahashi, H., Sakamoto, H., Nakajima, Y., and Imai, K. (1997) Decreased expression of biliary glycoprotein in hepatocellular carcinomas. *Int. J. Cancer* **74,** 15–19.
85. Thompson, J., Seitz, M., Chastre, E., et al. (1997) Down-regulation of carcinoembryonic antigen family member 2 expression is an early event in colorectal tumorigenesis. *Cancer Res.* **57,** 1776–1784.

86. Getz, G., Levine, E., and Domany, E. (2000) Coupled two-way clustering analysis of gene microarray data. *Proc. Natl. Acad. Sci. USA* **97,** 12,079–12,084.
87. Getz, G., Gal, H., Kela, I., and Dan, E. D. Coupled two-way clustering analysis of breast cancer and colon cancer gene experiments. Bioinformatics, in press.
88. Lin, Y. M., Furukawa, Y., Tsunoda, T., Yue, C. T., Yang, K. C., and Nakamura, Y. (2002) Molecular diagnosis of colorectal tumors by expression profiles of 50 genes expressed differentially in adenomas and carcinomas. *Oncogene* **21,** 4120–4128.
89. Shibata, D. and Aaltonen, L. A. (2001) Genetic predisposition and somatic diversification in tumor development and progression. *Adv. Cancer Res.* **80,** 83–114.
90. Huang, J., Qi, R., Quackenbush, J., Dauway, E., Lazaridis, E., and Yeatman, T. (2001) Effects of ischemia on gene expression. *J. Surg. Res.* **99,** 222–227.

10
Gene Expression Analysis of Prostate Carcinoma

William L. Gerald

INTRODUCTION AND CLINICAL ISSUES

Carcinoma of the prostate is the most common noncutaneous cancer of men in the U.S. and is expected to affect approx 198,000 individuals in 2001 *(1)*. It is estimated that more than 1 million men, over the age of 50 yr alive today, will die of this disease. The incidence of prostate cancer has increased sharply in the last decade, as serum prostate-specific antigen (PSA) testing has become widely available. This is primarily due to detection of clinically inapparent and early stage disease. Many early stage prostate cancers are relatively indolent, such that older men with disease often die of other causes. For example, the estimated lifetime risk of a man developing prostate cancer is 16%, however the risk of dying from the disease is about 3.4% *(2)*. In early stages, most prostate cancers are curable with local therapy, either surgery or radiation. On the other hand, more extensively invasive tumors and metastatic disease are much more aggressive and in many cases lethal. A critical challenge is to develop means to distinguish indolent cancers from those that are potentially lethal, so that therapeutic procedures can be tailored to an individual patient.

Androgens play a primary role in development and progression of prostate cancer, and surgical or medical androgen ablation is a mainstay in therapy. Response to hormonal therapy is variable and unpredictable. Virtually all patients so treated eventually develop androgen-independent prostate cancer. There is no therapy proven to cure hormone refractory prostate cancer, and most patients will eventually die from their disease *(3)*. The androgen-responsive biochemical pathways that drive prostate cancer and mechanisms of tumor resistance to androgen ablation therapy are unknown. A better understanding of androgen-independent tumors and more effective therapies are needed.

New high-throughput methods, such as comprehensive gene expression analysis, have potential to provide rapid advancement in characterization of neoplastic disease. This chapter summarizes gene expression studies relevant to prostate cancer, with a focus on attempts to identify diagnostic and prognostic markers, characterize the androgen response pathway, define mechanisms of androgen-independence, and identify new therapeutic targets. It is clear that even in these early days of expression profiling, there have been significant advances in techniques and methods of analysis,

From: *Expression Profiling of Human Tumors: Diagnostic and Research Applications*
Edited by: Marc Ladanyi and William L. Gerald © Humana Press Inc., Totowa, NJ

which suggest that these studies will further our understanding of the clinical biology of prostate cancer.

HISTOGENESIS, PATHOLOGY, AND EPIDEMIOLOGY OF PROSTATE CANCER

The vast majority of malignant tumors of the prostate are epithelial and termed adenocarcinomas. The prostate normally has several types of epithelial cells. Basal cells are located between the luminal cells and the basement membrane and form a continuous layer in the non-neoplastic gland. This cell layer may also contain a stem cell compartment that differentiates into luminal cells. Neuroendocrine cells are androgen-independent cells dispersed throughout the basal layer and are believed to provide paracrine signals that support the growth and function of luminal cells. The luminal cells are androgen-dependent and produce prostatic secretory proteins. Prostate adenocarcinomas have features of both basal and luminal cells, raising controversy as to the cell of origin (4). A likely possibility is that most cancers are derived from the ill-defined stem cell compartment.

The prostate develops through budding of epithelium from the urogenital sinus into the surrounding mesenchyme. Prostate formation is, therefore, a consequence of epithelial–mesenchymal interactions. Likewise, it is believed that the development of prostate carcinoma is associated with aberrant epithelial–stromal interactions. Specifically, aberrant growth factor signaling from stromal tissues may play an integral role in cancer progression (5,6). These paracrine mechanisms have not been well-defined, however they may be critically important in the early stages of development and play a significant role in providing a proper microenvironment for metastases (5).

The diagnosis of prostate cancer relies on histomorphologic assessment. Tumors evolve through a series of stages with increasing degrees of cytologic and architectural changes. Atypical and dysplastic luminal cell cytology, resembling that of cancer but associated with an intact basal cell layer, are referred to as prostatic intraepithelial neoplasia (PIN) (7). PIN is considered a precursor to invasive cancer based on epidemiologic, phenotypic, and molecular data (8). PIN is associated with abnormalities of molecular phenotype and genotype intermediate between normal epithelium and cancer. Invasive carcinoma lacks a basal cell layer.

Androgens play a primary role in normal prostate function and in the development and progression of prostate cancer (9–11). Prostate growth and maintenance of its structural and functional integrity require an adequate level of circulating androgen. Androgenic hormones bind to and activate the intracellular androgen receptor, which regulates specific gene expression. It is the expression of androgen-responsive genes that determines the balance between cell proliferation, cell death, and the differentiation of a normal prostatic epithelial cell. Prostate cancer cells are also typically androgen-dependent, and androgen ablation is the standard systemic therapy for this disease. Androgen deprivation induces programmed cell death in androgen-dependent normal, hyperplastic, preneoplastic, and malignant prostatic epithelial cells. Ablation can be achieved by surgical removal of the testes or chemically with gonadotropin-releasing hormone analogues, exogenous estrogens, progestational agents, anti-androgens, or adrenal enzyme synthesis inhibitors, such as ketoconazole and aminoglutethimide. Virtually all patients so treated respond, but eventually develop so-called androgen-independent prostate cancer, which is a serious clinical problem for which no consistently effective therapy exists. Time to therapy-resistant progression is variable and ranges between 1 and 3 yr.

Aging is the most significant risk factor for prostate cancer *(4)*. Clinically evident disease is usually not manifest prior to the age of 60 yr, although precursor lesions and low stage disease are detected much earlier. The incidence of prostate cancer is significantly higher in the U.S. than in many other countries, particularly Asian countries. Dietary and environmental factors are, therefore, presumed to play a significant but poorly understood role *(12)*. African-American men are more likely to develop and die from prostate cancer. Although heredity accounts for only a small percentage of prostate cancer, evidence for genetic susceptibility to prostate cancer is supported by epidemiological, twin studies, and segregation analyses *(13)*. Chromosomal regions that have been implicated to contain susceptibility loci include: 1q24-25, 1q42-43, Xq27-28, 1p36, 20q13, 17p11, 8p22-23, and 1p13. No single predisposition locus is by itself considered responsible for a large proportion of familial cancer. In addition, none of the susceptibility genes have been definitely identified.

EXPERIMENTAL BIOLOGY OF PROSTATE CANCER

Molecular Genetics

Several excellent reviews of the molecular biology of prostate cancer are available, and only a few of the more common molecular alterations and their potential significance are highlighted here *(4,12)*. The analysis of chromosomal alterations in cancer has identified many changes reflecting loss or gain of function of particular genes. Consistent allelic loss is expected to reflect the location of putative tumor suppressor genes. Loss of heterozygosity at chromosome arms 8p, 10q, 13q, and 17p are frequent events in prostate cancer, and losses at 6q, 7q, 16q, and 18q also occur. Gains of genetic material are expected to reflect the location of oncogenes. In prostate cancer, gains at 8q and 7 are fairly common. Individual genes at these loci have not been definitely assigned a role in prostate cancer, but several reasonable candidate genes have been proposed based on their location and functional properties.

One of the more common events in early prostate cancer development is loss of 8p *(14,15)*. Losses have been mapped to at least two regions: 8p12-21 and 8p22. Losses at 8p12-21 occur early in tumor development, and a candidate gene, *NKX3A*, maps to this region. *NKX3A* is highly expressed in the prostate, and disruption of this homeobox gene in mice leads to defects in prostate development and PIN-like lesions *(16,17)*. However, the gene is expressed in prostate cancers, and mutations of *NKX3A* have not been detected in human tumors.

Loss of 10q is frequent in prostate cancer and apparently more common in advanced disease. The *PTEN* gene at 10q23 encodes a lipid phosphatase, whose main substrate is phosphotidylinositol 3,4,5 triphosphate. Loss of *PTEN* leads to activation of *AKT* kinase activity and decreased sensitivity to cell death, but may also play a role in other cellular activities associated with neoplastic disease. *PTEN* mutations are associated with several autosomal dominant disorders, which include increased susceptibility to tumors. In addition, *PTEN* is mutated in several prostate cell lines and is homozygously deleted in about 10% of primary tumors *(18)*. Mutations, however, appear to be very rare in human prostate cancers. *PTEN* heterozygous mice develop dysplasia and carcinoma of several types, including prostatic epithelial dysplasias *(19)*. *MXI1* is another candidate tumor suppressor at 10q25 *(20)*. *MXI1* is a MYC-binding protein, and MYC overexpression is implicated in prostate cancer through amplification of 8q. Mutations of *MXI1* are apparently rare, however, *MXI1* mutant mice have been reported to develop prostatic hyperplasia and dysplasia *(21)*.

As in many other tumor systems, alterations in molecules that regulate cell cycle and apoptosis are likely to play a role in progression of prostate cancer. Loss of 17p occurs in advanced stages of prostate cancer, and deletions include the ubiquitous tumor suppressor gene *TP53*. Mutations of *TP53* are found in several prostate cancer cell lines, but occur infrequently in early stage disease *(22)*. They appear to be more common in advanced stages. Overexpression of *BCL2* occurs in some prostate cancers and is believed to reduce cell death. *BCL2* expression seems to be a relatively late event in prostate cancer and may be associated with progression to androgen-independent growth and resistance to chemotherapy *(23)*. Loss of activity of the CDK4 inhibitor, p27, and deletion of the locus at 12p12-13.1 occurs in prostate cancer, but the *CDKN1B* gene is rarely mutated. Instead, loss of activity occurs as a result of loss of expression, aberrant phosphorylation, and altered degradation. In the prostate, several studies have demonstrated that loss of p27 protein levels correlates with tumor grade and may provide a prognostic marker *(24)*.

Molecular Aspects of the Androgen Signaling Pathway and Androgen-Independent Prostate Carcinoma

The androgen signaling pathway plays a critical role in prostate cancer development and progression *(25)*. Ligand binding to androgen receptor results in conformational changes, homodimerization, dissociation from heat-shock proteins, phosphorylation, and subsequent interactions with specific androgen response elements in the promoter of androgen-target genes. These effector genes are expected to regulate important cellular processes. Transcriptional activity is dependent on ligand binding, release of repressors, and recruitment of coactivators. These critical interactions and multisubunit complexes are likely to be responsible for the diversity of androgen-regulated functions and cell context specificity. It is also likely that other growth factor-mediated signal transduction pathways influence androgen signaling.

Androgen ablation is a mainstay of therapy for high stage disease, but virtually all prostate cancers treated this way eventually become resistant to therapy. The critical molecular mechanisms by which prostate cancer cells become androgen-independent are largely unknown, but several possibilities are considered likely. A variety of genetic alterations are associated with androgen independence and include loss of heterozygosity at multiple sites of chromosome 8, amplifications of the long arm 8q, gains in Xp11-q13, and an increase in 7p and 5q *(4)*. Overexpression or mutation of the androgen receptor occurs in 20–30% of androgen-independent prostate tumors. Some mutant androgen receptors continue to respond to androgens, but are also stimulated by other androgen homologs *(26)*. An alternative model for androgen independence suggests that recruitment of nonsteroid receptor signal transduction pathways can activate the androgen pathway in the setting of clinical androgen deprivation *(27)*. The androgen receptor can be activated in a ligand-independent manner by insulin-like growth factor

(IGF)-1, epidermal growth factor (EGF), and keratinocyte growth factor (KGF), but the mechanistic details are not clear. In addition, the overexpression of the human epidermal growth factor receptor (HER)-2/neu receptor tyrosine kinase may be another mechanism of androgen-independent growth. Blocking the programmed cell death normally induced by androgen ablation may be one of the mechanisms for cells to become androgen-independent. Bcl-2 protein is undetectable in most androgen-dependent prostate cancers, but is expressed at high levels in some hormone-independent cancers. Expression of this gene inhibits the death rate of cancer cells. Further molecular studies may help define the biological basis of hormone-resistant disease and assist in identifying those patients likely to respond to sequential endocrine manipulations vs those who should receive alternative therapy.

Prostate Cancer Models

Prostate cancer research has been hampered by difficulties in generating permanent cell lines and xenografts for in vitro studies. Only a handful of lines are readily available (LNCaP, PC3, DU145, CWR22) and were isolated from metastatic tumors. Therefore, many studies are based on a small repertoire of cell lines of uncertain relevance to primary disease. An interesting spectrum of xenografts have been developed in recent years and, despite some limitations, provide very useful tools for study. The development of corresponding cell lines will further facilitate in vitro experimentation *(28)*. Several laboratories have undertaken the generation and use of genetic models of prostate cancer. One of the more interesting transgenic models results from the use of a prostate-specific promoter with a potent oncogene. The transgenic adenocarcinoma mouse prostate (TRAMP) carries a prostate-specific probasin promoter expressing simian virus 40 (SV40) large T and small t antigens *(29,30)*. These animals develop PIN and prostate carcinomas within 12 wk, which progress to metastasis in 30 wk. Several variants of this model have also been developed.

The *NKX3A* and *PTEN* knock-out mice as mentioned above are some of the few mutant mice to express a prostatic neoplasia phenotype. The *CDKN1B* knock-out mouse develops prostatic hyperplasia. Various crosses of these lines are being evaluated in an effort to model multistep tumor progression. For example, a cross resulting in *PTEN* heterozygous and *CDKN1B* homozygous loss contributes to progression of the neoplastic phenotype *(31)*. In addition, crosses between *NKX3A* and *PTEN* knockouts act synergistically in cancer progression *(17)*.

GENE EXPRESSION ANALYSIS

Technical Aspects

Given the limitations in our present understanding of the basic biology of prostate cancer, many investigators have turned to the use of high-throughput gene expression studies to provide a more complete characterization of this disease. Most commonly, these efforts have been designed to identify genes that participate in the process of prostate cancer development, progression, and androgen independence, or to identify genes that may serve as clinically useful markers for diagnosis or prognosis. These studies have many inherent technical and analytical challenges, however initial efforts are providing reason for optimism that these challenges can be met.

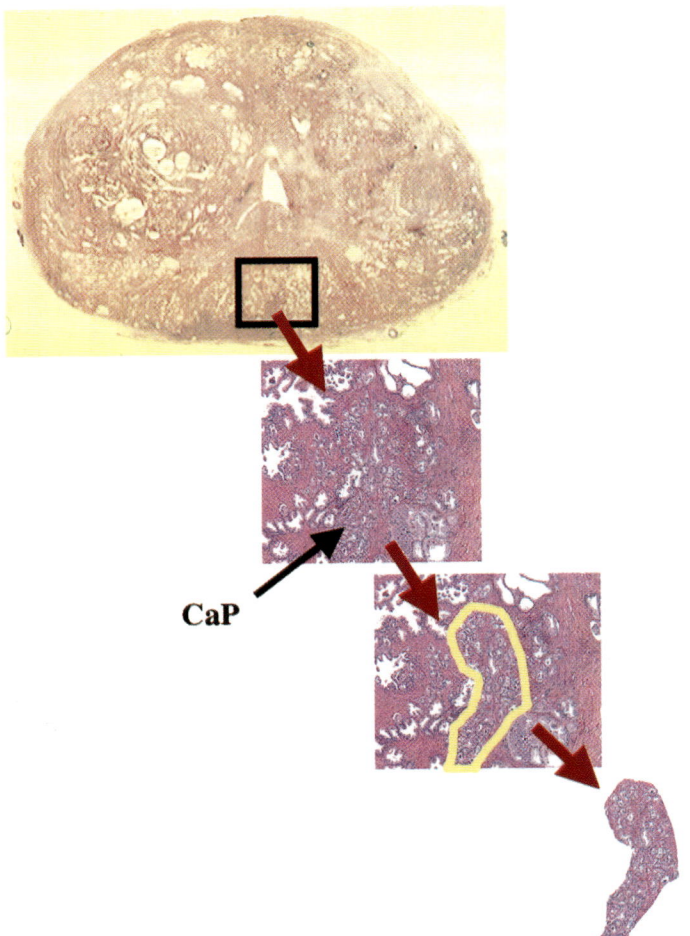

Fig. 1. Manual microdissection of prostate cancer to enrich for tumor cells. Photomicrographs depicting the complex histologic heterogeneity of prostate cancers and steps to manually dissect and enrich for tumor cells in samples for gene expression analysis. CaP designates prostate cancer.

The Impact of Tissue Heterogeneity on Gene Expression Analysis

Prostate cancers are heterogeneous and often multifocal. They infiltrate the gland, are difficult to identify grossly, and are intimately admixed with non-neoplastic epithelium and stroma (Fig. 1). Tumor cells are often associated with reactive stromal and inflammatory tissues. Even the neoplastic cells are heterogeneous and include varying patterns of *in situ* and invasive cancer. The histologic variability is so pervasive as to have been incorporated into the most commonly used grading system *(32)*. In addition, tumors that are distinctly located in the prostate may represent independent foci of neoplasia or intra-organ spread. Several studies have shown that topographically distinct lesions are often genetically and molecularly distinct *(33,34)*. This molecular and anatomic heterogeneity makes it difficult to obtain sufficient homogeneous material for study. Most studies to date have used some form of tissue dissection to enrich for cells of interest, however, even microdissection techniques that rely on morphologic assessment may not be adequate to address all these issues.

The interpretation of observed gene expression obtained from tissue samples is, therefore, dependent on understanding the complexity of the cellular content. A major challenge is identification of the cell types within a sample that contribute to the expression status of a specific gene. In heterogeneous samples, gene expression corresponding to very small, but biologically significant, components of tumors may not be evident using current methods. In addition, intersample comparisons can be limited by their lack of homogeneity. Few reports have addressed these issues, but the magnitude of the problem is illustrated in recent studies. Sequence analysis of cDNA libraries constructed from either cell sorted non-neoplastic luminal or basal epithelial cells demonstrated strong divergence in expressed sequences between these two prostate epithelial subtypes, even with relatively small numbers of sequences *(35)*. Sequences not detected in prior expression analysis of bulk prostate tissue were readily apparent in the purified individual cell types. This indicates that low to moderately abundant, but specific, transcripts from individual cell types, which make up only a small proportion of any sample, are still in the minority requiring very deep sequencing and sensitive hybridization techniques for detection. These cell-specific differences in gene expression are graphically evident by comparison of profiles obtained with samples enriched for either prostatic stroma, prostatic epithelium, or unselected bulk prostate (Fig. 2) *(36)*. It is readily apparent that the cellular composition of samples has a dramatic impact on the expression profile and must be accounted for in the analysis of human tumors.

Transcript Profiling Methods

Several methods have been used for high-throughput gene expression analysis in prostate carcinoma. Some rely on sequence analysis of cloned transcripts (differential display, subtractive hybridization, serial analysis of gene expression [SAGE], and sequencing of expressed sequence tags [ESTs]), while others use microarrayed sequence probes for quantitative hybridization studies. SAGE and ESTs require the conversion of mRNA to cDNA that is cloned and sequenced. Sequences are categorized and enumerated to provide readout of tag numbers per gene. These techniques can be very quantitative and precise, but are dependent on the depth of sequencing, accurate sequencing, and authentic mapping of sequence to gene. Low sequence tag counts are not very reliable.

Microarray-based hybridization studies involve labeling of transcript representations from cells and hybridization of the labeled target to individual sequence probes attached to a solid support. The bound label is proportional to the quantity of specific transcripts in the original RNA mixture. The probe arrays are miniaturized, such that thousands of individual probes can be attached to a very small substrate, providing high-throughput and experimental efficiency. This technique allows for the efficient analysis of virtually unlimited numbers of genes. The accuracy of measurements is dependent on specificity and sensitivity of individual probes. Only a relatively small number of probe sets on most arrays currently in use have been evaluated for these parameters, and accuracy

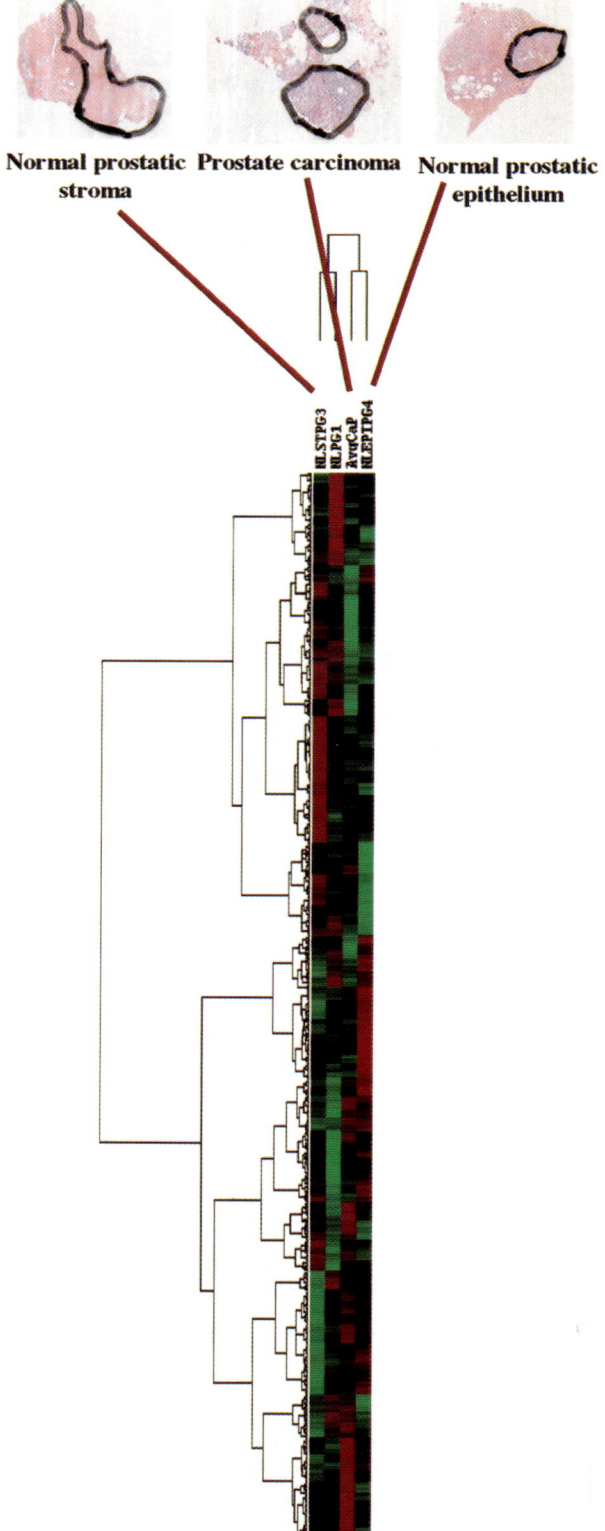

Fig. 2.

Data Analysis

The analysis of expression data is of primary importance and is thoroughly described in Chapter 6 in this text. A few critical issues are highlighted here. After the primary data has been evaluated to remove background and artifact, expression values for each gene can be determined. It should be kept in mind that most of these values have not been validated and are simply a measure of the signal obtained in a single experiment. Intra- and interexperiment reproducibility can often be very helpful in acceptance of signal intensity as a reflection of the true transcript level. Most studies are designed to identify genes that participate in a critical process or to classify samples into previously unrecognized biologically or clinically important subsets. The goal is to reduce the expansive datasets and identify the critically important genes. This can be accomplished using statistical or metric thresholds to identify genes whose expression varies significantly between samples of interest. A number of methods can then be used to identify samples or genes with the desired properties. This includes unsupervised algorithms, which search the data with few user-imposed restrictions in an effort to recognize molecular substructure and identify previously unrecognized classes of genes or samples, and supervised methods, which apply prior knowledge, such as histology, phenotype, stage or outcome, to the analysis. However, because of the large volume of data and relatively small number of samples, some associations are likely due to chance. It is imperative that the significance of correlations be established. The best approach is to test associations in independent sample sets, however, because studies frequently have small numbers of samples, permutation tests are often used as an alternative.

An important aspect of many techniques used in gene expression analysis is the capability to produce a visual display of the results, so that investigators are able to digest the large volume of data in an intuitive manner. It is often true that statistical significance does not always reflect biological significance, and low level but critical gene expression changes may often be lost with simply a mathematical approach. Visualization of the data facilitates the incorporation of investigator knowledge (and potentially bias) into interpretation of the data. Finally, the availability of the dataset allows continued analysis with different or new approaches, and public access to maximize use should be encouraged.

Fig. 2. *(previous page)* Differential gene expression in microdissected neoplastic and non-neoplastic prostate. Illustrated is a diagrammatic representation of expression for a cluster of genes that are differentially overexpressed in prostate carcinoma. An average linkage hierarchical clustering algorithm was used to group genes and samples based on similarity. Red represents overexpression, and green represents underexpression, relative to the scaled mean level of expression. NLEPIPG4, RNA from microdissected non-neoplastic prostatic epithelium as target; NLSTPG3, RNA from microdissected non-neoplastic prostatic stroma as target; NLPG1, RNA from bulk non-neoplastic prostate; Avg CaP, mean value of gene expression for RNA from 14 microdissected prostate carcinomas. The corresponding histologic section and design of microdissection is shown.

RESULTS OF GENE EXPRESSION ANALYSIS OF PROSTATE CANCER

The Prostate Transcriptome

The subset of genes expressed in normal and neoplastic prostate tissue has been estimated based on theoretical and practical considerations. These studies are beginning to define the prostate cancer transcriptome and a possible role for some genes in critical processes. The Cancer Genome Anatomy Project (CGAP) of the National Cancer Institute was designed to identify genes responsible for cancer by sequencing of cDNA clones representing RNA from normal, precancerous, and malignant cells (http://cgap.nci.nih.gov/). Currently, there are 17,040 genes identified in the 139,041 ESTs from prostate libraries listed in CGAP. These resources have been combined with independent data to identify 15,953 prostate-specific EST clusters in the prostate expression database (http://www.pedb.org/) *(37)*. This is believed to represent about 50–75% of the prostate transcriptome. An additional estimate, based on combining publicly available SAGE data, predicted 37,000 genes in the prostate transcriptome *(38)*. These estimates can be supplemented with microarray-based studies of gene expression. Using the comprehensive set of U95 oligonucleotide arrays (Affymetrix, Santa Clara, CA, USA) with 63,175 probe sets for gene/EST clusters from the unigene build 95, Latulippe et al. detected expression of genes corresponding to 5992 probe sets (absolute call of present based on Microarray Suite 4.0 with default settings) in all 23 primary prostate carcinomas studied, 34,518 probe sets detected gene expression in at least some cases, and 22,665 did not reliably detect expression (absolute call of absent) in any of these samples *(36)*. It is likely that these data represent all of the abundant and moderately expressed genes in the prostate, although low abundance genes, alternative forms, and those expressed only under special circumstances may not be well represented. The identification of a complete prostate transcriptome will define the lower limit of probes needed for prostate-specific cDNA arrays and provide insights into tissue-specific molecular characteristics and tissue-specific regulation of transcription.

Gene Expression Analysis of Neoplastic Transformation of the Prostate

A number of studies have focused on the identification of genes that are differentially expressed in the process of neoplastic transformation of prostatic epithelium. Differentially expressed genes are expected to represent molecular events that contribute to prostate cancer development and may serve as clinically useful markers for diagnosis. SAGE studies comparing 133,217 sequence tags representing 19,287 genes from pools of four prostate carcinomas and adjacent non-neoplastic prostate, identified 156 differentially expressed genes at a $p < 0.05$ (88 up-regulated and 68 down-regulated) *(38)*. Eighty-eight were changed at least fivefold in frequency. Among the most highly differentially expressed were *PRP4*, *DVL1*, *IFITM1*, and *PIN1*. Immunohistochemistry was used to validate expression of selected genes and demonstrated that some were derived from epithelial cells and others from stroma.

Several groups have compared expression differences between non-neoplastic prostate tissue and prostate cancer based on cDNA arrays. Elek et al. used a 588 gene commercial cDNA array and compared three prostate carcinoma samples to a single normal *(39)*. Semiquantitative autoradiography detected 19 differentially expressed genes (15 down-regulated and 4 up-regulated). One gene, *GSTT1*, was validated by reverse transcription polymerase chain reaction (RT-PCR) and found to also show low level of expression in some other tumor types. Using the same commercial array, Chaib et al. compared nonneoplastic tissue and cancer from a single patient. Several methods of analysis were evaluated *(40)*. Under the most stringent conditions, 5 genes were down-regulated and 10 were up-regulated. Relative changes in expression values were confirmed by RT-PCR and Northern blot, however, the various techniques were not quantitatively identical. There was little overlap in the genes identified as differentially expressed from these two studies, possibly due to the small number of samples. Using a 9984 gene cDNA array, Dhanasekaran et al. compared 24 benign prostate samples and 36 prostate carcinomas using two different reference RNA samples produced from non-neoplastic prostate tissue *(41)*. A scatter plot of the 200 highest ranked differentially expressed genes based on *t*-statistic and 200 highest ranked differentially expressed genes based on magnitude of the difference was used to identify genes highly ranked by both methods. Hepsin (*HPN*), *PIM1*, *LIM*, *TIMP2*, *HEVIN*, *RIG*, and *THBS*, among others, met these criteria. Luo et al. used a 6500 gene array with a reference composed of pooled RNA from two examples of benign prostatic hyperplasia (BPH) and compared expression between 16 prostate cancers and 9 BPH *(42)*. They used a weighted gene analysis to identify 210 significantly differentially expressed genes of the 3215 that were reliable measured (6.5%). Using these genes for classification by multidimensional scaling, they were able to reliably separate prostate cancers from BPH. *HPN* was the gene most highly ranked by this approach, and expression of this gene was validated by RT-PCR.

Similar studies have been carried out using oligonucleotide arrays (Affymetrix). Magee et al. studied 11 malignant and 4 benign samples on oligonucleotide arrays with probe sets for 7068 human transcripts *(43)*. They evaluated expression corresponding to 4712 probe sets that were scored present in at least one sample and used stringent criteria of at least threefold change in all 11 malignant samples compared to all 4 benign samples to identify differentially expressed genes. Only *HPN* was shown to be up-regulated, and no down-regulated genes were detected. Using relaxed criteria (threefold difference in expression values in 9 of 11 tumors compared to all 4 benign samples), they identified three additional genes: *HTR2B* and *CDK10* were both up-regulated, and *PGM5* was down-regulated. Welsh et al. used oligonucleotide arrays representing 8920 different genes to analyze 24 prostate cancers, 9 nonmalignant prostate tissues, and 21 cell lines *(44)*. Three thousand five hundred and thirty genes, which varied most between individual samples, were used for cluster analysis. Cell lines had distinct expression profiles from the tissue samples and, furthermore, malignant and nonmalignant samples were distinct. Using a metric that incorporated rank of each gene according to *t*-test, ratio, and absolute difference, they identified 20 highly ranked genes with distinctly different average expression in non-neoplastic and malignant tissues with nonoverlapping ranges. Among this group were *GA733-2*, *TACSTD1*, *FASN*, *MIC-1*, and *HPN*. Luo et al. used two different oligonucleotide arrays (Affymetrix Hu35k and U95A) to study 15 cancers and 15 non-neoplastic prostate samples *(45)*. They identified 84 genes that were differentially expressed at least twofold, with a *t*-test p value <0.05 and a minimum expression level of 500 in the group with highest expression. They examined 25 cDNAs by semiquantitative RT-PCR and found that about

80% had expression levels that were relatively the same as the chip result. They then used the same filtering method to identify 12 genes that were differentially expressed between tumors with capsular penetration and a group of normal and organ-confined tumors. *In situ* hybridization confirmed tumor cell expression for some genes. U95A oligonucleotide arrays with 12,600 probe sets were also used in experiments to determine differential gene expression in 52 prostate cancers compared to 50 nontumor samples *(46)*. Three hundred seventeen genes were up-regulated, and 139 were down-regulated in tumor samples with an adjusted *p* value of <0.001. K-nearest neighbor supervised machine learning was used to build predictors for classifying samples using small sets of genes. A model using 16 genes was 86% accurate using an independent data set. In a somewhat different approach, Stamey et al. compared nine high-grade prostate cancers (Gleason grade 4 or 5) to eight samples of prostatic hyperplasia using oligonucleotide arrays for 6800 genes *(47)*. Differentially expressed genes that were present in all cases and differentially expressed between the two groups with a *p* value of <0.0005 by student's *t*-test were selected. This resulted in 22 genes that were up-regulated and 64 that were down-regulated. These genes included many different functional categories. *HPN*, *SLC14A1*, *CYP3A7*, and prostate specific membrane antigen (*PSMA*) were included in the list.

Although some genes have been repeatedly detected as differentially expressed during prostate cancer development (e.g., *HPN*), it is interesting that the results of these many studies are in only partial agreement. It is likely that the variety of methods, samples, and experimental platforms contribute to the disparities.

Correlation of Gene Expression with Histologic Grade of Prostate Cancer

A critical issue in the care of prostate cancer patients is prognostically useful clinical classification of early stage disease. At this time, few studies have been performed with sufficient numbers of samples to establish strong clinical correlations, but some early results are intriguing. Welsh used a subgroup of 788 tumor-specific genes in cluster analysis and identified a tumor sample dichotomy, primarily resulting from differential expression of a group of ribosomal genes *(44)*. This division was statistically associated with Gleason's score. Analysis of variance was used to identify 20 genes that were differentially expressed between tumors subdivided into three classes of Gleason's score (5 or 6, 7, 8 or 9). Insulin-like growth factor binding protein (*IGFBP*)2 and *IGFBP5* were found to be highly expressed in high grade tumors. A similar finding was made by Singh et al. using oligonucleotide microarrays *(46)*. This group evaluated a large number of clinical variables for association with gene expression profiles. Although no expression correlates for age, serum PSA and local invasion were identified, there was a set of 29 genes that correlated with histologic grade (Gleason's score). The list did not include the two genes identified by Welsh et al. *(44)*. This same group developed a k-nearest neighbor approach to produce models using five genes to predict recurrence following prostatectomy. Several genes that were most commonly used for prediction were *ITPR3*, *SIAT1*, platelet-derived growth factor receptor-β (*PDGFR*β), and *CHRA*. Although these preliminary attempts require refinement with larger datasets, their success does support the concept that gene expression profiles may provide useful outcome measures.

Gene Analysis of Metastatic Progression of Prostate Carcinoma

Tumor metastasis is the most clinically significant event in prostate cancer patients. Development of metastases requires that cells from primary tumors detach, invade stromal tissue, and penetrate vessels by which they disseminate. They must then survive in the circulation to reach a secondary site. To form clinically significant tumors, metastatic cells must proliferate in the new microenvironment and recruit a blood supply. Those tumor cells growing at metastatic sites are then continually selected for growth advantage. This is a complex and dynamic process, which is expected to involve alterations in many genes and transcriptional programs.

Samples of metastatic prostate cancer are not abundant, and only a few studies have attempted to identify gene expression changes associated with this aggressive form of the disease. Hierarchical clustering of cDNA array data was used to identify coordinately expressed groups of genes that were specifically and differentially expressed in metastatic vs primary prostate cancer *(41)*. *IGFBP5*, *DAN1*, *FAT*, and *RAB5A* were some of the genes down-regulated in metastasis, and *MTA-1*, *MYBL2*, and *FLS353* were up-regulated. Magee et al. used oligonucleotide arrays and identified three genes that were down-regulated in all three metastases compared to all eight primary tumors (*GNA15, PDXK, DGKA*) *(43)*. With relaxed criteria, they found an additional six down-regulated and two up-regulated. LaTulippe et al. used oligonucleotide microarrays with 63,175 probe sets to analyze differential gene expression between 14 primary tumors from patients without recurrence and 9 metastatic prostate cancers *(36)*. An unsupervised analysis revealed a strong tendency for primary and metastatic tumors to have distinct expression profiles based on an average linkage hierarchical clustering algorithm (Fig. 3). As expected, some of the gene expression differences, which distinguished primary and metastatic tumors, were contributed by the small amount of contaminating non-neoplastic prostate tissue present in the primary tumor samples (Fig. 3, bottom panel). Three hundred ninety-one of the 63,175 probe sets detected tumor related differential gene expression of at least threefold with a student's t-test p value <0.001. These genes are expected to reflect the phenotype of these two cohorts and provide insight into the biology of prostate cancer progression. In keeping with this concept, 26 of the 100 most highly ranked characterized genes are believed to play a role in some aspect of cell cycle regulation, DNA replication and repair, or mitosis, including many genes that are known to be up-regulated in highly proliferative cells such as *RFC5*, *TOP2A*, *RFC4*, and *MAD2L1* (Table 1). This finding correlated with the increased proliferation index of metastatic tumors. Fifteen of the 100 highly ranked genes correspond to products potentially involved in signaling and signal transduction, and nine others may contribute to cell adhesion, cell migration, or extracellular matrix. These include hyaluronan-mediated motility receptor (*HMMR*), which encodes an extracellular matrix binding protein believed to play a role in cell motility through the Ras-ERK signaling pathway, and inositol polyphosphate 4-phosphatase type I (*INPP4A*), the substrates of which are intermediates in pathways regulated through the *AKT* proto-oncogene. A large proportion of the highly ranked differentially expressed genes are believed to be involved with the regulation of gene expression and gene product function. These findings suggest that the development and progression of prostate cancer metastases are associated with many gene expression changes related to

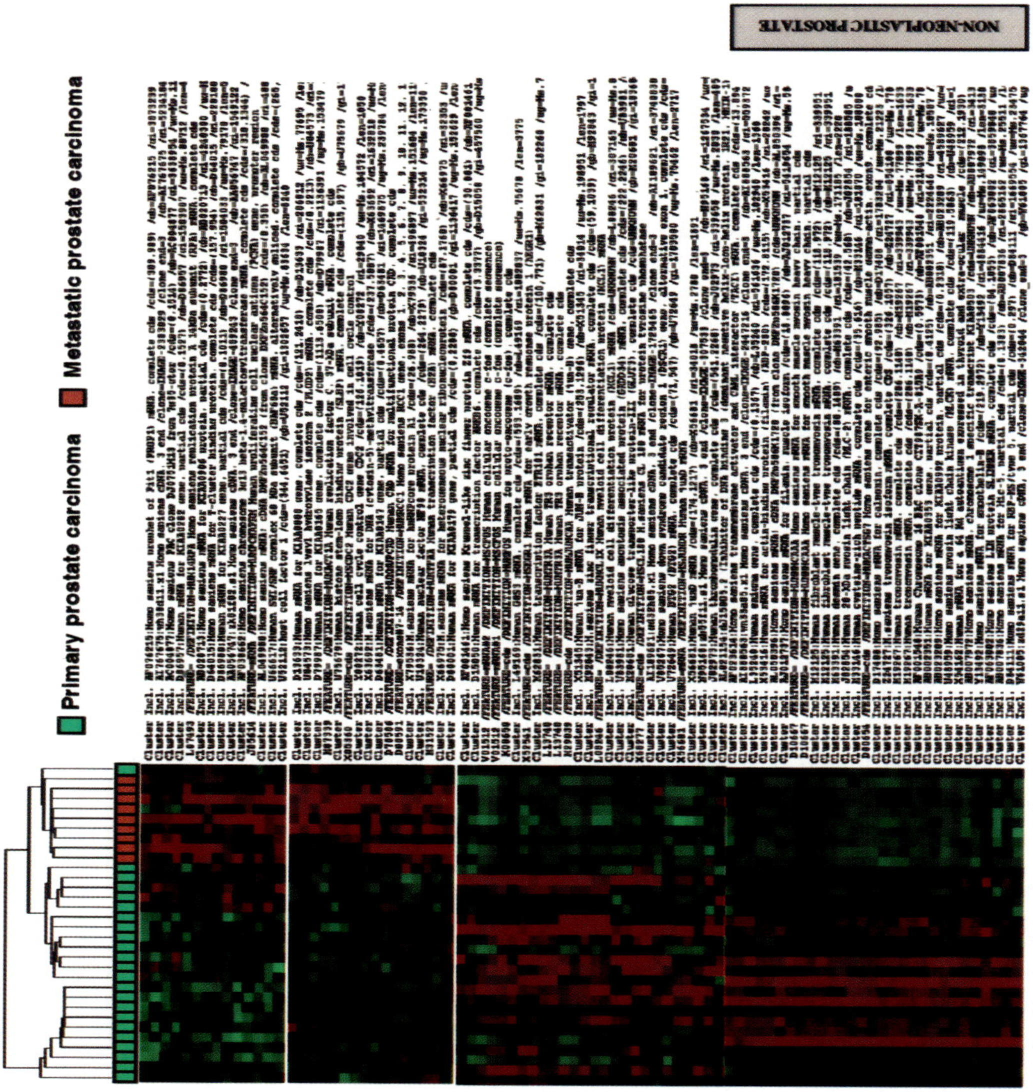

Fig. 3.

cell proliferation, interactions with the microenvironment, properties that might contribute to cell motility, activated signal transduction pathways, and regulation of gene product synthesis and function. Identification of genes, gene expression profiles, and biological pathways, which contribute to metastasis, may be of significant benefit to improved tumor classification and therapy.

Expression Analysis of the Androgen Response Pathway and Androgen-Independent Prostate Cancer

The prostate is dependent on androgens for normal development and function, and androgens play a role in prostate hyperplasia and neoplasia. The androgen response program is driven by ligand interaction with androgen receptor. The androgen receptor is a nuclear hormone transcription factor that modulates the expression of genes in response to androgen. Androgen-regulated genes are involved in a wide variety of functional roles including proliferation, differentiation, metabolism, and cell death. The critical role of the androgen response pathway in prostate disease makes it an important system for study in an effort to understand this complex network and identify specific events that are amenable to therapeutic intervention. Despite the importance of the androgen response pathway, relatively few downstream targets have been identified and characterized. Androgen deprivation therapy is effective for prostate cancer, but progression to androgen independence usually results in relapse within 2 yr. The molecular mechanisms underlying the clinically important transition from androgen dependence to androgen independence are poorly understood.

Controlled study of the complex processes associated with androgen signaling and androgen-independent progression in prostate cancer has proved difficult, because few models exist that reproducibly mimic the clinical course of the disease in men. LNCaP cells express a functional, although mutated, androgen receptor, proliferate in response to androgens, and produce PSA and other androgen-responsive genes in an androgen-dependent manner. A number of investigators have used this system for addressing the androgen response program. Using subtractive methods, several groups have identified potential androgen-responsive genes *(48,49)*. A combination of subtractive hybridization and reverse Northern analysis were used to identify genes that were largely prostate-specific and androgen-responsive in LNCaP *(50)*. These genes included *PSA*, *NKX3A*, *KLK4*, and seminogelin. Xu et al. analyzed 83,000 SAGE tags derived from androgen-stimulated and androgen-starved cells *(51)*. They identified 136 genes that were induced and 215 genes that were repressed by androgen. Waghray et al. analyzed 123,371 SAGE tags and identified 147 that were up-regulated

---

Fig. 3. *(previous page)* Cluster analysis of gene expression differences between primary and metastatic prostate cancer. Representative gene expression clusters enriched for genes differentially expressed between primary (green boxes) and metastatic (red boxes) prostate carcinomas. Clusters were selected from a hierarchical clustering dendrogram of gene expression data from all 12,559 probe sets of U95A array (rows) and 32 samples of prostate carcinoma (columns). Expression levels are pseudocolored red to indicate transcript levels above the median for that gene across all samples and green below the median. Color saturation is proportional to the magnitude of expression. A cluster of genes corresponding to high level expression in non-neoplastic prostate is labeled.

Table 1
100 Highest Ranked Differentially Expressed Genes in Metastatic Prostate Cancer and Functional Classification

Rank	Exp met		
	Cell cycle regulation, DNA replication and repair, and mitosis		
5	↑	DEEPEST	Deepest.
7	↑	KNTC1	Kinetochore-associated 1.
8	↑	FEN1	RAD2 (*S. pombe*) homolog, flap structure-specific endonuclease 1.
9	↑	TK1	Thymidine kinase.
11	↑	TOP2α	Topoisomerase (DNA) II α (170 kDa).
12	↑	CDKN3	Cyclin-dependent kinase inhibitor 3 (CDK2-associated dual specificity phosphatase).
16	↑	RFC5	Human replication factor C, 36-kDa subunit.
23	↑	RFC4	Replication factor C (activator 1) 4 (37 kDa).
28	↑	MAD2L1	MAD2 (mitotic arrest deficient, yeast, homolog)-like 1.
35	↑	KNSL2	Kinesin-like 2.
36	↑	CDC2	Cell division cycle 2, cell division control protein 2 homolog (ec 2.7.1.) (p34 proteinkinase) (cyclin-dependent kinase 1) (cdk1).
42	↑	MPHOSPH9	M-phase phosphoprotein 9.
44	↑	RNASEH1	Ribonuclease HI, large subunit.
50	↑	CCNE2	Cyclin E2.
54	↑	MCM7	Minichromosome maintenance deficient (*S. cerevisiae*) 7.
56	↑	BUB1β	Budding uninhibited by benzimidazoles 1 (yeast homolog), β.
57	↑	SMC4L1	SMC4 (structural maintenance of chromosomes 4, yeast)-like 1.
58	↑	STK15	Serine/threonine kinase 15.
59	↑	ZWINT	ZW10 interactor.
70	↑	MGC1780	DDA3: p53-regulated DDA3.
73	↑	CCNB2	Cyclin B2.
83	↑	TTK	TTK protein kinase.
84	↑	RPA3	Replication protein A3.
95	↑	CCNB1	Cyclin B1.
96	↑	CDC25B	Cell division cycle 25B.
	Signaling and signal transduction		
15	↑	STK11	Serine–threonine kinase 11 (Peutz-Jeghers syndrome).
17	↑	JAG1	Jagged 1.
21	↓	DUSP1	Dual specificity phosphatase 1.
26	↓	KIAA0135	Similar to PIM-1 Proto-oncogene.
34	↑	EDN3	Endothelin 3.
49	↑	CSNK1γ2	Casein kinase 1, γ 2.
51	↑	INPP4A	Inositol polyphosphate-4-phosphatase.
64	↑	TMPO	Thymopoietin.
67	↑	ESRRB	Estrogen-related receptor beta.

Table 1 (*continued*)

Rank	Exp met		
		Signaling and signal transduction	
76	↓	GPR68	Ovarian cancer G protein-coupled receptor 1; member of the G protein-coupled receptor family.
77	↓	LTBP1	Latent transforming growth factor β binding protein 1.
78	↓	IL8Rα	Interleukin 8 receptor, α.
79	↑	CIT	Citron (rho-interacting, serine/threonine kinase 21).
89	↓	MSMβ	Microseminoprotein β.
92	↓	PPP3Cβ	Protein phosphatase 3 (formerly 2B), catalytic subunit, β isoform (calcineurin A β).
		Transcriptional regulation, chromatin modification, RNA processing and protein synthesis, and modification	
2	↑	USP13	Ubiquitin specific protease 13 (isopeptidase T-3).
6	↑	SMARCD1	SWI/SNF related, matrix associated, actin dependent regulator of chromatin, subfamily d, member 1.
13	↑	MYBL2	v-myb avian myeloblastosis viral oncogene homolog-like 2.
18	↑	SF3A2	Splicing factor 3a, subunit 2, 66 kDa.
20	↑	HOXB5	HOX2, homeo box B5.
22	↓	ZFP36	Tristetraprolin.
24	↑	EZH2	Enhancer of zeste (*Drosophila*) homolog 2.
25	↓	SATB1	Special AT-rich sequence binding protein 1.
32	↓	FOS	v-fos FBJ murine osteosarcoma viral oncogene homolog.
33	↓	NR4A1	Nuclear receptor subfamily 4, group A.
43	↑	UBCH10	Ubiquitin carrier protein E2-C.
45	↑	PTTG1	Pituitary tumor-transforming 1, securin.
48	↑	CTRL	Chymotrypsin-like.
52	↑	EP300	E1A binding protein p300.
53	↑	E2EPF	Ubiquitin carrier protein.
62	↓	JUNB	jun B proto-oncogene.
68	↑	FOXM1	Forkhead box M1, hepatocyte nuclear factor.
72	↑	PROP1	Prophet of Pit1, paired-like homeodomain transcription factor.
75	↑	CNAP1	Chromosome condensation-related SMC-associated protein 1.
85	↑	HOXC5	Homeo box C5.
88	↑	U5-100K	prp28, U5 snRNP 100 kd protein.
91	↑	CGGBP1	CGG triplet repeat binding protein 1.
94	↓	CEBPD	ccaat/enhancer binding protein delta (c/ebp delta) (nuclear factor nf-il6-β).
99	↑	PLOD2	Procollagen-lysine, 2-oxoglutarate 5-dioxygenase.
		Cell adhesion, migration, cytoskeleton, and extracellular matrix	
1	↓	TAGLN	Transgelin actin, α 2.
3	↓	ACTA2	α actin.

(*continued*)

Table 1 (*continued*)

Rank	Exp met		
		Cell adhesion, migration, cytoskeleton, and extracellular matrix	
30	↓	FHL1	Four and a half LIM domains 1.
46	↓	ITGα8	Integrin, α 8.
47	↑	HMMR	Hyaluronan-mediated motility receptor (RHAMM).
61	↓	TPM3	Tropomyosin 3 (nonmuscle).
66	↓	ITGα7	Integrin, α 7.
80	↑	THBS2	Thrombospondin 2.
82	↓	RARRES1	Retinoic acid receptor responder (tazarotene-induced) 1.
97	↓	CEACAM7	Carcinoembryonic antigen-related cell adhesion molecule 7.
		Metabolism, biosynthesis, and molecular transport	
4	↓	ABCB1	P glycoprotein 1/multiple drug resistance 1.
10	↑	PGK1	Phosphoglycerate kinase 1.
19	↑	FTH1	Ferritin heavy chain 1; FTH1.
27	↑	NUP155	Nucleoporin 155 kDa.
29	↓	ABCA8	ATP-binding cassette, sub-family A (ABC1), member 8.
38	↓	FXYD3	FXYD domain-containing ion transport regulator 3.
41	↑	SLC29A1	Solute carrier family 29 (nucleoside transporters), member.
60	↑	FMO3	Flavin containing monooxygenase 3.
74	↑	PDK1	Pyruvate dehydrogenase kinase, isoenzyme 1.
86	↑	SCD	Stearoyl-CoA desaturase (δ-9-desaturase).
87	↑	KPNα2	Karyopherin α 2 (RAG cohort 1, importin α 1.
98	↓	APOD	Apolipoprotein D).
100	↑	ABCC4	ATP-binding cassette, sub-family C (CFTR/MRP), member 4.
		Unclassified	
14	↑	KIAA0101	None available.
31	↑	KIAA0906	None available.
37	↑	CGTHBA	Conserved gene telomeric to α globin cluster.
39	↑	KIAA0186	None available.
40	↑	KIAA0543	None available.
55	↓	DKFZP434N043	None available.
63	↑	R32184	None available.
65	↑	DKFZP434N093	None available.
69	↑	KIAA0008	None available.
71	↑	DKFZP564C152	None available.
81	↑	MEST	Mesoderm-specific transcript (mouse) homolog.
90	↑	DKFZP586L151	None available.
93	↓	EYA1	Eyes absent (*Drosophila*) homolog 1.

Rank, rank order of U95A probe sets based on *t*-statistic; exp met, mean expression value of metastatic prostate cancers above (↑) or below (↓) mean expression value of primaries.

and 204 that were repressed *(38)*. Both groups analyzed cells 24 h after androgen exposure, and many differentially expressed genes were identified in common. The differentially expressed genes represented many different functional categories and include both direct and indirect targets of the androgen receptor.

Nelson and coworkers used cDNA arrays composed of 1500 prostate-derived gene probes to characterize the androgen response program in a time course experiment with LNCaP cells *(52)*. They identified 20 genes that were up-regulated and none that were down-regulated. These included known (e.g., *PSA, KLK2, NKX3A*) and novel genes. There were distinct temporal patterns of gene expression, suggesting that some were direct and some indirect targets for the androgen receptor. Vaarala et al. used cDNA arrays with probes for 7075 human genes to analyze LNCaP sublines, which had been selected for expression of PSA as a surrogate marker of androgen-dependent–independent growth *(53)*. Some of the genes identified in this study were known to be androgen-responsive (e.g., *PSA, KLK2*). Northern blot and *in situ* hybridization were used to validate differential expression.

Using a different method to identify androgen-responsive genes, Wang et al. isolated 38 genes that were differentially expressed in the rat ventral prostate at 14 and 48 h following androgen replacement in castrated animals *(54)*. The majority of these were up-regulated following androgen replacement and were also prostate-specific. Six of the 11 up-regulated genes that could be identified were known to be androgen-regulated.

The CWR22 human xenograft is another androgen-responsive prostate cancer model that has also been used to investigate the development of androgen-independent disease. This model exhibits androgen-dependent growth, expression of a mutated androgen receptor, and expression of the androgen-responsive gene PSA. When deprived of androgen, tumors undergo involution, but recur as rapidly growing androgen-independent tumors in 3–12 mo. Bubendorf et al. used cDNA arrays with probes for 5184 prostate-derived genes to identify 37 that were increased by more than twofold and 135 that were reduced more than twofold in at least three out of four hormone refractory xenografts *(55)*. This group went on to study the progression of CWR22 using cDNA arrays with 6605 probes *(56)*. They profiled tumors before and during androgen ablation and after recurrence of hormone refractory tumors. Fifty-nine genes were repressed greater than threefold during androgen withdrawal, but 96.6% were reexpressed in the recurrent tumors, providing further support for the hypothesis that reactivation of androgen-responsive genes is involved in the growth of androgen-independent tumors. They also identified 164 genes that were differentially expressed between primary and recurrent tumors, including several that converge on the PI3K/AKT/FRAP pathway. Amler et al. used cDNA arrays with 9704 probes to study gene expression during androgen ablation and identified 122 genes with at least 2.5-fold decrease in expression at some time point after androgen withdrawal and 38 genes that were increased at least 2.5-fold in expression *(57)*. Comparison of the androgen-independent xenograft to the parental androgen-dependent CWR22 identified 13 genes that were increased and 44 genes that were decreased. In addition, 28 genes that were decreased after androgen withdrawal and in androgen-independent tumors were reexpressed on exposure to androgen, suggesting that only a subset were androgen-responsive genes. These time course experiments and relationship to androgen response suggested that there is only partial reactivation of the androgen response pathway in androgen-independent tumors in the absence of exogenous ligand. Six genes (*FKBP5, THRA, S100P, SDC1, NCOR1,* and *APELIN*) were found to be stably up-regulated in androgen-independent tumors and were not androgen-responsive, suggesting that they may serve as markers for androgen-independent disease.

This approach was extended to the analysis of gene expression changes in human prostate cancers during androgen ablation therapy and the development of androgen-independent disease *(58)*. Gene expression analysis was carried out for 23 untreated prostate cancers, 17 prostate cancers after 3 mo of androgen ablation therapy, and 3 prostate cancers that were progressing after long-term hormonal therapy (androgen-independent) using oligonucleotide arrays with 63,175 probe sets. Cluster analysis revealed that untreated cancers have expression profiles distinct from tumors following therapy. Interestingly, androgen-independent tumors tended to cluster with untreated cases (Fig. 4). Genes that were strongly differentially expressed during hormonal therapy were identified by using an algorithm that combined filters and ranking based on the mean and variance of expression levels for the two groups. A total of 659 of the 63,175 probe sets detected tumor-related differential gene expression of at least threefold with a p value <0.001. Two hundred ninety-five were up-regulated with androgen ablation therapy, and 364 were down-regulated. The expression levels for the majority of these genes also changed in the same direction in normal prostatic epithelium. Some of these genes are known to be androgen responsive (*PSA, KLK2,* and *KLK3,* among others). Of interest, about 97% of these genes did not demonstrate changes to the same extent in androgen-independent tumors growing in an androgen-deprived environment. In addition, the overall expression patterns for androgen-independent tumors more closely corresponded to that of the untreated androgen-dependent cancers, suggesting a reactivation of the androgen response pathway as predicted in the animal models. However, a unique set of differentially expressed genes was detected in androgen-independent tumors. For example, 145 of the 12,560 U95A probe sets detected tumor-related differential gene expression between primary androgen-dependent and metastatic androgen-independent tumor of at least threefold with a student's t-test p value <0.001. Of these, 50 were also differentially expressed threefold between androgen-independent and -dependent metastatic tumors. Twenty-five were up-regulated, and 25 were down-regulated. Included among the up-regulated genes were the androgen receptor and several involved in steroid biosynthesis. This further supports the concept that reactivation of androgen signaling is a common feature in prostate cancers that have become resistant to androgen ablation therapy and may occur through increased sensitivity and intracellular synthesis of ligand.

CONCLUDING REMARKS

It is clear that the combination of the entire genome sequence, microdissection, and comprehensive expression analysis will have a significant effect on research in cancer biology and oncology. It is also clear that these techniques should be used with discretion, common sense, and healthy skepticism to avoid over interpretation of the large volumes of data, much of which has yet to be validated as reliable. Nonetheless, a carefully controlled experimental design and rigorous data analysis could lead to new discoveries in gene function relationships, which might not be evident with traditional

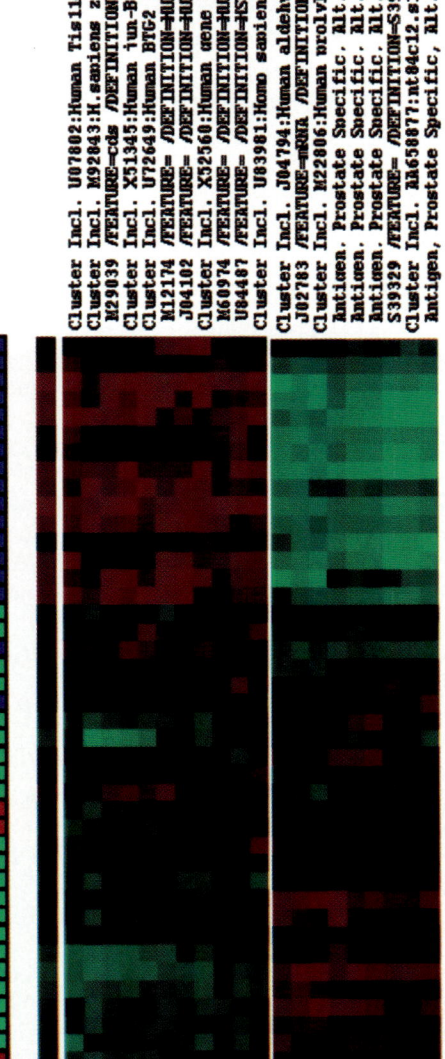

Fig. 4. Cluster analysis of gene expression during androgen ablation therapy and the development of androgen-independent disease. Representative gene expression clusters enriched for genes differentially expressed between primary untreated (green boxes), androgen-ablation treated (blue boxes), and metastatic androgen-independent (red boxes) prostate carcinomas. Clusters were selected from a hierarchical clustering dendrogram of gene expression data from all 12,559 probe sets of U95A array (rows) and 43 samples of prostate carcinoma (columns). Expression levels are pseudocolored red to indicate transcript levels above the median for that gene across all samples and green below the median. Color saturation is proportional to the magnitude of expression.

focused experiments. It is worth noting that there is only limited agreement between the studies described here with respect to expression of specific genes in association with clinically relevant subsets of prostate cancer. There are many reasons for this apparent lack of consensus, including the use of various analytical methods and criteria to identify the most significant genes for a particular process, different sources and methods of processing tissues and cells, variable methods of gene expression analysis, technical platforms, tissues compared, and genes interrogated. There is also the tendency in published results to discuss only the best characterized and most highly ranked genes, making comparisons of more inclusive lists difficult. Ultimately, the standardization of gene expression techniques and public availability of data will facilitate the combined analysis of large numbers of tissues, such that clinical correlative studies can be more efficiently and effectively performed. These novel technologies and approaches provide many opportunities for cancer biologists and clinicians, however, they will be most productively used in a cooperative manner that benefits from large numbers of samples and unlimited methods of analysis.

NOTE ADDED IN PROOF

Two recent publications present expression analysis of the androgen response program in LNCaP cells *(59,60)*.

REFERENCES

1. Greenlee, R. T., Hill-Hamon, M. B., Murray, T., and Thun, M. (2001) Cancer statistics. *CA Cancer J. Clin.* **51,** 15–36.
2. Barry, M. (2001) Prostate specific antigen testing for early diagnosis of prostate cancer. *N. Engl. J. Med.* **344,** 1373–1377.
3. Oesterling, J., Fuks, Z., Lee, C., and Scher, H. I. (1997) Cancer of the prostate, in *Cancer: Principles and Practice of Oncology, 5th ed.*, (Dvita, V. T., Hellman, S., and Rosenberg, S. A., eds.), Lippincott-Raven, Philadelphia, pp. 1322–1376.
4. Abate-Shen, C. and Shen, M. (2000) Molecular genetics of prostate cancer. *Genes Dev.* **14,** 2410–2434.
5. Hayward, S. W., Rosen, M. A., and Cunha, G. R. (1997) Stromal-epithelial interactions in the normal and neoplastic prostate. *Br. J. Urol.* **79,** 18–26.
6. Olumi, A. F., Grossfeld, G. D., Hayward, S. W., Carroll, P. R., Tlsty, T. D., and Cunha, G. R. (1999) Carcinoma-associated fibroblasts direct tumor progression of initiated human prostatic epithelium. *Cancer Res.* **59,** 5002–5011.
7. McNeal, J. E. and Bostwick, D. G. (1986) Intraductal dysplasia: a premalignant lesion of he prostate. *Hum. Pathol.* **17,** 64–71.
8. Bostwick, D. (1999) Prostatic intraepithelial neoplasia, in *Textbook of Prostate Cancer.* Martin Dunitz, LTD, London, pp. 35–50.
9. Klocker, H., Culig, Z., Kaspar, F., et al. (1994) Androgen signal transduction and prostatic carcinoma. *World J. Urol.* **12,** 99–103.
10. Logothetis, C. J., Hoosein, N. M., and Hsieh, J. T. (1994) The clinical and biological study of androgen independent prostate cancer (AI PCa). *Semin. Oncol.* **21,** 620–629.
11. Griffiths, K., Morton, M. S., and Nicholson, R. I. (1997) Androgens, androgen receptors, antiandrogens and the treatment of prostate cancer. *Eur. Urol.* **32(Suppl),** 24–40.
12. Dong, J.-T., Isaacs, W., and Isaacs, J. (1997) Molecular advances in prostate cancer. *Curr. Opin. Oncol.* **9,** 101–107.
13. Carter, B. S., Carter, H. B., and Isaacs, J. T. (1990) Epidemiologic evidence regarding predisposing factors to prostate cancer. *Prostate* **16,** 187–197.

14. Matsuyama, H., Pan, Y., Oba, K., et al. (2001) Deletions on chromosome 8p22 may predict disease progression as well as pathological staging in prostate cancer. *Clin. Cancer Res.* **7,** 3139–3143.
15. Swalwell, J., Vocke, C. D., Yang, Y., et al. (2002) Determination of a minimal deletion interval on chromosome band 8p21 in sporadic prostate cancer. *Genes Chromosomes Cancer* **33,** 201–205.
16. Bhatia-Gaur, R., Donjacour, A. A., Sciavolino, P. J., et al. (1999) Roles for Nkx3.1 in prostate development and cancer. *Genes Dev.* **13,** 966–977.
17. Kim, M. J., Cardiff, R. D., Desai, N., et al. (2002) Cooperativity of Nkx3.1 and Pten loss of function in a mouse model of prostate carcinogenesis. *Proc. Natl. Acad. Sci. USA* **99,** 2884–2889.
18. Steck, P. A., Pershouse, M. A., Jasser, S. A., et al. (1997) Identification of a candidate tumour suppressor gene, MMAC1, at chromosome 10q23.3 that is mutated in multiple advanced cancers. *Nat. Genet.* **15,** 356–362.
19. Di Cristofano, A., Pesce, B., Cordon-Cardo, C., and Pandolfi, P. P. (1998) Pten is essential for embryonic development and tumour suppression. *Nat. Genet.* **19,** 348–355.
20. Eagle, L. R., Yin, X., Brothman, A. R., Williams, B. J., Atkin, N. B., and Prochownik, E. V. (1995) Mutation of the MXI1 gene is prostate cancer. *Nat. Genet.* **9,** 249–255.
21. Shrieber-Agus, N., Meng, Y., Hoang, T., et al. (1998) Role of MSI1 in ageing organ systems and the regulation of normal and neoplastic growth. *Nature* **393,** 483–487.
22. Eastham, J., Stapleton, A., Gousse, A., et al. (1995) Association of p53 mutations with metastatic prostate cancer. *Clin. Cancer Res.* **1,** 111–118.
23. McDonnell, T., Navone, N., Troncoso, P., et al. (1997) Expression of bcl-2 oncoprotein and p53 protein accumulation in bone marrow metastases of androgen independent prostate cancer. *J. Urol.* **157,** 569–574.
24. Macri, E. and Loda, M. (1999) Role of p27 in prostate carcinogenesis. *Cancer Metastasis Rev.* **17,** 337–344.
25. Grossman, M., Huang, H., and Tindall, D. (2001) Androgen receptor signaling in androgen-refractory prostate cancer. *J. Natl. Cancer Inst.* **93,** 1687–1697.
26. Fenton, M. A., Shuster, T. D., Fertig, A. M., et al. (1997) Functional characterization of mutant androgen receptors from androgen-independent prostate cancer. *Clin. Cancer Res.* **3,** 1383–1388.
27. Lee, C. (1996) Role of androgen in prostate growth and regression: stromal-epithelial interaction. *Prostate* **6(Suppl),** 52–56.
28. van Weerden, W. M. and Romijn, J. C. (2000) Use of nude mouse xenograft models in prostate cancer research. *Prostate* **43,** 263–271.
29. Greenberg, N., DeMayo, F., Finegold, M., et al. (1995) Prostate cancer in a transgenic mouse. *Proc. Natl. Acad. Sci. USA* **92,** 3439–3443.
30. Gingrich, J., Barrios, R., Morton, R., et al. (1996) Metastatic prostate cancer in a transgenic mouse. *Cancer Res.* **56,** 4096–4102.
31. Di Cristofano, A., De Acetis, M., Koff, A., Cordon-Cardo, C., and Pandolfi, P. (2001) Pten and p27KIP1 cooperate in prostate cancer tumor suppression in the mouse. *Nat. Genet.* **27,** 222–224.
32. Gleason, D. F. (1992) Histologic grading of prostate cancer: a perspective. *Hum. Pathol.* **23,** 273–279.
33. Bostwick, D. G., Shan, A., Qian, J., et al. (1998) Independent origin of multiple foci of prostatic intraepithelial neoplasia: comparison with matched foci of prostate carcinoma. *Cancer* **83,** 1995–2002.
34. Macintosh, C. A., Stower, M., Reid, N., and Maitland, N. J. (1998) Precise microdissection of human prostate cancers reveals genotypic heterogeneity. *Cancer Res.* **58,** 223–228.
35. Liu, A. Y., Nelson, P., van den Engh, G., and Hood, L. (2002) Human prostate epithelial cell-type cDNA libraries and prostate expression patterns. *Prostate* **50,** 92–103.

36. LaTulippe, E., Satagopan, J., Smith, A., et al. (2002) Comprehensive gene expression analysis of prostate cancer reveals distinct transcriptional programs associated with metastatic disease. *Cancer Res.* **62**, 4499–4506.
37. Nelson, P. S., Clegg, N., Eroglu, B., et al. (2000) The prostate expression database (PEDB): status and enhancements in 2000. *Nucleic Acids Res.* **28**, 212–213.
38. Waghray, A., Schober, M., Feroze, F., Yao, F., Virgin, J., and Chen, Y. (2001) Identification of differentially expressed genes by serial analysis of gene expression in human prostate cancer. *Cancer Res.* **61**, 4283–4286.
39. Elek, J., Park, K., and Narayanan, R. (2000) Microarray-based expression profiling in prostate tumors. *In Vivo* **14**, 173–182.
40. Chaib, H., Cockrell, E., Rubin, M., and Macoska, J. (2001) Profiling and verification of gene expression patterns in normal and malignant human prostate tissues by cDNA microarray analysis. *Neoplasia* **3**, 43–52.
41. Dhanasekaran, S., Barrette, T., Ghosh, D., et al. (2001) Delineation of prognostic biomarkers in prostate cancer. *Nature* **412**, 822–826.
42. Luo, J., Duggan, D., Chen, Y., et al. (2001) Human prostate cancer and benign prostatic hyperplasia: molecular dissection by gene expression profiling. *Cancer Res.* **61**, 4683–4688.
43. Magee, J., Araki, T., Patil, S., et al. (2001) Expression profiling reveals hepsin overexpression in prostate cancer. *Cancer Res.* **61**, 5692–5696.
44. Welsh, J., Sapinoso, L., Su, A., et al. (2001) Analysis of gene expression identifies candidate markers and pharmacological targets in prostate cancer. *Cancer Res.* **61**, 5974–5978.
45. Luo, J.-H., Yu, Y., Cieply, K., et al. (2002) Gene expression analysis of prostate cancers. *Mol. Carcinog.* **33**, 25–35.
46. Singh, D., Febbo, P., Ross, K., et al. (2002) Gene expression correlates of clinical prostate cancer behavior. *Cancer Cell* **1**, 203–209.
47. Stamey, T., Warrington, J., Caldwell, M., et al. (2001) Molecular genetic profiling of Gleason grad4/5 prostate cancers compared to benign prostatic hyperplasia. *J. Urol.* **166**, 2171–2177.
48. Chang, G., Blok, L., Steenbeek, M., et al. (1997) Differentially expressed genes in androgen-dependent and independent prostate carcinomas. *Cancer Res.* **57**, 4075–4081.
49. Stubbs, A., Abel, P., Golding, M., et al. (1999) Differentially expressed genes in hormone refractory prostate cancer. *Am. J. Pathol.* **154**, 1335–1343.
50. Korkmaz, K., Korkmaz, C., Ragnhildstveit, E., Pretlow, T., and Saatcioglu, F. (2000) An efficient procedure for cloning hormone-responsive genes from a specific tissue. *DNA Cell Biol.* **19**, 499–506.
51. Xu, L., Su, Y., Labiche, R., et al. (2001) Quantitative expression profile of androgen-regulated genes in prostate cancer cells and identification of prostate-specific genes. *Int. J. Cancer* **92**, 322–328.
52. Nelson, P., Han, D., Rochon, Y., et al. (2000) Comprehensive analyses of prostate gene expression: convergence of expressed sequence tag databases, transcript profiling and proteomics. *Electrophoresis* **21**, 1823–1831.
53. Vaarala, M., Porvari, K., Kyllogen, A., and Vihko, P. (2000) Differentially expressed genes in two LNCaP prostate cancer cell lines reflecting changes during prostate cancer progression. *Lab. Invest.* **80**, 1259–1268.
54. Wang, Z., Tufts, R., Haleem, R., and Cai, X. (1997) Genes regulated by androgen in the rat ventral prostate. *Proc. Natl. Acad. Sci. USA* **94**, 12,999–13,004.
55. Bubendorf, L., Kolmer, M., Kononen, J., et al. (1999) Hormone therapy failure is human prostate cancer: analysis by complementary DNA and tissue microarrays. *J. Natl. Cancer Inst.* **91**, 1758–1764.
56. Mousses, S., Wagner, U., Chen, Y., et al. (2001) Failure of hormone therapy in prostate cancer involves systematic restoration of androgen responsive genes and activation of rapamycin sensitive signaling. *Oncogene* **20**, 6718–6723.

57. Amler, L., Agus, D., LeDuc, C., et al. (2000) Deregulated expression of androgen-responsive and nonresponsive genes in the androgen-independent prostate cancer xenograft model CWR22-R. *Cancer Res.* **60,** 6134–6141.
58. Holzberlein, J. LaTulippe, E., Satagopan, J., et al. Gene expression alterations associated with androgen ablation therapy of human prostate carcinoma. In preparation.
59. Deprimo, S. E., Diehn, M., Nelson, J. B., et al. (2002) Transcriptional programs activated by exposure of human prostate cancer cells to androgen. *Genome Biol.* **3,** 32.1–32.12.
60. Nelson, P. S., Clegg, N., Arnold, H., et al. (2002) The program of androgen-responsive genes in neoplastic prostate epithelium. *Proc. Natl. Acad. Sci. USA* **99,** 11,890–11,895.

11
Classification of Human Lung Carcinomas by mRNA Expression Profiling

Arindam Bhattacharjee and Matthew Meyerson

INTRODUCTION

Lung cancer deaths exceed the combined mortality from breast, prostate, and colorectal cancers (1). Every year more than 150,000 Americans die from this deadly disease. Lung carcinoma classes include small cell lung carcinomas (SCLC) and non-small cell lung carcinomas (NSCLC). The distinction of SCLC from NSCLC is important as the clinical course and treatments for the two diseases are different. SCLC patients initially respond well to chemotherapy, followed by regression of the tumor. However, in most cases, the patient relapses, develops chemo-resistance, and eventually dies from systemic dissemination of cancer. In contrast, in early stage NSCLC disease, treatment involves surgical removal of primary tumor (resection). However, 50% of those patients undergo relapse and eventually die from metastasis.

There exist a set of consensus criteria among pathologists for the SCLC vs NSCLC distinction. However, the histopathological distinction in tumor classes is not always clear. Occasionally, tumors are classified as adenosquamous carcinoma or combined SCLC with NSCLC features (2). Such subjective assessments may become even more difficult in poorly differentiated tumors (3). Thus, lung cancer classification can likely be improved by taking into consideration a molecular classification.

As the current distinction of SCLC from NSCLC rests on clinicopathological features, it may not reveal underlying alterations in genetic programs associated with the malignant process. In contrast, a biological classification of lung carcinoma based on large-scale transcriptional profiling, as well as other genomic and proteomic methods, offers a biological basis for subclassifying tumors, tumor class discovery, and predicting survival outcome. The recent development of targeted therapy against the Abl tyrosine kinase for chronic myeloid leukemia also illustrates the power of such biological knowledge from the standpoint of drug target discovery (4).

NSCLC accounts for 80% of lung cancer cases and is further subcategorized as adenocarcinoma, squamous cell lung carcinoma, and large cell carcinoma (5). Adenocarcinomas are the most common and represent 35% of the lung cancer cases, followed by squamous cell carcinomas, which account for 30% of the cases (2). Roughly 10% of lung cancer cases are large cell lung carcinomas (LCLC), typically diagnosed by

From: *Expression Profiling of Human Tumors: Diagnostic and Research Applications*
Edited by: Marc Ladanyi and William L. Gerald © Humana Press Inc., Totowa, NJ

exclusion of the other three types of lung cancers *(2)*. Neuroendocrine features, defined by microscopic morphology and immunohistochemistry, have been regarded as hallmarks of the high-grade SCLC that account for 18% of lung cancers, as well as intermediate–low-grade pulmonary carcinoid tumors *(5)*.

While the etiology of SCLC and squamous cell lung carcinomas, and most adenocarcinomas, are typically linked to tobacco smoking, the cause of some lung adenocarcinomas appears to be unclear *(6,7)*. Adenocarcinomas arise peripherally in the smaller airways, while SCLC and squamous cell tumors are centrally located in the larger airways. The tumor suppressor genes affected in lung cancer typically include the *pRb/p16^{INK4a}* pathway and *p53 (8,9)*. *p53* mutations and loss of heterozygosity (LOH) have been detected in greater than 50% of lung cancers *(9)* and are common genetic abnormalities in human cancers *(10)*. Overall, the most common activating oncogene mutations in lung adenocarcinomas are *K-ras* mutations *(11)*.

The histopathological subclassification of lung adenocarcinoma is also challenging. In one study, lung pathologists agreed on lung adenocarcinoma subclassification in only 41% of cases *(12)*. However, a favorable prognosis for bronchioloalveolar carcinoma (BAC), a histological subclass of lung adenocarcinoma, argues for refining such distinctions *(13,14)*. Metastases of non-lung origin are common in the lung and can be difficult to distinguish from primary lung adenocarcinomas *(15,16)*. As the clinical course of such a disease would differ from lung adenocarcinoma, confirmation of tumor origin prior to treatment is also important.

The only effective prognostic indicator in lung cancer currently in clinical use is surgical–pathological staging. Such staging is based on tumor size, nodal metastasis, and distant metastasis *(17)*. Although molecular markers such as *p16^{INK4a}*, *p53*, or *K-ras* status have been suggested as prognostic indicators *(8)*, they are of limited use. The simultaneous analysis of large numbers of independent clinical parameters may offer a more powerful insight into surgical–pathological staging. The development of microarray methods for large-scale analysis of gene expression *(18–21)* makes it possible to search systematically for molecular markers of cancer classification and outcome prediction in a variety of tumor types *(22–28)*. As we overcome technical hurdles in microarray analysis and as the cost of microarray experiments continues to decrease, use of expression analysis may become part of routine diagnostic practice.

In this chapter, we will focus on recent developments in lung cancer classification based on large-scale gene expression analysis. For this, we will compare and discuss gene expression analysis data obtained from over 250 human lung carcinomas, performed by three independent groups, including our own research group *(29–31)*.

LUNG TUMOR COLLECTION, PROCESSING, AND MODELS

Lung Tumor Specimen Source

Tumor collection is dictated to a large extent by clinical practice. For example, when SCLC is diagnosed, the disease has generally spread to distant organs (metastasis), and surgical resection is not performed. In such cases, the diagnostic biopsy or cytology specimen may be too limited for research tumor procurement. However, in limited stage (early stage) SCLC disease, when the tumor has not metastasized to distant organs, surgery is performed, and therefore, samples can be obtained. In contrast,

NSCLC is typically a slow growing tumor where surgical resection may benefit patients, making tumor collection quite feasible. Nevertheless, when comparing expression profiles of lung tumors one needs to be aware that differences observed could be due to the evolution of the tumor and disease progression, which may be altered following radiation or chemotherapy.

Classically, lung tumor specimens are either obtained as paraffin- or optimal cutting temperature (OCT)-embedded sections following surgery or bronchial biopsies. Other than primary tumors, there exist protocols that involve fine-needle biopsies from the mediastinal or thoracic nodes. There exist several lung tumor-derived cell lines from patients, which can also serve as a resource for examining the molecular profile of lung tumors. Finally, distant lymph node biopsy, blood, serum, and sputum samples may serve as additional sources of material that may prove to be valuable for developing and determining diagnostic applications. Such specimens may also be valuable for understanding the progression of lung tumors using gene expression analysis, immunohistochemistry, and DNA analysis.

Lung Tissue Handling

Integrity of lung tumor-derived RNA is a critical variable in mRNA-based expression profiling. Poor quality of RNA is generally observed when tumors are not frozen immediately following resection, which leads to nuclease-based degradation of RNA. Note that in Ramaswamy et al. *(3)*, at least 98 tumor samples were rendered unusable, as they did not meet RNA quality standards. This most likely reflects some tumors in which the RNA was degraded. In general, snap-frozen tissue is ideal for RNA extraction. Although suitable for DNA-based methods, paraffin-embedded tissue is not useful for mRNA-based expression profiling. In the current expression analysis studies *(29–31)*, snap-frozen tumor specimens were used.

Lung Tumor Heterogeneity

When profiling solid tumors, one question that arises is whether the profile generated is meaningful, as the tumor tissue consist also of lymphocytes, stroma, and other normal cells. To circumvent this problem, several research groups profiled crude tumors that are enriched in neoplastic tissue *(3,29,32)*. Crude tumors may also include particular types of tumors, such as adenosquamous tumors or SCLC with NSCLC features *(2)*, that may give a mixed expression profile. Under these circumstances, laser capture microdissection (LCM) may be a suitable approach that can be used for purification of tumor cells from contaminating cells in the tumor. Although current techniques of LCM and their application in expression profiling have not been extensively studied because of the limited quantities of mRNA obtained, LCM holds considerable promise in dissecting the transcriptional signature of tumor-specific cells from surrounding stroma. Both approaches of analyzing tumors have merits, and it may become necessary to use both to address a given problem in tumor classification.

Lung Cancer Models

Lung cancer models that are representative of human cancers are valuable tools to study tumor progression and discovery of therapeutic targets, provided they represent the tumor genotypically and not just phenotypically. While the significance of tumor suppressors ($p53$, pRB, $p16^{INK4a}$) and activated oncogenes, such as K-ras and c-myc, involved in lung cancer is well known *(8,9,11)*, the progression model and order of inactivation or activation mechanisms involving tumor suppressors and oncogenes is poorly recapitulated in current lung cancer models. However, some progress has been made in this nascent field. For example, overexpression of activated H-ras using the calcitonin–calcitonin gene-related peptide (CGRP) promoter induces pulmonary neuroendocrine (NE) cell hyperplasias and thyroid tumors *(33)*. This was surprising, as NE lung cancers such as SCLC typically do not have *ras* mutations. Recently, Linnoila et al. developed a mouse model of lung NE tumors *(34)*. In their model they demonstrated that *ASCL*, when expressed from a lung epithelial cell-specific CC10 (Clara cell) promoter created hyperplasia. When these mice were crossed to mice that expressed T antigen (TAg) from the CC10 promoter, the double transgenic mice gave rise to massive NE tumors that accelerated the onset of tumors generated by CC10/TAg mice or CC10/ASCL mice alone. As TAg sequesters pRB and p53, which are frequently dysfunctional in NSCLC and SCLC, this model recapitulates a NE lung tumor that could be similar to its human counterpart.

The recent development of *K-ras* transgenic mice that give rise to lung adenocarcinomas *(35–37)* could serve as an important basis for lung cancer models that reflect human lung cancers. The use of conditional expression together with viral infection *(35,38)* may be particularly useful. In order to recapitulate the human disease and to dissect lung tumor progression, one has to understand the set of genetic events that lead to lung cancer. Crosses can be made with mutants in tumor suppressor genes that are frequently mutated in lung cancer (e.g., *p16, p53*). Such mice may be useful as models for chemotherapeutic trials and to identify other mutations that may contribute to lung cancer pathogenesis.

Cell Culture

Cell lines in culture may acquire an altered genetic program due to artificial growing conditions and selection. Thus, it may be argued whether they represent the original tumor from which the cell line was derived. The use of tumor-derived cell lines in expression profiling is an alternative when sufficient cell numbers and corresponding mRNA is not readily obtainable from primary tumor material.

Lung cancer cell lines are very useful for drug target screens. Although lung cancer-derived cell lines have been evaluated for treatment responses by expression analysis *(39,40)*, the number of cell lines studied is limited, and a full-scale analysis of diverse lung tumor cell lines has not been performed to date to determine therapeutic targets. Such an analysis could begin with defining different classes of lung tumor cell lines using primary tumor signatures while maintaining prediction accuracy *(3,32)*. In addition to providing the foundation for classifying tumor cell lines, such an attempt allows for analyzing distinct classes of lung cancer cell lines that can be subsequently tested for response to different classes of chemical compounds (drugs) using techniques described by Scherf et al. *(39)*. This approach may be a vital link in discovery of therapeutic targets using cell lines that are amenable to high-throughput screens (HTS).

RESULTS FROM EXPRESSION PROFILING STUDIES OF LUNG CARCINOMAS REVEALS DISTINCT ADENOCARCINOMA SUBCLASSES

Transciptome Analysis in Lung Tumors

Although the SCLC and NSCLC distinction is widely accepted in the context of lung tumor classification, such a classification is based primarily on clinical and pathological attributes of the disease rather than similarity or dissimilarity at the level of biological circuitry. Thus, a basic purpose of transcriptional profiling of lung tumors aims at understanding the molecular view of lung tumors that is relevant from a diagnostic or therapeutic perspective. In that lung tissue is complex and is composed of a large number of cell types, it makes trancriptome analysis in lung tumors a daunting task. Few precursor cell types may give rise to lung tumors, as opposed to the diverse cell types that are known to make up the normal lung tissue. A particular cell type in the lung, such as the airway epithelial cell type 2 (AEC2, type II pneumocyte) is the dominant lung tissue cell type, while the pulmonary NE cell (PNEC) may represent less than 1% of lung cells. Thus, comparing a particular lung tumor (that arises from a particular lung cell type) to normal lung tissue may not be the ideal comparison, as it may not be the preneoplastic cell. This could explain, for example, why normal airway cell type-specific signatures are quite distinct from the signature of the A549 lung tumor cell line when compared using serial analysis of gene expression (SAGE) *(31)*. Similarly, lung NE tumor signatures may not be readily compared to the normal lung signatures, as the lung precursor cell for NE lung tumors are unknown or underrepresented. However, such a problem does not invalidate the rationale to study NE lung tumors. The NE tumor signature can be compared to other lung tumor types analyzed or to an independent dataset. It is also possible to compare NE tumors to diverse anatomical tumors and normal tissues studied by Ramaswamy et al. *(3)* and Su et al. *(32)*. A detailed understanding of genes involved in the transcriptional program of diverse lung tumors is helpful for understanding biological interactions, signaling pathways, and understanding the tumor's genetic program. Microarray analysis has already been shown to be useful for revealing transcriptional state and protein interaction *(41)*, autocrine or paracrine signaling *(42)*, and genetic network or pathway analysis *(43)* in model systems. These approaches may be readily applicable in genetic analysis of lung tumors.

Proof of Principle Study: Molecular Classification of Diverse Lung Tumors

It is of considerable interest to determine if a molecular classification recapitulates the existing histopathological classification of lung tumors and reveals previously undiscovered tumor classes. The existing classification of lung cancer as SCLC and NSCLC, and the subclassification of NSCLC, provides us with a framework for evaluating the significance of molecular classification of lung cancers *(29–31)*.

For example, we have used the Affymetrix (Santa Clara, CA, USA) oligonucleotide probe array method and applied hierarchical clustering *(44)* to classify a diverse group of 203 lung tumors and normal lung specimens using the 3312 most variably expressed transcripts *(29)*. Similarly, Garber et al. used cDNA microarrays to classify 67 lung tumors *(30)*. In contrast, Nacht et al. used SAGE to initially classify and train 9 tumors–cell lines and normal airway epithelial cells, which was followed by evaluation of marker genes on 43 tumors using quantitative polymerase chain reaction (PCR) and oligonucleotide probe arrays *(31)*. These studies used supervised and unsupervised clustering approaches for tumor classification, which included hierarchical clustering *(44)*, probabilistic clustering *(45)*, or multidimensional scaling *(31)*. Despite differences in sample acquisition, analytical methods, and analysis platforms, the resulting clusters from these studies have recapitulated the broadest existing distinctions between established histologic classes of lung tumors in a consistent manner, thus demonstrating that molecular classification can confirm existing clinicohistopathological classification. For example, in our 203 sample classification, pulmonary carcinoid tumors, SCLC, squamous cell lung carcinomas, and adenocarcinomas form distinct clusters, thus validating the experimental and analytic approach (Fig. 1). The genes that defined the cluster (groups) were in general agreement between the three study groups, and further details are described below.

Normal lung samples form a distinct group in our study *(29)* and that of Garber et al. *(30)*. However, it is also obvious that the normal lung specimen signature is more similar to adenocarcinomas. Although normal lung tissue is a good comparison to any class of lung tumor tissue, as indicated earlier, one needs to carefully consider that normal lung is composed of several cell types that contribute to the transcriptional fingerprint. For example, we observed marker genes that characterize normal lung samples but not lung tumors, including transformation growth factor β (TGFβ) receptor type II, tetranectin, and ficolin 3, (Fig. 1A). This was not unexpected, as elevated TGFβ receptor type II levels have been previously reported for normal bronchial and alveolar epithelium compared to lung carcinomas *(46)* and could mean that either the tumors observed have reduced expression or that the predominant TGFβ receptor type II expressing cell type in normal lung is absent in the tumors studied.

In our studies, SCLC and carcinoid tumors both showed high level expression of NE-specific genes (Fig. 1B), including known markers, such as insulinoma-associated gene 1 *(47)*, achaete-scute homolog 1 *(48)*, gastrin-releasing peptide, and chromogranin A. We also observed several previously undescribed markers for SCLC, such as thymosin-β expressed in neuroblastoma and the cell cycle inhibitor p18^{INK4c} (Fig. 1B). Similarly, Garber et al. *(30)* also observed expression of thymosin-β, the gene encoding 7B2, and glutaminyl cyclase in SCLC. In contrast, we observed expression of glutaminyl cyclase mostly in NE adenocarcinomas (*see* C2 tumors, below) and a few carcinoid tumors, while the expression of 7B2 was exclusively restricted to carcinoid tumors. Such differences could arise since the expression analysis platforms are different between the two studies. In addition, carcinoid tumors were not included in the study conducted by Garber et al. *(30)*, which could explain the differences in expression patterns observed. In general, carcinoid tumors, examined in our study *(29)*, appeared to be the most distinct form of lung tumors, which is consistent with previous reports *(49)*. Only a few markers were shared between SCLC and carcinoids, while a distinct group of genes defined carcinoid tumors *(29)*.

Squamous cell lung carcinomas, for which one diagnostic feature is keratinization *(2)*, formed a discrete expression cluster characterized by high levels of transcripts for multiple keratin types (keratin 5, 13, 14, 15, to 17), S100 calcium binding protein A2 and the keratinocyte-specific protein stratifin (Fig. 1C) in all three studies *(29–31)*. The squamous tumors also showed overexpression of p63, a p53-related protein essential

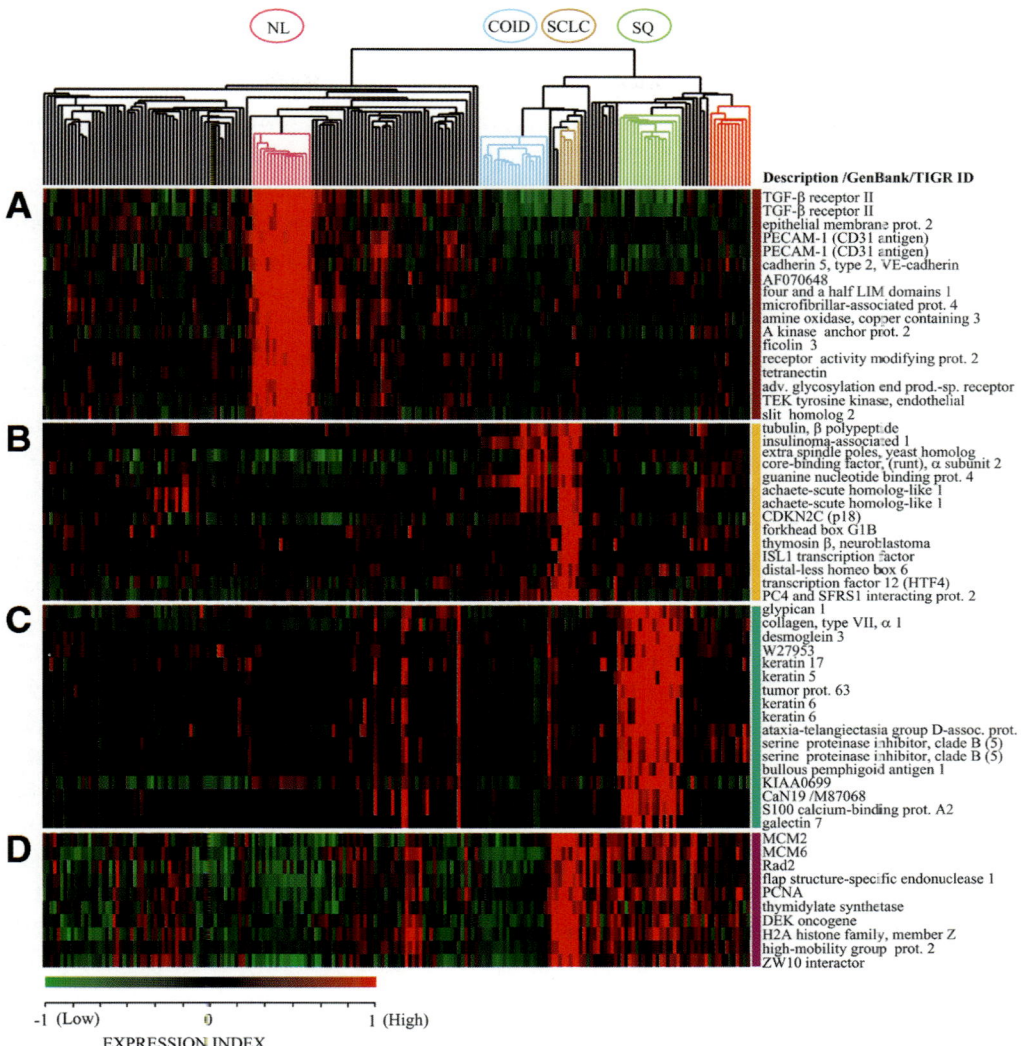

Fig. 1. Hierarchical clustering defines subclasses of lung tumors. Two-dimensional hierarchical clustering of 203 lung tumors and normal lung samples was performed with 3312 genes. The normalized expression index for each gene (rows) in each sample (columns) is indicated by a color code. **(A)** Clusters of genes with high relative expression in normal lung (NL, pink branch). **(B)** NE tumors: SCLC (gold branch) and pulmonary carcinoids (COID, light blue branch). **(C)** Squamous cell lung carcinomas with keratin markers (SQ, light green branch). **(D)** Proliferation-related markers. Black branches are adenocarcinomas, and a subset of adenocarcinomas suspected as colon metastases (CM, red branch) is indicated. Reprinted with permission from *Proc. Natl. Acad. Sci. USA* **98,** 13790–13795.

for the formation of squamous epithelia *(50)*. Several adenocarcinomas that show high expression of squamous-associated genes (Fig. 1C), also displayed histological evidence of squamous features in our data *(29)* and is consistent with previous histological findings *(2)*. Finally, expression of proliferative markers, such as proliferating cell nuclear antigen (*PCNA*), thymidylate synthase, minichromosome maintenance (*MCM*)2 and *MCM6*, was highest in SCLC, which is known to be the most rapidly dividing lung tumor followed by squamous cell carcinomas (Fig. 1D). The overexpression of PCNA and other proliferative markers, although not reported, were noted in the data of Garber et al. *(30)*. However, unlike the other major lung tumor classes shown in our studies, lung adenocarcinomas were not defined by a unique set of marker genes in all three studies *(29–31)*. Even in global tumor classification studies, the lung adenocarcinomas have few marker genes that accurately distinguish lung adenocarcinomas from other anatomically distinct primary epithelial tumors, such as prostate cancers *(3,32)*.

Identification of Adenocarcinomas Metastatic to the Lung

Transcriptional fingerprinting by microarray analysis of human lung tumors offers a wealth of information quite rapidly that was previously difficult to obtain. A key issue in lung tumor diagnosis is the discrimination of a primary lung tumor from a distant metastasis to the lung as the clinical course or treatment of the disease may differ from primary lung cancer. In our study *(29)*, microarray analysis readily defined extra-pulmonary metastasis with non-lung expression signatures among putative lung adenocarcinomas, suggesting that expression analysis may serve as a diagnostic tool to confirm and identify metastases to the lung. However, the datasets of Nacht et al. *(31)* and Garber et al. *(30)* did not have tumor samples that represented lung metastases from extrapulmonary primaries.

Briefly, we identified one distinct hierarchical cluster of 12 samples that most likely represent metastatic adenocarcinomas from the colon *(29)*. These tumors show expression of galectin-4, *CEACAM1*, and liver-intestinal cadherin 17, as well as overexpression of *c-myc*, which is common in colon carcinoma (Figs. 1, CM, and 2A). Marker gene selection for these 12 tumors included cdx1, cdx2, and cytokeratin 20 among others, which were also observed by Giordano et al. to indicate colonic origin *(51)*. Ramaswamy et al. observed that cdx1 transcription factor was over expressed in colorectal tumors *(3)*. Cdx1 is a target of the Wnt-1/β-catenin pathway that is frequently mutated in colorectal tumors. Of the 10 samples in the colon expression signature group in our study, for which clinical history and/or histopathologic information was available, only seven samples had been previously diagnosed as metastases of colonic origin. Several other adenocarcinomas also showed non-lung signatures that correlated with breast-associated markers (estrogen receptor and mammaglobin-1) and were associated with a clinical history and histopathology consistent with breast metastasis *(29)*. Thus, clustering identified suspected metastases of extra-pulmonary origin, as well as putative metastases that escaped previous diagnosis, suggesting a pivotal role of gene expression analysis in the diagnosis of lung tumors.

Class Discovery Among Lung Adenocarcinomas

Finding novel tumor classes or redefining existing tumor classes using microarray data is a challenge. Successful class discovery depends on clues unrelated to molecular signature. For example, Golub et al. chose acute leukemias as a test case in which variability in clinical outcome, subtle differences in nuclear morphology, enzyme-based histochemical analyses (periodic acid-Schiff and myeloperoxidase staining), and antibodies recognizing either lymphoid or myeloid cell surface molecules were existent, and their clinical impact was well-understood. They were able to use molecular profile

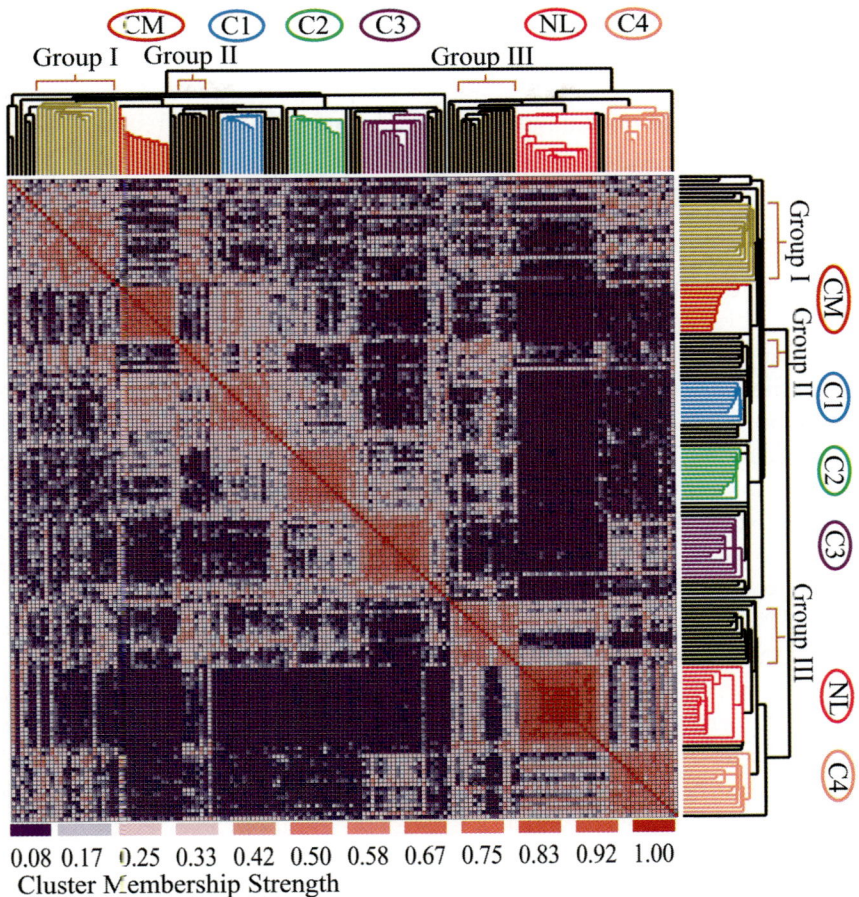

Fig. 2. Clustering defines adenocarcinoma subclasses. Comparison of classifications derived by hierarchical clustering (dendrogram) and probabilistic clustering (colored matrix) algorithms. The two-dimensional colored matrix is a visual representation of a corresponding numerical matrix, whose entries record a normalized measure of association strength between samples. Strong association approaches a value of 1 (red) and poor association is close to 0 (blue). CM, colon metastasis; NL, normal lung; C1 through C4 are adenocarcinoma clusters, and I, II, and III additional groups with weaker association. Reprinted with permission from *Proc. Natl. Acad. Sci. USA* **98,** 13790–13795.

to rediscover the classification *(25)*. However, in solid tumors such as lung cancers, especially adenocarcinomas, such robust classification is nonexistent. Therefore, discovery of new tumor classes can raise questions and needs to be validated by independent datasets.

Also, strong signatures in diverse lung tumor classes may obscure the successful subclassification of lung adenocarcinoma, a class of tumors that is believed to be rather heterogeneous. Therefore, in our study *(29)*, we also used hierarchical clustering to subclassify a dataset restricted to lung adenocarcinomas (Fig. 2). To avoid spurious variations contributing to the clustering process, we selected genes whose expression levels were most highly reproducible in a randomly chosen set of duplicate adenocarcinoma samples, yet whose expression varied widely across the chosen sample set. To reduce potential classification-bias due to choice of clustering method and to clarify cluster boundaries, we also used a model-based probabilistic clustering method *(45)*. Out of 200 bootstrapped resampled iterations, we measured the frequency with which two samples appeared in the same cluster as a reflection of the overall strength of each pair-wise association. Thereafter, we displayed the strength of pair-wise association according to the tree structure obtained from hierarchical clustering that define clusters corresponding to normal lung, putative colon metastases, and four subclasses of lung adenocarcinoma (C1 to C4) (see Fig. 2 in ref. 29). Several smaller and/or less robust groups were also observed.

Comparison of our classification data (that was restricted to adenocarcinomas) when compared to the data of Garber et al. *(30)* showed striking similarities for at least two of their classes discovered based on signature genes. A fourth group discovered by Garber et al. *(30)* corresponds primarily to LCLC, which was excluded from our study. Garber et al. analyzed lung tumors by hierarchical clustering including all histological groups of lung samples, including adenocarcinomas, SCLC, squamous carcinomas, LCLC, etc. *(30)*. When we performed a similar analysis, we observed clusters C2, C3, and C4 as coherent groupings within the hierarchical clustering of the larger set of tumors using the 3312 gene set, while cluster C1 was split into different groups *(29)*. Thus, tumor selection is a variable that appears to affect a subset of cluster boundaries. Nevertheless, the reproducibility of several adenocarcinoma subclasses across multiple datasets and clustering methods supports the validity of the adenocarcinoma clusters and their boundaries.

In order to identify marker genes that define each of the proposed clusters in such studies, a supervised approach such as a K-nearest neighbor can be employed to extract marker genes from the entire set of 12,600 transcript sequences *(25)*. For each cluster, selected genes can include those that are preferentially expressed in the cluster relative to all other samples, using the signal-to-noise metric *(25)*. The genes whose expression correlates best with each class may serve as markers for class prediction in future studies. Similar statistical approaches were used by Garber et al. *(30)* and Nacht et al. *(31)* for gene selection.

Molecular Signature of Lung Adenocarcinoma Subclasses

Hierarchical clustering approaches define samples that are defined by co-expressed transcripts. As described above, we defined four distinct subclasses of primary lung adenocarcinomas in our study *(29)* that were arbitrarily named C1 to C4. Similarly, Garber et al. *(30)* observed three adenocarcinoma classes labeled AC1 to AC3.

Tumors in the C1 cluster expressed high levels of genes associated with cell division and proliferation (Fig. 2B), which are also expressed in the squamous cell lung carcinoma and SCLC samples described earlier (Fig. 1D). Relatively high level expression of proliferation-associated genes was also seen in cluster C2. Although similar genes were observed in dataset of Garber et al. *(30)*, they were mostly elevated in SCLC and squamous cancers. As our adenocarcinoma classification was restricted to adenocarcinomas (Fig. 3), the C1 cluster was evident. However, when we mapped the C1 cluster members to a hierarchical clustering that involved all samples, the members of this category were distributed in smaller groups with adenocarcinoma clusters and squa-

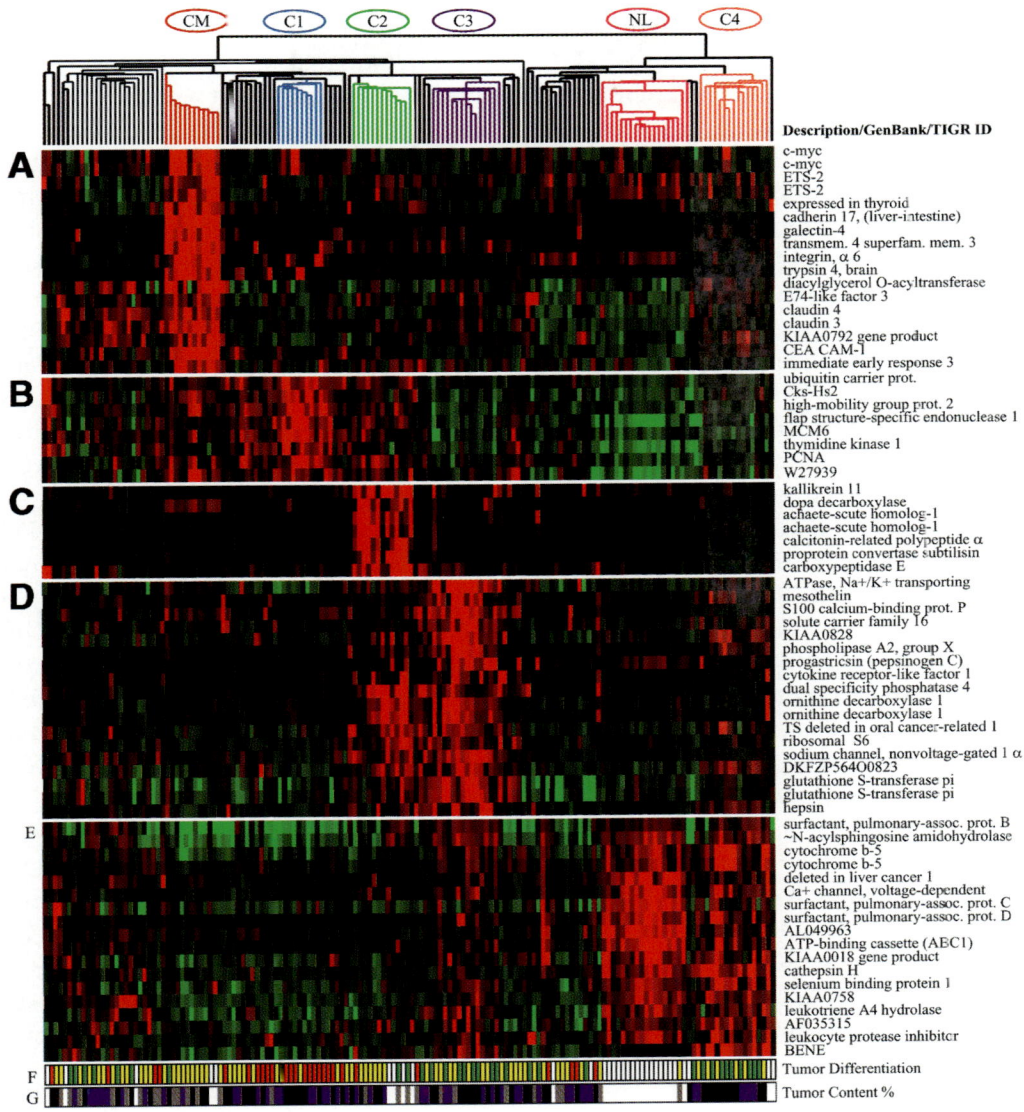

Fig. 3. Gene expression clusters and histologic differentiation within lung adenocarcinoma subclasses. Genes expressed at high levels in specific subsets of adenocarcinomas. **(A)** Colon metastases. **(B)** Proliferation-related gene expression (C1). **(C)** NE gene expression (C2). **(D)** ODC and surfactant gene expression (C3) and C2. **(E)** Type II pneumocyte gene expression (C4), C3, and normal lung. **(F)** Histopathological degree of differentiation (red, poor; yellow, moderate; green, well; white, not available or irrelevant). **(G)** Estimated nucleated tumor content (white, not determined; grey, 30–40%; blue, 40–70%; black, >70%. Reprinted with permission from *Proc. Natl. Acad. Sci. USA* **98**, 13790–13795.

mous clusters. As C1 tumors are poorly differentiated tumors, they are probably difficult to classify and may not express a set of well-differentiated molecular markers.

In our dataset *(29)*, high expression of NE markers were observed in cluster C2 (Fig. 2C). Several of these markers, such as dopa decarboxylase and achaete-scute homolog 1, were also expressed in SCLC and pulmonary carcinoids. The serine protease, kallikrein 11, was uniquely expressed in the NE C2 adenocarcinomas, but not in other NE lung tumors. Only a single adenocarcinoma tumor that closely segregated with SCLC in the dataset of Garber et al. *(30)* showed a molecular signature comparable to C2 tumors.

C3 tumors were defined by high level expression of two sets of genes in our dataset *(29)*. Expression of one gene cluster, including ornithine decarboxylase 1 (ODC1) and glutathione S-transferase pi (Fig. 2D), was shared with the NE C2 cluster. Expression of the second set of genes was shared with cluster C4 and with normal lung (Fig. 2E). The C3 adenocarcinoma subclass is similar to AC2 tumor class observed by Garber et al. *(30)*. In both datasets, this class of tumors express ODC1, citron, DUSP4, and hepsin.

Highest expression of type II alveolar pneumocyte markers, such as thyroid transcription factor 1 (TTF1) and surfactant protein B, C, and D genes, was seen in cluster C4, followed by normal lung and C3 cluster (Fig. 3E). These markers and others, such as cytochrome b5, cathepsin H, and epithelial mucin 1, are also expressed in AC1 tumor class of Garber et al. *(30)*.

The expression of genes such as amphiregulin, epiregulin, and vascular endothelial growth factor (VEGF)-C was seen in AC3 and was also observed in C0 tumors in our study *(29)*. The cluster representing LCLC analyzed by Garber et al. *(30)* shared a set of genes with the AC3 type tumors, which included amphiregulin, epiregulin, VEGF-C, plasminogen activator, and Dickkopf-1. In the dataset of Garber et al. *(30)* large cell tumors showed down-regulation of several genes that included claudin 4 and 7, epithelial-specific ETS factor, discoidin domain receptor, *PAX-8*, and *CATX-8* compared to other tumors. Based on these findings, Garber et al. *(30)* suggested that loss of PAX-8 expression and expression of Dickkopf-1 suggests a mesenchymal transitional state for large cell cancers.

All three studies *(29–31)* observed expression of genes associated with detoxification and antioxidant properties that were highly expressed in a set of adenocarcinomas and mostly in squamous cell tumors. These genes include glutathione peroxidase, glutathione S-transferase, carboxylesterase, and aldo-keto reductase. Their presence in squamous cell lung cancers, which are usually centrally located in the lung and associated with tobacco smoking, may reflect a response by the bronchial epithelium to carcinogenic insults.

Relation Between Gene Expression Tumor Classes and Histological Analysis

We noticed in our study *(29)* that cluster C1 tumors primarily contained poorly differentiated tumors, while C3 and C4 contained predominantly well-differentiated tumors. This is consistent with Garber et al. *(30)*, as tumors in cluster AC1 (corresponding to our C4 tumors) and AC2 tumors (corresponding to C3 tumors) are well-differentiated. Cluster C2 and the other adenocarcinomas were moderately differentiated (Fig. 2F). Note that C2 type tumors in the dataset of Garber et al. *(30)* is probably represented by the adenocarcinoma segregating with SCLC tumors. In our study, 10 of the 14 C4 tumors had been identified as BACs by at least one out of three pathologists who examined the tumors. The presence of type II pneumocyte markers, and the high fraction of well-differentiated tumors diagnosed as BACs suggest that cluster C4 is likely to be a gene expression counterpart of histologically-defined BAC.

It is possible that correlation of gene expression with histopathology is a result of differential tumor content. Although microscopic analysis in our dataset *(29)* indicated that our samples varied in homogeneity (Fig. 2), contamination of normal lung cells does not seem to have overwhelmed the expression signatures. The degree to which tumors clustered with normal samples did not reflect the percentage of tumor cells in a sample in most cases. Class C4 is most similar to normal lung in both hierarchical and probabilistic clustering, yet these tumors all contain at least 50% estimated tumor nuclei and, in the majority of the samples, over 80%. In contrast, classes C2 and CM contained tumors with as few as 30% estimated tumor nuclei, but are sharply distinguishable from the normal lung.

Validation of Lung Tumor Expression Profiling Results

Validation of expression profiling results is needed at different levels. At one level, it is necessary to know if the profiling data agrees with actual transcript levels in the tissue studied. In the case of lung cancer classification data, Nacht et al. noted that SAGE data were comparable to oligonucleotide probe array and quantitative PCR results for several genes *(31)*. As expression data are "noisy," it may be necessary for lung expression profiling data to be evaluated with quantitative PCR.

The use of other datasets to validate classifications based on one expression dataset is also necessary to determine unifying concepts and arrive at a consensus hypothesis. In our own experience, we have observed that experimental noise can be a dominant feature in the data structure and should be properly evaluated by determining the various levels of experimental errors. One way of evaluating and distinguishing experimental noise from biological signal is to determine the performance or classifying accuracy of biological replicates, as was determined in our dataset and that of Garber et al. *(29,30)*.

Validation of a classification scheme on an independent dataset may be performed to determine the ability of predicting a biological tumor class. This process is necessary and increases the significance of the observed or newly discovered tumor class. However, the prediction and its relevance are even more difficult when misclassification cannot be accurately evaluated. For example, since lung adenocarcinoma tumor classification and biology have not been well defined, it is difficult to understand the extent of misclassification. Once we are better able to understand the biological networks in adenocarcinomas and/or their impact on clinical end points, we may be closer to validating lung adenocarcinoma profiling results.

Novel Insights Related to Tumor Progression and Metastasis

Supervised analysis may delineate gene sets that correlate with metastatic potential. In our dataset *(29)*, C4 tumors did not metastasize, and this observation is consistent with histopathological definition of BAC. Tumor profile analysis also indicated that C3 tumors (AC2 in ref. *30*) express markers shared with the C2 type tumors. Indeed, the C3 tumors have a relatively better prognosis in the dataset of Garber et al. *(30)*, while the C2 tumors tend to have a worse prognosis in our dataset. Thus, it would be interesting to compare the genes in these sets of tumors that correlate with metastasis or survival.

Garber et al. *(30)* also report results that address tumor progression. When comparing two intrapulmonary metastatic tumors from a patient, the tumor expression profiles fell into AC group 1 and AC group 3. When they performed comparative genome hybridization (CGH) analysis, they were indeed independent samples derived from the same primary tumor and may be viewed as tumor progression and or metastasis formation. This is intriguing, as Garber et al. *(30)* propose that large cell tumors and AC3 tumors may be transitional mesenchymal-like tumors. More examples will have to be evaluated to confirm this statement. We have also noted that C4 and C0 tumors are most similar to AC1 and AC3, share expression for a set of genes, and are more similar to each other (Fig. 2). These correspond mostly to BACs or tumors with BAC-like features, respectively.

Novel Insights from Tumor Profiling

A histological review can readily identify cases of lung adenocarcinomas, squamous cell lung carcinomas, and SCLC. However, histopathological subdivision within a particular subtype and its clinical significance are not well understood. For example, adenocarcinomas are histologically classified into the following four classes: acinar, papillary, BAC, and solid adenocarcinoma with mucin formation *(2)*. However, in the case of acinar or papillary adenocarcinoma, the distinction is not well correlated with biomarkers or with clinical outcome. To this end, our data and those of Garber et al. *(30)* have a large set of adenocarcinomas and have identified several distinct subclasses within the lung adenocarcinomas with distinct signatures that are likely to be more reproducible than histological appearance of tumors. For example, the C2 adenocarcinoma subclass, defined by NE gene expression, was associated with a less favorable outcome in our dataset, and current histopathological criterion may not be able to identify this group unless specific immunohistochemical analysis are performed. While adenocarcinomas with NE features have been described previously *(52,53)*, unique markers that precisely define this group have not been described.

Our current analysis uncovered putative NE markers, such as kallikrein 11, that may discriminate the C2 tumors readily from all other lung tumors. The C2 marker kallikrein 11, which is related to the vasodepressor renal kallikrein, may be of clinical interest, given the unexplained observation of orthostatic hypotension in some lung cancer patients *(54)*. Kallikrein 11 may be useful for diagnosis of NE lung adenocarcinomas in serum or sputum samples. In the future, such an approach may become important for diagnosis and therapeutic decisions.

Expression analysis may be used for informed selection and design of rational therapies. Highly expressed genes could become important therapeutic targets. Expression analysis have shown overexpression of several receptor tyrosine kinases, such as discoidin domain receptors, RET and VEGF-C in subclasses of lung cancer *(29,30)*. Small molecule inhibitors, blocking antibodies, vaccines, and antisense drugs designed against such targets in combination with existing chemotherapy regimens could become valuable for treating lung cancer *(37)*. Several agents that target one or more members of receptor tyrosine kinases are currently undergoing clinical investigation for lung cancers. For example, ZD1839 (Iressa), an epidermal growth factor receptor (EGFR) tyrosine kinase inhibitor, which blocks signal transduction pathways implicated in the proliferation, survival, and growth of cancer cells, has recently entered phase III clini-

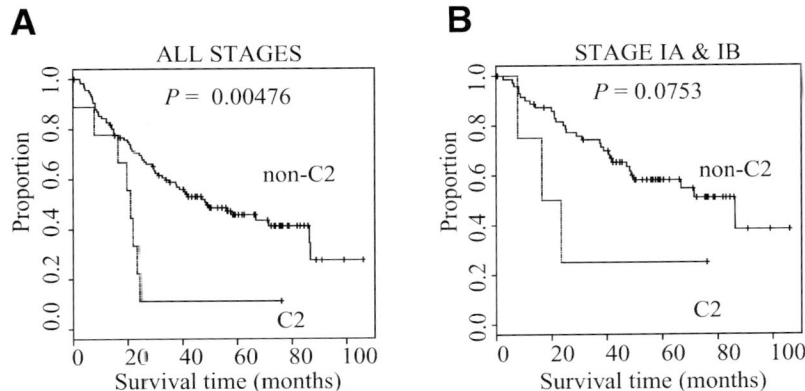

Fig. 4. Survival analysis of NE C2 adenocarcinomas. Kaplan-Meier curves for C2 vs all other adenocarcinomas. (A) All patients. n = 9 for C2, n = 117 for others. (B) Patients with stage I tumors only. n = 4 for C2, n = 72 for others. Reprinted with permission from *Proc. Natl. Acad. Sci. USA* **98**, 13790–13795.

cal trials for lung cancer *(55)*. Alternatively, cells that harbor mutations in tumor suppressors or have oncogenic mutations can also be targeted separately from normal cells. A virus targeted to *p53*-defective cells is currently being evaluated for lung cancer and squamous tumors as well as lung adenocarcinomas that have mutations in *p53 (56)*. Similarly, peptide antibody against mucin-1 (BIOMIRA, Edmonton, Alberta, Canada), which is overexpressed in C3/C4 type tumors *(29,31)*, might be used to target mucin-1 positive lung tumors.

NOVEL INSIGHTS FROM COMPARING TUMOR PROFILING WITH CLINICAL RESPONSE: CORRELATION OF PATIENT OUTCOME WITH PUTATIVE ADENOCARCINOMA CLASSES

Clustering can classify tumors into distinct groups. Garber et al. *(30)* identified clusters based on the similarity of transcriptional profile *(30)*. We also identified clusters using a similar approach as well as on the basis of stable clusters identified by combining clustering methods and bootstrapping of tumor samples *(29)*. With our dataset, we asked whether lung cancer patient outcome correlated with the subclasses of lung adenocarcinomas defined. However, such clusters may or may not correlate with clinical end points. For example, in our dataset, there was no detectable difference in prognosis between the primary lung adenocarcinomas and the metastases to the lung of colonic origin. However, in our dataset *(29)*, the NE C2 adenocarcinomas were associated with a less favorable survival outcome than all other adenocarcinomas (Figs. 3A and 4B). The median survival for C2 tumors was 21 mo compared to 40.5 mo for all non-C2 tumors ($p = 0.00476$). When only stage I tumors were considered, the median survival for patients with C2 tumors was 20 mo compared to 47.8 mo for patients with non-C2 tumors. In contrast, the median survival for patients with C4 tumors was 49.7 mo, while the median survival for patients with non-C4 tumors was 33.2 mo ($p = 0.049$). Garber et al. *(30)* also compared patient outcome to adenocarcinoma classes. They observed that AC2 tumors (which were similar to C3 tumors) had the best survival (up to 50 mo), while the AC3 tumors (which were similar to C0 tumors) had the worst prognosis. The AC1 tumors (similar to C4 tumors) had intermediate survival. It is evident that, in either study, the correlation of survival with expression-based classification has revealed distinct results. While the results are not contradictory, they are not in perfect agreement. The classification methods in both studies are distinct and may be a factor in the observed differences in clinical end points. It can also be argued that, in either classification, adenocarcinomas are classified by molecular markers that do not correlate or are relevant to clinical endpoints in both datasets *(29,30)*. Therefore, supervised approaches should be more effective in revealing the markers associated with clinical end points, such as survival or recurrence of tumors. Further cross-validation studies will be necessary to determine the set of genetic and other patient information that best describe clinical endpoints.

FUTURE

The preliminary data on lung cancer studies from three independent groups suggests that classification by expression profiling may be indicative of prognosis. However, prognosis may correlate with a small set of expressed genes independent of the larger gene sets that define sample clusters. Determination of biological clusters or classes based on hierarchical clustering shows the set of genes that are co-expressed providing information on how a tumor's biological network operates. This may reveal novel autocrine or paracrine signaling pathways *(42)* significant in tumor proliferation and metastasis.

Another issue is how should we compare different lung cancer datasets. Cross platform or intra- and interplatform comparisons are currently not feasible. Even within the same platform, we have observed that experimental noise can dampen biological signals. For now, we have to evaluate datasets independently. Issues also include platform-specific problems, such as cross-hybridization of target (mRNA). Eventually, expression analysis will be integrated with other approaches that examine DNA amplifications and losses in cancers (such as comparative genome hybridization or single nucleotide polymorphisms), protein expression and protein interaction (proteomics), and the consensus information from all such databases to paint a picture of tumor classes and tumorigenesis.

SUMMARY

Comparison of our results *(29)* to independent studies performed with different set of tumors and expression-profiling platforms *(30,31)*, reveals a number of similarities. Genes characteristic for previously defined tumor classes, such as SCLC and squamous cell lung carcinoma, overlap heavily between analyses. Furthermore, two of the adenocarcinoma classes we have identified, those with type II pneumocyte gene expression (C4) and those with ODC/surfactant gene expression (C3), have counterparts in the data of Garber et al. *(30)*. However, other findings are, so far, unique to individual data sets. Differences between the studies indicate that the number of samples in a single study alone is probably too small to allow for the generation of a classification scheme that fully represents the complexity of lung cancer.

Cumulative evidence already suggests that gene expression patterns can predict the clinical behavior and therapeutic response of cancers *(39,40)*. Together, the generation of gene expression-based classification of lung cancer and a subclassification of lung adenocarcinoma serves as the first step towards a new molecular taxonomy that should provide new molecular targets for prognosis and rational therapy.

ACKNOWLEDGMENTS

We offer special thanks to David Botstein, Mitch Garber, and Jin Jen for their assistance with this review. Special thanks are due to David Livingston, who has stimulated and coordinated the lung cancer classification project at DFCI/WI/MIT funded by U01 CA84995 from the National Cancer Institute, and to our collaborators Todd Golub, Jane Staunton, Cheng Li, Stefano Monti, Wing Wong, William Richards, David Sugarbaker, and Bruce Johnson. Part of this work was also supported in part by Millennium Pharmaceuticals, Affymetrix, and Bristol-Myers Squibb. M.M is a Pew Scholar in the Biomedical Sciences.

REFERENCES

1. Greenlee, R. T., Hill-Harmon, M. B., Murray, T., and Thun, M. (2001) Cancer statistics, 2001. *CA Cancer J. Clin.* **51,** 15–36.
2. Travis, W. D., Colby, T. V., Corrin, B., Shimosato, Y., and Brambilla, E. (1999) Histological typing of lung and pleural tumors, in *World Health Organization International Histological Classification of Tumors.* Springer Verlag, Berlin.
3. Ramaswamy, S., Tamayo, P., Rifkin, R., et al. (2001) Multiclass cancer diagnosis using tumor gene expression signatures. *Proc. Natl. Acad. Sci. USA* **98,** 15,149–15,154.
4. Druker, B. J., Talpaz, M., Resta, D. J., et al. (2001) Efficacy and safety of a specific inhibitor of the BCR-ABL tyrosine kinase in chronic myeloid leukemia. *N. Engl. J. Med.* **344,** 1031–1037.
5. Travis, W. D., Travis, L. B., and Devesa, S. S. (1995) Lung cancer. *Cancer* **75,** 191–202.
6. Bennett, W. P., Hussain, S. P., Vahakangas, K. H., Khan, M. A., Shields, P. G., and Harris, C. C. (1999) Molecular epidemiology of human cancer risk: gene-environment interactions and p53 mutation spectrum in human lung cancer. *J. Pathol.* **187,** 8–18.
7. Hainaut, P., Olivier, M., and Pfeifer, G. P. (2001) TP53 mutation spectrum in lung cancers and mutagenic signature of components of tobacco smoke: lessons from the IARC TP53 mutation database *Mutagenesis* **16,** 551–553.
8. Kim, D. H., Nelson, H. H., Wiencke, J. K., et al. (2001) p16(INK4a) and histology-specific methylation of CpG islands by exposure to tobacco smoke in non-small cell lung cancer. *Cancer Res.* **61,** 3419–3424.
9. Hollstein, M., Sidransky, D., Vogelstein, B., and Harris, C. C. (1991) p53 mutations in human cancers. *Science* **253,** 49–53.
10. Levine, A. J., Mcmand, J., and Finlay, C. A. (1991) The p53 tumour suppressor gene. *Nature* **351,** 453–456.
11. Gazdar, A. F. (1994) The molecular and cellular basis of human lung cancer. *Anticancer Res.* **14,** 261–267.
12. Sorensen, J. B., Hirsch, F. R., Gazdar, A., and Olsen, J. E. (1993) Interobserver variability in histopathologic subtyping and grading of pulmonary adenocarcinoma. *Cancer* **71,** 2971–2976.
13. Breathnach, O. S., Ishibe, N., Williams, J., Linnoila, R. I., Caporaso, N., and Johnson, B. E. (1999) Clinical features of patients with stage IIIB and IV bronchioloalveolar carcinoma of the lung. *Cancer* **86,** 1165–1173.
14. Breathnach, O. S., Kwiatkowski, D. J., Finkelstein, D. M., et al. (2001) Bronchioloalveolar carcinoma of the lung: recurrences and survival in patients with stage I disease. *J. Thorac. Cardiovasc. Surg.* **121,** 42–47.

15. Flint, A. and Lloyd, R. V. (1992) Pulmonary metastases of colonic carcinoma. Distinction from pulmonary adenocarcinoma. *Arch. Pathol. Lab. Med.* **116,** 39–42.
16. Shirakusa, T., Tsutsui, M., Motonaga, R., Ando, K., and Kusano, T. (1988) Resection of metastatic lung tumor: the evaluation of histologic appearance in the lung. *Am. Surg.* **54,** 655–658.
17. Mountain, C. F. (2000) The international system for staging lung cancer. *Semin. Surg. Oncol.* **18,** 106–115.
18. Chee, M., Yang, R., Hubbell, E., et al. (1996) Accessing genetic information with high-density DNA arrays. *Science* **274,** 610–614.
19. Lockhart, D. J., Dong, H., Byrne, M. C., et al. (1996) Expression monitoring by hybridization to high-density oligonucleotide arrays. *Nat. Biotechnol.* **14,** 1675–1680.
20. Schena, M., Shalon, D., Davis, R. W., and Brown, P. O. (1995) Quantitative monitoring of gene expression patterns with a complementary DNA microarray. *Science* **270,** 467–470.
21. Schena, M., Shalon, D., Heller, R., Chai, A., Brown, P. O., and Davis, R. W. (1996) Parallel human genome analysis: microarray-based expression monitoring of 1000 genes. *Proc. Natl. Acad. Sci. USA* **93,** 10,614–10,619.
22. Alizadeh, A. A., Eisen, M. B., Davis, R. E., et al. (2000) Distinct types of diffuse large B-cell lymphoma identified by gene expression profiling. *Nature* **403,** 503–511.
23. Alon, U., Barkai, N., Notterman, D. A., et al. (1999) Broad patterns of gene expression revealed by clustering analysis of tumor and normal colon tissues probed by oligonucleotide arrays. *Proc. Natl. Acad. Sci. USA* **96,** 6745–6750.
24. Bittner, M., Meltzer, P., Chen, Y., et al. (2000) Molecular classification of cutaneous malignant melanoma by gene expression profiling. *Nature* **406,** 536–540.
25. Golub, T. R., Slonim, D. K., Tamayo, P., et al. (1999) Molecular classification of cancer: class discovery and class prediction by gene expression monitoring. *Science* **286,** 531–537.
26. Khan, J., Wei, J. S., Ringner, M., et al. (2001) Classification and diagnostic prediction of cancers using gene expression profiling and artificial neural networks. *Nat. Med.* **7,** 673–679.
27. Perou, C. M., Sorlie, T., Eisen, M. B., et al. (2000) Molecular portraits of human breast tumours. *Nature* **406,** 747–752.
28. Welsh, J. B., Zarrinkar, P. P., Sapinoso, L. M., et al. (2001) Analysis of gene expression profiles in normal and neoplastic ovarian tissue samples identifies candidate molecular markers of epithelial ovarian cancer. *Proc. Natl. Acad. Sci. USA* **98,** 1176–1181.
29. Bhattacharjee, A., Richards, W. G., Staunton, J., et al. (2001) Classification of human lung carcinomas by mRNA expression profiling reveals distinct adenocarcinoma subclasses. *Proc. Natl. Acad. Sci. USA* **98,** 13,790–13,795.
30. Garber, M. E., Troyanskaya, O. G., Schluens, K., et al. (2001) Diversity of gene expression in adenocarcinoma of the lung. *Proc. Natl. Acad. Sci. USA* **98,** 13,784–13,789.
31. Nacht, M., Dracheva, T., Gao, Y., et al. (2001) Molecular characteristics of non-small cell lung cancer. *Proc. Natl. Acad. Sci. USA* **98,** 15,203–15,208.
32. Su, A. I., Welsh, J. B., Sapinoso, L. M., et al. (2001) Molecular classification of human carcinomas by use of gene expression signatures. *Cancer Res.* **61,** 7388–7393.
33. Sunday, M. E., Haley, K. J., Sikorski, K., et al. (1999) Calcitonin driven v-Ha-ras induces multilineage pulmonary epithelial hyperplasias and neoplasms. *Oncogene* **18,** 4336–4347.
34. Linnoila, R. I., Zhao, B., DeMayo, J. L., et al. (2000) Constitutive achaete-scute homologue-1 promotes airway dysplasia and lung neuroendocrine tumors in transgenic mice. *Cancer Res.* **60,** 4005–4009.
35. Jackson, E. L., Willis, N., Mercer, K., et al. (2001) Analysis of lung tumor initiation and progression using conditional expression of oncogenic K-ras. *Genes Dev.* **15,** 3243–3248.
36. Johnson, L., Mercer, K., Greenbaum, D., et al. (2001) Somatic activation of the K-ras oncogene causes early onset lung cancer in mice. *Nature* **410,** 1111–1116.
37. Meuwissen, R., Linn, S. C., van der Valk, M., Mooi, W. J., and Berns, A. (2001) Mouse model for lung tumorigenesis through Cre/lox controlled sporadic activation of the K-Ras oncogene. *Oncogene* **20,** 6551–6558.

38. Fisher, G. H., Wellen, S. L., Klimstra, D., et al. (2001) Induction and apoptotic regression of lung adenocarcinomas by regulation of a K-Ras transgene in the presence and absence of tumor suppressor genes. *Genes Dev.* **15,** 3249–3262.
39. Scherf, U., Ross, D. T., Waltham, M., et al. (2000) A gene expression database for the molecular pharmacology of cancer. *Nat. Genet.* **24,** 236–244.
40. Staunton, J. E., Slonim, D. K., Coller, H. A., et al. (2001) Chemosensitivity prediction by transcriptional profiling. *Proc. Natl. Acad. Sci. USA* **98,** 10,787–10,792.
41. Ge, H., Liu, Z., Church, G. M., and Vidal, M. (2001) Correlation between transcriptome and interactome mapping data from *Saccharomyces cerevisiae*. *Nat. Genet.* **29,** 482–486.
42. Graeber, T. G. and Eisenberg, D. (2001) Bioinformatic identification of potential autocrine signaling loops in cancers from gene expression profiles. *Nat. Genet.* **29,** 295–300.
43. Roberts, C. J., Nelson, B., Marton, M. J., et al. (2000) Signaling and circuitry of multiple MAPK pathways revealed by a matrix of global gene expression profiles. *Science* **287,** 873–880.
44. Eisen, M. B., Spellman, P. T., Brown, P. O., and Botstein, D. (1998) Cluster analysis and display of genome-wide expression patterns. *Proc. Natl. Acad. Sci. USA* **95,** 14,863–14,868.
45. Cheeseman, P. and Stutz, J. (1996) Bayesian classification (autoclass): theory and results, in *Advances in Knowledge Discovery and Data Mining* (Fayyad, U. M., Piatetsky-Shapiro, G., Smyth, P., and Uthurasamy, R., eds.), MIT Press, Cambridge.
46. Kang, Y., Prentice, M. A., Mariano, J. M., et al. (2000) Transforming growth factor-beta 1 and its receptors in human lung cancer and mouse lung carcinogenesis. *Exp. Lung Res.* **26,** 685–707.
47. Lan, M. S., Russell, E. K., Lu, J., Johnson, B. E., and Notkins, A. L. (1993) IA-1, a new marker for neuroendocrine differentiation in human lung cancer cell lines. *Cancer Res.* **53,** 4169–4171.
48. Ball, D. W., Azzoli, C. G., Baylin, S. B., et al. (1993) Identification of a human achaete-scute homolog highly expressed in neuroendocrine tumors. *Proc. Natl. Acad. Sci. USA* **90,** 5648–5652.
49. Anbazhagan, R., Tihan, T., Bornman, D. M., et al. (1999) Classification of small cell lung cancer and pulmonary carcinoid by gene expression profiles. *Cancer Res.* **59,** 5119–5122.
50. Yang, A., Schweitzer, R., Sun, D., et al. (1999) p63 is essential for regenerative proliferation in limb, craniofacial and epithelial development. *Nature* **398,** 714–718.
51. Giordano, T. J., Shedden, K. A., Schwartz, D. R., et al. (2001) Organ-specific molecular classification of primary lung, colon, and ovarian adenocarcinomas using gene expression profiles. *Am. J. Pathol.* **159,** 1231–1238.
52. Berendsen, H. H., de Leij, L., Poppema, S., et al. (1989) Clinical characterization of non-small-cell lung cancer tumors showing neuroendocrine differentiation features. *J. Clin. Oncol.* **7,** 1614–1620.
53. Skov, B. G., Sorensen, J. B., Hirsch, F. R., Larsson, L. I., and Hansen, H. H. (1991) Prognostic impact of histologic demonstration of chromogranin A and neuron specific enolase in pulmonary adenocarcinoma. *Ann. Oncol.* **2,** 355–360.
54. Vio, C. P., Olavarria, V., Gonzalez, C., Nazal, L., Cordova, M., and Balestrini, C. (1998) Cellular and functional aspects of the renal kallikrein system in health and disease. *Biol. Res.* **31,** 305–322.
55. Ciardiello, F. and Tortora, G. (2001) A novel approach in the treatment of cancer: targeting the epidermal growth factor receptor. *Clin. Cancer Res.* **7,** 2958–2970.
56. Nemunaitis, J., Cunningham, C., Buchanan, A., et al. (2001) Intravenous infusion of a replication-selective adenovirus (ONYX-015) in cancer patients: safety, feasibility and biological activity. *Gene Ther.* **8,** 746–759.

12
Molecular Profiling of Bladder Cancer Using High-Throughput DNA Microarrays

Marta Sánchez-Carbayo and Carlos Cordon-Cardo

INTRODUCTION

Understanding the biology of tumorigenesis and tumor progression of bladder cancer is essential for improving our capacity to diagnose and treat this group of diseases. Unraveling the biological complexity underlying these processes is expected to provide novel tools of predictive nature and identify therapeutic targets. The further characterization of the regulatory mechanisms and pathways controlling cellular homeostasis that are altered in bladder tumors will be achieved, at least in part, by global analyses of gene expression. Until recently, the ability to identify and analyze gene expression patterns has been technically limited to relatively few genes per study. This limitation is being overcome by the development of a number of methods that allow for a more comprehensive analysis of these patterns. Some of the most powerful methods include differential display *(1)*, serial analysis of gene expression (SAGE) *(2)*, massively parallel signature sequencing (MPSS) *(3)*, as well as protein composition-based approaches, such as protein microarrays *(4)* and combined two-dimensional gel followed by mass spectral analysis *(5)*.

Concurrent with the development of these techniques has been the tremendous increase in the DNA sequence information available for a range of organisms through genome sequencing efforts *(6,7)*. An important component of the mouse and human sequencing efforts has been the determination, by cDNA sequencing, of both full and partial sequences for tens of thousands of genes. In addition to increasing the utility of the techniques described above, this expansion of gene and genome information has served as the basis for development of one of the most powerful new techniques available to provide both static and dynamic views of gene expression patterns in cells and tissues, the so-called DNA microarrays *(8,9)*.

DNA microarrays can be used to "interrogate" complex mixtures of a myriad of nucleic acid sequences for both presence and abundance of known genes. Hybridization-based analysis and the microarray format together constitute an extremely versatile technology. There is an increasingly broad range of applications for microarrays, including genotyping polymorphisms and mutations *(10,11)*, determining the binding sites of DNA-binding proteins *(12)*, and identifying structural alterations using arrayed comparative genome hybridization (CGH) approaches *(13)*. The most widespread use of this technology to date has been the analysis of gene expression.

In this chapter, we describe the potential role of this technology in the context of uroepithelial neoplasms, discuss some practical considerations for the use of DNA microarray platforms, mainly when utilizing clinical material, and summarize applications of this technology to the study of bladder cancer.

CLINICAL AND MOLECULAR PATHOLOGIC FEATURES OF BLADDER CANCER

Molecular epidemiology studies of bladder cancer have contributed to the modern viewpoint of tumorigenesis mainly based on associations with environmental exposures and DNA damaging agents. Bladder cancer was one of the first neoplastic diseases to be found associated with an industrial chemical exposure *(14)*. It has also been related to genotypic characteristics of individual polymorphisms, such as the fast acetylator phenotype *(15)*. In addition, bladder cancer is one of the first tumors in which early stage disease is treated with immunotherapy *(16)*. The early diagnosis of bladder cancer is increasing, particularly in superficial preinvasive stages due to simplified procedures such as the use of flexible cystoscopy.

Pathologically, most bladder tumors are transitional cell carcinomas. There is, however, increasing recognition of the prognostic importance associated with the metaplastic variants displaying squamous and glandular differentiation as part of their clonal evolution. Bladder tumors are pathologically stratified based on stage, grade, tumor size, presence of concomitant carcinoma *in situ*, and multicentricity. The power of these histopathological variables in defining the clinical subtypes of bladder cancer and predicting the clinical outcome of individual patients has certain limitations. Within each stage, it has been very difficult to clinically identify useful parameters that can predict risk of disease recurrence or progression. It is for these reasons that many groups of investigators have examined additional molecular characteristics of bladder cancer that may be of predictive value. Phenotypic features associated with tumor aggressiveness include cell cycle and apoptosis regulators.

Bladder cancer, including some superficial lesions, has been reported to carry a significant number of genetic alterations at the time of diagnosis *(17)*. If genetic changes occur randomly in tumor cells, then selective advantage must supply the drive for some changes to become stabilized, such as critical activation of specific oncogenic events or inactivation of certain tumor suppressor genes.

A substantial body of work has suggested that superficial papillary tumors (pTa) differ from flat carcinoma *in situ* lesions (TIS) and muscle invasive tumors in their molecular pathogenesis and pathways of progression. The number of genetic alterations is substantially higher in the invasive lesions, but there also appears to be a difference in the specific alterations present. Based on data from several groups, it appears that there are at least two major molecular pathways of bladder tumor development and evolution that can be followed (Fig. 1). The first, represented by papillary superficial tumors, is associated with chromosome 9 losses, including inactivation of cyclin-dependent kinase inhibitor 2A (CDKN2A) (p16) on 9p and still unknown genes associated with telomeric 9q loci *(18,19)*. The second pathway includes inactivation of

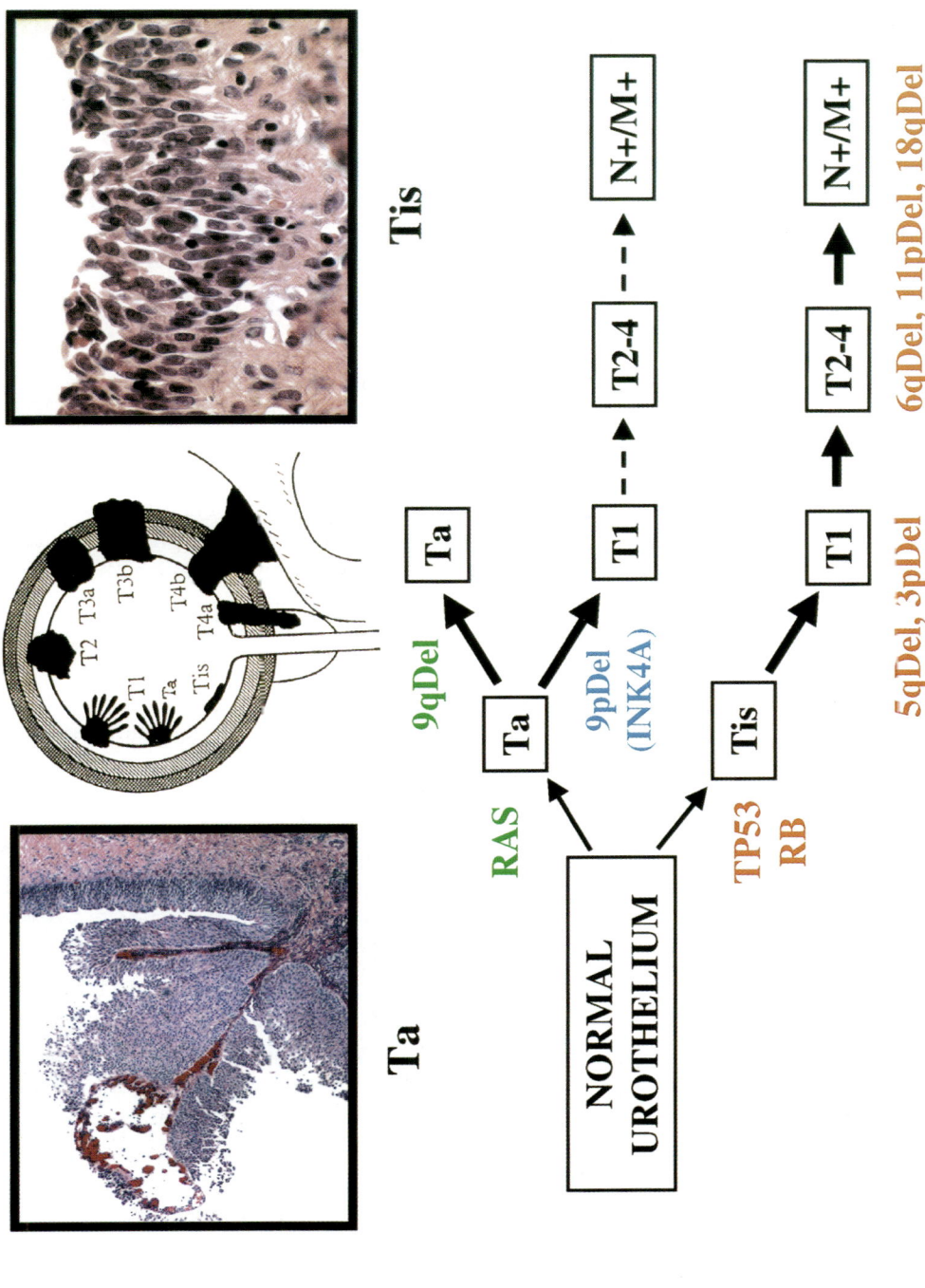

Fig. 1. Overview of the main genetic alterations associated with the two major progression pathways described for bladder cancer. T, N, M refers to tumor-node-metastasis (TNM) staging nomenclature.

p53 on chromosome 17 (17p11.3) and retinoblastoma (RB) on 13q14, seen in flat carcinoma *in situ* and pT1 tumors *(20–22)*.

PRACTICAL CONSIDERATIONS IN THE USE OF DNA MICROARRAYS FOR BLADDER CANCER ANALYSIS

Selection of the Type of Arrays

There is a variety of commercially available oligonucleotide, glass, or nylon cDNA microarrays. Alternatively, some researchers have opted to manufacture their own spotted cDNA microarrays. Issues of cost, setup time, personnel available, flexibility, and product range will influence this decision. The oligonucleotide and spotted cDNA formats each have unique advantages and disadvantages and thus offer investigators a distinct choice *(23–25)*.

Sample Preparation

Several types of samples are available to study bladder cancer by expression profiling. Normal urothelia and tumor tissues can be obtained by transurethral resection, cystectomy, or cystoprostatectomy. Due to the close monitoring of bladder cancer patients, sequential biopsies obtained over time allow addressing critical issues related to tumor progression and response to treatment. Optimal results are achieved by handling tissue promptly and either extracting RNA immediately from fresh aliquots or deep freezing in liquid nitrogen in either tubes or using cryomolds and embedding medium. This latter format allows verification of histopathological characteristics, since it represents a frozen tissue block. It also provides adequate samples for tissue microdissection if required.

Bladder cancer offers an additional source of material for tumor profiling studies based on direct access to exfoliated tumor cells through urine samples and bladder washes. This approach has not been reported to date, but represents an alternative strategy mainly by the amount and purity of tumor cells collected.

A critical issue in microarray technology is the quantity and quality of the RNA from which the hybridization sample is prepared. This is particularly critical when the RNA is isolated from clinical specimens. Potential solutions to the quantitative limitation include probe labeling protocols that increase sensitivity through label signal amplification using dendrimers *(24)*, probe amplification protocols that reduce the amount of RNA required through the use of highly efficient phage RNA polymerases or polymerase chain reaction (PCR) amplification *(25–27)*, and posthybridization amplification methods in which the target–probe duplexes are detected enzymatically *(27)*. Currently, probe amplification is the most frequently used method to address issues dealing with limiting starting material and to improve the detection of low abundance gene transcripts. Although there are studies showing favorable data *(27,28)*, care must be taken when amplifying probe material, as this may introduce bias, such that the hybridization probe does not accurately reflect the transcript representation in the original RNA sample *(28)*.

Problems of RNA quality usually are severe when working with archived samples, as many fixing and embedding protocols utilize aldehyde-based products, which damage RNA integrity *(29,30)*. Nonetheless, there have been studies that indicate that the RNA quality is not diminished, at least for certain tumors, following certain fixation protocols *(31)*.

Tumor heterogeneity is a critical issue in tumor expression profiling, in general, and particularly in the case of bladder cancer. Tumor cells are surrounded by normal connective tissue and inflammatory infiltrates. In addition to manual microdissection of tissue sections from embedded frozen tumor blocks, laser capture microdissection appears to be a technique capable of isolating relatively pure cancer cells populations from clinical specimens *(32,33)*. The degree of the effect of the laser beam on the quality of the RNA obtained is still controversial, although there are reports showing good results using different microdissecting platforms *(31)*. Flow cytometry sorting is also possible using bladder washes or disaggregated tissue preparations as previously described *(34)*. The establishment of suitable tissue banks that are well annotated and procured following appropriated guidelines is a logical component of any translational program *(35)*.

EXPRESSION PROFILING AND THE STUDY OF BLADDER CANCER

Microarray-based gene expression profiling has found a number of important applications in the study of carcinogenesis and cancer biology. Broadly speaking, these applications can be described as: gene and pathway discovery, functional classification of genes, and tumor classification.

Gene and pathway discovery is mainly based on functional association of changes in gene expression between different cell states or phenotypes. This approach, which is associating a change in the expression of a gene with a change in physiological state, is one of the simplest ways in which gene expression profiling can be used to suggest or predict gene function. Another way in which expression profiling can be used in the functional classification of genes is often referred to as "guilt by association." This method is based on the observation that genes with related expression patterns, genes that presumably are co-regulated, are likely to be functionally related and involved in the control of the same biological processes or physiological pathways. When genes with similar expression profiles are grouped, a process referred to as clustering, novel genes (usually expressed sequence tags [ESTs]) are often found mixed with genes of known function. A tentative activity for the novel genes can be inferred by this grouping. Moreover, new functions can be ascribed to known genes when they are grouped with genes that have a distinct functional classification. Similarly, previously unknown functions can be ascribed to pathways when expression changes in the genes that are part of the pathway correlate with changes in the physiological state of the cell.

Functional classification of genes is a traditional approach to assigning a functional role to a gene when overexpressed and observing the effect(s) of its expression on known pathways or processes. This approach has been especially useful in identifying the downstream targets of transcription factors. The genes identified as either up- or down-regulated in these experiments are likely to play important roles in the functional pathways controlled by the gene under investigation.

In these experiments, it is often critical to be able to tightly control expression of the gene under study. Another potential problem, particularly in experiments in which cells are transiently transfected, can be in restricting the analysis to only those cells that actually express the gene of interest. This can be avoided using expression constructs in which the gene of interest is fused to a tag, such as a green fluorescent protein, that can be used to select and enrich for cells expressing the gene.

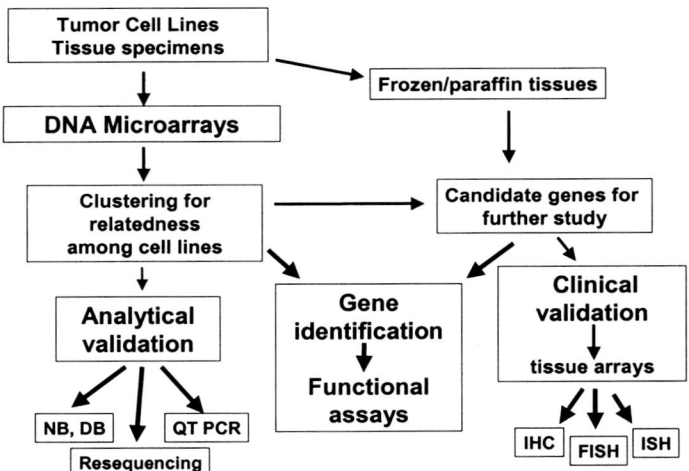

Fig. 2. The general procedure of a tumor expression profiling experiment includes RNA isolation from tumor biopsy and control samples, preparation of the hybridization probe, hybridization with the DNA microarray, data acquisition and analysis, and verification of the results using, for example, tissue microarrays. NB, Northern blotting; DB, dot blotting; QT-PCR, quantitative PCR; IHC, immunohistochemistry; ISH, *in situ* hybridization; FISH, fluorescent ISH.

Tumor classification is one of the most exciting and potentially most powerful applications of expression profiling with DNA microarrays. Major goals for improving cancer treatment include the early and accurate diagnosis of tumor type and determining the extent of the disease. The traditional approach to tumor classification is based on clinicopathological criteria. It is expected that the integration of gene expression patterns, as determined by DNA microarrays, will provide a better means for classifying tumors into biologically meaningful and clinically useful categories. In addition, expression profiling of well-curated tumor specimens has the potential of identifying target genes for novel diagnostic, prognostic, or therapeutic approaches. Although the true clinical utility of expression profile-based tumor classification is controversial and still unproven, early results are encouraging *(36–41)*. A general outline of the process of microarray expression profiling for classification of tumor specimens, including the important step of target gene validation using several procedures, such as high-throughput tissue microarrays, is illustrated in Fig. 2.

APPLICATION OF DNA MICROARRAYS TO THE STUDY OF BLADDER CANCER

The main advantage of DNA arrays is that they allow the study of the multiple transcriptional events that take place when normal urothelium is transformed into tumor tissue in single experiments. Expression profiling using cell lines has been used to gain an insight into the molecular events associated with the disease. An example of how the technology can be applied to the discovery of gene functions and pathways in bladder cancer is provided by the following study. Tumor cell growth inhibition mediated by genistein was induced in the susceptible bladder tumor line TCCSUP. Expression profiling was analyzed at various time points using cDNA chips containing 884 sequence-verified known human genes. Transient induction of early growth response protein-1 (EGR-1) was found, relating this event to its proliferation and differentiation effects. The study reported many groups of genes with distinct expression profiles, most of them encoding proteins that regulate cell growth or the cell cycle *(42)*.

The following study provides an example of the functional classification of genes applied to bladder cancer, relating the expression patterns of p53-mediated apoptosis in resistant tumor cell lines vs sensitive tumor cell lines. The ECV-304 bladder carcinoma cell line was selected for resistance to p53 by repeated infections with a p53 recombinant adenovirus Ad5-cytomegalovirus (CMV)-p53. Its expression pattern using cDNA arrays containing 5730 genes was compared with p53-sensitive ECV-304 cells. A number of potential p53 transcription or related targets were identified playing roles in cell cycle regulation, DNA repair, redox control, cell adhesion, apoptosis, and differentiation. Proline oxidase, a mitochondrial enzyme involved in the proline–pyrroline-5-carboxylate redox cycle, was up-regulated in sensitive, but not in resistant cells. Further experiments with pyrroline-5-carboxylate (P5C), a proline-derived metabolite generated by proline oxidase, inhibited the proliferation and survival of resistant and sensitive cells, inducing apoptosis in both cell lines. These results showed the implication of proline oxidase and the proline/P5C pathway in p53-induced growth suppression and apoptosis *(43)*.

Expression analysis has also been utilized to monitor in vitro the effect of therapies under preclinical and clinical trials, such as DNA methylation inhibitors, of drugs attempting to reactivate silenced genes in human cancers. High-density oligonucleotide gene expression microarrays were used to examine the effects of 5-aza-2'-deoxycytidine treatment on a human bladder tumor cell line (T24) as compared to human fibroblast cells (LD419). Data obtained 8 d after recovery from this treatment showed that more genes were induced in tumorigenic cells (61 genes induced) than nontumorigenic cells (34 genes induced). Approximately 60% of induced genes did not have CpG islands within their 5' regions, suggesting that some genes activated by this treatment may not result from the direct inhibition of promoter methylation. It was also shown that a high percentage of genes activated in both cell types belonged to the interferon (IFN) signaling pathway, confirming previous reports from other tumor cell types *(44)*.

EXPRESSION PROFILING FINDINGS USING HUMAN BLADDER CANCER SPECIMENS

Gene expression patterns may vary sufficiently to complicate the task of sample classification based upon expression profiles. A significant challenge lies in choosing the best groups of genes with which to identify the biologically related tumor subclasses. Certain groups or "clusters" of genes vary consistently in tumor samples, and these genes can frustrate attempts at subclassification *(36,37)*. Tumor classifications based upon these types of "dominant" gene clusters are unlikely to identify useful tumor subclasses. Much care must be placed in the choice of genes used to subclassify tumors, and the choices must be thoroughly and rigorously verified and validated, both statistically and clinically.

There have been few reports dealing with molecular classification of bladder cancer expression profiling using DNA microarrays. The most extensive one has monitored the expression patterns of superficial and invasive tumor cell suspensions prepared from 36 normal and 29 bladder tumor biopsies using oligonucleotide microarrays containing 6500 genes. This study also analyzed pools of cells made from normal urothelium as well as pools of tumors of different stages such as pTa grade I and II and pT2 grade III and IV bladder cancer specimens (34). The pooling approach may smooth out individual differences, but on the other hand, it can dilute strong intensities of relevant genes that may differentiate specific groups with different prognosis. Single cell suspensions were prepared from cooled biopsies immediately after surgery following a procedure previously used for the preparation of bladder tumors for flow cytometry (45). Single-cell suspensions can be inspected under the microscope to ensure the presence of enriched urothelial cells, avoiding samples with peripheral blood contamination. These pools can be produced with similar numbers of tumor cells from each specimen. In addition, RNA-preserving guanidium thiocyanate, immediately disrupts the cells and inactivates RNases.

Hierarchical clustering of gene expression levels grouped bladder cancer specimens according to stage and grade. By organizing genes with similar expression patterns into clusters, several functionally related genes were identified. The most significant were obtained by examining log-fold change of expression and included genes involved in cell cycle, cell growth, immunology, cell adhesion, transcription, and proteinase genes clustering into separate groups. Superficial papillary tumors showed increased transcription factor and ribosomal levels, as well as up-regulation of proteinase encoding genes. In the invasive tumors, increased levels of cell cycle-related transcripts were observed, which might reflect the increased level of growth factor and oncogene transcripts found. A loss of cellular adhesion proteins was found in invasive tumors and may be related to tissue invasion and metastasis. The invading tumor cells seem to challenge the immune system as reflected by an increase in immunological proteins (34).

The use of a common reference sample is not possible with oligonucleotide microarrays, as only a single sample can be tested per array. Instead, comparative scaling and normalization tools are used to allow comparisons to be made among samples. The study mentioned above represent an example of this technology for sample classification. In these cases, the mean expression level of a gene across all samples is calculated, the change for each sample relative to this mean is determined, and the size of the changes across all the samples are then compared (38). In the classification studies using spotted cDNA arrays described below, the choice of the common reference to which the experimental samples will be compared is a very important issue. Among the strategies that can be adopted, a pooled common reference sample from closely related or multiple related cell types has frequently been used (36,37).

The expression profiling of nine bladder cancer cell lines, including T24, J82, 5637, HT-1376, RT4, SCaBER, TCCSUP, UMUC-3, and HT1197, has been compared against a pool containing equal RNA quantities of each of them using cDNA arrays containing 8976 genes (46,47). Hierarchical clustering classified these tumor cells according to the histopathological characteristics of the tumors they were derived. The squamous carcinoma cell line SCaBER was distinguished from the other cell lines obtained from transitional carcinomas. Moreover, cell lines from invasive lesions clustered together and were segregated from cell lines obtained from a metastatic and a papillary superficial tumor. We focused on identifying potential targets that differenti-

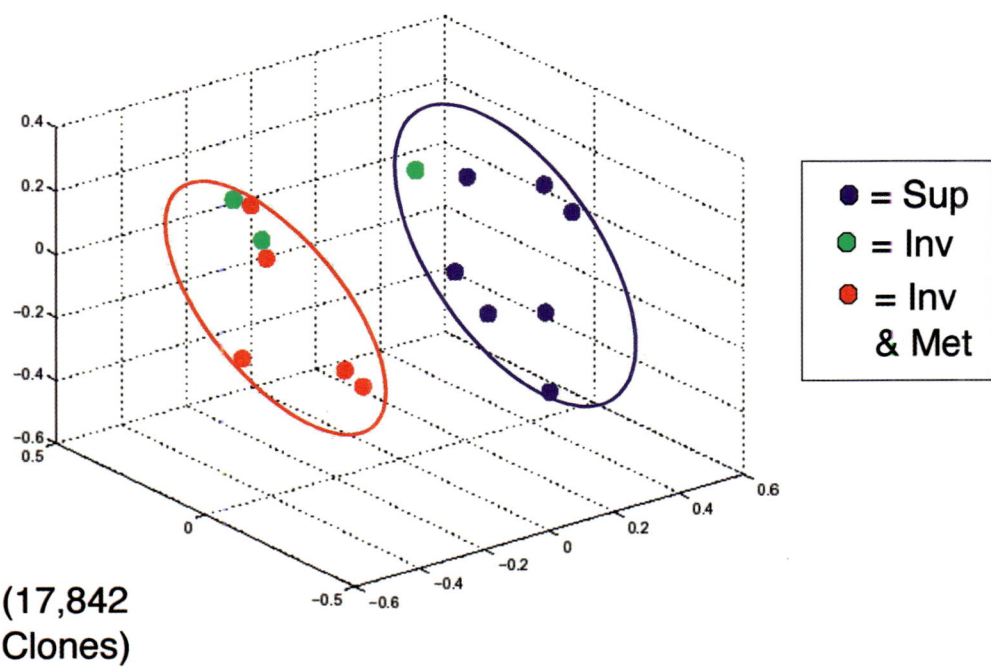

Fig. 3. Superficial and invasive bladder tumors showed differential expression profiles using cDNA microarrays containing 17,842 known genes and ESTs.

ated squamous features within bladder cancer based on the genes that were differentially expressed in SCaBER. Caveolin-1 and keratin 10 were differentially expressed in SCaBER and certain invasive tumor cell lines when compared to RT4 cells, which are derived from a papillary superficial bladder tumor. We further characterized the expression patterns of keratin 10 and caveolin-1 as potential markers of squamous differentiation in primary bladder tumors using tissue microarrays. Interestingly, the expression of these genes in the tumors analyzed was significantly associated with squamous differentiation, histopathological stage, and tumor grade. Moreover, when a bootstrapping resampling technique was applied, the cells clustered based on their p53, RB, and INK4A status. E-cadherin, zyxin, and moesin were identified as genes differentially expressed in these clusters. Interestingly, the expression of these genes was significantly associated with histopathological stage and tumor grade as well *(46)*. These results revealed that molecular profiling clustered bladder cancer based on histopathogenesis and biological criteria.

In a preliminary study, the expression profiling of 15 bladder tumors has been analyzed against a pool of bladder cancer cell lines using cDNA microarrays containing 18,609 known genes and ESTs *(47,48)*. The application of bootstraps and multidimensional scaling methods into the hierarchical clustering allowed the classification of superficial bladder tumors vs the invasive lesions. Using a high number of clones has allowed to validate critical known targets involved in bladder cancer progression, such as p21 or cyclin E *(49)*, as well as to identify novel molecular targets which require further analytical, in vitro, and clinical validation *(47,48)* (Fig. 3). The application of tissue microarrays represents a high-throughput approach for validation of novel potential markers for bladder cancer by immunohistochemistry or fluorescence *in situ* hybridization in paraffin blocks *(49–51)*. Frozen tissue microarrays may enable further characterization of novel genes by *in situ* hybridization of ESTs and known genes for which specific antibodies are not available.

Comparisons between expression profiles of tumor and normal tissue instead of purified primary cancer cells will also provide insight into the biology of the malignancy, as well as information concerning its cellular composition. It is also possible to identify the contribution of the different cellular components of the sample, cancerous and noncancerous, to the overall expression profile of a histologically complex tumor by using laser microdissection *(32,33)*. Furthermore, knowledge of the expression patterns of untransformed cells from which the malignancy has potentially developed can greatly assist in assigning a cellular origin to the tumor under study *(36,37)*. An experimental justification for this latter practice derives from the observation that the expression profiles of tumor cell lines often correlate with the profiles of their tissue of origin *(39)*.

OTHER APPLICATIONS OF MICROARRAYS IN BLADDER CANCER RESEARCH

In addition to expression profiling studies, specific oligonucleotide microarrays have been applied to the study of DNA variation in clinical material. The short length of oligonucleotide targets gives them the ability to discriminate between multiple probes that differ in sequence at a single base. Because of this, oligonucleotide microarrays have been developed to identify simple polymorphisms and allelic variations in DNA *(10)*. The primary applications of these types of microarrays have been in the genotyping of single nucleotide polymorphisms (SNPs) and in the identification of mutations in medically important genes *(39)*.

Oligonucleotide microarrays have also been used to analyze gene mutations in tumor samples for the presence of mutations in the TP53 tumor suppressor gene, which is a valuable predictor for bladder cancer outcome *(34)*. The traditional manual dideoxy sequencing has been compared with the much faster microarray sequencing on a commercially available chip and the concordance between methods was 92%. DNA samples extracted from 140 human bladder tumors were subjected to a multiplex-PCR before loading onto the p53 GeneChip (Affymetrix, Santa Clara, CA, USA). Each of the 1464 gene chip positions corresponded to an analyzed nucleotide in the p53 gene sequence. If there is a heterozygous or homozygous mutation, the oligonucleotide probes will hybridize differently from the wild-type sequence and reveal a signal that detects this particular base change. It is possible to detect either a wild-type sequence or a mixture of wild-type and mutant alleles. The system is almost free from interference by mixtures of templates from nonpathological and pathological tissue. Even 1% content of target from the diseased tissue can be detected in the presence of 99% wild-type content. This is in contrast to the capabilities of classical sequencing methodologies, which usually require at least 30% tumor tissue for accurate detection of mutations. Although conventional and microchip sequencing techniques were not in complete agreement, the speed and ease of the microarray analysis provided an argument for their use in this application on a larger scale. A drawback of use of microarrays in mutation screening is the inherent inability of this technique to detect previously unidentified mutations not represented on the array, or frameshift mutations, namely insertions or deletions. This is certainly an important limitation, because it has already been established that such frameshift mutations represent 5–10% of all known p53 mutations *(52)*. The p53 GeneChip fails to detect such mutations because it is not designed to do so, with the exception of single-base deletions.

Recently, genome-wide screening using SNP arrays has been applied to DNA isolated from microdissected superficially invasive T1 and muscle invasive T2–4 bladder tumors. These oligonucleotide commercial microarrays contain 1494 biallelic polymorphic sequences and can detect loss or gain of at least one allele. The authors reported that out of a genotype of 1204 loci, 343 were heterozygous in the bladder tumors under study. Allelic imbalance was detected in known areas of imbalance on chromosomes 6, 8, 9, 11, and 17, and a new area of imbalance was detected on the p arm of chromosome 6. Microsatellite analysis of T2–4 tumors and Ta tumors showed that allelic imbalance was more frequent in T2–4 tumors than in Ta or T1 tumors. However, when pairs of T1 and T2–4 tumors were analyzed from eight patients, 68% of imbalances detected in T1 tumors (146 imbalances) occurred in the subsequent T2–4 tumors (99 imbalances). Moreover, the authors report that homozygous TP53 mutations were more often associated with high allelic imbalance than with low allelic imbalance. In this study, SNP arrays were shown to be feasible for high-throughput genome-wide scanning for allelic imbalances in bladder cancer in a faster manner than comparable microsatellite-based analyses. Not only did they confirm known areas of chromosomal losses *(18–22)*, but they also identified areas with common allelic imbalances that could harbor potential new tumor suppressors in bladder cancer *(53)*. Although these data are restricted to the polymorphic areas contained in the arrays, in the future, it should be possible to fabricate high-density SNP microarrays for other predefined chromosomal locations, which could make noninformative areas informative. In addition, high-throughput CGH arrays may be of interest to confirm the alterations found at the genomic level in a detailed manner *(13)*. The microarray is a convenient platform for assays involving biomolecules other than nucleic acids. Arrays of tissues, peptides, antibodies, proteins, and even cells have been developed *(4,49–51,54–57)*. This is further evidence of the strength and versatility for high-throughput screening. These should provide means of rapidly validating at the protein level, the genes identified by expression profiling using DNA microarrays.

FINAL COMMENTS

The area of data analysis and management deserves special comment. Expression profiling experiments can produce data sets that are, at least by the standards of cell biology, extremely large. A variety of numerical analysis approaches and algorithms have been used to cluster genes based upon their expression patterns. Various statistical approaches have likewise been used to perform the class predictions that identify expression patterns that correlate with phenotypic characteristics, such as tumor type. Still, there exists a significant need to develop additional bioinformatic methods to extract all the information contained in these very rich, deep expression pattern data sets and to integrate it with other forms of biological information (e.g., the information contained in the published literature [58]). In addition, there is the hope that the raw data from published microarray experiments will be made available to the scientific community with unrestricted access in a uniform format. An international effort is underway to develop guidelines and consensus on data handling and annotation.

Despite the relative youth of the technology, microarray use has already had a broad and significant impact on the study of cancer. As illustrated here, it has been possible to assign potential functional roles to novel genes in both the signaling pathways controlling and the phenotypic changes associated with cancer development. Microarray studies will continue to correlate changes in the expression of specific genes and groups of genes with cancer and cancer-related phenotypes. Following the biological validation of these expression–phenotype correlations, the result will be a more complete list of the genes controlling cancer development and progression. From this, a clearer view should emerge of the principles and pathways controlling bladder cancer physiology *(59,60)*.

The early studies summarized here and elsewhere indicate that expression profiling of a relatively small number of genes may provide a molecular means of defining clinically important tumor subtypes not identified using standard methods. Moreover, these subtypes may identify specific subgroups of patients that will benefit from distinct treatment regimes. Currently needed are carefully controlled large-scale expression profiling studies on large numbers of clinically well-annotated cases before the final clinical utility of this technique can be accurately judged. It is clear that the results of these studies will add to our understanding of the mechanisms of carcinogenesis and may also improve our ability to diagnose and treat the disease. As interest in microarrays and their use in the study of cancer continues to increase, so does the likelihood that their use will have an important clinical application.

REFERENCES

1. Liang, P. and Pardee, A. B. (1992) Differential display of eukaryotic messenger RNA by means of the polymerase chain reaction. *Science* **257,** 967–971.
2. Velculescu, V. E., Zhang, L., Vogelstein, B., and Kinzler, K. W. (1995) *Science* **270,** 484–487.
3. Brenner, S., Williams, S. R., Vermaas, E. H., et al. (2000) In vitro cloning of complex mixtures of DNA on microbeads: physical separation of differentially expressed cDNAs. *Proc. Natl. Acad. Sci. USA* **97,** 1665–1670.
4. Haab, B. B., Dunham, M. J., and Brown, P. O. (2001) Protein microarrays for highly parallel detection and quantitation of specific proteins and antibodies in complex solutions. *Genome Biol.* **2,** 1–4.
5. Pandey, A. and Mann, M. (2000) Proteomics to study genes and genomes. *Nature* **405,** 837–846.
6. International Human Genome Sequencing Consortium. (2001) *Nature* **409,** 860–921.
7. Venter, J. C. (2000) Remarks at the human genome announcement. *Funct. Integr. Genomics* **1,** 154–155.
8. Duggan, D. J., Bittner, M., Chen, Y., Meltzer, P., and Trent, J. M. (1999) Expression profiling using cDNA microarrays. *Nat. Genet.* **21,** 10–14.
9. Lipshutz, R. J., Fodor, S. P., Gingeras, T. R., and Lockhart, D. J. (1999) *Nat. Genet.* **21,** 20–24.
10. Hacia, J. G. and Collins, F. S. (1999) Mutational analysis using oligonucleotide microarrays. *J. Med. Genet.* **36,** 730–736.
11. Fan, J. B., Chen, X., Halushka, M. K., et al. (2000) Parallel genotyping of human SNPs using generic high-density oligonucleotide tag arrays. *Genome Res.* **10,** 853–860.

12. Iyer, V. R., Horak, C. E., Scafe, C. S., Botstein, D., Snyder, M., and Brown, P. O. (2001) Genomic binding sites of the yeast cell-cycle transcription factors SBF and MBF. *Nature* **409,** 533–538.
13. Pinkel, D., Segraves, R., Sudar, D., et al. (1998) High resolution analysis of DNA copy number variation using comparative genomic hybridization to microarrays. *Nat. Genet.* **20,** 207–211.
14. Davies, J. M., Somerville, S. M., and Wallace, D. M. (1976) Occupational bladder tumour cases identified during ten years' interviewing of patients. *Br. J. Urol.* **48,** 561–566.
15. Roberts, D. W., Benson, R. W., Groopman, J. D., et al. (1988) Immunochemical quantitation of DNA adducts derived from the human bladder carcinogen 4-aminobiphenyl. *Cancer Res.* **48,** 6336–6342.
16. Mack, D. and Frick, J. (1995) Low-dose bacille Calmette-Guerin (BCG) therapy in superficial high-risk bladder cancer: a phase II study with the BCG strain Connaught Canada. *Br. J. Urol.* **75,** 185–187.
17. Neiman, P. E. and Hartwell, L. H. (1991) Malignant instability. Workshop on genetic instability and its role in carcinogenesis sponsored by the Programs in Molecular Medicine of the Fred Hutchinson Cancer Research Center and the University of Washington, Seattle, WA, USA, January 11–12, 1991. *New Biol.* **3,** 347–351.
18. Yeager, T., Stadler, W., Belair, C., Puthenveettil, J., Olopade, O., and Reznikoff, C. (1995) Increased p16 levels correlate with pRb alterations in human urothelial cells. *Cancer Res.* **55,** 493–497.
19. Balazs, M., Carroll, P., Kerschmann, R., Sauter, G., and Waldman, F. M. (1997) Frequent homozygous deletion of cyclin-dependent kinase inhibitor 2 (MTS1, p16) in superficial bladder cancer detected by fluorescence in situ hybridization. *Genes Chromosom. Cancer* **19,** 84–89.
20. Reuter, V. E. and Melamed, M. R. (1989) The lower urinary tract, in *Diagnostic Surgical Pathology*, (Sternberg, S. S., ed.), Raven Press, New York, pp. 1355–1392.
21. Grossman, H. B., Liebert, M., Antelo, M., et al. (1998) p53 and RB expression predict progression in T1 bladder cancer. *Clin. Cancer Res.* **4,** 829–834.
22. Cote, R. J., Dunn, M. D., Chatterjee, S. J., et al. (1998) Elevated and absent pRb expression is associated with bladder cancer progression and has cooperative effects with p53. *Cancer Res.* **58,** 1090–1094.
23. Chen, J. J., Wu, R., Yang, P. C., et al. (1998) Profiling expression patterns and isolating differentially expressed genes by cDNA microarray system with colorimetry detection. *Genomics* **51,** 313–324.
24. Cheung, V. G., Morley, M., Aguilar, F., Massimi, A., Kucherlapati, R., and Childs, G. (1999) Making and reading microarrays. *Nat. Genet.* **21,** 15–19.
25. Granjeaud, S., Bertucci, F., and Jordan, B. R. (1999) Expression profiling: DNA arrays in many guises. *Bioessays* **21,** 781–790.
26. Stears, R. L., Getts, R. C., and Gullans, S. R. (2000) A novel, sensitive detection system for high-density microarrays using dendrimer technology. *Physiol. Genomics* **3,** 93–99.
27. Eberwine, J., Yeh, H., Miyashiro, K., et al. (1992) Analysis of gene expression in single live neurons. *Proc. Natl. Acad. Sci. USA* **89,** 3010–3014.
28. Zhao, N., Hashida, H., Takahashi, N., Misumi, Y., and Sakaki, Y. (1995) High-density cDNA filter analysis: a novel approach for large-scale, quantitative analysis of gene expression. *Gene* **156,** 207–213.
29. Takahashi, N., Hashida, H., Zhao, N., Misumi, Y., and Sakaki, Y. (1995) High-density cDNA filter analysis of the expression profiles of the genes preferentially expressed in human brain. *Gene* **164,** 219–227.
30. Wang, E., Miller, L. D., Ohnmacht, G. A., Liu, E. T., and Marincola, F. M. (2000) High-fidelity mRNA amplification for gene profiling. *Nat. Biotechnol.* **18,** 457–459.

31. Klimecki, W. T., Futscher, B. W., and Dalton, W. S. (1994) Effects of ethanol and paraformaldehyde on RNA yield and quality. *Bio Techniques* **16,** 1021–1023.
32. Cerroni, L., Arzberger, E., Ardigo, M., Putz, B., and Kerl, H. (2000) Monoclonality of intraepidermal T lymphocytes in early mycosis fungoides detected by molecular analysis after laser-beam-based microdissection. *J. Invest. Dermatol.* **114,** 1154–1157.
33. Specht, K., Richter, T., Muller, U., Walch, A., Werner, M., and Hofler, H. (2001) Quantitative gene expression analysis in microdissected archival formalin-fixed and paraffin-embedded tumor tissue. *Am. J. Pathol.* **158,** 419–429.
34. Thykjaer, T., Workman, C., Kruhoffer, M., et al. (2001) Identification of gene expression patterns in superficial and invasive human bladder cancer. *Cancer Res.* **61,** 2492–2499.
35. Cerroni, L., Minkus, G., Putz, B., Hofler, H., and Kerl, H. (1997) Laser beam microdissection in the diagnosis of cutaneous B-cell lymphoma. *Br. J. Dermatol.* **136,** 743–746.
36. Perou, C. M., Sorlie, T., Eisen, M. B., et al. (2000) Molecular portraits of human breast tumours. *Nature* **406,** 747–752.
37. Alizadeh, A. A., Eisen, M. B., Davis, R. E., et al. (2000) Distinct types of diffuse large B-cell lymphoma identified by gene expression profiling. *Nature* **403,** 503–511.
38. Welsh, J. B., Zarrinkar, P. P., Sapinoso, L. M., et al. (2001) Analysis of gene expression profiles in normal and neoplastic ovarian tissue samples identifies candidate molecular markers of epithelial ovarian cancer. *Proc. Natl. Acad. Sci. USA* **98,** 1176–1181.
39. Risch, N. J. (2000) Searching for genetic determinants in the new millennium. *Nature* **405,** 847–856.
40. Ahrendt, S. A. and Sidransky, D. (1999) The potential of molecular screening. *Surg. Oncol. Clin. N. Am.* **8,** 641–656.
41. Golub, T. R., Slonim, D. K., Tamayo, P., et al. (1999) Molecular classification of cancer: class discovery and class prediction by gene expression monitoring. *Science* **286,** 531–537.
42. Chen, C. C., Shieh, B., Jin, Y. T., et al. (2001) Microarray profiling of gene expression patterns in bladder tumor cells treated with genistein. *J. Biomed. Sci.* **8,** 214–222.
43. Maxwell, S. A. and Davies, G. E. (2000) Differential gene expression in p53-mediated apoptosis-resistant vs. apoptosis-sensitive tumor cell lines. *Proc. Natl. Acad. Sci. USA* **97,** 13,009–13,014.
44. Liang, G., Gonzales, F. A., Jones, P. A., Orntoft, T. F., and Thykjaer, T. (2002) Analysis of gene induction in human fibroblasts and bladder cancer cells exposed to the methylation inhibitor 5-aza-2'-deoxycytidine. *Cancer Res.* **62,** 961–966.
45. Zarbo, R. J., Visscher, D. W., and Crissman, J. D. (1989) Two-color multiparametric method for flow cytometric DNA analysis of carcinomas using staining for cytokeratin and leukocyte-common antigen. *Anal. Quant. Cytol. Histol.* **11,** 391–402.
46. Sánchez-Carbayo, M., Socci, N. D., Charytonowicz, E., et al. (2002) Molecular profiling of bladder cancer using cDNA microarrays: defining histogenesis and biological phenotypes. *Cancer Res.* **62,** 6973–6980.
47. Sánchez-Carbayo, M., Capodieci, P., and Cordon-Cardo, C. (2003) Tumor suppressor role of KiSS-1 in bladder cancer: loss of KiSS-1 expression is associated with bladder cancer progression and clinical outcome. *Am J. Pathol.* **162,** 609–618.
48. Sánchez-Carbayo, M., Socci, N. D., Lozano, J. J., Li, W., Belbin, T. J., Prystowsky, M. B., Ortiz, A. R., Childs, G., and Cordon-Cardo, C. Gene discovery in bladder cancer progression using cDNA microarrays, submitted.
49. Richter, J., Wagner, U., Kononen, J., et al. (2000) High-throughput tissue microarray analysis of cyclin E gene amplification and overexpression in urinary bladder cancer. *Am. J. Pathol.* **157,** 787–794.
50. Kononen, J., Bubendorf, L., Kallioniemi, A., et al. (1998) Tissue microarrays for high-throughput molecular profiling of tumor specimens. *Nat. Med.* **4,** 844–847.

51. Nocito, A., Bubendorf, L., Maria Tinner, E., et al. (2001) Microarrays of bladder cancer tissue are highly representative of proliferation index and histological grade. *J. Pathol.* **194,** 349–357.
52. Hollstein, M. (1999) New approaches to understanding p53 gene tumor mutation spectra. *Mutat. Res.* **431,** 199–209.
53. Primdahl, H., Wikman, F. P., von der Maase, H., Zhou, X. G., Wolf, H., and Orntoft, T. F. (2002) Allelic imbalances in human bladder cancer: genome-wide detection with high-density single-nucleotide polymorphism arrays. *J. Natl. Cancer Inst.* **94,** 216–223.
54. Reineke, U., Volkmer-Engert, R., and Schneider-Mergener, J. (2001) Applications of peptide arrays prepared by the SPOT-technology. *Curr. Opin. Biotechnol.* **12,** 59–64.
55. MacBeath, G. and Schreiber, S. L. (2000) Printing proteins as microarrays for high-throughput function determination. *Science* **289,** 1760–1763.
56. Ziauddin, J. and Sabatini, D. M. (2001) Microarrays of cells expressing defined cDNAs. *Nature* **411,** 107–110.
57. Sánchez-Carbayo, M. (2003) Use of high-throughput DNA microarrays to identify biomarkers for bladder cancer. *Clin. Chem.* **49,** 23–31.
58. Jenssen, T. K., Laegreid, A., Komorowski, J., and Hovig, E. (2001) A literature network of human genes for high-throughput analysis of gene expression. *Nat. Genet.* **28,** 21–28.
59. Knowles, M. A. (2001) What we could do now: molecular pathology of bladder cancer. *Mol. Pathol.* **54,** 215–221.
60. Adam, B. L., Vlahou, A., Semmes, O. J., and Wright, G. L. Jr. (2001) Proteomic approaches to biomarker discovery in prostate and bladder cancers. *Proteomics* **1,** 1264–1270.

13
Gene Expression Profiling of Renal Cell Carcinoma and its Clinical Implications

Masayuki Takahashi and Bin Tean Teh

INTRODUCTION

Renal cell carcinoma (RCC) is the most common malignancy arising in the adult kidney, representing 2% of all malignancies and 2% of cancer-related deaths. It is the 10th most common cancer in the U.S., where it causes more than 12,000 deaths per year. Its incidence has been increasing, a phenomenon that cannot be accounted for by the wider use of imaging procedures (1). RCC is more common in men than women, especially in men over 55 yr of age. Risk factors include genetic predisposition (2), hypertension, obesity (3), and occupational exposures (4). RCC may be as small as 1 cm in diameter when discovered (usually incidentally), or as bulky as several kilograms. It most often manifests with pain, as a palpable mass, as hematuria, or as another paraneoplastic syndrome.

HISTOLOGICAL SUBTYPE OF RCC AND GENETIC EVENTS

RCC may be well circumscribed, or it may invade the perirenal adipose tissue or the renal vein. Some tumors are predominantly cystic, and cystic degeneration is common. It is a clinicopathologically heterogeneous disease, subdivided into clear cell, granular, chromophobe, spindle, papillary, and collecting-duct subtypes based on morphological features (5). Originating in the proximal renal tubule, clear cell RCC (ccRCC) or conventional RCC is the most common adult renal neoplasm (75–80% of all renal neoplasms). Other types of RCC, by descending frequency, are: papillary (10–15%), chromophobe (4–6%), collecting duct (<1%), and those forms that are yet to be classified (<2%). Both sarcomatoid and cystic RCC are not considered independent entities. Sarcomatoid RCC, characterized by prominent spindle cell features, is thought to represent the high-grade end of the cytologic spectrum in all of the subgroups.

With recent advances in molecular genetics, the subtypes of RCC have been associated with distinct genetic abnormalities. This association has led to a proposal for molecular diagnosis (6). For example, most of the clear cell RCC exhibit a loss of chromosome 3 and inactivating mutations of the von Hippel-Lindau (VHL) gene, whereas papillary renal cell carcinomas (PRCC) are frequently associated with trisomy of chromosomes 3q, 7, 12, 16, 17, and 20, and loss of the Y chromosome. Recently,

From: *Expression Profiling of Human Tumors: Diagnostic and Research Applications*
Edited by: Marc Ladanyi and William L. Gerald © Humana Press Inc., Totowa, NJ

it has been proposed that, even in the absence of prominent papillae, these aberrant chromosomal features could support the diagnosis of PRCC. Conversely, kidney cancers that do not possess these genetic characteristics should not be designated as PRCCs, even when papillary structures are prominent *(7)*. Frequent loss of sex chromosomes and of chromosomes 1 and 14 have been found in renal oncocytoma, a benign entity composed of large eosinophilic cells in an acinar arrangement *(8)*. The concept of molecular diagnosis has been further advanced by the use of microarray gene expression profiles, which basically expand the genomic information to include gene expression or transcription profiles.

MANAGEMENT OF RCC

Surgical removal remains the mainstay therapy for patients with localized primary tumors. In metastatic RCC, surgery has also been recommended for the primary tumor, and surgery combined with postoperative adjunctive use of interferon-α offers longer survival than interferon therapy alone *(9)*. Metastatic RCC are resistant to chemotherapy (e.g., 5-fluorouracil [5-FU] *[10]*, paclitaxel *[11]*, vinblastine *[12]*), radiation, and hormonal therapy (e.g., tamoxifen *[13,14]*). On the other hand, immunotherapy with interferon-α or interleukin-2 (IL-2) has resulted in response rates of 10–20%, and a few patients with metastatic RCC exhibit complete responses *(15)*. Many combination therapies based on these agents have been attempted, but to date, no significant survival benefits have been found. The combination therapy of IL-2 and interferon-α has led to no better a response rate than that for IL-2 alone *(16,17)*. However, several studies with triple-drug combinations of interferon-α, IL-2, and 5-FU have shown higher response rates *(13,18)*. More recently, nonmyeloablative allogeneic stem-cell transplantation, which employs donor T cells to combat host RCC tissues, has been found effective in sustaining regression of metastatic disease *(19)*. In addition to this new promising immunotherapy, discovering and understanding the underlying genetic alterations in RCC may give us novel information leading to the discovery of new drugs. Individualized treatment based on the gene expression profiles of each tumor may lead to improved response rates in the future.

PROGNOSTIC FACTORS OF RCC

Overall, approx 30% of patients with macroscopically complete resection of RCC have a recurrence after radical nephrectomy *(20,21)*, and patients with metastatic RCC have a life expectancy of about 12 mo *(15,22)*. It is, therefore, important to identify prognostic factors that can predict patients' outcomes and, thereby, influence decisions regarding their treatment. To date, a number of prognostic factors have been proposed, which are mainly divided into patient-related factors and tumor-related factors. According to the College of American Pathologists Working Classification for Prognostic Markers, prognostic factors that are currently used in patient management include existence of symptoms, weight loss, performance status, erythrocyte sedimentation rate (ESR), anemia, serum calcium and serum alkaline phosphatase *(23)*. Surgical margins, the number and location of metastases, tumor, node, metastases (TNM) staging, and histological grading are well-accepted tumor-related prognostic factors *(24–26)* . Biomolecular factors, such as proliferation markers (e.g., *Ki-67*, proliferating cell nuclear antigen [*PCNA*]), apoptosis markers (e.g., *p53*, *bcl-2*, *p21*), growth factors, cell adhesion molecules, angiogenesis promoters, tumor suppressor genes, oncogenes, and cytokines have also been examined, but none has been sufficiently well-established to gain wide clinical use. Microarray analysis may help us to identify new biomolecular markers, prognostic factors, and new pathways because of its comprehensive nature.

MICROARRAY ANALYSIS AS COMPREHENSIVE GENE EXPRESSION PROFILING

The clinical challenges mentioned above point to a need to understand in a comprehensive fashion the underlying molecular mechanisms of kidney cancers. One recent biomedical breakthrough is the use of high-throughput microarray technology, which allows comprehensive gene expression profiling of tumors. These gene expression profiles can serve as the molecular signatures of particular tumors, and they may be used to distinguish among histological subtypes and novel distinct subtypes that are correlated with clinical outcome. This distinction may reflect the heterogeneity in transformation mechanisms, cell types, or aggressiveness among tumors. For example, approx 100 genes were identified as differentially expressed by serous as compared to mucinous ovarian cancers (27). Other studies have identified distinct gene sets that distinguish between acute myeloid leukemia and acute lymphoblastic leukemias (28), between hereditary breast cancer with *BRCA1* and *BRCA2* mutations (29), and between hepatitis-B and hepatitis-C positive hepatocellular carcinomas (30). Furthermore, several studies have identified prognostic sets of genes in various cancer types that may underlie the heterogeneity in tumor aggressiveness. These include diffuse large B cell lymphomas (31), breast cancers (32), lung cancers (33), and central nervous system embryonal tumors (34), some of which are reviewed elsewhere in this book. Importantly, the identification of such genes may lead to the discovery of a number of new potential targets for cancer diagnosis and therapy.

RCC cDNA MICROARRAY STUDIES

The first paper that sought to examine gene expression profiles of RCC with cDNA microarray technology was published in 1999 (35). In this report, RNA extracted from a kidney cancer cell line (CRL-1933) and normal kidney tissue were radioactively labeled and hybridized to a membrane containing 5184 cDNA spots. A total of 89 genes or expressed sequence tags (ESTs) were identified as differently expressed in the RCC cell line compared to the normal kidney tissues. Thirty-eight sequences were up-regulated, including 26 known genes and 12 ESTs, whereas 51 sequences were down-regulated, including 25 known genes and 26 ESTs. Vimentin, a cytoplasmic intermediate filament, was specifically selected for further immunohistochemical study using a tissue array containing 532 RCC cases, because of previous reports regarding its expression in each RCC subtype (36,37). Subsequently, vimentin expression was found to be an independent prognostic factor for clear cell RCC. However, because the study was based on only a single cancer cell line, certain gene expression changes may have resulted from the culture conditions *per se* or through multiple culture passages, which are all factors that could make it more difficult to choose a new molecular target for further study.

To date, several papers on gene expression profiling of primary RCC have been published. One of them sought to find the molecular markers of different pathological subtypes *(38)*. In this paper, four ccRCC, one chromophobe RCC, and two oncocytomas were subjected to microarray analysis containing 7075 genes. Poly(A)-rich RNA was extracted from the RCC samples and matched normal kidney tissues and reverse-transcribed with incorporation of Cy3 and Cy5 dyes. These reverse transcripts were competitively hybridized to microarray spots, and the ratios of expression levels in tumor to those in normal kidney were determined. Using at least a twofold alteration of transcript expression in at least two tumors as the criterion for significant differential expression, a total of 189 genes were identified. Hierarchical average linkage analysis *(39)* revealed two distinct classes of tumor expression patterns, which were for ccRCC and chromophobe RCC/oncocytoma. Each histopathological group was clustered together using all 4906 expressed genes, as well as the 189 differentially expressed genes. This distinction was also confirmed by another analysis method called quality threshold (QT) clustering algorithm *(40)*. Chromophobe RCC and oncocytomas were found to have similar gene expression profiles, although only two oncocytomas and one chromophobe RCC were analyzed. Interestingly, among the common expression data of chromophobe RCC and oncocytoma, distal nephron markers, such as *parvalbumin (41)* and *β-defensin-1 (42)* were up-regulated. *Parvalbumin* expression was also confirmed by immunohistochemistry, indicating its potential usefulness as a new marker for differential diagnosis of RCC. This study also found that chromophobe RCC and oncocytoma exhibited up-regulation of genes related to mitochondrial biology and oxidative phosphorylation, including the *mitochondrial adenine nucleotide translocator*, *proton-transporting ATP synthase subunits*, *carbonic anhydrase isoenzymes*, and *cytochrome c oxidase subunits*. We have obtained the gene expression data for a chromophobe RCC that is related to Birt-Hogg-Dube syndrome *(43)* and found that these mitochondria-related genes as well as *parvalbumin* were also up-regulated (unpublished). Obviously, additional microarray studies involving larger series of tumors are warranted to elucidate and confirm the most reliable molecular markers for each RCC subtype.

DIFFERENTIALLY EXPRESSED GENES IN CLEAR CELL RCC

It is critical to obtain a large number of samples for comprehensive gene expression profiling. Recently, two papers on gene expression profiling were published, each evaluating about 30 ccRCC cases *(44,45)*. The backbone of these studies lies in the identification of differentially expressed genes, i.e., genes that are either up- or down-regulated in the tumors when compared with matched normal renal tissues. Some of the differentially expressed genes play a normal physiological role in normal kidney, and their functions may be impaired by the replacement of normal cells by cancerous tissue. When the altered genes in our data set were compared with those reported in other RCC microarray papers *(38,44)*, around 80% of the altered genes showed similar expression patterns. Mislabeling and contamination of cDNA clones and various extent of RNA degradation may account for some discrepancies between studies. Moreover, the differences in the reported genes may result from a different selection of probes spotted on microarray slides or from different selection criteria for altered genes. Some up-regulated genes in the tumors may point to the involvement of oncogenes or enhancers of cancer formation and progression. It may be possible to develop inhibi-

**Table 1
Commonly Up-Regulated Genes in ccRCC**

GenBank® accession no.	Gene name	Average fold	% of RCC
H86554	Ceruloplasmin.	16.9	96.2
R00332	ESTs, highly similar to growth factor-responsive protein.	14.1	96.4
T72235	Nicotinamide N-methyltransferase.	13.5	96.6
W72051	Fatty acid binding protein 7, brain.	13.2	87.5
W70343	Lysyl oxidase.	11.2	95.8
W30988	ESTs, highly similar to angiopoietin-related protein.	11.1	100
H99075	ESTs.	10.7	95.7
W93163	Tumor necrosis factor, α-induced protein 6.	10.5	100
T54298	ESTs, highly similar to angiopoietin-related protein.	8.1	100
AA598601	Insulin-like growth factor binding protein 3.	7.6	96.6
AA678335	Phosphodiesterase I/nucleotide pyrophosphatase 3.	7.6	84.0
AA164819	ESTs.	7.1	96.3
AA485896	ESTs.	6.8	96.4
N26171	ESTs.	6.2	87.5
AA487787	Von Willebrand factor.	6.2	100
AA450189	Enolase 2, (γ, neuronal).	6.0	96.4
R62612	Fibronectin 1.	5.6	93.1
H20872	Fc fragment of IgG, low affinity IIIa, receptor for (CD16).	5.5	85.7
W72293	ESTs.	5.5	93.1
AA055835	Caveolin 1, caveolae protein 22 kDa.	5.4	92.9
AA873159	Apolipoprotein C-1.	5.3	88.9
AA017544	Regulator of G-protein signaling 1.	5.2	85.7
R19956	Vascular endothelial growth factor.	5.1	96.4
H99816	Procollagen-lysine, 2-oxoglutarate 5-dioxygenase 2.	5.1	96.4
R49597	ESTs.	4.6	95.8
AA405000	Homo sapiens ribonuclease 6 precursor.	4.5	96.2
H58873	Solute carrier family 2, member 1.	4.5	93.1
T62491	Chemokine, receptor 4.	4.4	89.7
AA443899	CD36 antigen-like 1.	4.2	89.3
AI004331	Human MHC class II HLA-DQ-β.	4.2	85.7
AA488892	ESTs, weakly similar to Gag-Pol polyprotein.	4.0	85.7

Thirty-one genes or ESTs that are at least threefold up-regulated in at least 75% of RCC ($n = 29$). Average fold reflects average ratios of all analyzable spots. Percentage of RCC is the fraction of tumors that have at least twofold up-regulation.

Ig, immunoglobulin; MHC, major histocompatibility factor; HLA, human leukocyte antigen.

tors or down-regulators of these genes as potential therapeutic agents. Some of the down-regulated genes, on the other hand, may be tumor suppressors that could be potential therapeutic targets if their function could be restored or enhanced.

Up-Regulated Genes in Clear Cell RCC

The list of up-regulated genes for clear cell RCC is shown in Table 1. Interestingly, some of the genes that have the highest differential expression ratio in the tumors have no known or little-known association with ccRCC. For example, *ceruloplasmin*, a protein involved in iron and copper homeostasis, has the highest increase in expression in the ccRCC. Its serum level increases markedly in anemia of iron deficiency, hemorrhage, renal failure, sickle cell disease, pregnancy, and inflammation. To date, only one study has reported its secretion by RCC *(46)*, and another study described its elevation in the serum of RCC patients *(47)*. Another copper-related protein, *lysyl oxidase*, was also up-regulated in ccRCC. It is an extracellular enzyme involved in the connective tissue maturation pathway. It is highly expressed in invasive breast cancer cell lines *(48)*, but it has heretofore not been studied in RCC. *Caveolin-1* is an integral protein of caveolae, which is involved in signal transduction and lipid transport. Its overexpression has been reported to be a relatively common feature of breast cancer and advanced prostate cancer *(49)*, but it has not previously been associated with RCC.

Additional microarray studies with more RCC samples and further protein studies will be needed to validate the involvement of these and other genes. In addition, in order to understand their roles in tumorigenesis of RCC, functional studies together with in vitro and in vivo studies are needed.

Down-Regulated Genes in RCC

The list of down-regulated genes for ccRCC is shown in Table 2. Some of the many down-regulated genes may be involved in the tumorigenesis of ccRCC. Those that are highly down-regulated are *kininogen, fatty acid binding protein 1, phenylalanine hydroxylase, epidermal growth factor* (*EGF*), *plasminogen*, and *aldolase B*. Most strikingly, *kininogen* was found to be more than 27-fold down-regulated in the tumors. *Kininogen*, a molecule involved in the activation of the cellular contact system, has recently been shown to be an inhibitor of angiogenesis *(50)*. Its down-regulation may concur with up-regulation of *vascular endothelial growth factor* (*VEGF*), resulting in hypervascularization, which is a characteristic of ccRCC. We also found the *metallothionein* (*MT*) family to be coordinately down-regulated. *MT* is known to modulate the release of gaseous mediators, such as hydroxyl radical or nitric oxide, apoptosis, and the binding and exchange of heavy metals such as zinc, cadmium or copper. Differential expression of this family of genes has been reported in many cancers *(51)*, and several subtypes (*MT-1A*, *MT-1G*, and *MT-1H*) were reported to be down-regulated in RCC *(52,53)*. Our study supports these reports and additionally found *MT-1L* and *MT-1E* to be down-regulated. *Aldolase B*, one of the three aldolase glycolytic enzyme catalyzing the reversible conversion of fructose-1, 6-bisphosphate to glyceraldehyde 3-phosphate and dihydroxyacetone phosphate, is another gene that is down-regulated in RCC. It has been found in abundance in normal renal cortex compared with RCC, suggesting that it plays a physiological role in normal kidney *(54)*. The relatively lower expression in the cancer, as confirmed by our study, may be due to displacement of normal tissue. One of the heparan sulfate proteoglycans, *glypican 3*, also stood out in our analysis, but has never been associated with kidney cancer. Its down-regulation has been reported in mesotheliomas, ovarian cancer, and breast cancer *(55)*. Glypican 3-deficient mice have been shown to exhibit several clinical features, including developmental overgrowth and dysplastic kidneys *(56)*.

Table 2
Commonly Down-Regulated Genes in ccRCC

GenBank accession no.	Gene name	Average fold	% of RCC
R89067	Kininogen.	27.2	100
T53220	Fatty acid binding protein 1, liver.	22.8	95.8
AA682293	Phenylalanine hydroxylase.	20.4	96.0
AA705692	ESTs.	18.0	100
AA954947	Epidermal growth factor (β-urogastone).	15.0	100
H72098	Aldolase B, fructose-biphosphate.	13.6	100
AA411988	EST.	13.3	100
T73187	Plasminogen.	12.0	100
T51617	Solute carrier family 22, member 3.	11.8	96.4
AA777384	ESTs.	11.0	96.2
H53340	Metallothionein 1G.	10.0	100
AA844930	Glycoprotein 2 (zymogen granule membrane).	9.6	100
AA101792	Phosphatidylinositol glycan, class F.	9.4	96.6
AA858026	Protein C inhibitor (plasminogen activator inhibitor III).	9.4	100
H18950	ESTs, highly similar to hepatocyte nuclear factor 4 γ.	9.2	100
AA405769	Phosphoenolpyruvate carboxykinase 1 (soluble).	8.9	96.6
AA040387	X-prolyl aminopeptidase 2, membrane-bound.	8.8	96.4
H77766	Metallothionein 1H.	8.4	96.6
W16424	ESTs.	8.4	92.6
H88329	Calbindin 1 (28 kDa).	8.1	100
N62179	Methylmalonate-semialdehyde dehydrogenase.	7.9	100
AA775872	Glypican 3.	7.9	100
AA457718	Homo sapiens mRNA; cDNA DKFZp564B076.	7.8	95.7
R24266	Growth factor receptor-bound protein.	7.1	80.8
R54778	Collagen, type XVI, α 1.	7.1	100
AA702640	Dopa decarboxylase.	7.0	96.3
N55459	RNA helicase-related protein.	6.9	96.6
AA664180	Glutathione peroxidase 3.	6.6	92.9
AA227594	Mal, T cell differentiation protein.	6.3	100
H68509	UDP glycosyltransferase 2 family, polypeptide B10.	6.1	95.5
AA676466	Argininosuccinate sythetase.	6.1	96.4
H96140	Acyl-coenzyme A dehydrogenase, short–branched chain.	6.0	96.0
H11346	Aldehyde dehydrogenase 4.	6.0	92.9
AA862999	Calcium-sensing receptor.	6.0	100
AA497001	ESTs, weakly similar to BcDNA.GH02901.	6.0	96.3
AA449780	EST.	5.9	88.9
AA704995	Putative glycine-N-acyltransferase.	5.6	92.9
T94781	Potassium inwardly-rectifying channel, subfamily J.	5.6	92.9
N89673	ESTs.	5.6	92.6
H37880	ESTs, moderately similar to ALU SUBFAMILY SP.	5.6	96.3
AA663884	Synaptosomal-associated protein, 25 kDa.	5.5	95.7
R25818	Aldehyde dehydrogenase 9.	5.5	100
AA700604	Sorbitol dehydrogenase.	5.4	92.6
W95082	Hydroxysteroid (11-β) dehydrogenase 2.	5.4	96.6

(continued)

Table 2 *(continued)*

GenBank accession no.	Gene name	Average fold	% of RCC
AA677655	*Klotho.*	5.4	92.3
N80129	*Metallothionein 1L.*	5.3	86.2
AA402915	*Aminoacylase 1.*	5.3	96.3
AA863424	*Dipeptidase 1* (renal).	5.2	93.1
H72722	ESTs, highly similar to *METALLOTHEIONEIN-1B*.	5.2	86.2
N78083	*Hydroxysteroid dehydrogenase.*	5.1	96.4
R06601	ESTs, moderately similar to *METALLOTHIONEIN-II*.	5.1	82.8
AA131240	ESTs.	5.0	92.0
AA485965	*Succinate-CoA ligase, GDP-forming, α-subunit.*	4.9	92.9
AA196287	ESTs, weakly similar to alternative spliced product using exon 13A.	4.9	96.6
R61229	*Glycine amidinotrasferase* (L-arginine: glycine amidinotrasferase).	4.8	82.8
AA872383	*Metallothionein 1E* (functional).	4.8	82.8
N23898	*G protein-coupled receptor kinase 2* (Drosophila)-like.	4.8	92.9
AA699427	*Fructose-biphosphatase 1.*	4.7	93.1
T68892	*Secreted frizzled-related protein 1.*	4.7	96.2
AA873355	*ATPase,* Na+/K+ transporting, α-1 polypeptide.	4.7	100
AI000188	*UDP glycosyltransferase 2 family,* polypeptide B7.	4.6	85.7
AA459197	Sodium channel, nonvoltage-gated 1 α.	4.6	89.7
T65482	*L-3-hydroxyacyl-coenzyme A dehydrogenase,* short chain.	4.4	96.2
AA457374	*DKFZP586B0319.*	4.3	91.7
R33037	ESTs.	4.3	92.0
AA437099	ESTs.	4.3	85.2
W01011	*SA (rat hypertension-associated) homolog.*	4.2	89.7
AA863449	*Oviductal glycoprotein 1,* 120 kDa.	4.0	92.9
N53031	*UDP glycosyltransferase 2 family,* polypeptide B4.	4.0	86.2
AA458884	*S100 calcium-binding protein A2.*	4.0	92.9
AA608575	*Propionyl coenzyme A carboxylase,* α polypeptide.	3.8	89.7
H18608	*Solute carrier family 22 (organic anion transporter),* member 8.	3.6	89.3

Seventy-two genes or ESTs that are at least threefold down-regulated in at least 75% of RCC ($n = 29$). Average fold reflect average ratios of all analyzable spots. Percentage of RCC is the fraction of tumors that have at least twofold down-regulation.

The Classification of Altered Genes in Human RCC Specimens by Gene Ontology

Boer et al. *(44)* in particular sought to understand gene expression profiling of RCC by putting expression data into the categories of biological pathways and cellular components, a concept proposed by the Gene Ontology Consortium *(57)*. According to the results of their categorization of biological pathways, a large number of up-regulated genes in RCC belonged to the cell adhesion category, e.g., *fibronectin 1* and *laminin-*

α4. They also found that the genes of several signal transduction pathways, such as *guanosine-5'-triphosphate (GTP)*-binding protein and kinases, had a tendency toward up-regulation. Other biological categories that encompassed a large proportion of up-regulated genes were nucleotide and nucleic acid metabolism, protein metabolism and modification, cell shape and cell size, and immune response. We also examined our data using three different criteria: at least a twofold alteration in gene expression by 60% or more of the tumors, at least a threefold alteration in 60% or more of the tumors, or at least a twofold alteration in 80% or more of the tumors. Results from applying these different criteria showed a similar trend. We found that the genes of immune response and cell adhesion were up-regulated in a large percentage of tumors, findings consistent with the data of Boer et al. The up-regulation of the immune response-related genes, such as several types of major histocompatibility complex, *monokine induced by γ-interferon, interferon-induced protein 17, interferon-γ-inducible protein 16, CD32,* and *CD53,* might indicate why some RCC respond to immunotherapy, although some of them could simply reflect the inflammatory and immune reactions taking place in the tumor environment. In contrast with the study by Boer et al., we could not find any clear differences between the number of up- and down-regulated genes in the following categories: nucleotide and nucleic acid metabolism, protein metabolism and modification, cell shape and cell size.

With regard to down-regulated genes, Boer et al. described that the genes of transport, ion homeostasis, and electron transport were mostly down-regulated. Our data showed that, besides transport, the genes of carbohydrate metabolism (especially gluconeogenesis), lipid metabolism, energy pathways, biosynthesis, isoprenoid catabolism, amino acid and derivative metabolism, cytoplasm organization, and biogenesis were mostly down-regulated. It seems that many metabolic pathways of normal kidney cells would be disrupted in RCC cells. The presence of many down-regulated genes of lipid metabolism, such as *fatty acid coenzyme A ligase, aldehyde dehydrogenase, acyl-coenzyme A dehydrogenase,* may explain why ccRCC have abundant cytoplasm, which contains cholesterol, cholesterol ester, phospholipids, and glycogen. Some discrepancies between our data and Boels' data may be due to several factors. First, they used rather different criteria to select altered genes, such as a different ratio-voting criterion of at least a 3.5-fold alteration in at least 30% of the tumors. They also applied an adapted sign test that could provide higher sensitivity and better control over the rate of false positives. Another source of discrepancy between our data sets may be the differences in cDNA spots on the microarray slides. Finally, some discrepancies could be related to the source of ontology data, since we used the Celera database as well as that provided by The Gene Ontology Consortium. We further note that most gene ontology information is assigned computationally, but not reviewed or confirmed by further experiments. It remains a challenge to assign a RCC-specific category of genes with altered expression by using computationally assigned ontology categorization, and more sophisticated software is needed to more clearly categorize these data. It is also essential to compare the RCC ontology data with that of other cancers to identify RCC-specific pathways.

We also examined genes according to the categorization of molecular function. The genes encoding defense–immunity proteins were mostly up-regulated, similar to the results of biological process categorization. Structural proteins, such as several kinds of ribosomal protein, *laminin*, *keratin 4*, *fibronectin 1*, *collagen type IV*, *chondroitin sulfate proteoglycan 2*, and *caveolin* were up-regulated. In the category of glycosaminoglycan binding proteins, the encoding genes showed a tendency toward up-regulation. The up-regulated genes categorized as enzymes were kinases, such as *serum-inducible kinase*, and *phosphorylase kinase-α2*, and phosphatases, such as *protein tyrosine phosphatase*, *autotaxin* and *phosphatase* and *tensin homolog*. On the other hand, the genes encoding heavy metal binding were mostly down-regulated. Among them, several subtypes of *metallothionein (MT-1L, MT-1H, MT-1G, MT-1E)* and *membrane metallo-endopeptidase* are related to zinc binding. *Phenylalanine hydroxylase* and *stearol-CoA desaturase* are related to iron binding. In addition, genes involved in electron transfer, steroid binding, carbohydrate metabolism, lipid metabolism, and amino acid and derivative metabolism were mostly down-regulated. Again, while the altered expression of genes in those molecular function categories would seem to be due to decreased normal metabolism, some of the down-regulated genes could be related to RCC tumorigenesis or invasiveness.

IDENTIFICATION OF THE GENE SET SPECIFIC TO PARTICULAR CLINICAL SUBSETS OF RCC

In our study, the availability of 29 ccRCC frozen tissue specimens with up to 12 yr of follow-up information allowed us to identify expression signatures of ccRCC that have different clinical outcomes, as well as to assess their clinical implications. First, we used hierarchical clustering *(39)* to look at the variation in gene expression among the tumors. The clustering algorithm grouped both genes and tumors based on similarities among their expression patterns. The 3184 genes were selected for clustering based on the total gene expression profiling. The selected genes were those analyzable in at least 75% of the tumors and with expression ratios that varied at least twofold in at least two experiments. Overall, there was a great variation in gene expression among the tumors. The tumors clustered into two main groups (Fig. 1), which were correlated with cause-specific survival at 5 yr, with only two tumors that did not cluster by that parameter. We used the program Cluster Identification Tool (CIT) *(58)* to identify and rank subclusters of genes that distinguish between the two defined sample groups. Briefly, the expression ratios of tumor to matched normal kidney of all the genes within each subcluster or node of a dendrogram were averaged for each patient, and the averages were placed into two groups based on user-defined criteria. The mean (μ) and standard deviation (σ) for the averaged expression ratios of all patients in each group are calculated. A discrimination score (DS) for each subcluster was calculated

---

Fig. 1. *(opposite page)* Clustering of 29 ccRCC with the selected 3148 genes. Rows represent individual cDNAs and columns represent individual RCC mRNA samples. The color of each square represents the median-polished normalized ratio of gene expression in a tumor relative to patient-matched normal kidney tissue. Red indicates gene expression above the median; green, below the median; black, equal to the median; and gray, inadequate or missing data. The color saturation indicates the degree of divergence from the median. Red patient's number indicates cancer death within 5 yr after surgery, blue indicates alive, and black indicates short follow-up. This dendrogram shows the structure of similarity in relationships between the gene expression profiles.

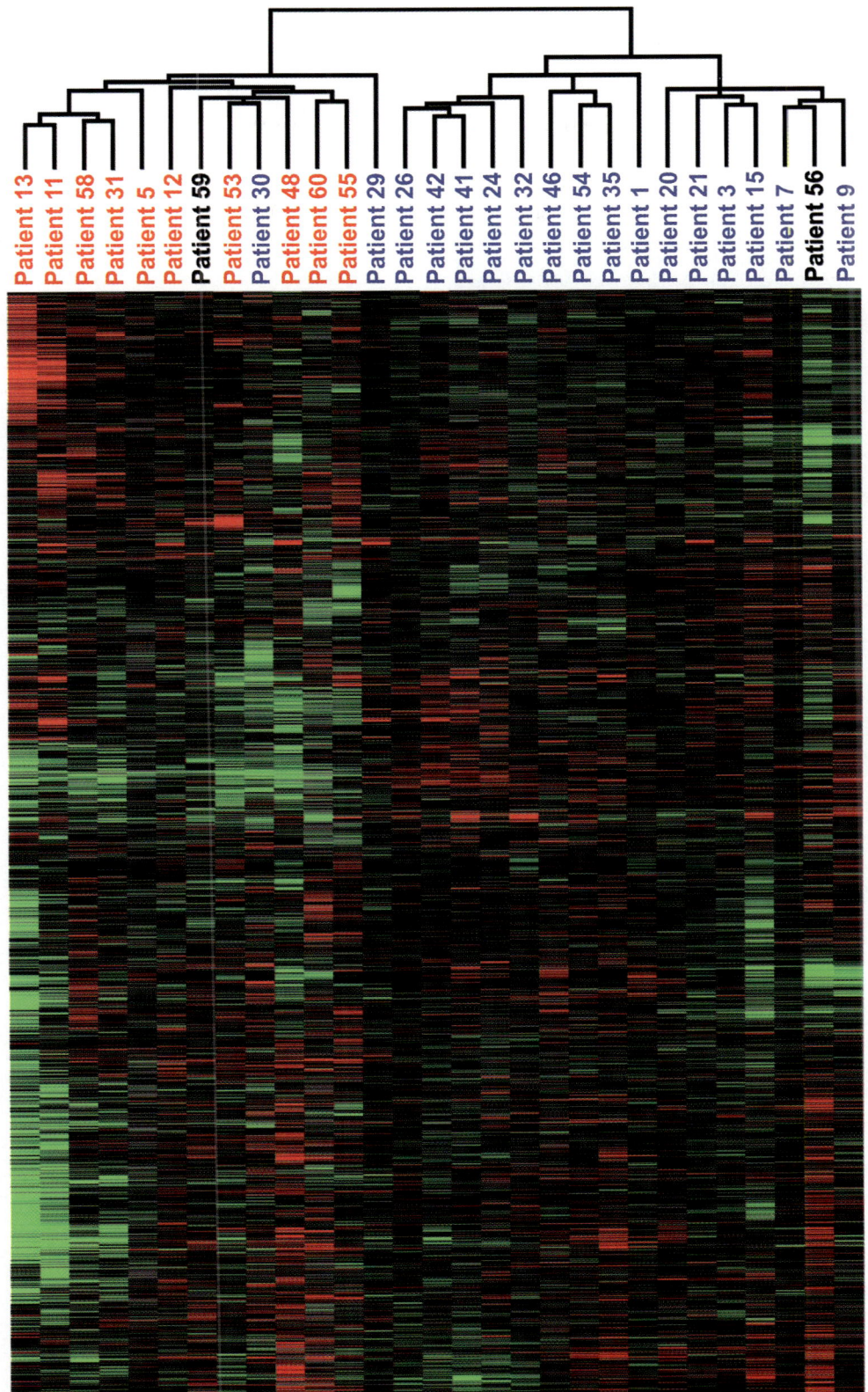

Fig. 1.

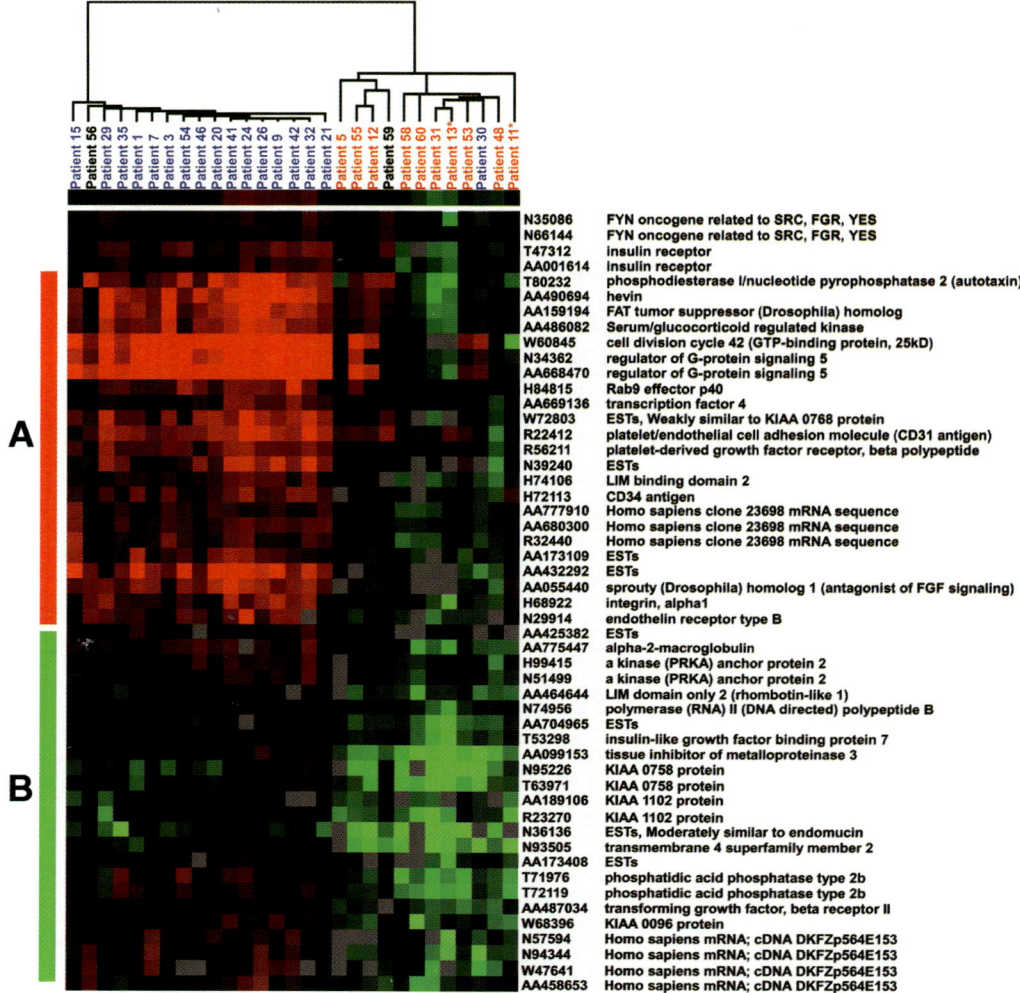

Fig. 2. Clustering of the 51 genes of cluster 1281 using nonmedian centered values. In this case, the color of each square corresponds to actual normalized gene expression level relative to normal kidney tissue, using the same scale as in Fig. 1. **(A)** Genes mostly up-regulated in tumors with the good outcome. **(B)** Genes mostly down-regulated in tumors with the poor outcome.

as $DS = |\mu_1 - \mu_2| / (\sigma_1 + \sigma_2)$, where the subscripts referred to the sample group. A large DS indicated that the genes in that cluster exhibited great variation between the two groups, but low variation within each group (28). A DS of 1.0 would approximate a significantly discriminating cluster ($\alpha = 0.05$). By using the CIT, we sought to identify particular subsets of genes that most strongly defined two distinct groups of patients by patient outcome or other clinicopathological findings. Since tumor staging is often used to determine the prognosis of ccRCC (24), we tested whether stage grouping has any specific gene expression pattern, and whether the gene expression pattern could possibly be valuable as a predictor of cancer progression. However, no gene clusters significantly distinguished between stage I and II and stage III and IV tumors (data not shown).

When we applied the CIT to a tumor grouping based on cause-specific survival at 5 yr, multiple clusters with a high DS were found. Among them, one cluster and its parent cluster had the highest DS (1.70). The tumors were reclustered based on these gene sets with highest DS. Clustering based on the genes of this cluster grouped the tumors by outcome, except for patient 30. This patient showed an expression pattern similar to that of the poor outcome group, but had no evidence of disease after 5 yr. The dendrogram for this cluster showed a strict segregation of patients by outcome with a high correlation score (0.839).

Then a permutation t-test was used to calculate the ability of individual genes to distinguish between two groups (29). Patients were randomly permuted into two groups 10,000 times, and for each gene, a t-statistic was calculated. The distribution of t-statistics defined a 99.9% significance threshold ($\alpha = 0.001$). If a gene's t-statistic for the user-defined patient grouping passed the 99.9% significance threshold, the gene was considered to significantly distinguish the two groups. The 94% of genes from the best cluster significantly ($\alpha < 0.001$) differentiated two groups based on survival at 5 yr.

The tumors were reclustered based on the genes of the best node using nonmedian centered expression values, so that the colors represented the actual gene expression level of the tumor relative to normal, but the grouping by cause-specific survival remains distinct. This cluster is comprised of genes up-regulated in the tumors with a good clinical outcome and genes down-regulated in tumors with a poor clinical outcome (Fig. 2). The diversity in the gene expression profiles largely defined two patient groups that were distinguishable by cause-specific survival at 5 yr. These findings may reflect the existence of distinct subclasses of clear cell RCC that differ in clinical behavior. We showed that while no statistically significant clusters of genes correlated with random groupings of tumors or with staging of tumors, multiple clusters consisting of dozens of genes with high statistical significance correlated with cause-specific survival at 5 yr. This result showed that only certain groupings of patients have distinguishing gene expression signatures, likely only the groupings that have an underlying biological basis. Therefore, the two groups of ccRCC identified by gene expression profiling may represent two classes of ccRCC, an aggressive and a nonaggressive class, that have distinct molecular bases for distinct mechanisms of progression.

CLINICAL SIMULATION TEST WITH THE PROGNOSTIC SET OF GENES

To verify that this prognostic set of genes is robust, we simulated its likely clinical use for each tumor sample. A "test" tumor was removed from the group, a new set of predictive genes was generated from the remaining 28 tumors using CIT, and the test tumor was clustered with the other samples using the predictive gene set. The test tumor was classified as high risk or low risk depending on whether it clustered with the poor outcome or good outcome tumors, respectively. This process was repeated for each tumor. The prognostic classification of each tumor was considered "correct" if it corresponded to the actual outcome. The predictive gene set was slightly different at each simulation, but on average 95% of the genes were conserved.

Table 3 presents the results of the simulation. Prognostic classification by gene expression profiling was a better predictor than staging in five patients (patients 35, 9,

Table 3
Patient Clinical Data and Corresponding Prognosis Classifications

				Follow-up		Prognosis group	
Patient no.	Grade	Stage	Outcome	(mo)	Outcome	Staging	Gene expression
46	G1	S1	NED	62.6	L	L	L
42	G1	S1	NED	77.3	L	L	L
41	G1	S1	NED	80.3	L	L	L
30	G2	S3	NED	87.1	L	H*	H*
7	G1	S1	NED	92.1	L	L	L
26	G1	S1	NED	96.0	L	L	L
24	G1	S1	NED	97.3	L	L	L
15	G1	S1	OCD	100.4	L	L	L
32	G1	S2	OCD	110.4	L	L	L
1	G1	S1	NED	111.6	L	L	L
21	G1	S1	NED	114.6	L	L	L
20	G1	S1	NED	115.8	L	L	L
35	G1	S3	NED	120.5	L	H*	L
9	G1	S3	NED	120.9	L	H*	L
3	G1	S1	NED	137.2	L	L	L
29	G3	S3	AWC	89.4	L	H*	L
54	G1	S4	AWC	105.6	L	H*	L
13	G3	S4	Death	3.2	H	H	H
48	G2	S4	Death	4.9	H	H	H
11	G3	S3	Death	18.8	H	H	H
60	G3	S4	Death	20.8	H	H	H
31	G3	S3	Death	22.6	H	H	H
53	G3	S4	Death	26.2	H	H	H
5	G2	S4	Death	31.7	H	H	H
12	G2	S4	Death	33.8	H	H	H
55	G2	S2	Death	55.8	H	L*	H
58	G3	S4	Death	24.0	H	H	H
56	G3	S4	AWC	27.8	U	H	L
59	G2	S3	NED	41.1	U	H	H

Patient clinical data and corresponding prognostic classifications. Grade and stage information (columns 2 and 3) corresponds to the primary tumor. Outcomes (column 4) are: no evidence of disease (NED), alive with cancer (AWC), other cause of death (OCD), and cancer death. Follow-up (column 5) is the duration between nephrectomy and latest outcome assessment. Outcome group (column 6) is the risk group based on actual patient outcome that was used for predictive gene set generation (L, low risk; H, high risk; U, unknown). Pathology prognosis group (column 7) is based on staging (L, stage I/II; H, stage III/IV). Gene expression prognosis group (column 8) is based on a gene expression prognosis test based on the selected genes. An asterisk (*) indicates a deviation in outcome from the predicted risk group.

29, 54, and 55). Patient 29, who had a grade 3 tumor invading into the renal vein at operation (high risk by staging), had a low risk gene expression profile and now has been alive for 7.5 yr. Patient 55, who had a stage II, grade 2 tumor (low risk by staging), had a high risk gene expression profile and died of ccRCC at 4.6 yr after the operation. Patient 54, classified as low risk by gene expression profiling, had bone metastases at initial diagnosis, but is still alive with stable bone metastases after 8.8 yr. One patient was misgrouped by both staging and gene expression profiling (patient 30), who presented with stage III tumor and had a high risk gene expression profile, but 7 yr later, has no evidence of disease.

PROGNOSTIC SET OF GENES IN RCC

Figure 2 shows the gene set that is considered the best for discriminating between patients with good clinical outcome and poor outcome based on subclusters generated by hierarchical clustering, instead of gene by gene. Some of the individual genes that were not selected may still have prognostic value. However, in our selected set of genes, many of them provide insights into the biology of the two groups of ccRCC. For example, *sprouty*, the mammalian homolog of the *Drosophila melanogaster* angiogenesis inhibitor, was up-regulated exclusively in the good outcome group, suggesting that failure to properly inhibit angiogenesis may contribute to the aggressive form of ccRCC. *The regulator of G-protein signaling 5* was exclusively up-regulated in the good outcome tumors and may be important for the proper control of cancer progression. *Transforming growth factor-β receptor II (TGFβRII)* and its downstream effector, *tissue inhibitor of metalloproteinase 3 (TIMP-3)*, were exclusively down-regulated in the poor outcome group. Loss of the *TGFβII* signaling pathway previously has been shown to be important for the development of aggressive cancers *(59)*, and loss of *TIMP-3* expression by promoter methylation has been shown to increase tumorigenicity due to unregulated matrix metalloproteinases (MMPs) *(60)*. A recent study demonstrated the inhibition of invasion in melanoma cell lines by overexpressing *TIMP-3* by adenovirus-mediated gene delivery *(61)*. Again, the identification of this pathway, which is down-regulated in aggressive ccRCC, presents numerous potential targets for intervention. Progress in this direction may supplement the still low response rate to current adjuvant therapies, such as interferon-α and IL-2.

SURVIVAL ANALYSES BASED ON THE EXPRESSION PROFILE OF THE PROGNOSTIC SET OF GENES

We used Kaplan-Meier survival analysis (Fig. 3) to further compare the significance of the prognostic classifications determined by stage, grade, and gene expression profile and tested by the log-rank test. Three patients were excluded from the statistic analyses because they had <5 yr of clinical follow-up. Classification by grade ($p < 0.0001$) was better than that by stage ($p = 0.0036$). Just as significant as grade, gene expression profiling is also more accurate in predicting clinical outcome than staging alone. Correlation of histological grade or stage with the gene expression profile was analyzed as the Spearman correlation coefficient by the exact test with the SAS/STAT analysis package (version 8.0; SAS Institute, Cary, NC, USA). It turned out that histological grade and gene expression classification were highly correlated (correlation coefficient = 0.7703; $p < 0.0001$), indicating that grading is the phenotype resulting from gene expression profiling. Surprisingly, within the high risk group defined by staging (stage III and IV), gene expression profiling significantly distinguished two groups of patients with different outcomes. Multivariate analysis of these parameters was also attempted, but was prohibited by the sample size of our cohort and strong correlation of grading and gene expression profiling. A larger cohort of patients would

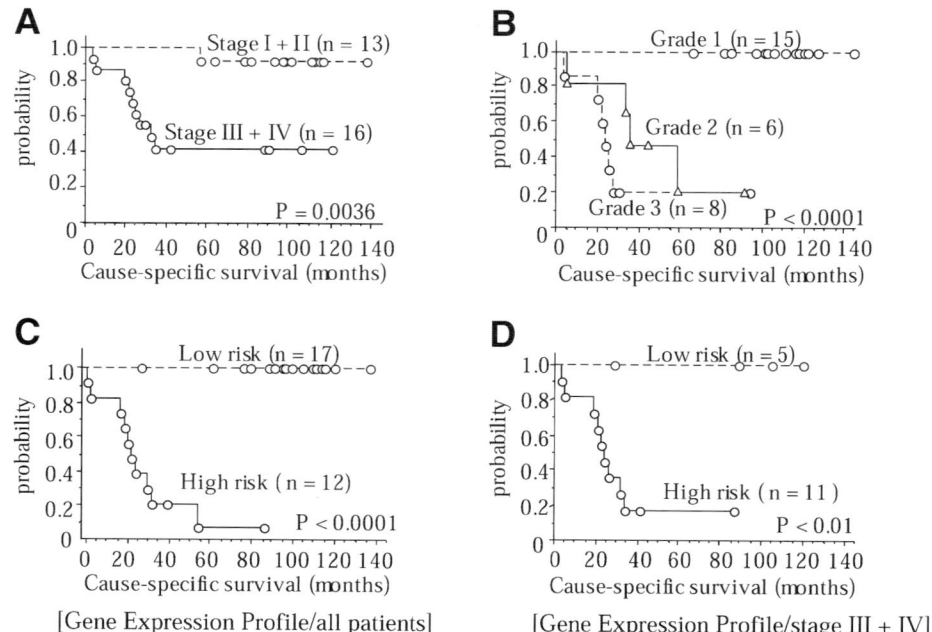

Fig. 3. Cause-specific survival curves based on staging (**A**), histological grading (**B**), gene expression profiling in all patients (**C**), and that in patients with stage III/IV (**D**) by Kaplan-Meier method.

allow multivariate analysis. This analysis may lead to the development of more accurate prognostic methods for ccRCC patients.

FUTURE cDNA MICROARRAY EXPRESSION STUDIES IN RCC

To date, promising new data have been accumulating and molecular signatures of RCC have been uncovered. However, there are still many issues to be resolved in understanding the molecular signature of each cell type of RCC and to realize the potential clinical benefits. For instance, it is important to define the specific gene set for each cell type of RCC with a large number of samples to better understand tumorigenesis of each cell type of RCC, and to differentiate accurately RCC cases that are difficult to diagnose histopathologically. Second, the prognostic gene set that we have identified warrants further study with a large number of new RCC cases. More information can be obtained by using microarrays containing larger numbers of genes and improved analytic methods. From these analyses, a better and more specific gene set to differentiate the different outcome groups or metastasis-related gene sets may be discovered. Moreover, it may be interesting to examine and compare the gene expression profiles of RCC cases that have complete or partial response to immunotherapy and those that are immunotherapy-resistant. This may lead to the discovery of new drugs that could prove effective in immunotherapy-resistant RCC.

Microarray-related technology will continue to develop and evolve. Its cost will likely decrease, and hence, more research groups will undertake such studies. As mentioned above, there is generally a tremendous need for more powerful and

POST-GENE EXPRESSION PROFILING STUDIES

High-throughput screening technologies have identified an enormous number of potential molecular targets that are related to cancer. However, the studies are solely based on gene expression, which may or may not reflect certain critical genetic alterations. For example, the *VHL* tumor suppressor gene is found inactivated in approx 70% of ccRCC. However, its differential expression in the tumors does not stand out and, therefore, did not appear on the list of down-regulated genes. It is well known that, besides *VHL*, a number of tumor suppressor genes (e.g., *MENIN* and *WT1*) do not have the expected low expression in tumors, for reasons still not well understood.

Therefore, the very obvious challenge is to prioritize and select the targets that could be essential to tumorigenesis or tumor aggressiveness. This will first require a more complete understanding of the functions of these genes and how they interact with other genes in contributing to cancer formation and progression. Actually, a majority of genes identified in RCC studies remain either completely unknown or very poorly understood with regard to their functions. While we expect that more microarray studies will be done to complement and validate the present studies, more functional studies are essential. Perhaps these studies can be designed and executed in unconventional ways. For example, rather than based on single genes, these studies could be more pathway-based, i.e., all components related to a pathway could be studied at the same time.

The potential targets identified require validation in in vitro and in animal model systems that mimic critical aspects of disease progression and response to therapy. To date, there is only one RCC model, that of the Eker rat, which develops ccRCC. It is caused by a tuberous sclerosis type 1 (TSC1) mutation that is rarely involved in human RCC samples *(62)*. With the identification of potential targets through microarray studies, it may be feasible to create new target-specific animal models that caricaturize the pathophysiology of human RCC.

ACKNOWLEDGMENT

We thank Dr. Rick Hay for critically reviewing this manuscript.

REFERENCES

1. Chow, W. H., Devesa, S. S., Warren, J. L., and Fraumeni, J. F., Jr. (1999) Rising incidence of renal cell cancer in the United States. *JAMA* **281**, 1628–1631.
2. Takahashi, M., Kahnoski, R., Gross, D., Nicol, D., and Teh, B. T. (2002) Familial adult renal neoplasia. *J. Med. Genet.* **39**, 1–5.
3. Chow, W. H., Gridley, G., Fraumeni, J. F., Jr., and Jarvholm, B. (2000) Obesity, hypertension, and the risk of kidney cancer in men. *N. Engl. J. Med.* **343**, 1305–1311.
4. Moyad, M. A. (2001) Review of potential risk factors for kidney (renal cell) cancer. *Semin. Urol. Oncol.* **19**, 280–293.
5. Mostofi, F. K. and Davis, C. J. (1998) *International Histological Classification of Tumors*. Springer, Berlin.

6. Bugert, P. and Kovacs, G. (1996) Molecular differential diagnosis of renal cell carcinomas by microsatellite analysis. *Am. J. Pathol.* **149,** 2081–2088.
7. Storkel, S., Eble, J. N., Adlakha, K., et al. (1997) Classification of renal cell carcinoma: Workgroup No. 1. Union Internationale Contre le Cancer (UICC) and the American Joint Committee on Cancer (AJCC). *Cancer* **80,** 987–989.
8. Presti, J. C., Jr., Moch, H., Reuter, V. E., Huynh, D., and Waldman, F. M. (1996) Comparative genomic hybridization for genetic analysis of renal oncocytomas. *Genes Chromosom. Cancer* **17,** 199–204.
9. Flanigan, R. C., Salmon, S. E., Blumenstein, B. A., et al. (2001) Nephrectomy followed by interferon alfa-2b compared with interferon alfa-2b alone for metastatic renal-cell cancer. *N. Engl. J. Med.* **345,** 1655–1659.
10. Kish, J. A., Wolf, M., Crawford, E. D., et al. (1994) Evaluation of low dose continuous infusion 5-fluorouracil in patients with advanced and recurrent renal cell carcinoma. A Southwest Oncology Group Study. *Cancer* **74,** 916–919.
11. Einzig, A. I., Gorowski, E., Sasloff, J., and Wiernik, P. H. (1991) Phase II trial of taxol in patients with metastatic renal cell carcinoma. *Cancer Invest.* **9,** 133–136.
12. Fossa, S. D., Droz, J. P., Pavone-Macaluso, M. M., Debruyne, F. J., Vermeylen, K., and Sylvester, R. (1992) Vinblastine in metastatic renal cell carcinoma: EORTC phase II trial 30882. The EORTC Genitourinary Group. *Eur. J. Cancer* **28A,** 878–880.
13. Atzpodien, J., Kirchner, H., Illiger, H. J., et al. (2001) IL-2 in combination with IFN- alpha and 5-FU versus tamoxifen in metastatic renal cell carcinoma: long-term results of a controlled randomized clinical trial. *Br. J. Cancer* **85,** 1130–1136.
14. Schomburg, A., Kirchner, H., Fenner, M., Menzel, T., Poliwoda, H., and Atzpodien, J. (1993) Lack of therapeutic efficacy of tamoxifen in advanced renal cell carcinoma. *Eur. J. Cancer* **29A,** 737–740.
15. Minasian, L. M., Motzer, R. J., Gluck, L., Mazumdar, M., Vlamis, V., and Krown, S. E. (1993) Interferon alfa-2a in advanced renal cell carcinoma: treatment results and survival in 159 patients with long-term follow-up. *J. Clin. Oncol.* **11,** 1368–1375.
16. Vogelzang, N. J., Lipton, A., and Figlin, R. A. (1993) Subcutaneous interleukin-2 plus interferon alfa-2a in metastatic renal cancer: an outpatient multicenter trial. *J. Clin. Oncol.* **11,** 1809–1816.
17. Atkins, M. B., Sparano, J., Fisher, R. I., et al. (1993) Randomized phase II trial of high-dose interleukin-2 either alone or in combination with interferon alfa-2b in advanced renal cell carcinoma. *J. Clin. Oncol.* **11,** 661–670.
18. Ellerhorst, J. A., Sella, A., Amato, R. J., et al. (1997) Phase II trial of 5-fluorouracil, interferon-alpha and continuous infusion interleukin-2 for patients with metastatic renal cell carcinoma. *Cancer* **80,** 2128–2132.
19. Childs, R., Chernoff, A., Contentin, N., et al. (2000) Regression of metastatic renal-cell carcinoma after nonmyeloablative allogeneic peripheral-blood stem-cell transplantation. *N. Engl. J. Med.* **343,** 750–758.
20. Levy, D. A., Slaton, J. W., Swanson, D. A., and Dinney, C. P. (1998) Stage specific guidelines for surveillance after radical nephrectomy for local renal cell carcinoma. *J. Urol.* **159,** 1163–1167.
21. Ljungberg, B., Alamdari, F. I., Rasmuson, T., and Roos, G. (1999) Follow-up guidelines for nonmetastatic renal cell carcinoma based on the occurrence of metastases after radical nephrectomy. *BJU Int.* **84,** 405–411.
22. Hermanek, P. and Schrott, K. M. (1990) Evaluation of the new tumor, nodes and metastases classification of renal cell carcinoma. *J. Urol.* **144,** 238–242.
23. Henson, D. E., Fielding, L. P., Grignon, D. J., et al. (1995) College of American Pathologists Conference XXVI on clinical relevance of prognostic markers in solid tumors. Summary. Members of the Cancer Committee. *Arch. Pathol. Lab. Med.* **119,** 1109–1112.

24. Tsui, K. H., Shvarts, O., Smith, R. B., Figlin, R. A., deKernion, J. B., and Belldegrun, A. (2000) Prognostic indicators for renal cell carcinoma: a multivariate analysis of 643 patients using the revised 1997 TNM staging criteria. *J. Urol.* **163,** 1090–1095.
25. Gettman, M. T., Blute, M. L., Spotts, B., Bryant, S. C., and Zincke, H. (2001) Pathologic staging of renal cell carcinoma: significance of tumor classification with the 1997 TNM staging system. *Cancer* **91,** 354–361.
26. Moch, H., Gasser, T., Amin, M. B., Torhorst, J., Sauter, G., and Mihatsch, M. J. (2000) Prognostic utility of the recently recommended histologic classification and revised TNM staging system of renal cell carcinoma: a Swiss experience with 588 tumors. *Cancer* **89,** 604–614.
27. Ono, K., Tanaka, T., Tsunoda, T., et al. (2000) Identification by cDNA microarray of genes involved in ovarian carcinogenesis. *Cancer Res.* **60,** 5007–5011.
28. Golub, T. R., Slonim, D. K., Tamayo, P., et al. (1999) Molecular classification of cancer: class discovery and class prediction by gene expression monitoring. *Science* **286,** 531–537.
29. Hedenfalk, I., Duggan, D., Chen, Y., et al. (2001) Gene-expression profiles in hereditary breast cancer. *N. Engl. J. Med.* **344,** 539–548.
30. Okabe, H., Satoh, S., Kato, T., et al. (2001) Genome-wide analysis of gene expression in human hepatocellular carcinomas using cDNA microarray: identification of genes involved in viral carcinogenesis and tumor progression. *Cancer Res.* **61,** 2129–2137.
31. Alizadeh, A. A., Eisen, M. B., Davis, R. E., et al. (2000) Distinct types of diffuse large B-cell lymphoma identified by gene expression profiling. *Nature* **403,** 503–511.
32. Sorlie, T., Perou, C. M., Tibshirani, R., et al. (2001) Gene expression patterns of breast carcinomas distinguish tumor subclasses with clinical implications. *Proc. Natl. Acad. Sci. USA* **98,** 10,869–10,874.
33. Bhattacharjee, A., Richards, W. G., Staunton, J., et al. (2001) Classification of human lung carcinomas by mRNA expression profiling reveals distinct adenocarcinoma subclasses. *Proc. Natl. Acad. Sci. USA* **98,** 13,790–13,795.
34. Pomeroy, S. L., Tamayo, P., Gaasenbeek, M., et al. (2002) Prediction of central nervous system embryonal tumour outcome based on gene expression. *Nature* **415,** 436–442.
35. Moch, H., Schraml, P., Bubendorf, L., et al. (1999) High-throughput tissue microarray analysis to evaluate genes uncovered by cDNA microarray screening in renal cell carcinoma. *Am. J. Pathol.* **154,** 981–986.
36. Dierick, A. M., Praet, M., Roels, H., Verbeeck, P., Robyns, C., and Oosterlinck, W. (1991) Vimentin expression of renal cell carcinoma in relation to DNA content and histological grading: a combined light microscopic, immunocytochemical and cytophotometrical analysis. *Histopathology* **18,** 315–322.
37. Beham, A., Ratschek, M., Zatloukal, K., Schmid, C., and Denk, H. (1992) Distribution of cytokeratins, vimentin and desmoplakins in normal renal tissue, renal cell carcinomas and oncocytoma as revealed by immunofluorescence microscopy. *Virchows Arch. A Pathol. Anat. Histopathol.* **421,** 209–215.
38. Young, A. N., Amin, M. B., Moreno, C. S., et al. (2001) Expression profiling of renal epithelial neoplasms: a method for tumor classification and discovery of diagnostic molecular markers. *Am. J. Pathol.* **158,** 1639–1651.
39. Eisen, M. B., Spellman, P. T., Brown, P. O., and Botstein, D. (1998) Cluster analysis and display of genome-wide expression patterns. *Proc. Natl. Acad. Sci. USA* **95,** 14,863–14,868.
40. Heyer, L. J., Kruglyak, S., and Yooseph, S. (1999) Exploring expression data: identification and analysis of coexpressed genes. *Genome Res.* **9,** 1106–1115.
41. Bindels, R. J., Timmermans, J. A., Hartog, A., Coers, W., and van Os, C. H. (1991) Calbindin-D9k and parvalbumin are exclusively located along basolateral membranes in rat distal nephron. *J. Am. Soc. Nephrol.* **2,** 1122–1129.

42. Valore, E. V., Park, C. H., Quayle, A. J., Wiles, K. R., McCray, P. B., Jr., and Ganz, T. (1998) Human beta-defensin-1: an antimicrobial peptide of urogenital tissues. *J. Clin. Invest.* **101,** 1633–1642.
43. Khoo, S. K., Bradley, M., Wong, F. K., Hedblad, M. A., Nordenskjold, M., and Teh, B. T. (2001) Birt-Hogg-Dube syndrome: mapping of a novel hereditary neoplasia gene to chromosome 17p12-q11.2. *Oncogene* **20,** 5239–5242.
44. Boer, J. M., Huber, W. K., Sultmann, H., et al. (2001) Identification and classification of differentially expressed genes in renal cell carcinoma by expression profiling on a global human 31,500-element cDNA array. *Genome Res.* **11,** 1861–1870.
45. Takahashi, M., Rhodes, D. R., Furge, K. A., et al. (2001) Gene expression profiling of clear cell renal cell carcinoma: gene identification and prognostic classification. *Proc. Natl. Acad. Sci. USA* **98,** 9754–9759.
46. Saito, K., Saito, T., Draganac, P. S., et al. (1985) Secretion of ceruloplasmin by a human clear cell carcinoma maintained in nude mice. *Biochem. Med.* **33,** 45–52.
47. Pejovic, M., Djordjevic, V., Ignjatovic, I., Stamenic, T., and Stefanovic, V. (1997) Serum levels of some acute phase proteins in kidney and urinary tract urothelial cancers. *Int. Urol. Nephrol.* **29,** 427–432.
48. Kirschmann, D. A., Seftor, E. A., Nieva, D. R., Mariano, E. A., and Hendrix, M. J. (1999) Differentially expressed genes associated with the metastatic phenotype in breast cancer. *Breast Cancer Res. Treat.* **55,** 127–136.
49. Thompson, T. C. (1998) Metastasis-related genes in prostate cancer: the role of caveolin-1. *Cancer Metastasis Rev.* **17,** 439–442.
50. Zhang, J. C., Claffey, K., Sakthivel, R., et al. (2000) Two-chain high molecular weight kininogen induces endothelial cell apoptosis and inhibits angiogenesis: partial activity within domain 5. *FASEB J.* **14,** 2589–2600.
51. Janssen, A. M., van Duijn, W., Oostendorp-Van De Ruit, M. M., et al. (2000) Metallothionein in human gastrointestinal cancer [in process citation]. *J. Pathol.* **192,** 293–300.
52. Nguyen, A., Jing, Z., Mahoney, P. S., et al. (2000) In vivo gene expression profile analysis of metallothionein in renal cell carcinoma. *Cancer Lett.* **160,** 133–140.
53. Izawa, J. I., Moussa, M., Cherian, M. G., Doig, G., and Chin, J. L. (1998) Metallothionein expression in renal cancer. *Urology* **52,** 767–772.
54. Zhu, Y. Y., Takashi, M., Miyake, K., and Kato, K. (1991) An immunochemical and immunohistochemical study of aldolase isozymes in renal cell carcinoma. *J. Urol.* **146,** 469–472.
55. Xiang, Y. Y., Ladeda, V., and Filmus, J. (2001) Glypican-3 expression is silenced in human breast cancer. *Oncogene* **20,** 7408–7412.
56. Cano-Gauci, D. F., Song, H. H., Yang, H., et al. (1999) Glypican-3-deficient mice exhibit developmental overgrowth and some of the abnormalities typical of Simpson-Golabi-Behmel syndrome. *J. Cell Biol.* **146,** 255–264.
57. Ashburner, M., Ball, C. A., Blake, J. A., et al. (2000) Gene ontology: tool for the unification of biology. The Gene Ontology Consortium. *Nat. Genet.* **25,** 25–29.
58. Rhodes, D. R., Miller, J. C., Haab, B. B., and Furge, K. A. (2002) CIT: identification of differentially expressed clusters of genes from microarray data. *Bioinformatics* **18,** 205–206.
59. Engel, J. D., Kundu, S. D., Yang, T., et al. (1999) Transforming growth factor-beta type II receptor confers tumor suppressor activity in murine renal carcinoma (Renca) cells. *Urology* **54,** 164–170.
60. Bachman, K. E., Herman, J. G., Corn, P. G., et al. (1999) Methylation-associated silencing of the tissue inhibitor of metalloproteinase-3 gene suggest a suppressor role in kidney, brain, and other human cancers. *Cancer Res.* **59,** 798–802.

61. Ahonen, M., Baker, A. H., and Kahari, V. M. (1998) Adenovirus-mediated gene delivery of tissue inhibitor of metalloproteinases-3 inhibits invasion and induces apoptosis in melanoma cells. *Cancer Res.* **58,** 2310–2315.
62. Parry, L., Maynard, J. H., Patel, A., et al. (2001) Analysis of the TSC1 and TSC2 genes in sporadic renal cell carcinomas. *Br. J. Cancer* **85,** 1226–1230.

14
Expression Profiling of Pancreatic Ductal Adenocarcinoma

Christine A. Iacobuzio-Donahue and Ralph H. Hruban

INTRODUCTION

Pancreatic cancer is a uniquely challenging cancer. First, it is a deadly cancer. Pancreatic cancer is the fourth leading cause of cancer death in men and in women, and, despite advances in the treatment of other types of cancer, pancreatic cancer continues to have one of the highest mortality rates of any malignancy. Each year in the U.S. approx 29,000 patients are diagnosed with pancreatic cancer, and approx 29,000 will die of their disease (1). The poor prognosis for patients with pancreatic cancer is, in large part, due to the fact that almost all patients are diagnosed at an advanced stage of disease, as no known tumor markers exist that could be used to screen for pancreatic cancer at an earlier, potentially curative stage. This is a particular problem for those patients with a strong familial history of pancreatic cancer, who may have up to a 57-fold greater risk of developing pancreatic cancer in their lifetime (2,3). Second, even when a mass caused by a pancreatic cancer is identified, it can be very difficult to establish a definitive diagnosis. Deadly infiltrating adenocarcinomas of the pancreas can be so well differentiated that it can be difficult, and even at times impossible, to distinguish cancer from reactive changes histologically. Third, even when the diagnosis can be firmly established, pancreatic cancer simply does not respond to current chemotherapeutic or radiation therapies. Perhaps more than any other tumor type, a better understanding of the gene expression of pancreatic cancer is urgently needed.

CURRENT STATE OF MOLECULAR BIOLOGY OF PANCREATIC CANCER

A brief review of the current understanding of the genetic alterations associated with pancreatic cancer will help set the stage for a more detailed discussion of gene expression in pancreatic cancer. The last decade has seen a dramatic increase in our understanding of the molecular biology of pancreatic cancer, with pancreatic cancer now considered one of the better characterized neoplasms at the genetic level. Recent advances include the identification of the precursor lesions of invasive pancreatic carcinoma, known as pancreatic intraepithelial neoplasias (PanINs) (4–6). Just as there is a progression in the colorectum from normal epithelium, to adenoma, to infiltrating carcinoma, so too is there a genetic progression in the pancreas from normal duct epithelium, to PanINs, to invasive duct adenocarcinoma (7). This progression has been

From: *Expression Profiling of Human Tumors: Diagnostic and Research Applications*
Edited by: Marc Ladanyi and William L. Gerald © Humana Press Inc., Totowa, NJ

shown to be associated with the accumulation of multiple genetic alterations, including activating point mutations in the K-*ras* gene, telomere shortening, and inactivation of the *p16*, *p53*, *DPC4*, and *BRCA2* tumor suppressor genes *(4,5,8–11)*. Other mechanisms also contribute to carcinogenesis of the pancreas, such as overexpression of growth factors and their receptors, changes in activity of signal transduction pathways *(12–14)*, and alteration of low frequency mutational targets *(15)*. Recently, the interesting possibility that islet cells undergo transformation and progression to pancreatic adenocarcinomas has been hypothesized *(16)*, while others have suggested that the cell type that undergoes transformation in the pancreas is a stem cell *(17)*.

While significant advances in our understanding of pancreatic cancer genetics have been made, it is clear that if we are to impact on patient outcome, much remains to be learned regarding the fundamental changes in gene expression that occur in adenocarcinoma of the pancreas. Whereas many of the advances in the understanding of the genetics of pancreatic cancer have centered on those events that occur in the development and early genetic progression of pancreatic adenocarcinoma, other aspects of pancreatic cancer, such as tumor invasion, metastasis, or chemotherapeutic resistance, are much less well understood. Studies using global gene expression methodologies provide a unique opportunity to better understand this lethal tumor and to have a significant impact on patient care.

GENERAL ASPECTS OF THE PANCREAS TRANSCRIPTOME

Before we discuss recent developments in global analyses of gene expression in pancreatic cancer, it would be worthwhile to review what is known about gene expression in pancreatic cancer at the present time. A number of genes expressed in pancreatic ductal adenocarcinomas have been identified over the past 20 yr using a variety of techniques. These genes are summarized in Table 1.

Normal pancreatic duct epithelial cells express the cytokeratins 7, 8, 18, and 19 *(18)*. Occasional expression of cytokeratin 4 may also be noted. Infiltrating ductal adenocarcinomas express the same set of cytokeratins as the normal duct epithelium (cytokeratins 7, 8,18, and 19). More than 50% of ductal adenocarcinomas also express cytokeratin 4, and ductal adenocarcinomas are usually negative for cytokeratin 20, which is a cytokeratin commonly expressed in other lower gastrointestinal tract neoplasms. Normal acinar cells, on the other hand, express cytokeratins 8 and 18 only, and islet cells express cytokeratins 8, 18, and occasionally also 19. Acinar cell carcinomas and islet cell neoplasms express similar cytokeratins to their normal cell counterparts. Thus, the patterns of cytokeratin expression can be helpful in distinguishing nonductal-type pancreatic neoplasms (i.e., acinar and islet cell neoplasms) and other gastrointestinal neoplasms (i.e., colorectal neoplasms) from ductal adenocarcinoma.

Ductal adenocarcinomas generally do not express vimentin nor do they usually express the endocrine markers synaptophysin and chromogranin, which are two proteins commonly expressed by islet cells and islet cell neoplasms. Ductal adenocarcinomas also generally do not express the pancreatic enzymes trypsin, chymotrypsin, and lipase, all of which are highly expressed genes of acinar cells and acinar cell neoplasms. Thus, the presence of these genes in gene expression data of pancreatic cancers likely represents trapped islets or atrophic acini within the infiltrating mass and not the up-regulation of expression by the neoplastic cells themselves.

Table 1
Immunohistologic Markers of Pancreatic Normal and Neoplastic Tissues

Marker	Normal duct epithelium	Normal acini	Normal islets	Duct adenoCa	Acinar Ca	Islet neoplasm
Cytoskeletal proteins						
CK4	±	–	–	±	–	–
CK7	+	–	–	+	–	–
CK8	+	+	+	+	+	+
CK18	+	+	+	+	+	+
CK19	+	–	±	+	–	±
CK20	–	–	–	±	–	±
Vimentin	–	–	±	–	–	±
Endocrine markers						
Synaptophysin	–	–	+	–	–	+
Chromogranin	–	–	+	–	±	+
Trypsin	–	+	–	–	+	–
Chymotrypsin	–	+	–	–	±	–
Lipase	–	+	–	–	+	–
Cell surface molecules						
MUC1	+	±	–	+	ND	–
MUC3	±	–	–	+	ND	–
MUC4	–	–	–	+	ND	–
MUC5/6	–	ND	–	+	ND	–
CEA	–	–	–	+	±	–
CA19–9	±	ND	–	+	ND	–

Data from refs. *18–21* and (www.immunoquery.com).
+, positive in ≥50% of cases; ±, positive in <50% of cases; –, negative.

Duct adenocarcinomas typically also express a variety of cell surface glycoproteins *(19,20)*. For example, duct adenocarcinomas are known to express several sulfated (acidic) mucins, such as MUC1, MUC3, MUC4, and MUC5/6 *(19,21)*. The cell surface molecule carcinoembryonic antigen (CEA) is also expressed in nearly all duct adenocarcinomas, although the amount of expression seen tends to be greatest in well-differentiated tumors and least in poorly differentiated tumors. In contrast, CEA is typically not expressed in the normal pancreas and in chronic pancreatitis *(20)*. The cell surface antigen CA19-9 is also expressed in the vast majority of ductal adenocarcinomas. However, it also is expressed in normal duct epithelium, particularly in samples of chronic pancreatitis *(20)*. Therefore, although serum levels of CA19-9 are often elevated in patients with pancreatic cancer, these elevations are not specific for cancer, and CA19-9 cannot be used as a population-based screening test *(22)*. Other markers with similar expression profiles to MUC1, CEA, and CA19-9 in duct adenocarcinomas are DuPan2, Span1, and TAG72 *(18)*. While these patterns of known gene expression in pancreatic cancer provide insight into the biology of pancreatic cancer, none of these genes has proven to be useful for screening tests nor have they proven to be useful therapeutic targets. A more complete knowledge of the gene expression patterns in pancreatic ductal adenocarcinomas is, therefore, needed.

THE CHALLENGE OF STUDYING SPECIMENS FROM THE PANCREAS

While it has frequently been assumed that global expression profiling would be impossible in pancreatic tissues because of the high levels of RNases and other enzymes in the pancreas and the low neoplastic cellularity of most pancreatic cancers, in fact, these hurdles can be overcome. Two sample types have proven quite useful for gene expression profiling in the pancreas. These include cultured cell lines (normal and malignant) and surgically resected pathologic tissue specimens, both of which have inherent advantages and disadvantages for gene expression studies.

Pancreatic cell lines are very useful because they are pure populations of epithelial cells. One can, therefore, obtain an undiluted view of gene expression patterns. Recently, short-term cultures of non-neoplastic pancreatic ductal cells have also become available, allowing one to easily identify groups of genes differentially expressed by the cancer cell lines as compared to normal duct epithelial cells. Additionally, neoplastic cell lines are particularly useful for evaluating the response of the neoplastic cells to various treatment strategies, delineating signaling cascades or cellular functions that may be altered by various experimental conditions. While these cell lines are clearly useful, one must also appreciate their limitations. Cell lines are grown in artificial conditions that can result in the dysregulation of gene expression, particularly the down-regulation of gene expression related to the normal interactions of epithelial cells with their surrounding extracellular matrix components *(23,24)*. While this feature of cell lines may not affect some directed gene expression studies, it is nonetheless important to be aware of this limitation in interpretation of gene expression data based solely on the analysis of cell lines.

Surgically resected tissue specimens, because they represent the neoplasm in its "native" state, are also essential for gene expression studies. However, two concerns exist regarding the use of surgically resected pancreatic tissue samples, i.e., the predominance of non-neoplastic stromal cells within the tumor tissue specimens and the extent of mRNA degradation in pancreatic tissues. Typically, resected pancreatic cancers are composed of a minor population of infiltrating neoplastic epithelial cells surrounded by a predominance of dense fibrous (or desmoplastic) non-neoplastic stroma. This stroma contains proliferating fibroblasts, small endothelial-lined vessels, inflammatory cells, and trapped residual atrophic parenchymal components of the organ invaded (Fig. 1A) *(24,25)*. A consistently low ratio of the infiltrating neoplastic epithe-

---

Fig. 1. *(opposite page)* **(A)** Infiltrating pancreatic duct adenocarcinoma (×160). In this typical example of an infiltrating pancreatic duct adenocarcinoma, the non-neoplastic host stromal response (desmoplasia) accounts for the majority of the cellularity of the mass and is composed of proliferating fibroblasts, small vessels, and inflammatory cells. The neoplastic epithelium forms both glandular structures and individual cells that infiltrate this stromal reaction. **(B)** Normal pancreas (×160). Normal pancreatic tissue is composed predominantly of acinar cells and islets (upper left), with a minority of the cellularity accounted for by the duct epithelial cells (lower right) from which most pancreatic cancers are believed to arise.

Fig. 2. *(opposite page)* Immunohistochemical labeling for mesothelin **(A)** and PSCA **(B)**. Note the intense labeling of the infiltrating cancer (panel A, right, and panel B, top) and the absence of labeling of the normal pancreatic ducts (panel A, left, and panel B, bottom). Adapted with permission from Argani et al. *(61,62)*.

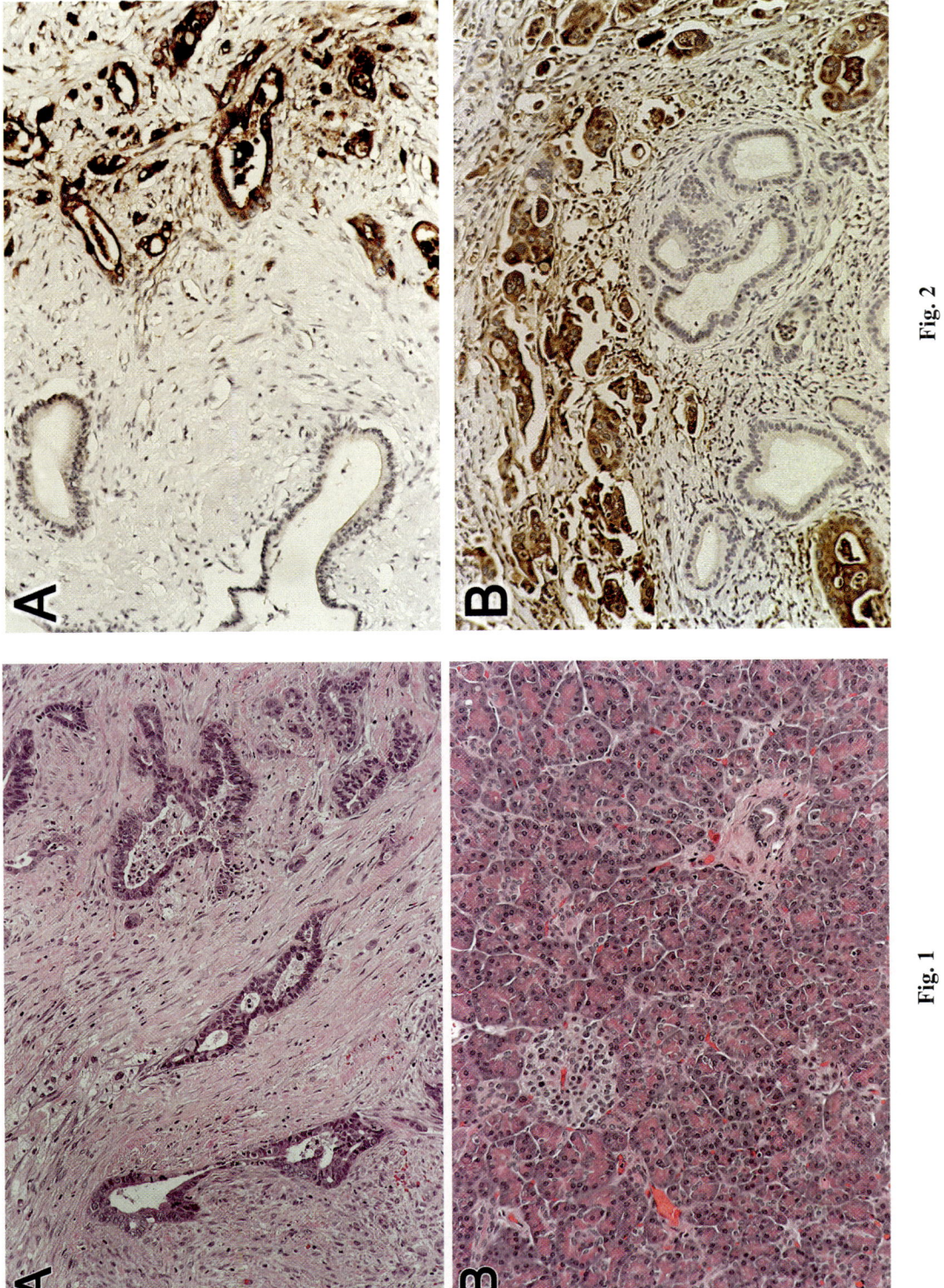

Fig. 1

Fig. 2

261

lial cells to this abundant non-neoplastic desmoplastic response is rather unique to duct adenocarcinomas of the pancreas, in contrast to infiltrating carcinomas arising in other organ or tissue types (26). Microdissection or other methods of purification of the epithelial component may be used to overcome this perceived obstacle, however, we have found that tissue samples in their native state can, nonetheless, be quite informative in gene expression studies. For example, we have obtained considerable information about the nature of gene expression related to tumor–stromal interactions by use of these stromal-rich tissue samples (24,25). Furthermore, we have found that robust patterns of neoplastic epithelial gene expression can be detected in tissue samples of pancreatic cancer, despite the predominance of the stromal cell component.

Another common perception regarding the use of pancreatic tissues is that they contain a large amount of endogenous RNases, which can potentially interfere in the extraction of mRNA for gene expression studies. RNases are a major secretory product of normal pancreatic acinar cells. However, there is commonly a significant loss of acinar cells within infiltrating pancreatic cancers due to atrophy or destruction of the gland by the neoplasm, thus facilitating the study of mRNA expression patterns within these cancer tissues. We have found that with careful technique, adequate amounts of mRNA can be extracted from quickly frozen surgically resected samples.

Normal pancreatic tissues contain a great predominance of acini and islet cells (Fig. 1B). It is, therefore, also important to recognize that ductal cells comprise only a small proportion of bulk normal pancreas. Microdissection of the duct epithelium provide one method to enrich the normal samples, whereas the use of non-neoplastic duct epithelial cells in culture is another alternative that we have used in several studies (25,27). For example, extensive gene expression data on two normal ductal epithelial cell lines is available in the on-line serial analysis of gene expression (SAGE) database (www.ncbi.nlm.nih.gov/SAGE).

While each of these sample types alone can provide limited information on the gene expression patterns in pancreatic cancer, the analysis of these different sample types together can provide a comprehensive view of the gene expression patterns in pancreatic cancer and can account for both in vivo and in vitro expression-specific patterns. Samples of normal pancreas can aid in identifying the contributions of trapped residual acinar and islet cells to the gene expression profiles detected in resected primary tumor tissues, whereas pancreas cancer cell lines, when studied together with resected pancreatic cancer tissues, can be used to identify those genes specifically expressed by the neoplastic epithelium. Likewise, genes found to be expressed solely within resected pancreas cancer tissues likely highlight those genes whose expression relates to the non-neoplastic stromal elements, but can also highlight those genes expressed by the neoplastic epithelium due to tumor–stromal interactions (24).

GLOBAL ANALYSES OF GENE EXPRESSION

Four new technologies have revolutionized our ability to study gene expression in pancreatic cancer. These include SAGE, cDNA microarrays, oligonucleotide arrays, and proteomics. Typically, the most differentially expressed genes (i.e., those with the most robust expression patterns) may be commonly identified by each of these methods, whereas those genes specifically identified by only one of these methods may reflect the sensitivity of that particular system.

SAGE

SAGE is a recently developed technique that allows one to obtain a quantitative and comprehensive profile of gene expression. In SAGE, cellular mRNA transcripts are converted to cDNA and then cleaved at specific sites by restriction enzymes into small fragments of 10–14 bases, which are known as tags *(28)*. These tags are linked together, amplified, and then sequenced. The abundance of each tag provides a quantitative measure of the number of transcripts from which the tag was derived in the total mRNA sample, therefore allowing expression levels for a particular tag to be compared between different samples. The ability to quantitate gene expression is a major advantage of SAGE, as compared to other gene expression methodologies.

Normal pancreas and pancreatic cancers were among the first tissues to be studied by SAGE *(29)*. These studies have identified a number of genes differentially expressed by pancreatic cancer, and they have revealed much about the biology of this tumor type. Ryu et al. *(27)* studied six pancreatic cancer cell lines, two short-term cultures of normal pancreatic duct cells and two surgically resected tissue samples of invasive duct adenocarcinoma by SAGE, and identified 86 genes that were differentially expressed by the cancers (*see* Table 2). Forty-nine of these genes were overexpressed by the cancers as compared to normal cells, while 37 were underexpressed by the cancers. The 49 overexpressed genes included secretory (e.g., *HE4*), cell-surface (e.g., mesothelin), transmembrane (e.g., *CEACAM6*), and tight junction protein coding genes (e.g., claudin 4), possibly corresponding to altered cellular attachments and cell surface architecture, and resulting in aberrant cell–cell interactions that are characteristic of cancer cells. A number of genes related to calcium homeostasis were also identified and included genes such as *S100A4*, *S100A10*, *Trop-2*, and *ALG-2 (30)*.

SAGE analyses have also provided insight into the process of tissue invasion in pancreatic cancers. Using principal component analysis (PCA) of SAGE data derived from pancreatic cancer cell lines and primary pancreatic cancer tissues, Ryu et al. *(25)* identified a large cluster of "invasion-specific" genes of infiltrating pancreatic cancer. These 74 known genes and 16 expressed sequence tags (ESTs) were expressed in surgically resected pancreatic cancer tissue samples, but were not seen in normal tissues nor were they seen in cultured pancreatic cancer cell lines. The genes identified within this "invasion-specific" cluster included collagens type $1\alpha1$, $1\alpha2$, connective tissue growth factor (*CTGF*), and hevin, reflecting the cellular components of the host stromal response in tissue specimens, whereas cell lines would not be expected to express these genes. This invasion-specific cluster was, therefore, thought to be specific to the desmoplastic response. Also of note is that significant numbers of the expressed genes that were invasion-specific for pancreatic cancer samples were also invasion-specific in other tumor types. For example, in this study, insulin-like growth factor binding protein 5 (IGFBP5) was found to be elevated in both resected pancreatic and breast cancers (*see* Table 3).

Because the spatial localization of gene expression was not determined initially for these invasion-specific tags, their cellular origin within the primary tumor remained unclear (neoplastic epithelium, vasculature, or stroma), as well as their role in the invasive process. Therefore, in a related study *(24)*, we used *in situ* hybridization to evaluate 12 of these invasion-specific genes in two samples of resected pancreatic cancer tissue. These genes were selected to represent different cellular functions, such as

Table 2
Representative Differentially Expressed Genes Identified by SAGE in Pancreatic Cancers as Compared to Normal Ductal Epithelium

Tag	Gene/ESTs	Function
Up-regulated in cancer		
GACATCAAGT	Keratin 19	Cytoskeletal and microfibrillar.
ATGTGTAACG	S100A4 (Mst1)	Calcium-binding protein.
ATCGTGGCGG	Claudin 4	Tight junction barrier function.
GCCTACCCGA	Trop-2	Tumor-associated calcium signal transducer.
CAAACCATCC	Keratin 18	Cytoskeletal and microfibrillar.
CCTGCTTGTC	HE4	Secreted protease inhibitor.
GCCCAGCATT	PSCA	GPI-anchored glycoprotein/prostate-specific.
GGAACTGTGA	Tetraspan 1	Possible interconnecting cell surface molecules.
AAGGATAAAA	CEACAM6	CEA-related cell adhesion molecule.
CCCCCTGCAG	Mesothelin	GPI-anchored/cell adhesion/mesothelioma and ovarian cancer antigen.
CTCGCGCTGG	Claudin 3	Tight junction barrier function.
AGCAGATCAG	S100A10	Calcium-binding protein.
TGCCTTACTT	ALG-2	Ca^{2+}-binding protein required for T cell receptor-, Fas-, and glucocorticoid-induced cell death.
Down-regulated in cancer		
GGTTATTTTG	PAI, type I	Serine (or cysteine) proteinase inhibitor.
CAAACTGGTC	Stanniocalcin 1	Stimulates renal phosphate reabsorption.
CTAACGCAGC	AP-1 (proto-oncogene c-Jun)	Transcription factor.
GAGAAGGGCA	Matrilin 1	Major component of extracellular matrix.
	Sphingosine kinase 1	Kinase.
ATCCGGACCC	GADD34	Apoptosis-associated /growth arrest and DNA damage-induced.
GAAAGTGGCT	A novel transmembrane protein	Have two follistatin modules and an epidermal growth factor (EGF) domain.

Data from ref. 27.
GPI, glycosylphosphatidylinositol.

growth factors (CTGF), signal transduction (β-catenin), cellular adhesion (β-catenin, intercellular adhesion molecule-1 [ICAM-1]), extracellular matrix remodeling (matrix metalloproteinase [MMP]-2, -11, and -14), markers of specific cell types (hevin, endothelium and thrombospondin-1, extracellular matrix), as well as genes whose role in neoplasia is unknown (apolipoproteins C-1 and D, α-2 macroglobulin, and α-2 macroglobulin receptor).

Detectable expression of all 12 genes were observed in both neoplasms, thus confirming their expression in pancreatic cancer tissues. However, for each gene, detect-

Table 3
The Pancreatic Cancer Invasion Cluster

SAGE tag	Gene	Cellular function[a]
tttgcacctt	Connective tissue growth factor	EM,A
gacctatctc	Palladin	
tgcacttcaa	Hevin	A
tcttgattta	α2 Macroglobulin	
ccctacctg	Apolipoprotein D	
gatagcacag	IFGBP5	
gtttatggat	Matrix G1a protein	A
caggagaccc	MMP-11/preg-spec β1GP9	EM,A
aagatcaaga	Actin α1 or α2 or α1	
gggaggggtg	MMP-14/HMG1 and Y	EM
cggggtggcc	Cartilage matrix protein	EM
gatgaggaga	Collagen 1α2	EM,A
atgtgaagag	Osteonectin (secreted)	EM,A
gaccagcaga	Collagen 1α1	EM
ggaaatgtca	MMP-2	EM,A
aggtcttcaa	Thrombospondin 1/EST	A
taagtagcaa	Integral membrane protein 2B	
tggccccagg	Apolipoprotein C-1	
aaatagatcc	β-Catenin	

Data from refs. 24 and 25.
[a]HMG, high mobility group. EM, extracellular matrix; A, angiogenesis.

able expression was found to localize to one or more of four distinct compartments of the invasive tumors, i.e., the neoplastic epithelium, the angioendothelium, the juxtatumoral stroma (those stromal cells immediately adjacent to the invasive neoplastic epithelium), or the panstromal compartment (all stromal tissue within the host response). Eight of these 12 genes localized to one architectural compartment. For example, CTGF, ICAM-1, β-catenin, and MMP-14 all localized to the neoplastic epithelium, while hevin localized to endothelial cells of small capillaries or venules. Apolipoprotein C-1, apolipoprotein D, and MMP-11 expression localized to the juxtatumoral stroma, whereas the neoplastic epithelium and endothelium were negative for these genes. Four genes were localized to two or more regions of the infiltrating carcinomas and showed more complex patterns of expression. For example, α-2 macroglobulin localized to endothelial cells and juxtatumoral stroma, while expression of α-2 macroglobulin receptor and thrombospondin-1 both showed expression within the neoplastic epithelium and the panstromal compartments. MMP-2 expression localized to the panstromal and angioendothelium of the tumors.

The identification of these distinct compartments helped to identify a highly organized, structured, and coordinated process of tumor invasion in the pancreas and highlighted aspects of pancreatic cancer biology previously unrecognized. First, the data indicate that although these compartments of gene expression are distinct, potential lines of communication between different compartments may exist. To provide one example, we found α-2 macroglobulin to be expressed by the juxtatumoral stroma, while the receptor for this gene product, α-2 macroglobulin receptor, is expressed by the neoplastic epithelium. Thus, the host stromal response and the juxtatumoral stroma, in particular, may play an active role in the invasive process. Second, our data indicate that important differences in gene expression between in vivo and in vitro systems exist. The finding of invasion-specific genes identified by SAGE that are highly expressed in the neoplastic epithelium of infiltrating pancreatic cancers (*CTGF*, *ICAM-1*, β-catenin, and *MMP-14*), but not in passaged cell lines derived from pancreatic cancer, supports this possibility. Third, and perhaps most interesting, we found that regional differences in the host stromal response to infiltrating pancreatic cancer exist. Our identification of the juxtatumoral stroma and its associated specific gene expression as compared to the remaining host stromal response offers new possibilities for the study of the desmoplastic response.

cDNA Microarrays

Alternative methods of gene expression analyses, such as cDNA microarrays, have also shed light on the nature of pancreatic cancer. Unlike SAGE, cDNA microarray analyses involve the competitive hybridization of cDNAs derived from experimental and control samples to cDNA microarray chips, thus permitting the identification of relationships of global gene expression patterns among different sample types.

In collaboration with Patrick O. Brown Ph.D. at Stanford, we have used cDNA microarray to analyze samples of infiltrating pancreatic duct adenocarcinoma, pancreatic cancer cell lines, and normal pancreatic tissues *(31)*. We identified several clusters that could be related to the biological or histological features of the samples. The largest cluster identified contained cDNAs whose expression was increased in both pancreatic cancer cell lines and in primary pancreatic tumors as compared to normal pancreatic tissues. Genes included in this large pancreas cancer-specific cluster spanned a variety of classes of gene function and were characterized by those involved in cell membrane junctions (claudins 3, 4, and 7) *(32)*, cell–matrix interactions (integrin-α 3 and -α 6) *(33)*, cytoskeletal assembly (keratins 7, 17, and 19) *(34)*, cell cycle regulation (p21, cyclin D) *(35)*, transcription factors (T cell factor [TCF]7) *(36)*, calcium homeostasis (S100A10 and S100A11) *(37)*, and proteolytic processing (urokinase plasminogen activator and MMP-24) *(12,38)*. A partial list of the genes identified using cDNA microarrays is given in Table 4.

Several smaller yet informative clusters of genes were also found and were associated with cellular proliferation and invasion-specific gene expression related to both stromal cells and neoplastic epithelium. The proliferation cluster included chromosome remodeling genes (e.g., structural maintenance chromosome [SMC]4-like 1), cell cycle regulating genes (e.g., cyclin A2), and genes associated with cytoskeletal remodeling (e.g., myosin heavy polypeptide 1), whereas the invasion-specific cluster included genes such as collagen 1α1, fusin, and *IGFBP7*.

Oligonucleotide Arrays

Oligonucleotide arrays provide yet another method to study global gene expression patterns. With this method, cDNAs are hybridized in a noncompetitive manner to oligonucleotide arrays, with the intensity of signal reflecting in a linear fashion the amount of mRNA expression present in the original sample.

Table 4
Representative cDNAs Identified as Overexpressed in Pancreatic Cancer by Microarray Analysis in Pancreatic Cancer Cell Lines, Normal Pancreas, and Resected Pancreatic Cancer Tissues

Known gene name	Cellular function
Pancreas cancer-specific cluster	
Claudin 3	Cell junction component
Claudin 4	Cell junction component
Integrin-α3	Cell adhesion
Cytokeratin 7	Cytoskeleton
Cytokeratin 17	Cytoskeleton
Cytokeratin 19	Cytoskeleton
p21	Cell cycle regulation
Cyclin D	Cell cycle regulation
TCF7	Transcription factor
S100A10	Calcium homeostasis
S100A11	Calcium homeostasis
UPA	Proteolytic processing
MMP-24	Extracellular matrix processing
Pancreas cancer tissue-specific cluster (desmoplasia)	
Collagen 1α1	Extracellular matrix component
Fusin	Lymphocyte cytokine
IGFBP7	Cell growth regulation
Galectin 4	Cell adhesion
Bone marrow stromal cell antigen 2	Cell growth regulation
Pancreas cancer cell line-specific cluster (proliferation)	
SMC4-like 1	Chromosome remodeling
Cyclin A2	Cell cycle regulation
Myosin heavy polypeptide 1	Cytoskeletal remodeling

Data from ref. *31*.

In collaboration with Grace L. Shen-Ong, Ph.D., and GeneLogic, Inc. (Gaithersburg, MD, USA), we have used oligonucleotide arrays to identify genes differentially expressed in pancreatic cancer, using cDNAs prepared from normal pancreas, normal gastrointestinal mucosa, resected pancreas cancer tissues, and pancreas cancer cell lines. GeneChips® (Affymetrix) were utilized, and the genes that were overexpressed in the pancreatic adenocarcinoma tumor tissues or cell lines were compared to all normal tissues and were identified *(39)*. One hundred eighty fragments were found to be expressed at least 5-fold greater in pancreatic cancer samples as compared to normal tissues, 12 of which were expressed greater than 10-fold. The level of significance for each gene fragment ranged from less than $p = 0.00001$ to $p = 0.01$ (modified Welch *t*-test).

Characterization of the 180 fragments identified revealed that 56 fragments corresponded to ESTs, and 124 fragments corresponded to known genes. Among these 124 fragments, 10 genes were represented by two or more fragments, resulting in 107 known genes identified as expressed at least 5-fold or greater in pancreatic cancers as compared to normal. Ten genes were identified as having high levels of expression in the SAGE analyses of normal pancreatic duct epithelium. These genes were excluded, leaving 97 remaining differentially expressed genes.

Of the 97 genes analyzed, 28 genes were previously reported to be associated with pancreatic cancer, whereas 69 genes were not. Of these 69 genes not identified in this PubMed search as having been reported in pancreatic cancer, 21 have been reported before in association with tumor types other than pancreatic cancer, while 48 genes have not been reported in association with any neoplasm. These 97 candidate tumor markers of pancreatic cancer represented a variety of cellular functions (a partial list is provided in Table 5). Genes identified included those involved in cell membrane junctions (claudin 1, connexin 26) *(40,41)*, signal transduction (ras GTPase-activating protein-like) *(42,43)*, calcium homeostasis (Trop-2, S100 calcium-binding protein P) *(44)*, cytoskeletal assembly (fascin, keratin 7, rabkinesin6, and pleckstrin) *(45–48)*, cell surface adhesion and recognition (integrin β-like 1) *(49)*, DNA transcription (topoisomerase II-α, transcription factor brain-muscle-Arnt-like-protein [BMAL]2, and acute myelogenous leukemia [AML]1) *(50–52)*, DNA repair (ataxia telangiectasia group D-complementing [ATDC]) *(53)*, or extracellular matrix remodeling and function (collagens 1α1, 1α2, and X1α1, heat-shock protein 47, MMP-14, and MMP-7) *(24,54,55)*. The cellular localization of the corresponding gene products was also determined using the on-line database On-line Mendelian Inheritance in Man (OMIM) available through the National Center for Biotechnology Information (NCBI) Web site (http://www.ncbi.nlm.nih.gov/entrez/query). Genes were found to encode membrane-bound proteins (prostate stem cell antigen, OB-cadherin), cytoplasmic proteins (fascin, ATDC), nuclear proteins (topoisomerase II-α, paraneoplastic antigen MA1), as well as extracellular proteins, such as those involved in extracellular matrix homeostasis (heat-shock protein 47, thrombospondin 2) or secreted protein products (osteopontin). Each of these genes identified represents a potential tumor marker for the development of pancreas cancer screening tests and novel chemotherapeutic modalities.

Protein Chips

Proteomic methods have provided an alternative approach to the study of pancreatic cancer and the identification of novel biomarkers. In this method, small amounts of protein are directly applied to biochips coated with specific chemical matrices (hydrophobic, cationic, anionic, normal phase, etc.) and analyzed by mass spectrometry to obtain a protein "fingerprint" of the sample.

Using ProteinChip® (Ciphergen) SELDI technology, Rosty, Goggins, and coworkers screened for differentially expressed proteins in pancreatic juice samples from patients both with and without pancreatic duct adenocarcinoma *(56)*. A 16.5-kDa protein peak was identified in 10 out of 15 (67%) of the patients with pancreatic adenocarcinoma, but in only 1 out of 7 (17%) of the patients with other pancreatic diseases. This protein was identified as hepatocarcinoma–intestine–pancreas/pancreatitis-associated–protein-1 (HIP/PAP-1), which is a protein released from pancreatic acini during acute pancreatitis and overexpressed in hepatocellular carcinoma. The quantification of HIP/PAP-1 amounts in pancreatic juice and serum samples by enzyme linked immunosorbent assay (ELISA) confirmed the significantly elevated amounts of this protein in the samples from patients with pancreatic adenocarcinoma. Furthermore, patients with pancreatic juice HIP/PAP-1 levels greater or equal to 20 µg/mL were

Table 5
Representative Highly Expressed Genes Identified by Oligonucleotide Arrays in Pancreatic Cancer Cell Lines and Tissues

Known gene name	Fold change	SAGE normal tags[a]	Reported in pancreas	Ref.	Cellular location[b]
Ataxia-telangiectasia group D-associated protein	5.21	0,0	No		C
Cadherin 11, type 2, OB-cadherin (osteoblast)	5.93	0,0	No		M
Claudin 1	5.61	0,0	No		M
Collagen, type I, α-2	8.84	4,2	Yes	(25,27)	EM
Collagen, type XI, α-1	6.88	0,0	Yes	(25,27)	EM
Cyclin-dependent kinase inhibitor 2A (p16)	5.87	0,0	Yes		C
Drebin 1	5.24	0,1	No		
Gap junction protein, β 2, 26 kDa (connexin 26)	7.32	0,0	Yes	(63)	M
Integrin, β-like 1 (with EGF-like repeat domains)	7.49	0,0	No		M
Interleukin 8	6.53	0,0	Yes		C
Lipocalin 2 (oncogene 24p3)	8.86	5,3	No		
Keratin 7	10.77	2,4	Yes	(46)	C
Matrix metalloproteinase 14 (membrane-inserted)	7.27	0,0	Yes	(24)	M
Matrix metalloproteinase 7 (matrilysin, uterine)	8.79	0,0	Yes	(25)	S
Paraneoplastic antigen MA1	5.49	0,1	No		N
Pleckstrin homology-like domain, family A, member 1	14.66	5,2	No		
Prostate stem cell antigen	5.34	0,0	Yes	(61)	M
RAB6 interacting, kinesin-like (rabkinesin6)	5.09	0,0	No		C
ras GTPase activating protein-like	7.01	4,4			
Runt-related transcription factor 1 (aml1 oncogene)	5.92	0,4	No		N
S100 calcium-binding protein P	8.73	0,0	No		N
Secreted phosphoprotein 1 (osteopontin)	7.98	0,0	No		S
Heat shock protein 47	6.41	1,4	No		
Singed (*Drosophila*)-like (sea urchin fascin homolog like)	13.31	1,1	No		C
Thrombospondin 2	9.92	0,0	No		EM
Topoisomerase (DNA) II α (170 kDa)	5.28	1,0/2,0	No		N
Transcription factor BMAL2	7.03	0,0	No		N
Transmembrane, prostate androgen induced RNA	9.54	1,2	No		
Trop-2	9.29	0,0	Yes	(27)	M

Data from ref. *39*.

[a]Total tags present in normal duct epithelial cell SAGE libraries HX and H126, available through the SAGEmap database (www.ncbi.nlm.nih.gov/SAGE).

[b]C, cytoplasmic; M, cell membrane; EM, extracellular matrix; N, nuclear; S, secreted.

21.9 times more likely to have pancreatic adenocarcinoma than patients with levels less than 20 µg/ml. Immunohistochemical labeling of paraffin-embedded samples of pancreatic duct adenocarcinoma confirmed the strong expression of HIP/PAP-1 in acinar cells immediately adjacent to the cancers, with only rare expression in the neoplastic epithelium, indicating the main source of this protein is acini undergoing atrophy or destruction. Based on this early screen identifying HIP/PAP-1, Dr. M. Goggins and collaborators are actively using this technology to screen for additional markers.

Summary

Our initial attempts to study pancreatic cancer using these global expression methodologies have revealed a wealth of information. Interestingly, these various methodologies have revealed similar findings regarding the genes or cellular processes most highly up-regulated in pancreatic cancers, i.e., calcium homeostasis (S100A10, Trop-2), cell–cell interactions (claudins 3,4), cell–matrix interactions (integrin-$\alpha 3$, integrin-β-like 1), extracellular matrix remodeling (MMP-7, MMP-14, collagens $1\alpha 1$ and $1\alpha 2$), as well as genes such as prostate stem cell antigen (PSCA), which has been identified both by SAGE and oligonucleotide arrays, but whose role in pancreatic cancer is unknown. These findings not only provide novel insight into the biology of pancreatic cancer, but serve to generate new hypotheses for the study of pancreatic cancer.

VALIDATION OF GENE EXPRESSION

While global gene expression analyses can reveal much about the biological processes of pancreatic cancer, each of the candidate genes identified should be validated to confirm that gene expression differences are real. Perhaps the most commonly used methods for validation are those which confirm mRNA expression, such as reverse transcription polymerase chain reaction (RT-PCR) and *in situ* hybridization, or those which confirm protein expression, i.e., immunohistochemistry and Western blotting.

RT-PCR not only provides confirmation of gene expression, but also allows a semiquantitative determination of gene expression in samples. This method is perhaps best for confirming gene expression in cell lines or microdissected samples in which the cell of origin is known. Alternatively, *in situ* hybridization can be used to both confirm mRNA expression and also localize mRNA expression in tissues, which is an issue of particular relevance in stromal-rich pancreatic cancers. *In situ* hybridization can be performed on fresh frozen or paraffin-embedded tissues, which is an advantage for validating mRNA expression in archival tissue specimens. Furthermore, in our experience, formalin-fixed paraffin-embedded pancreatic cancers have adequate amounts of mRNA for the reliable detection of gene expression *(24)*.

Validation of gene expression by immunohistochemical labeling is yet another option with several advantages. Immunohistochemical detection of protein expression can be used to determine the cell type of origin of expression, similar to *in situ* hybridization. However, unlike *in situ* hybridization, immunohistochemistry can also provide information on the cellular localization of the protein product (e.g., the cell membrane, nucleus, or cytoplasm), as well as demonstrate translation of the gene product studied. One disadvantage, however, is that the ability to perform immunohistochemistry is often limited by the availability of suitable primary antibodies specific for the protein of interest, unlike *in situ* hybridization, in which specific probes can be easily generated to any transcript of interest.

APPLICATIONS

These initial gene expression studies of pancreatic cancer provide a host of new and exciting genes with potential as novel diagnostic aids, tumor markers, or targets for therapeutic development.

The identification of genes significantly overexpressed in infiltrating pancreatic ductal carcinomas has immediate diagnostic potential. Overexpression of novel tumor markers of pancreatic cancer can be used to differentiate infiltrating pancreatic duct adenocarcinoma from chronic pancreatitis, particularly in small tissue samples or cytologic material *(57)*. For example, immunohistochemical labeling of a small biopsy for a panel of genes specifically up-regulated in invasive cancer could be used to establish the diagnosis of cancer.

Novel markers of pancreatic cancer also have the potential for the development of new screening tests for pancreatic cancer. For example, several of the genes we have identified have been found to be membranous or secreted proteins, suggesting that they may be shed into the blood or pancreatic secretions. If so, these proteins may also serve as screening markers, not only for identification of primary pancreatic cancers at an earlier stage, but also for the identification of recurrent disease at an earlier phase when it may be more responsive to adjuvant therapies. In addition, whereas the use of any one marker individually may have a limited sensitivity or specificity in detecting pancreatic cancer, the development of a panel of markers may significantly increase the specificity of detecting clinically inapparent pancreatic cancers without decreasing the sensitivity. For example, tissue inhibitor of metalloproteinase 1 (TIMP-1) was identified by SAGE as overexpressed in pancreatic cancer *(29)*. Serum TIMP-1 levels, when combined with other markers of pancreatic cancer, provided a reasonably sensitive and specific screening test for pancreatic cancer. The development of tagged antibodies to one or more of these genes may also be useful in the diagnostic radiologic imaging of small primary pancreatic cancers or metastases before they become clinically apparent.

Novel markers of pancreatic cancer may also have important therapeutic applications. For example, Jaffee et al. *(58)* have recently shown that cell-mediated immunotherapy can be both safe and effective in patients with pancreatic cancer. The identification of genes highly overexpressed in pancreatic cancer may, therefore, represent potential targets for the development of cell-mediated vaccines. Similarly, genes identified that encode for cell surface proteins hold promise for the development of antibody-based immunotherapy against pancreatic carcinoma *(59,60)*. Finally, in some cases, small molecules that specifically block the function of the identified protein could be developed.

The value of these new markers can be demonstrated with a few specific examples. Comparisons of SAGE libraries derived from pancreatic duct adenocarcinomas to SAGE libraries derived from non-neoplastic tissues have identified the SAGE tags for both mesothelin and PSCA as highly expressed in SAGE libraries derived from pancreatic carcinomas *(61,62)*. The tag for mesothelin was present in seven of eight pancreatic cancer cell line SAGE libraries, but not in SAGE libraries derived from normal duct epithelial cells. Interestingly, 60 of 60 duct adenocarcinomas were strongly immunoreactive for the mesothelin protein (Fig. 2A), and many of these carcinomas had a membranous pattern of staining, suggesting a potential target for immune-based therapies. PSCA was also identified using this approach. PSCA is a recently discovered gene initially thought to be restricted in expression to prostate basal cells and prostate cancers. PSCA mRNA expression was confirmed in 14 of 19 pancreatic cancer cell lines by RT-PCR, and protein overexpression was confirmed immunohistochemically in 36 of 60 archival paraffin-embedded primary pancreatic adenocarcinomas (Fig. 2B) *(61)*. Importantly, the intense labeling of infiltrating pancreatic cancers with these markers could be used to easily distinguish infiltrating neoplastic cells from benign ducts and immune therapies targeting both of these antigens are under development *(62)*.

SUMMARY

The development of global gene expression methodologies have resulted in a virtual explosion of information in the study of human cancers. The use of these various techniques in the study of pancreatic cancer exemplifies this phenomenon. Compared to only 5 yr ago, we are now aware of hundreds of genes with potential importance in the biology of pancreatic cancer.

While the gene expression studies we have discussed represent encouraging "first steps" on the road to the cure of pancreatic cancer, much more remains to be learned about this tumor. Potential areas for future study include, but are not limited to, the gene expression patterns associated with invasion and metastasis of pancreatic cancer, the genes expressed in incipient pancreatic cancers and their associated genetic alterations, and the identification and development of biologic markers for early detection or therapy. These and other aspects of pancreatic cancer remain areas of enormous potential for research opportunities that can greatly impact on the survival of patients with this highly lethal tumor.

ACKNOWLEDGMENTS

We thank Dr. N. Volkan Adsay for his helpful suggestions regarding the immunohistochemical markers of normal and neoplastic pancreatic tissues.

REFERENCES

1. Greenlee, R. T., Hill-Harmon, M. B., Murray, T., and Thun, M. (2001) Cancer Statistics, 2001. *CA Cancer J. Clin.* **51,** 16–36.
2. Tersmette, A. C., Petersen, G. M., Offerhaus, G. J. A., et al. (2001) Increased risk of incident pancreatic cancer among first-degree relatives of patients with familial pancreatic cancer. *Clin. Cancer Res.* **7,** 738–744.
3. Giardiello, F. M., Welsh, S. B., Hamilton, S. R., et al. (1987) Increased risk of cancer in the Peutz-Jeghers syndrome. *N. Engl. J. Med.* **316,** 1511–1514.
4. Wilentz, R. E., Geradts, J., Maynard, R., et al. (1998) Inactivation of the *p16 (INK4A)* tumor-suppressor gene in pancreatic duct lesions: Loss of intranuclear expression. *Cancer Res.* **58,** 4740–4744.
5. Wilentz, R. E., Iacobuzio-Donahue, C. A., Argani, P., et al. (2000) Loss of expression of Dpc4 in pancreatic intraepithelial neoplasia: evidence that *DPC4* inactivation occurs late in neoplastic progression. *Cancer Res.* **60,** 2002–2006.

6. Hruban, R. H., Adsay, N. V., Albores-Saavedra, J., et al. (2001) Pancreatic intraepithelial neoplasia (PanIN): a new nomenclature and classification system for pancreatic duct lesions. *Am. J. Surg. Pathol.* **25,** 579–586.
7. Hruban, R. H., Wilentz, R. E., and Kern, S. E. (2000) Genetic progression in the pancreatic ducts. *Am. J. Pathol.* **156,** 1821–1825.
8. Redston, M. S., Caldas, C., Seymour, A. B., et al. (1994) *p53* mutations in pancreatic carcinoma and evidence of common involvement of homocopolymer tracts in DNA microdeletions. *Cancer Res.* **54,** 3025–3033.
9. Schutte, M., Hruban, R. H., Geradts, J., et al. (1997) Abrogation of the *Rb/p16* tumor-suppressive pathway in virtually all pancreatic carcinomas. *Cancer Res.* **57,** 3126–3130.
10. Hruban, R. H., van Mansfeld, A. D. M., Offerhaus, G. J. A., et al. (1993) K-*ras* oncogene activation in adenocarcinoma of the human pancreas. A study of 82 carcinomas using a combination of mutant-enriched polymerase chain reaction analysis and allele-specific oligonucleotide hybridization. *Am. J. Pathol.* **143,** 545–554.
11. Wilentz, R. E., Su, G. H., Dai, J. L., et al. (2000) Immunohistochemical labeling for Dpc4 mirrors genetic status in pancreatic: a new marker of *DPC4* inactivation. *Am. J. Pathol.* **156,** 37–43.
12. Ellenrieder, V., Hendler, S. F., Ruhland, C., Boeck, W., Adler, G., and Gress, T. M. (2001) TGF-beta-induced invasiveness of pancreatic cancer cells is mediated by matrix metalloproteinase-2 and the urokinase plasminogen activator system. *Int. J. Cancer* **93,** 204–211.
13. Ebert, M., Yokoyama, M., Friess, H., Kobrin, M. S., Buchler, M. W., and Korc, M. (1995) Induction of platelet-derived growth factor A and B chains and over-expression of their receptors in human pancreatic cancer. *Int. J. Cancer* **62,** 529–535.
14. Friess, H., Yamanaka, Y., Buchler, M., et al. (1993) Enhanced expression of the type II transforming growth factor beta receptor in human pancreatic cancer cells without alteration of type III receptor expression. *Cancer Res.* **53,** 2704–2707.
15. Goggins, M., Schutte, M., Lu, J., et al. (1996) Germline BRCA2 gene mutations in patients with apparently sporadic pancreatic carcinomas. *Cancer Res.* **56,** 5360–5364.
16. Pour, P. M. and Schmied, B. (1999) The link between exocrine pancreatic cancer and the endocrine pancreas. *Int. J. Pancreatol.* **25,** 77–87.
17. Rao, M. S., Yeldandi, A. V., and Reddy, J. K. (1990) Stem cell potential of ductular and periductular cells in the adult rat pancreas. *Cell Differ. Dev.* **29,** 155–163.
18. Solcia, E., Capella, C., and Klöppel, G. (1997) *Atlas of Tumor Pathology: Tumors of the Pancreas. 3rd ed.* Armed Forces Institute of Pathology, Washington, D.C.
19. Balague, C., Audie, J. P., Porchet, N., and Real, F. X. (1995) In situ hybridization shows distinct patterns of mucin gene expression in normal, benign, and malignant pancreas tissues. *Gastroenterology* **109,** 953–964.
20. Shimizu, M., Saitoh, Y., Ohyanagi, H., and Itoh, H. (1990) Immunohistochemical staining of pancreatic cancer with CA19-9, KM01, unabsorbed CEA, and absorbed CEA. *Arch. Pathol. Lab. Med.* **114,** 195–200.
21. Terada, T., Ohta, T., Sasaki, M., Nakanuma, Y., and Kim, Y. S. (1996) Expression of MUC apomucins in normal pancreas and pancreatic tumours. *J. Pathol.* **180,** 160–165.
22. Ritts, R. E. and Pitt, H. A. (1998) CA19-9 in pancreatic cancer. *Surg. Clin. N. Am.* **7,** 93–101.
23. Ross, D. T., Scherf, U., Eisen, M. B., et al. (2000) Systematic variation in gene expression patterns in human cancer cell lines. *Nat. Genet.* **24,** 227–235.
24. Iacobuzio-Donahue, C. A., Ryu, B., Hruban, R. H., and Kern, S. E. (2002) Exploring the host desmoplastic response to pancreatic carcinoma: gene expression of stromal and neoplastic cells at the site of primary invasion. *Am. J. Pathol.* **160,** 91–99.

25. Ryu, B., Jones, J., Hollingsworth, M. A., Hruban, R. H., and Kern, S. E. (2001) Invasion-specific genes in malignancy: serial analysis of gene expression comparisons of primary and passaged cancers. *Cancer Res.* **61,** 1833–1838.
26. Hahn, S. A., Seymour, A. B., Hoque, A. T. M. S., et al. (1995) Allelotype of pancreatic adenocarcinoma using xenograft enrichment. *Cancer Res.* **55,** 4670–4675.
27. Ryu, B., Jones, J., Hollingsworth, M. A., Hruban, R. H., and Kern, S. E. (2002) Relationships and identification of differentially expressed genes among pancreatic cancers examined by large-scale serial analysis of gene expression. *Cancer Res.* **62,** 5351–5357.
28. Velculescu, V. E., Zhang, L., Vogelstein, B., and Kinzler, K. W. (1995) Serial analysis of gene expression. *Science* **270,** 484–487.
29. Zhou, W., Sokoll, L. J., Bruzek, D. J., et al. (1998) Identifying markers for pancreatic cancer by gene expression analysis. *Cancer Epidemiol. Biomarkers Prev.* **7,** 109–112.
30. Rosty, C., Ueki, T., Argani, P., et al. (2002) Overexpression of S100A4 in pancreatic ductal adenocarcinomas is associated with poor differentiation and DNA hypomethylation. *Am. J. Pathol.* **160,** 45–50.
31. Iacobuzio-Donahue, C. A., Maitra, A., Olsen, M., et al. (2003) Exploration of global gene expression of pancreatic adenocarcinoma by cDNA microarray analysis. *Am. J. Pathol.*, in press.
32. Heiskala, M., Peterson, P. A., and Yang, Y. (2001) The roles of claudin superfamily proteins in paracellular transport. *Traffic* **2,** 93–98.
33. Kikkawa, Y., Sanzen, N., Fujiwara, H., Sonnenberg, A., and Sekiguchi, K. (2000) Integrin binding specificity of laminin-10/11: laminin-10/11 are recognized by alpha 3 beta 1, alpha 6 beta 1 and alpha 6 beta 4 integrins. *J. Cell Sci.* **113,** 869–876.
34. Schussler, M. H., Skoudy, A., Ramaekers, F., and Real, F. X. (1992) Intermediate filaments as differentiation markers of normal pancreas and pancreas cancer. *Am. J. Surg. Pathol.* **140,** 559–568.
35. DiGiuseppe, J. A., Redston, M. S., Yeo, C. J., Kern, S. E., and Hruban, R. H. (1995) p53-independent expression of the cyclin-dependent kinase inhibitor p21 in pancreatic carcinoma. *Am. J. Pathol.* **147,** 884–888.
36. Sparks, A. B., Morin, P. J., Vogelstein, B., and Kinzler, K. W. (1998) Mutational analysis of the APC/beta-catenin/Tcf pathway in colorectal cancer. *Cancer Res.* **58,** 1130–1134.
37. Ruse, M., Lambert, A., Robinson, N., Ryan, D., Shon, K. J., and Eckert, R. L. (2001) S100A7, S100A10, and S100A11 are transglutaminase substrates. *Biochemistry* **40,** 3167–3173.
38. Ellenrieder, V., Alber, B., Lacher, U., et al. (2000) Role of MT-MMPs and MMP-2 in pancreatic cancer progression. *Int. J. Cancer* **85,** 14–20.
39. Iacobuzio-Donahue, C. A., Maitra, A., Shen-Ong, G. L., et al. (2002) Discovery of novel tumor markers of pancreatic cancer using global gene expression technology. *Am. J. Pathol.* **160,** 1239–1249.
40. Morita, K., Furuse, M., Fujimoto, K., and Tsukita, S. (1999) Claudin multigene family encoding four-transmembrane domain protein components of tight junction strands. *Proc. Natl. Acad. Sci. USA* **96,** 511–516.
41. Carrio, M., Romagosa, A., Mercade, E., et al. (1999) Enhanced pancreatic tumor regression by a combination of adenovirus and retrovirus-mediated delivery of the herpes simplex virus thymidine kinase gene. *Gene Ther.* **6,** 547–553.
42. Ripani, E., Sacchetti, A., Corda, D., and Alberti, S. (1998) Human Trop-2 is a tumor-associated calcium signal transducer. *Int. J. Cancer* **76,** 671–676.
43. Namima, M., Takeuchi, K., Watanabe, Y., et al. (1998) Localization of GTPase-activating protein-(GAP) like immunoreactivity in mouse cerebral regions. *Mol. Chem. Neuropathol.* **35,** 157–172.
44. Donato, R. (2001) S100: a multigenic family of calcium-modulated proteins of the EF-hand type with intracellular and extracellular functional roles. *Int. J. Biochem. Cell Biol.* **33,** 637–668.

45. Tseng, Y., Fedorov, E., McCaffery, J. M., Almo, S. C., and Wirtz, D. (2001) Micromechanics and ultrastructure of actin filament networks crosslinked by human fascin: a comparison with alpha-actinin. *J. Mol. Biol.* **310,** 351–366.
46. Goldstein, N. S. and Bassi, D. (2001) Cytokeratins 7, 17, and 20 reactivity in pancreatic and ampulla of vater adenocarcinomas. Percentage of positivity and distribution is affected by the cut-point threshold. *Am. J. Clin. Pathol.* **115,** 695–702.
47. Lai, F., Fernald, A. A., Zhao, N., and Le Beau, M. M. (2000) cDNA cloning, expression pattern, genomic structure and chromosomal location of RAB6KIFL, a human kinesin-like gene. *Gene* **248,** 117–125.
48. Sato, T. K., Overduin, M., and Emr, S. D. (2001) Location, location, location: membrane targeting directed by PX domains. *Science* **294,** 1881–1885.
49. Schwartz, M. A. (2001) Integrin signaling revisited. *Trends Cell Biol.* **11,** 466–470.
50. Tanner, M., Jarvinen, P., and Isola, J. (2001) Amplification of HER-2/neu and topoisomerase IIalpha in primary and metastatic breast cancer. *Cancer Res.* **61,** 5345–5348.
51. Ikeda, M., Yu, W., Hirai, M., et al. (2000) cDNA cloning of a novel bHLH-PAS transcription factor superfamily gene, BMAL2: its mRNA expression, subcellular distribution, and chromosomal localization. *Biochem. Biophys. Res. Commun.* **275,** 493–502.
52. Downing, J. R. (2001) AML1/CBFbeta transcription complex: its role in normal hematopoiesis and leukemia. *Leukemia* **15,** 664–665.
53. Hosoi, Y. and Kapp, L. N. (1994) Expression of a candidate ataxia-telangiectasia group D gene in cultured fibroblast cell lines and human tissues. *Int. J. Radiat. Biol.* **66(6 Suppl),** S71–S76.
54. Gress, T. M., Muller-Pillasch, F., Lerch, M. M., et al. (1994) Balance of expression of genes coding for extracellular matrix proteins and extracellular matrix degrading proteases in chronic pancreatitis. *Z. Gastroenterol.* **32,** 221–225.
55. Dafforn, T. R., Della, M., and Miller, A. D. (2001) The molecular interactions of heat shock protein 47 (Hsp47) and their implications for collagen biosynthesis. *J. Biol. Chem.* **276,** 49,310–49,319.
56. Rosty, C., Christa, L., Kuzdzal, S., et al. (2002) Identification of hepatocarcinoma-intestine-pancreas/pancreatitis-associated-protein I (HIP/PAP-1) as a biomarker for pancreatic adenocarcinoma by protein biochip technology. *Cancer Res.* **62,** 1868–1875.
57. Tascilar, M., Offerhaus, G. J., Altink, R., et al. (2001) Immunohistochemical labeling for the Dpc4 gene product is a specific marker for adenocarcinoma in biopsy specimens of the pancreas and bile duct. *Am. J. Clin. Pathol.* **116,** 831–837.
58. Jaffee, E. M., Hruban, R. H., Biedrzycki, B., et al. (2001) Novel allogeneic granulocyte-macrophage colony-stimulating factor-secreting tumor vaccine for pancreatic cancer: a phase I trial of safety and immune activation. *J. Clin. Oncol.* **19,** 145–156.
59. Jaffee, E. M., Schutte, M., Gossett, J., et al. (1998) Development and characterization of a cytokine-secreting pancreatic adenocarcinoma vaccine from primary tumors for use in clinical trials. *Cancer J. Sci. Am.* **4,** 194–203.
60. McDevitt, M. R., Ma, D., Lai, L. T., et al. (2001) Tumor therapy with targeted atomic nanogenerators. *Science* **294,** 1537–1540.
61. Argani, P., Rosty, C., Reiter, R. E., et al. (2001) Discovery of new markers of cancer through serial analysis of gene expression (SAGE): prostate stem cell antigen (PSCA) is overexpressed in pancreatic adenocarcinoma. *Cancer Res.* **61,** 4320–4324.
62. Argani, P., Iacobuzio-Donahue, C., Ryu, B., et al. (2001) Mesothelin is expressed in the vast majority of adenocarcinomas of the pancreas: identification of a new cancer marker by serial analysis of gene expression (SAGE). *Clin. Cancer Res.* **7,** 3862–3868.
63. Carrio, M., Mazo, A., Lopez-Iglesias, C., Estivill, X., and Fillat, C. (2001) Retrovirus-mediated transfer of the herpes simplex virus thymidine kinase and connexin26 genes in pancreatic cells results in variable efficiency on the bystander killing: implications for gene therapy. *Int. J. Cancer* **94,** 81–88.

15
Gene Expression in Ovarian Carcinoma

Garret M. Hampton

INTRODUCTION

Cancer of the ovary accounts for the highest tumor-related mortality among women diagnosed with gynecological malignancy. Of the 23,400 new cases of ovarian cancer estimated by the American Cancer Society (ACS) in 2001, roughly 60% will die from their disease. The remarkably high mortality rate is due, in large part, to the late stage at which patients with ovarian cancer are diagnosed, prompting sizable efforts to identify diagnostic molecules for early stage detection. A growing body of literature is emerging on gene transcription in normal and malignant ovarian tissues, revealing genes whose aberrant transcription may contribute to the neoplastic phenotype, as well as pinpointing novel entry points for therapeutic intervention and identifying secreted molecules that could be detected by diagnostic assays. This chapter will focus on recent results obtained by expression profiling in ovarian cancers, with an emphasis on the genes identified, the emergence of molecular signatures of the disease, as well as the discovery of molecular distinctions between ovarian cancers of varying histology and grade.

OVARIAN CANCER: EPIDEMIOLOGY, PATHOLOGY, AND GENETIC ALTERATIONS

Of the 625,000 new cases of cancer anticipated by the ACS to afflict females in the U.S. in 2001, approx 80,300 will involve the female genitalia with 51,100 arising in the uterus, including the uterine cervix, 23,400 arising in the ovary, 3600 and 2100 arising in the vulva, vagina, and other genitalia, respectively *(1)*. Cancer-related mortality resulting from tumors of the female genitalia is most pronounced for ovarian cancer, estimated by the ACS to account for 13,400 deaths in females of all ages, and accounting for the single largest number of cancer-related deaths in females between the ages of 60–79 yr (approx 7100 deaths per annum) *(1)*. Based on a relatively steady incidence rate of ovarian cancer over the past 10 yr *(1)*, these figures suggest that roughly 60% of patients with ovarian cancer will die from their disease. In contrast, the mortality from uterine cancers, which are the most commonly diagnosed genital cancers, is considerably lower at 22% *(1)*. The high ratio of death–incidence for patients with ovarian carcinomas is largely due to late stage diagnosis, at a time when the disease has

From: *Expression Profiling of Human Tumors: Diagnostic and Research Applications*
Edited by: Marc Ladanyi and William L. Gerald © Humana Press Inc., Totowa, NJ

typically spread beyond the ovary. The 5-yr survival rate for patients with stage 1 disease, which is confined to the ovary, is >90%, whereas the 5-yr survival rate for patients diagnosed with stage III or IV disease is 15–20%. Late stage diagnosis of ovarian carcinomas can be attributed to the fact that the disease is relatively "asymptomatic" in its early stages, and that the symptoms of late stage disease, such as abdominal discomfort, weight loss, diarrhea or constipation, vaginal bleeding, and shortness of breath, are nonspecific complaints.

Although there are a number of suspected risk factors for ovarian carcinoma, such as the number of ovulations a woman experiences during her lifetime (which may allow the accumulation of genetic alterations though cycles of epithelial rupture and repair at sites of ovulation), the most established risk factor is a family history of ovarian cancer. It is estimated that as many as 5–10% of ovarian carcinomas are associated with an inherited predisposition to the disease *(2,3)*. There are two key familial syndromes in which inherited mutations play a causative role: hereditary nonpolyposis colon carcinoma (HNPCC) or Lynch type II syndrome, which results from mutations in DNA mismatch repair genes, and hereditary breast–ovarian cancer, which is linked to mutations in the *BRCA1* gene and, to a lesser extent, *BRCA2* and other unidentified loci. In families with an established risk of breast cancer (mostly *BRCA*-related), there is an almost 50% increased risk of developing ovarian cancer.

The majority of ovarian cancers are epithelial in origin, arising from the surface epithelial lining of the ovary (ovarian surface epithelium [OSE]), or from the lining of cortical inclusion cysts, which are formed by invagination of the OSE into the superficial ovarian cortex. Histologically, ovarian carcinomas can be subdivided into several categories: serous papillary adenocarcinomas, which represent a majority of carcinomas arising in the ovary (approx 80%), and mucinous, endometrioid, clear-cell, and transitional cell carcinomas, which constitute the remaining cases. Borderline serous tumors (termed tumors of low malignant potential [LMP]) encompass a special category of tumors, in that available molecular evidence (see below) suggests that they represent a separate entity. These tumors are typically noninvasive and have very little potential for aggressive biological behavior.

Since most cases of ovarian carcinoma are diagnosed when the disease has spread from the ovary, our knowledge of how these cancers arise and progress is limited to circumstantial histological observations, such as the co-existence of dysplasia or ovarian intraepithelial neoplasia (OIN) adjacent to invasive carcinoma, as well as evidence derived from the study of OSE and carcinoma-derived cells in culture *(4)*. Thus, unlike other gynecologic neoplasms, such as those that arise in the cervix, there is no clear delineation of the genetic events that accompany the emergence of precursor lesions, and progression to frank malignancy and metastatic spread *(5)*.

Genetic analyses of ovarian cancers have, therefore, largely been confined to relatively late-stage disease (reviewed in refs. 4–6, and references therein). Invasive serous adenocarcinomas are characterized by frequent mutation of p53, and consistent losses of chromosomes 1, 6q, 11, 13q, 17p, and 17q, possibly pinpointing the existence of tumor suppressor loci whose loss may contribute to tumor progression. In contrast, the profile of genetic alterations in mucinous tumors is typified by mutation of *k*-ras, which is a rare event in serous tumors. The genetic alterations that characterize serous LMP lesions appear to be more typical of an invasive mucinous histology, with a low frequency of p53 mutations, a high rate of k-ras mutations, and frequent replication-error (RER+) mutations, detectable as microsatellite instability. It is noteworthy that mutations of the p53 or K-ras genes in LMP tumors are typically different from those detected in subsequent carcinomas isolated from the same patient *(7)*. Thus, LMP tumors are unlikely to be the precursor of invasive serous adenocarcinomas.

TISSUES PROCUREMENT AND PROCESSING

The OSE, from which ovarian carcinomas arise, represents a minute fraction of cells in the ovary (for e.g., *see* Fig. 2 in ref. *4*). This contrasts with carcinomas that arise in other organs, such as the prostate, where normal epithelial cells constitute a greater fraction of the tissue. The small amount of OSE poses special challenges for comparing the transcription of genes in the "normal" state, where there is a limited amount of RNA available, vs transcription in carcinomas, where cancerous epithelial cells tends to be more plentiful, particularly in late stage disease. Many studies have not taken these features of ovarian biology into account and have simply used whole ovarian tissue as a "normal" control to compare transcriptional profiles in ovarian carcinomas. Clearly, the results of these studies must be interpreted cautiously. A more biologically relevant approach has been to isolate OSE "brushings" from the surface of freshly procured ovaries. Samples are literally brushed, or lightly scraped, and the cells deposited in cell culture media prior to RNA preparation *(8)*. The epithelial character of these cell isolates can be verified by cytokeratin staining *(8)*, and sufficient RNA can be generated by this method by pooling individual isolates. A third approach has been to extend the proliferative capacity of the OSE by placing freshly procured ovaries in epithelial cell-selective media and expanding the OSE cells in short-term cultures *(9)*. A review of the procedures for OSE isolation and the properties of these cells in culture are presented in ref. *(4)*. Ovarian carcinoma cells can also be isolated in a similar manner *(9)* and then directly compared to normal short-term cultures.

Our own laboratory has taken a more crude strategy, scraping the surface of fresh-frozen normal ovaries, thereby enriching, to some extent, for OSE cells. We have also enriched for stromal cells in some of the same specimens and then directly compared the gene expression profiles of these samples to determine if this approach is tenable. We have identified several clusters of expressed genes (*see* Fig. 1 in ref. *10*) whose identities collectively suggest that we can distinguish tissue samples with enriched cellular populations *(10)*. However, the use of gene annotation alone as a tool to decipher cell-specific transcription is not an exact science, and the reliability of this approach is only based on a few reports (*see* ref. *11*).

Ideally, techniques such as laser capture microdissection (LCM) should enable the selective procurement of OSE cells, as well as enriched populations of carcinoma cells (see section on technical aspects in Chapter 1 of this text). Although these techniques have not yet been applied to the analysis of gene expression in ovarian carcinoma, there is emerging evidence from the examination of other cancers *(12–14)*, that the integration of LCM with RNA amplification and microarray hybridization may prove reliable in profiling the expression of genes from relatively pure cellular populations, while maintaining the relative representation of individual transcripts in a complex RNA population.

These issues also pertain to other analytical techniques, such as serial analysis of gene expression (SAGE), which typically requires sizable amounts of starting RNA. However, recent studies have shown that the SAGE method can be adapted to smaller amounts of RNA, requiring as few as 100,000 cells, or 500–5000 times less than the typical starting material of conventional SAGE protocols *(15)*. Indeed, these modifications have recently been applied to SAGE analysis of a single cell-derived colony of mesenchymal stem cells *(16)*, suggesting that these methods may make feasible the analysis of relatively pure OSE cell populations.

DIFFERENTIAL GENE EXPRESSION IN OVARIAN CARCINOMAS AND CELL LINES

Differential expression of genes in ovarian carcinomas of varying histology, grade and stage form the basis of several reports published over the past 3 yr *(8–10,17–22)*. Table 1 lists the key reports dealing with tissue samples and cell cultures, as well as some of the features of the studies, such as the method(s) used, the numbers of genes interrogated, and the types of tumors or cells analyzed. Table 2 lists a series of 21 up-regulated and 3 down-regulated genes that have been independently discovered by different profiling studies of ovarian carcinomas or validated in the same study by independent methods (e.g., reverse transcription polymerase chain reaction [RT-PCR], Northern blot analysis, or immunohistochemistry [IHC]). The genes listed in Table 2 represent only a small fraction of the hundreds of genes that have been reported as differentially expressed in ovarian cancer. As such, it is important to remember that these genes do not encompass the diversity of transcriptional differences between normal and cancerous tissue[1]. However, the fact that they have been identified, in some cases repeatedly, suggests that they represent true biological differences between the normal OSE and ovarian carcinomas, particularly in those cases where a corresponding difference in protein expression has been shown by IHC.

Several of the genes identified in these studies are reported as differentially expressed in many other carcinomas, such as CD9, GA733-2, and Muc-1. In contrast, others, such as mesothelin and pax8, appear to have a more restricted profile. From a biomedical perspective, it is notable that many of the gene products are predicted to encode cell surface proteins (e.g., GA733-2, mesothelin, muc-1) or are secreted from expressing cells (e.g., HE4, matrix metalloproteinase [MMP]-7). Not surprisingly, therefore, a number of these gene products are the subjects of translational diagnostic and therapeutic research. For example, mesothelin is being pursued as an ovarian

---

[1]There are several reasons why the list of genes is small. First, the criteria used to identify the most "significantly" differentially expressed genes are typically different from study to study. Second, most studies, with a few exceptions, use carcinomas of varying histology in the same normal versus tumor comparisons. Thus, it is difficult, and somewhat misleading, to directly compare studies in which one report may have focused solely on serous adenocarcinomas, while another may have used a combination of serous, endometrioid and clear cell carcinomas. Third, for the reasons outlined in the section entitled Tissue Procurement and Processing, the choice or source of normal tissue likely has a profound impact on the ability to identify differential gene expression. Lastly, different expression platforms have inherently different abilities to monitor transcripts in cells. SAGE is arguably the most comprehensive method, in that one can interrogate a majority of expressed transcripts. In contrast, microarrays are limited by the numbers and diversity of genes printed or imprinted on them.

Table 1
Studies of Differential Expression

Study (reference)	Gene expression platform	Genes/elements queried	Cancer tissue histology	Normal tissue/ cell sample baseline
In ovarian carcinoma tissues				
Wang et al. (18)	cDNA microarray (nylon filter)	5766	Serous, mucinous, clear cell, endometrioid.	Commerical ovarian RNA (nonenriched).
Schummer et al. (17)	cDNA microarray (nylon filter)	21,500	Mostly serous adenocarcinomas.	OSE cells, normal ovarian tissue (mostly stroma); fetal ovary pool; other normal tissues.
Ono et al. (19)	cDNA microarray (glass)	9121	Serous, mucinous.	Adjacent normal tissues from the same patients (nonenriched).
Hough et al. (20)	SAGE	6800	Serous adenocarcinomas.	OSE cells (HOSE-4; IOSE29).
Welsh et al. (10)	Oligonucleotide array (glass)		Serous adenocarcinomas (variable grade, mostly stage III).	Normal ovary, macro OSE enrichment.
Shridhar et al. (8)	cDNA microarray (nylon filter)	25,000	Serous, clear cell, enodmetrioid (high grade, variable stage).	OSE brushings.
In tumor and normal ovarian-derived cell cultures				

Study (reference)	Gene expression platform	Genes/elements queried	Cancer cells	Normal cells
Wong et al. (21)	cDNA microarray (glass)	2400	In-house developed cell lines; SK-OV-3.	OSE short-term culture.
Ismail et al. (9)	Subtractive hybridization/ cDNA microarray (nylon filter)	255	In-house cancer cell culture.	OSE short-term culture.
Tonin et al. (22)	Oligonucleotide array (glass)	6416	In-house developed cell lines.	OSE short-term culture.

Table 2
Differentially Expressed Genes in Ovarian Cancers Identified by Microarray and SAGE

			Up-regulated genes			
Up-regulated genes	Fold change	Citation	Microarray/ SAGE validation	PCR validation	Northern validation	IHC validation
CD9	4.34	(18)	(10)			
Mesothelin	6.84	(18)	(20)	(18,49)	(18)	
HE4	7.26	(18)	(8,10,19,20)	(8,10,17)	(17)	
Keratin 8	8.37	(18)	(8,10)			
Matrilysin/MMP-7	8.58	(18)	(8)	(8)		
GA733-2/EpCAM	13.87	(18)	(10,21,49)	(49)		(20)
SLC7A5/E16	4.4[a]	(17)	(21)	(17)		
Mucin1	5.5[a]	(17)	(8,10,49)	(17)		
14.3.3 Sigma	3.4–5.4[a]	(17)		(17)		
Breast epithelial antigen BA46	4.1[a]	(17)		(17)		
B-actin	4.4[a]	(17)		(17)		
Progesterone binding protein	5.9[a]	(17)		(17)		
Ryudocan	4.2[a]	(17)		(17)		
Keratin 18	NR	(19)	(10)			
Lutheran blood group antigen/B-CAM	17	(20)	(10)	(10)		
Kunitz serine protease type 2	34	(20)	(10)			
Ceruloplasmin/ferroxidase	79	(20)	(8)	(49)		
Claudin 4	109	(20)	(8)	(49)		(20)
ApoJ	39	(20)		(49)		(20)
CD24	NR	(10)	(21)	(10)		
Uncoupling protein 2	NR	(10)	(8)			
pax8	NR	(10)	(8)			
Keratin 7	NR	(10)	(8)			

Down-regulated genes

Down-regulated genes	Fold change	Citation	Microarray/SAGE validation	PCR validation
Nexin	0.23	(18)	(8)	
Amphiregulin	>10	(8)		(8)
Integral membrane protein A2	>10	(8)		(8)

Twenty-six genes from six independent studies are tabulated for which initial observations of differential expression on microarrays or by SAGE have been verified by additional array/SAGE studies, or by other independent methods (RT-PCR, Northern, or IHC). Genes are listed in chronological order of their discovery, from 1999 through 2001. The study in which the genes were first identified is cited first; confirmatory array or SAGE studies are cited secondarily. The identification of genes for which other studies confirm their differential expression is based largely the nomenclature of the gene annotation, rather than a rigorous analysis of the accession numbers published along with the gene annotation (not provided in some cases). Thus, some genes for which confirmatory data may have been published are not included because of disparate nomenclature. Additionally, genes for which no functional annotation was available at the time of publication are omitted for clarity. The interested reader is therefore encouraged to examine the original papers in detail. Genes were identified from Table 2 in (18); Table 2 in (17); Fig. 2 in (19); Table 3 in (20); Fig. 2 in (10); and Table 4 in (8).

[a]Fold change is calculated as the average of the individual fold changes reported for cases in which ratios were >2.5.

NR, not reported.

carcinoma target using high affinity endotoxin A-conjugated single-chain Fv antibodies *(23)*, as well as toxin-conjugated mouse anti-mesothelin monoclonal immunoglobulin (Ig)G antibodies *(24)*. Enzyme-linked immunosorbent assay (ELISA) analysis of MMP-7 (matrilysin) shows that it is more highly elevated in the cyst fluid of patients with serous adenocarcinomas (seven out of eight cases) vs the cyst fluid of serous adenomas (four out of fourteen) *(25)*, suggesting its potential utility for the detection of invasive disease. Broad-spectrum inhibitors of MMPs, including MMP-7, have also recently been described. One such small molecule, Bristol-Myers Squibb (BMS)-275291, inhibits the protease activity of MMP-1, -2, -7, -9, and -14 at low nM levels and has been shown to inhibit lung metastases, as well as tumor angiogenesis in vivo *(26)*. Since it is speculated that increased MMP-7 expression reflects increasing invasiveness of ovarian carcinomas *(27)*, inhibition of MMP-7 protease activity may have some clinical benefit.

A number of investigators have used short- or long-term ovarian carcinoma-derived cell lines as the starting point for large-scale differential expression studies *(9,21,22)* (Table 1). Typically, transcript profiles of these cells are compared to short-term cultures of OSE cells, some of which have been immortalized by transfection with simian virus 40 (SV40) T antigen (see ref. *4*). Wong et al. *(21)* have identified genes such as GA733-2 and CD24 as highly up-regulated in long-term cultures of ovarian carcinoma-derived cells compared to short-term OSE cultures (Table 2). Tonin and coworkers have also reported elevation of these same genes, as well as HE4, MUC-1, and keratins 8 and 18 in cells derived from tumors with aggressive courses *(22)*. Differential expression of these genes is highly concordant with several studies on primary tumors *(10,17,18)*. However, as with interstudy comparisons of tumor tissue, there is considerable diversity among the genes identified by cell–cell comparisons, and a rigorous comparison is not currently feasible.

The results of these cell studies and their relationship to those obtained in carcinomas highlight a broadly interesting question: To what extent do tumor-derived cell lines reflect their ostensible tumor of origin? Figure 1 depicts the expression levels of 50 genes that we find to be most highly elevated (by average fold change) in a series of serous papillary adenocarcinomas when compared to "normal" ovarian tissue and for which we have also examined their expression in a series of commonly used ovarian carcinoma-derived cell lines and one OSE line, IOSE-80, which was stably transfected with SV40 T antigen (a gift from Dr. Nelly Auersperg).

This simple analysis illustrates several facets of the use of cell lines as models of the carcinomas from which they are derived. Most remarkably, a majority (47 out of 50) of the genes we find to be overexpressed in ovarian carcinomas compared to normal tissue are also overexpressed in at least one of the ovarian carcinoma-derived cell lines when compared to IOSE-80 cells. The exceptions are insulin-like growth factor 2 (IGF-2), which is reported as overexpressed in a fraction of serous adenocarcinomas through loss of imprinting (LOI) *(28)*, properdin, which is reported as overexpressed by independent cDNA microarray analysis *(19)*, and kallikrein 11 (hK11), whose protein product is elevated and secreted into the serum of patients with ovarian cancer *(29)*. Second, although the overexpression of these 47 genes is considerably less uniform in cell lines than in primary tumors, each of the genes can nonetheless be found overexpressed in at least one line. For genes that have been reported by several studies, such as GA733-2,

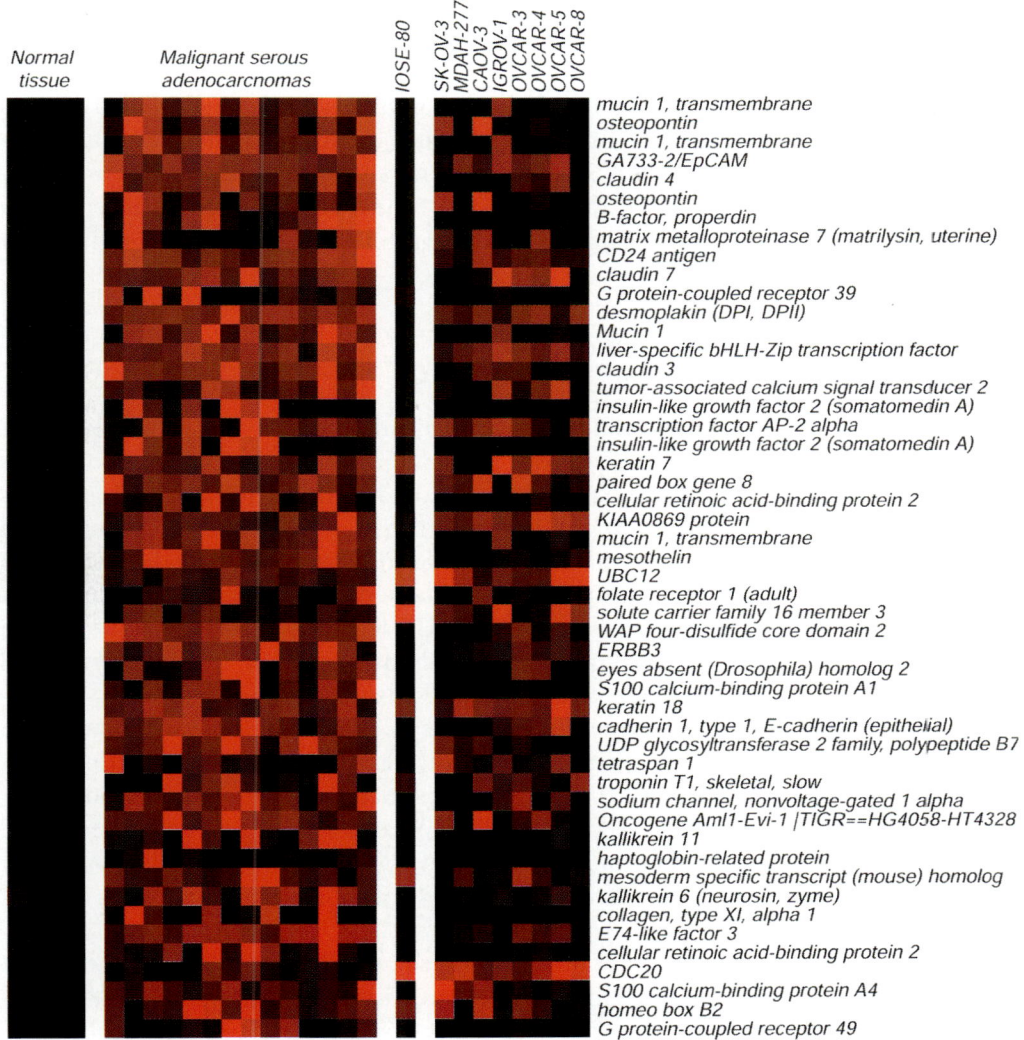

Fig. 1. Differentially expressed genes in ovarian carcinomas and cell lines. Genes differentially expressed between normal ovarian tissue (Normal tissue) and tumor tissue (Malignant serous adenocarcinomas) were selected based on the magnitude of the fold differences in hybridization intensities between the average of each tissue type. Shown are the 50 genes with the highest fold change. The levels of expression of these 50 genes are depicted in a series of nine cell lines. IOSE-80 represents normal OSE cells (a gift from Dr. N. Auersperg, BC, Canada); carcinoma-derived cell lines were obtained from ATCC (Manassas, VA, USA) or from Dr. J. Reed (Burnham Institute, CA, USA). The relative levels of gene expression (depicted in each row) across all samples (columns) in were median-centered and normalized by 'Cluster' and output in 'Treeview' *(50)*. Red, increased gene expression; blue, decreased expression; black, median level of gene expression. The color intensity is proportional to the hybridization intensity of a gene from its median level across all samples.

expression appears to be more uniform across multiple cell lines. While there is no clear "best" representative cell line, the use of a sufficiently large number of cell lines is likely to capture a majority of important genes. These data suggest that at least one cell-based model may be available for following up the biological consequences of gene-specific down-regulation (e.g., by RNA interference, antibody therapeutics, or small molecules).

In contrast to the tumor data, we also find comparable levels in IOSE-80 and ovarian carcinoma-derived cells for a small number of genes, such as keratin 7, 18, ubiquitin-conjugating enzyme E2M, and CDC20. The latter two cell cycle-related genes are likely up-regulated as a consequence of rapid growth of these cells in culture, whereas the uniform expression of keratins suggests that our normal samples are not sufficiently enriched for epithelial cells.

Validation of Differential Gene Expression by RT-PCR, IHC, and Gene Transfer

Identification of differentially expressed genes by microarray hybridization or SAGE is, in essence, "descriptive genomics", since the output is merely a series of observations. While validation of these observations by independent RNA-based analyses (e.g., RT-PCR) is certainly important, these confirmatory experiments do not reveal whether the transcribed gene is translated or, indeed, whether elevated levels of an encoded protein are functional in the context of tumorigenesis or progression. Thus, for any one gene identified by microarray or SAGE techniques for which the known biology may be interesting in the context of the disease, there are many experimental steps required to validate its involvement in the disease process.

Few studies to date have corroborated increases in gene expression with that of the encoded protein in ovarian cancers. Exceptions to this include GA733-2 (EpCAM), ApoJ (clusterin), and claudin 4 *(20)* (Table 2), for which increases in the encoded proteins observed by IHC on tissue sections reflect increases at the RNA level. Since IHC can detect protein expression at the single cell level, this method is perhaps the best way to validate lower levels or absence of expression in the small numbers of OSE cells of the ovary.

Thus far, functional validation of differential gene expression is lacking for almost all microarray-based studies, including those specifically focused on ovarian carcinoma. Few examples exist where a candidate gene of interest has been introduced into the appropriate precursor cells to mimic up-regulated expression or antagonized in fully malignant cells to mimic decreased expression. In melanoma, the introduction of RhoC into cells with limited metastatic potential increased their ability to metastasize, whereas dominant-negative RhoC constructs inhibited metastasis *(30)*. These observations reflect the increased expression of RhoC seen on oligonucleotide microarrays following selection of metastatic variants from precursor melanoma cells with limited metastatic potential *(30)*. Recently Lee et al. *(31)* have identified thymosin β-10 as down-regulated in four out of five ovarian carcinomas profiled on nylon arrays imprinted with a selection of 588 genes (Atlas Array 1; Clontech Laboratories, Palo Alto, CA, USA). Introduction of thymosin β-10 into ovarian cancer cell lines SKOV-3 and PA-1 *via* adenoviral infection led to markedly decreased cell growth rates, as well as a high rate of cell death in these cell lines. Such studies serve to validate the potential of microarrays, as they clearly link a descriptive genomic observation with a functional role for a gene in tumor development.

MOLECULAR CLASSIFICATION, CLASS DISCOVERY, AND CORRELATES OF DISEASE HISTOLOGY

Using Molecular Classification to Identify Genes Preferentially Elevated in Ovarian Carcinomas

We, and others, have begun to evaluate whether gene expression technologies can facilitate the identification of genes whose transcription typifies tumors of distinct anatomic sites *(32–34)*. The primary motivation for these efforts is the objective identification of a tumor's anatomic site of origin, which in combination with classical histopathology, may lead to robust tumor diagnosis *(35,36)*. Although histological examination of a solid tumor, in concert with clinical information, typically leads to a correct diagnosis in the vast majority of cases, there remains a subset of tumors (including some metastases of undetermined primary origin *[37]*) for which a correct diagnosis may be difficult. Thus, molecular methods might provide an independent and objective tool to determine the anatomic or tumor origin of these cases.

An ancillary goal of these molecular studies is the identification of genes whose transcription is elevated in a particular type of cancer vs many other cancer types. These genes are likely to expose aspects of tumor type-specific biology, which, to date, have been unattainable. Figure 2 depicts a small series of genes that we recently identified as highly predictive of 11 different classes of human tumors, 10 of which arise at distinct anatomic sites *(33)*. Collectively, we found that this set of genes was capable of correctly "predicting" the anatomic origin of approx 85% of 75 "blinded" tumors with high confidence. This blinded set of tumors was representative of many of the carcinomas used to build the prediction algorithm, as well as a small selection of 12 metastatic lesions (for which we correctly predicted the tumor origin in 9 out of 12 cases) *(33)*. Thus, we consider these classifier genes "characteristic" of the tumor. In a similar study, Ramaswamy et al. *(34)* identified genes whose transcription typified (and were predictive of) 14 different tumor classes, many of which overlap with our own tumor collection. A substantial number of the classifier genes identified in both studies are concordant, as well as overlapping with some of those identified by Giordano and colleagues *(32)*, who reported classifier genes characteristic of ovarian, lung, and colon carcinomas. The results of these studies indicate that different computational methods can uncover the same fundamental sets of tumor type-specific genes. Some of the differences between studies are likely attributable to the use of different oligonucleotide microarrays, as well as the fact that the studies documented by Su et al. *(33)* and Ramaswamy et al. *(34)* attempted to find tumor type-specific genes in the context of more than 10 tumor classes, whereas the study by Giordano and colleagues focused on 3 tumor classes *(32)*. In the context of a limited number of tumor classes, the identification of the estrogen receptor (ER) as predictive of ovarian carcinomas, for example, is reasonable (*see* Table 1 in ref. *32*); however, such a gene is not likely to be predictive of ovarian cancer in the context of tumor classes that include ER-positive carcinomas of the breast *(33,34)*.

We have asked to what extent "classifier" genes, which were found to be characteristic of ovarian tumors, reflect features of the carcinomas or the ovarian tissue from which the tumors arose. A simple comparison of normal and tumor ovarian tissues revealed that 18 out of 28 genes (which were individually at least 92% predictive of

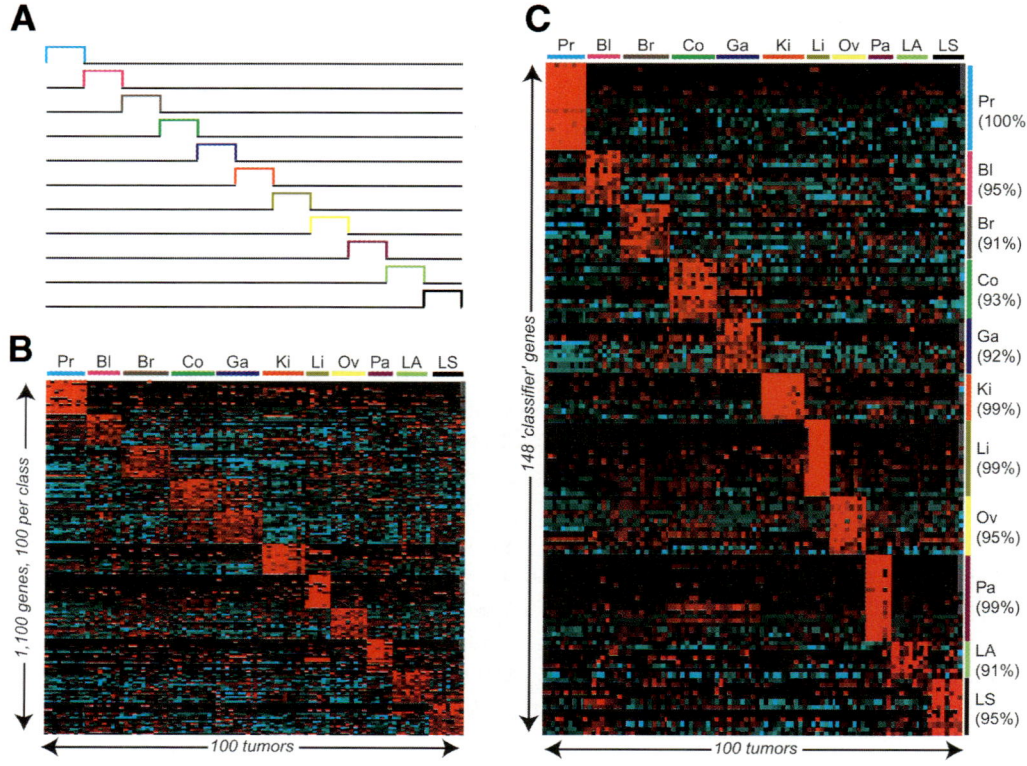

Fig. 2. Molecular signatures of carcinomas from diverse anatomic sites. To identify tumor class-specific classifiers, we sought genes whose expression was uniformly high among carcinomas of a specific anatomic site, and uniformly low among carcinomas of all other anatomic sites or histologies (i.e., one-vs-all; depicted in panel A). This was achieved using the Wilcoxon rank sum test, which tests the null hypothesis that gene expression in one tumor class is no different from gene expression in any other tumor class. The genes in each class that had significant p scores represent those that dispute the null hypothesis and define those that are most different among tumor classes. One hundred of the Wilcoxon-selected genes from each tumor class, depicted in panel B, were subjected to a prediction accuracy test, in which each of the genes was tested for its ability to discriminate one tumor class from all other tumor classes, using a support vector machine (SVM)-learning algorithm. Leave-out-one cross validation (LOOCV) was used to blind ourselves sequentially to each of the 100 tumor samples, and the SVM was trained on the remaining samples and then used to predict the class of the blinded sample depicted in panel C. The accuracy of the tumor class-specific classifiers is shown to the right of the panel in C (*see* ref. *33* for detailed methods). Pr, prostate; Bl, bladder; Br, breast; Co, colorectal; Ga, gastroesophageal; Ki, kidney; Li, liver; Ov, ovary; Pa, pancreas; LA, lung adenocarcinomas; LS, lung squamous cell carcinoma. Levels of gene expression are presented as described in Figure 1. Red, increased gene expression; blue, decreased expression; black, median level of gene expression. The color intensity is proportional to the hybridization intensity of a gene from its median level across all samples. The list of classifier genes identified by the SVM/LOOCV method are available from our Web site at (www.gnf.org/cancer/epican). Reprinted with kind permission from Cancer Research.

ovarian carcinoma by cross-validation) were highly elevated in the carcinomas, including genes such as mesothelin, that have been previously identified by more traditional methods as well as by microarray analyses (Fig. 3). Expression of the other 10 genes was essentially identical in normal and cancerous tissue samples, implying that they form part of a characteristic transcriptional program of that tissue.

Ovarian Cancer Profiling

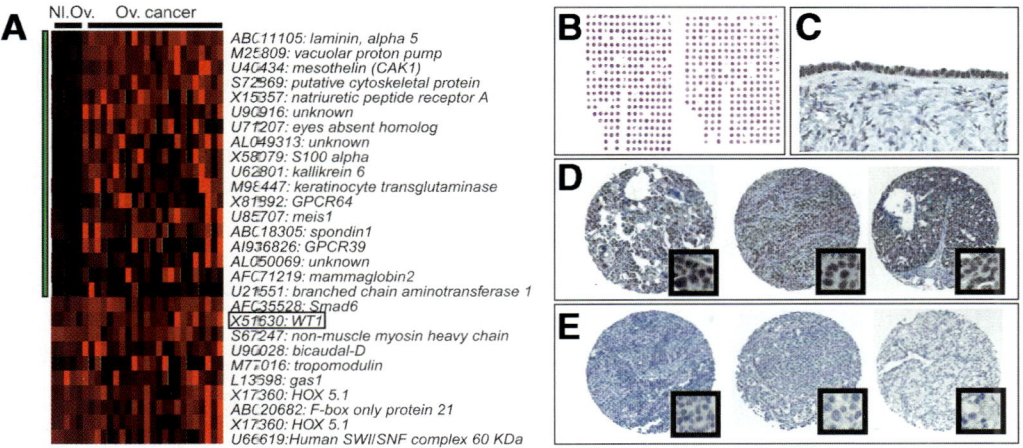

Fig. 3. Genes and proteins predictive for serous papillary adenocarcinomas of the ovary. Expression levels of highly predictive classifier genes in normal and malignant samples of the ovary are depicted in panel A. Green bars represent differentially expressed genes where the mean level of expression in tumor samples >3 times the mean level of expression in normal tissues and where $p < 0.01$ by an unpaired t-test. Gene expression was normalized and output in Treeview as described in the legends to Figs. 1 and 2. (Panel B) Visualization of a tissue microarray containing 36 normal epithelial tissues and 229 carcinomas representative of the 10 anatomic sites of the tumors profiled in the study stained with hematoxylin and eosin. Tissue microarrays were stained with an antibody specific to the WT protein. Panel C depicts the normal serous lining of the ovary positive for WT. Panel D illustrates three serous papillary carcinomas of the ovary that were positive for WT, whereas other tumors, such as breast, lung, and kidney carcinomas were uniformly negative for WT (immunoperoxidase technique) as shown in panel E. Insets show magnified view of nuclei. Reprinted with kind permission from Cancer Research.

An immediate clinical utility of the information gathered from these analyses is the derivation of antibodies against some of the proteins encoded by the classifier genes for use in tumor diagnosis. To test this idea, we chose several genes whose protein products are detectable by commercial antibodies and performed IHC on medium density tissue microarrays containing most of the carcinomas profiled in the original study. An example of one of these IHC experiments for the Wilm's tumor gene product (WT-1) protein is shown in Fig. 3. IHC revealed positive nuclear staining in 18 of 20 ovarian carcinomas, with no visible nuclear staining in any of the other carcinomas, consistent with the expression results obtained by microarray hybridization. Therefore, in the context of the 11 tumor classes that were investigated, WT-1 was found to be highly predictive of ovarian carcinomas. A long-term goal arising from these studies is the derivation of antibodies against secreted proteins encoded by genes such as mammaglobin-2 (MGB-2) (Fig. 3), which may have potential utility as serum diagnostics. Proof-of-concept that this approach may be successful comes from the observation that the protein product of kallikrein 6 (hK6), which we also identified as elevated in ovarian carcinoma (Fig. 3), is detected at elevated levels in the serum of a significant fraction of patients with ovarian carcinoma *(38)*.

Molecular Class Discovery

The molecular classification methods discussed above were designed to identify genes that typify cancers of a specific anatomic origin and, therefore, transcend differences in stage, grade, and other phenotypic features of a particular type of cancer. While this approach has provided possible new ways in which to augment traditional histopathology for primary diagnosis, the larger problems in cancer, specifically, response to therapy, disease-free interval, recurrence, and overall survival, are likely rooted in the molecular heterogeneity of specific types of cancers, highlighting the potential benefits of molecular stratification. Recent evidence suggests that expression profiling can identify molecular subgroups of cancers, which, on occasion, correlate with diverse biological behaviors. Class discovery methods, which range from visual identification of consistent subgroups by hierarchical clustering, to more thorough statistical analysis, have led to the molecular classification of leukemias *(39)* and lymphomas *(40)*, melanomas *(41)*, breast carcinomas *(42)*, and lung adenocarcinomas *(43,44)*.

Few studies have evaluated the existence of molecular subgroups among histological types of ovarian carcinomas. My own laboratory's work in this area has focused on serous papillary adenocarcinomas, and our preliminary study of these carcinomas identified a potential molecular distinction between noninvasive LMP and low-grade invasive carcinomas vs invasive carcinomas of higher grade *(10)*. This distinction appears to be predominantly based on increased expression of ribosomal genes in LMP/low-grade carcinomas, as well as increased expression of multiple cell cycle genes in mostly high-grade tumors. We have since expanded this study by profiling a larger number of carcinomas. Using a variety of computational approaches, we are led to the same general conclusion that noninvasive borderline and invasive low-grade carcinomas are molecularly very similar and easily distinguished from moderate and high-grade carcinomas (to be described in detail elsewhere).

Molecular Correlates of Tumor Histology

There is now clear evidence that tumors of different histology from specific anatomic sites may often be discerned by their molecular profiles. For example, in lung cancer, clustering algorithms readily enable the distinction between squamous, large cell, small cell, and adenocarcinoma *(43,44)*. Ono et al. *(19)* have assessed the molecular profiles of ovarian carcinomas with mucinous or serous papillary histologies using cDNA arrays comprised of 9121 distinct human genes. Using the Mann-Whitney test, the authors reported significant differential expression of 115 genes between histological subtypes. The identity of the genes do not readily pin-point a major underlying biological theme related to the distinction between the histological subtypes, and it is somewhat surprising that genes known to be differentially expressed, such as the mucin, MUC-2 *(45)*, were not identified by this analysis. However, additional studies with these genes and their encoded proteins may reveal biologically relevant insights. The molecular distinction between serous tumors of varying grade has also recently been addressed by analysis of gene expression. Using the Atlas Human Cancer Array™, Tapper et al. have specifically compared the expression of 588 genes in well differentiated and poorly differentiated carcinomas *(46)*. Of note, they reported increased expression of several extracellular matrix proteins, including COL3A1, fibronectin, and biglycan, in poorly differentiated carcinomas, which is an observation that we have consistently made in our own tumor collection (unpublished data). Other interesting genes identified as up-regulated in poorly differentiated carcinomas include proliferating cell nuclear antigen (PCNA) and epidermal growth factor receptor (EGFR), consistent with increased cell cycle activity and growth factor receptor signaling.

TRANSLATIONAL RESEARCH: NEW MARKERS FOR OVARIAN CANCER DIAGNOSIS

Although we, and others, have identified many highly expressed genes encoding secreted proteins in ovarian carcinomas, we collectively lack the evidence at this time that these proteins are detectable in the serum of patients with ovarian cancer. Mok and coworkers have recently described one of the first examples of this type of translational diagnostic research, in which their observation of an overexpressed gene in ovarian carcinoma has led to the identification of the encoded protein in the serum of ovarian cancer patients *(47)*. The authors previously described the overexpression of prostasin, a secreted protease originally identified in human seminal fluid, in ovarian carcinoma-derived cell-lines compared to those obtained from normal ovarian epithelium. IHC showed that the prostasin protein is expressed in ovarian tissues, albeit at significantly higher levels in the cytoplasm of ovarian carcinoma cells. Consistent with this observation, the authors demonstrated by ELISA that the prostasin protein is detectable at elevated levels in patients with ovarian carcinoma (mean level of 13.7 µg/mL) compared to control subjects (7.5 µg/mL). The correlation between serum levels of prostasin and CA125 in a subset of patients was low, suggesting that measurement of a combination of both proteins might provide additional information than either alone. Whereas the sensitivity of CA125 and prostasin was approx 65 and 51%, respectively, the combination of both molecules had an overall sensitivity of detection of 92%. Additional studies are needed to establish the diagnostic potential of prostasin, since a significant fraction of the screened cases (>50%) were stage III carcinomas. It will also be important to see how prostasin is correlated to other histological parameters, as well as its potential usefulness in following patients postoperatively. Nonetheless, the translation of a descriptive genomic observation to a potentially useful biomedical application, such as the diagnosis of a proportion of patients with ovarian carcinomas, demonstrates the potential of these gene expression platforms in providing candidate reagents for major unmet biomedical needs. A number of the genes identified by expression profiling, which encode secreted proteins, such as HE4 *(8,17,18,20,48)*, KLK6 *(33)*, and MGB-2 *(33)*, hold some promise for the emergence of a new class of molecular markers that may, in combination, identify a majority of females with ovarian carcinomas.

CONCLUSIONS

Although significant progress has been made in using large-scale gene expression platforms to decipher the genes involved in ovarian malignancies, a substantial amount of effort is needed to delineate the molecular distinctions between ovarian carcinomas of varying histology, grade, and stage, as well as the underlying molecular distinctions that correlate with a tumor's clinical behavior and response to chemotherapeutics. Molecular class discovery methods are continually improving; incorporating "reliability" measures to judge the likelihood that a gene's expression is correlated with the biological features in question. One assumes that the application of these methods to ovarian carcinomas may eventually help to stratify patients according to their tumor's molecular profile, leading to better management of the patient.

Because a majority of ovarian cancers are diagnosed at late stage, relatively little is known about how these tumors arise. Reliable methods that allow expression profiles to be generated from small amounts of RNA are beginning to emerge, which, in combination with microdissection of putative precursor lesions, either alone or adjoining frankly malignant tumors, will allow an assessment of the changes in gene expression that accompany the transition from a premalignant state to an invasive carcinoma.

Functional characterization of the genes presented in Table 2, and others yet to be discovered by genome-wide analyses, are required to understand how they contribute to the neoplastic phenotype. Central to these experiments are the appropriate model systems in which such studies can be carried out. The isolation of parental "normal" OSE cells and their cancerous counterparts in vitro are now well established, providing an excellent resource by which to assess the functional effects of gene transfer or transcriptional antagonism. Likewise, many of the common and widely available ovarian carcinoma-derived cell lines can be established as xenografts in athymic or severe combined immunodeficient (SCID) mice, facilitating interrogation of gene action in vivo. As illustrated in Fig. 1, we are beginning to accumulate the information required to make rational decisions regarding the choice of carcinoma cell line to perform these experiments, ensuring that the resultant observations are as relevant as possible.

ACKNOWLEDGMENTS

I would like to thank my many colleagues at GNF for their support and encouragement, and Drs. Quinn Deveraux and Henry Frierson, Jr. for critically reading the manuscript.

REFERENCES

1. Greenlee, R. T., Hill-Harmon, M. B., Murray, T., and Thun, M. (2001) Cancer Statistics, 2001. *CA Cancer J. Clin.* **51,** 15–36.
2. Boyd, J. (1998) Molecular genetics of hereditary ovarian cancer. *Oncology* **12,** 399–406.
3. Lynch, H. T., Casey, M. J., Lynch, J., White, T. E., and Godwin, A. K. (1998) Genetics and ovarian carcinoma. *Semin. Oncol.* **25,** 265–280.
4. Auersperg, N., Wong, A. S. T., Choi, K.-C., Kang, S. K., and Leung, P. C. (2001) Ovarian surface epithelium: biology, endocrinology and pathology. *Endocr. Rev.* **22,** 255–288.
5. Shelling, A. N. and Foulkes, W. (2001) Molecular genetics of ovarian cancer. *Mol. Biotechnol.* **19,** 13–27.
6. Feeley, K.M. and Wells, M. (2001) Precursor lesions of ovarian epithelial malignancy. *Histopathology* **38,** 87–95.
7. Ortiz, B. H., Ailawadi, M., Colitti, C., et al. (2001) Second primary or recurrence? Comparative patterns of p53 and K-ras mutations suggest that serous borderline ovarian tumors and subsequent serous carcinomas are unrelated tumors. *Cancer Res.* **61,** 7264–7267.
8. Shridhar, V., Lee, J., Pandita, A., et al. (2001) Genetic analysis of early-versus late-stage ovarian tumors. *Cancer Res.* **61,** 5895–5904.
9. Ismail, R. S., Baldwin, R. L., Fang, J., et al. (2000) Differential gene expression between normal and tumor-derived ovarian epithelial cells. *Cancer Res.* **60,** 6744–6749.
10. Welsh, J. B., Zarrinkar, P. P., Sapinoso, L. M., et al. (2001) Analysis of gene expression profiles in normal and neoplastic ovarian tissue samples identifies candidate molecular markers of epithelial ovarian cancer. *Proc. Natl. Acad. Sci. USA* **98,** 1176–1181.
11. Ross, D. T., Scherf, U., Eisen, M. B., et al. (2000) Systematic variation in gene expression patterns in human cancer cell lines. *Nat. Genet.* **24,** 227–235.

12. Ohyama, H., Zhang, X., Kohno, Y., et al. (2000) Laser capture microdissection-generated target sample for high-density oligonucleotide array hybridization. *BioTechniques* **29**, 530–536.
13. Alevizos, I. M., Zhang, M., Ohyama, X., et al. (2001) Oral cancer in vivo gene expression profiling assisted by laser capture microdissection and microarray analysis. *Oncogene* **20**, 6196–6204.
14. Kitahara, O., Furukawa, Y., Tanaka, T., et al. (2001) Alterations of gene expression during colorectal carcinogenesis revealed by cDNA microarrays after laser-capture microdissection of tumor tissues and normal epithelia. *Cancer Res.* **61**, 3544–3549.
15. Datson, N. A., van der Perk-de Jong, J., van den Berg, M. P., de Kloet, E. R., and Vreugdenhil, E. (1999) MicroSAGE: a modified procedure for serial analysis of gene expression in limited amounts of tissue. *Nucleic Acids Res.* **27**, 1300–1307.
16. Tremain, N., Korkko, J., Ibberson, D., Kopen, G. C., DiGirolamo, C., and Phinney, D. G. (2001) MicroSAGE analysis of 2,353 expressed genes in a single cell-derived colony of undifferentiated human mesenchymal stem cells reveals mRNAs of multiple cell lineages. *Stem Cells* **19**, 408–418.
17. Schummer, M., Ng, W. V., Bumgarner, R. E., et al. (1999) Comparative hybridization of an array of 21,500 ovarian cDNAs for the discovery of genes overexpressed in ovarian carcinomas. *Gene* **238**, 375–385.
18. Wang, K., Gan, L., Jeffery, E., et al. (1999) Monitoring gene expression profile changes in ovarian carcinomas using cDNA microarray. *Gene* **229**, 101–108.
19. Ono, K., Tanaka, T., Tsunoda, T., et al. (2000) Identification by cDNA microarray of genes involved in ovarian carcinogenesis. *Cancer Res.* **60**, 5007–5011.
20. Hough, C. D., Sherman-Baust, C. A., Pizer, E. S., et al. (2000) Large-scale serial analysis of gene expression reveals genes differentially expressed in ovarian cancer. *Cancer Res.* **60**, 6281–6287.
21. Wong, K.-K., Cheng, R. S., and Mok, S. (2001) Identification of differentially expressed genes from ovarian cancer cells by MICROMAX cDNA microarray system. *BioTechniques* **30**, 670–675.
22. Tonin, P. N., Hudson, T. J., Rodier, F., et al. (2001) Microarray analysis of gene expression mirrors the biology of an ovarian cancer model. *Oncogene* **20**, 6617–6626.
23. Chowdhury, P., Viner, J. L., Beers, R., and Pastan, I. (1998) Isolation of a high-affinity stable-chain Fv specific fir mesothelin from DNA-immunized mice by phage display and construction of a recombinant immunotoxin with anti-tumor activity. *Proc. Natl. Acad. Sci. USA* **95**, 669–674.
24. Hassan, R., Viner, J. L., Wang, Q. C., Margulies, I., Kreitman, R. J., and Pastan, I. (2000) Anti-tumor activity of K1-LysPE38QQR, an immunotoxin targeting mesothelin, a cell-surface antigen overexpressed in ovarian cancer and malignant mesothelioma. *J. Immunother.* **23**, 473–479.
25. Furuya, M., Ishikura, H., Ogawa, Y., et al. (2000) Analyses of matrix metalloproteinases and their inhibitors in cyst fluid of serous ovarian tumors. *Pathobiology* **68**, 239–244.
26. Naglich, J. G., Jure-Kunkel, M., Gupta, E., et al. (2001) Inhibition of angiogenesis and metastasis in two murine models by the matrix metalloproteinase inhibitor, BMS-275291. *Cancer Res.* **61**, 8480–8485.
27. Tanimoto, H., Underwood, L. J., Shigemasa, K., et al. (1999) The matrix metalloprotease pump-1 (MMP-7, matrilysin): a candidate marker/target for ovarian cancer detection and treatment. *Tumour Biol.* **20**, 88–98.
28. Chen, C. L., Ip, S. M., Cheng, D., Wong, L. C., and Ngan, H. Y. (2000) Loss of imprinting of the IGF-II and H19 genes in epithelial ovarian cancer. *Clin. Cancer Res.* **6**, 474–479.
29. Diamandis, E. P., Okui, A., Mitsui, S., et al. (2002) Human kallikrein 11: a new biomarker of prostate and ovarian carcinoma. *Cancer Res.* **62**, 295–300.
30. Clark, E. A., Golub, T. R., Lander, E. S., and Hynes, R. O. (2000) Genomic analysis of metastasis reveals an essential role for RhoC. *Nature* **406**, 532–535.

31. Lee, S. H., Zhang, W., Choi, J. J., et al. (2001) Overexpression of the thymosin beta-10 gene in human ovarian cancer cells disrupts F-actin stress fiber and leads to apoptosis. *Oncogene* **20,** 6700–6706.
32. Giordano, T. J., Shedden, K. A., Schwartz, D. R., et al. (2001) Organ-specific molecular classification of primary lung, colon, and ovarian adenocarcinomas using gene expression profiles. *Am. J. Pathol.* **159,** 1231–1238.
33. Su, A. I., Welsh, J. B., Sapinoso, L. M., et al. (2001) Molecular classification of human carcinomas by use of gene expression signatures. *Cancer Res.* **61,** 7388–7393.
34. Ramaswamy, S., Tamayo, P., Rifkin, R., et al. (2001) Multiclass cancer diagnosis using tumor gene expression signatures. *Proc. Natl. Acad. Sci. USA* **98,** 15,149–15,154.
35. Gerald, W. L. (2000) A practical approach to the differential diagnosis of small round cell tumors of infancy using recent scientific and technical advances. *Int. J. Surg. Pathol.* **8,** 87–97.
36. Ladanyi, M., Chan, W., Triche, T. J., and Gerald, W. L. (2001) Expression profiling of human tumors: the end of surgical pathology? *J. Mol. Diagn.* **3,** 92–97.
37. Hillen, H. F. (2000) Unknown primary tumors. *Postgrad. Med. J.* **76,** 690–693.
38. Diamandis, E. P., Yousef, G. M., Soosaipillai, A. R., and Bunting, P. (2000) Human kallikrein 6 (zyme/protease M/neurosin): a new serum biomarker of ovarian carcinoma. *Clin. Biochem.* **33,** 579–583.
39. Golub, T. R., Slonim, D. K., Tamayo, P., et al. (1999) Molecular classification of cancer: class discovery and class prediction by gene expression monitoring. *Science* **286,** 531–537.
40. Alizadeh, A. A., Eisen, M. B., Davis, R. E., et al. (2000) Distinct types of diffuse large B-cell lymphoma identified by gene expression profiling. *Nature* **403,** 503–511.
41. Bittner, M., Meltzer, P., Chen, Y., et al. (2000) Molecular classification of cutaneous malignant melanoma by gene expression profiling. *Nature* **406,** 536–540.
42. Sorlie, T., Perou, C. M., Tibshirani, R., et al. (2001) Gene expression patterns of breast carcinomas distinguish tumor subclasses with clinical implications. *Proc. Natl. Acad. Sci. USA* **98,** 10,869–10,874.
43. Bhattacharjee, A., Richards, W. G., Staunton, J., et al. (2001) Classification of human lung carcinomas by mRNA expression profiling reveals distinct adenocarcinoma subclasses. *Proc. Natl. Acad. Sci. USA* **98,** 13,790–13,795.
44. Garber, M. E., Troyanskaya, O. G., Schluens, K., et al. (2001) Diversity of gene expression in adenocarcinoma of the lung. *Proc. Natl. Acad. Sci. USA* **98,** 13,784–13,789.
45. Hanski, C., Hofmeier, M., Schmitt-Graff, A., et al. (1997) Overexpression or ectopic expression of MUC2 is the common property of mucinous carcinomas of the colon, pancreas, breast, and ovary. *J. Pathol.* **182,** 385–391.
46. Tapper, J., Kettunen, E., El-Rifai, W., Seppala, M., Andersson, L. C., and Knuutila, S. (2001) Changes in gene expression during progression of ovarian carcinoma. *Cancer Genet. Cytogenet.* **128,** 1–6.
47. Mok, S. C., Chao, J., Skates, S., et al. (2001) Prostasin, a potential serum marker for ovarian cancer: identification through microarray technology. *J. Natl. Cancer Inst.* **93,** 1458–1464.
48. Welsh, J. B., Sapinoso, L. M., Su, A. I., et al. (2001) Analysis of gene expression identifies candidate markers and pharmacological targets in prostate cancer. *Cancer Res.* **61,** 5974–5979.
49. Hough, C. D., Cho, K. R., Zonderman, A. B., Schwartz, D. R., and Morin, P. J. (2001) Coordinately up-regulated genes in ovarian cancer. *Cancer Res.* **61,** 3869–3876.
50. Eisen, M. B., Spellman, P. T., Brown, P. O., and Botstein, D. (1998) Cluster analysis and display of genome-wide expression patterns. *Proc. Natl. Acad. Sci. USA* **95,** 14,863–14,868.

16
Classification of Pediatric Tumors Using DNA Microarrays

Javed Khan and Marc Ladanyi

INTRODUCTION

Expression profiling is of particular interest in pediatric tumors for both clinical and scientific reasons. Clinically, pediatric tumors are often histologically primitive, leading to difficulties in diagnosis, and tumors of similar morphology may have markedly different behavior and treatment responses. Despite the improving trend in survival among pediatric patients with cancer over the last 25 yr (1), several fundamental problems remain in the management of pediatric cancers. First, the choice of chemotherapeutic agents for treatment of cancer is primarily empirical in nature, based on their efficacy in clinical trials and not on targeting specific genes, proteins, or pathways known to be active in that cancer. Second, the majority of these drugs target all dividing cells, including those in normal bone marrow and mucosa, which often leads to severe dose-limiting toxicity. Third, other idiopathic sometimes fatal toxicities, such as cardiomyopathy, may occur as maximal tolerated doses are reached. Fourth, there is currently no cure for the 35% of patients with the most aggressive disease, including those with metastatic disease and those with poor prognostic molecular markers, such as gene amplification (e.g., *MYCN* in neuroblastoma). Finally, despite very careful clinical and pathological prognostic stratifications, 30% of patients with apparently "low-risk" cancers will die from their cancer, while a similar percentage in the "high risk" groups will survive. Therefore, there is a need for more accurate markers of prognosis and treatment response to be delineated. For these reasons, there has been an increasing emphasis on using global genomic approaches to determine the biological and molecular features of high risk cancers, correlating these with diagnosis and prognosis, and identifying new targets for therapy (Fig. 1).

Biologically, most pediatric tumors are considered developmental or embryonal tumors, and their analysis may, therefore, be of scientific interest with respect to normal development, differentiation, and apoptosis. Because the key genetic lesions in many pediatric tumors result in chimeric and/or amplified transcription factors (e.g., EWS-FLI1, PAX7-FKHR, MYCN), the gene expression profiles in these tumors are likely to reveal some of the direct effects of these aberrant oncogenic proteins, thereby providing important clues to their essential mechanisms and pathways.

From: *Expression Profiling of Human Tumors: Diagnostic and Research Applications*
Edited by: Marc Ladanyi and William L. Gerald © Humana Press Inc., Totowa, NJ

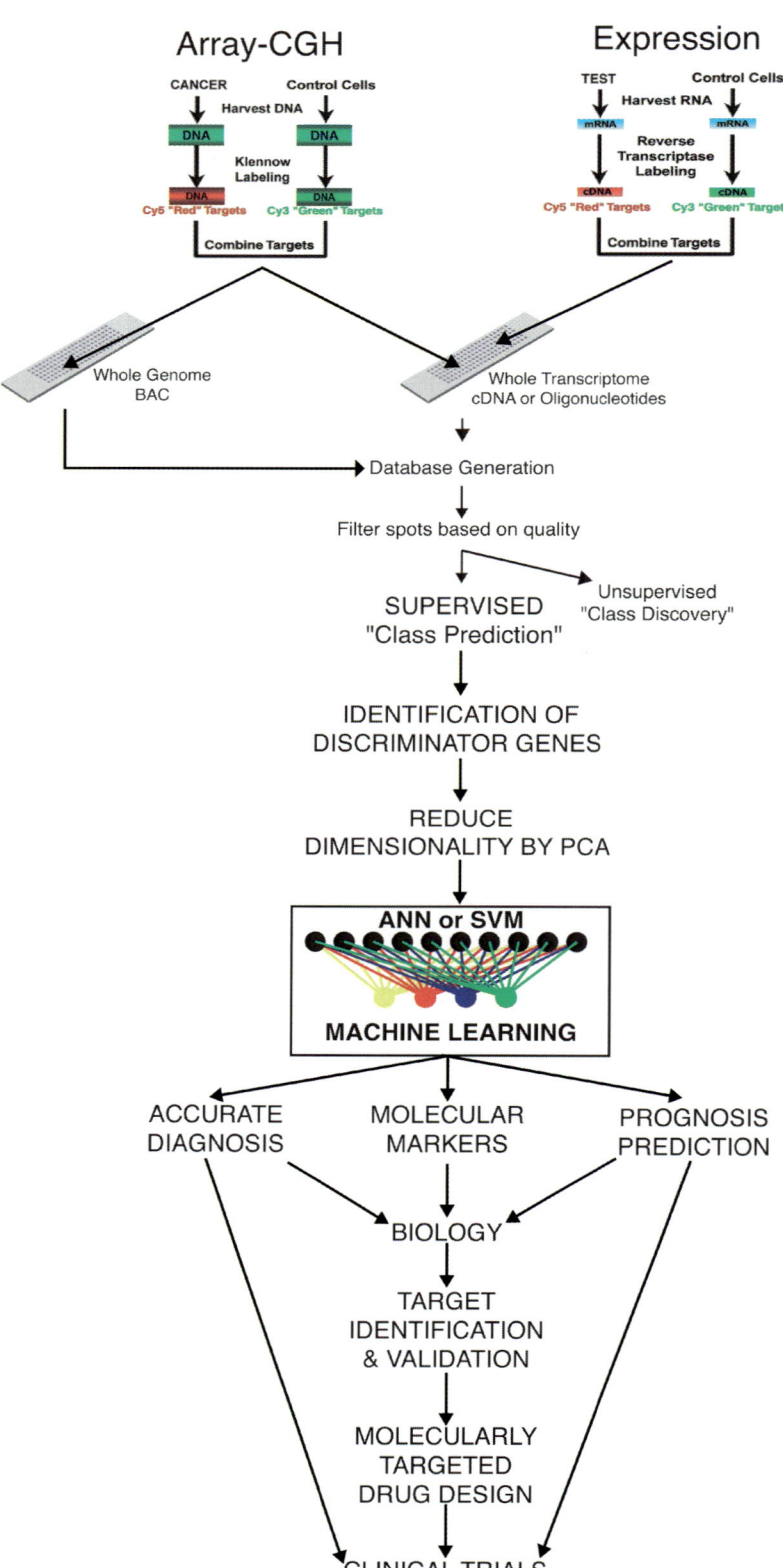

Fig. 1.

Relatively few reviews of expression profiling studies of pediatric cancers are available *(2)*. Here, we summarize the data obtained so far on neuroblastomas (NB), rhabdomyosarcomas (RMS), Ewing's sarcomas (EWS), Burkitt's lymphomas (BL), brain tumors, acute leukemias, and Wilms tumor (WT).

NEUROBLASTOMAS AND SMALL ROUND BLUE CELL SARCOMAS

Cancers that belong to the category of small round blue cell tumors (SRBCT) of childhood include NB, RMS, EWS, and BL, and are classical examples of cancers that may pose diagnosis difficulties in clinical practice. Their name comes from their uniform appearance on routine histologic examination, which can make them difficult to distinguish from one another. Accurate diagnosis of the SRBCTs is essential, as the treatment options, response to therapy, and prognosis varies widely depending on the diagnosis. Several diagnostic techniques are utilized to diagnose them, including cytogenetics, interphase fluorescence *in situ* hybridization, reverse transcription polymerase chain reaction (RT-PCR), and immunohistochemistry.

We recently used expression profiling to approach the diagnostic problem of undifferentiated SRBCTs *(3)*. A machine learning artificial intelligence system (artificial neural network [ANN]) was trained using microarray data from 63 training samples (23 EWS, 20 RMS, 12 NB, and 8 BL) hybridized to 6567 gene cDNA microarrays. To determine which genes were most important for the classification, the calibrated ANNs were analyzed, and the genes were ranked according to how sensitive the output was with respect each gene's expression level. A set of 96 genes was capable of discriminating between these 4 tumor types, the ANN correctly classified 25 additional test samples. We also identified additional lineage-associated genes in specific tumor types, i.e., muscle-specific genes in RMS and neural-specific genes in EWS. For instance, we found *FGFR4*, a gene that is expressed during myogenesis and prevents terminal differentiation in myocytes *(4,5)*, to be highly expressed only in RMS, but not in normal muscle. Additionally, as a receptor protein kinase, it represents a possible new target for therapy. The relatively strong differential expression of FGFR4 in RMS was confirmed by immunostaining of tissue microarrays (Fig. 2A). Likewise, the prominent expression of some neural lineage proteins previously not studied in EWS, such as EphrinB1 and NPYY1 *(6)*, was also detected by the cDNA microarray analysis and later demonstrated using appropriate antibodies on tissue microarrays of EWS cases

---

Fig. 1. *(previous page)* Potential roles of DNA microarrays in pediatric cancers. With the current expansion in the number of available clones or sequences for both gene expression (cDNA and oligonucleotides) and allelic imbalance profiling (using bacterial artificial chromosomes [BACs]), it is possible to perform array-based comprehensive genomic studies on pediatric cancers. The goal of this type of research is to identify expression or genomic imbalance (amplification or deletion) signatures that correlate with molecular markers, diagnosis and prognosis. There has also been an explosion in tools available to perform supervised clustering using standard statistics and machine learning algorithms. The primary aims of these studies are to guide therapy by diagnostic and risk of relapse predictions, as well as to decipher the biological processes involved in the oncogenic process. It should also be possible to use these techniques to identify specific novel targets for therapy, from which rational molecularly targeted drugs can be designed and eventually evaluated in patients in the context of clinical trials.

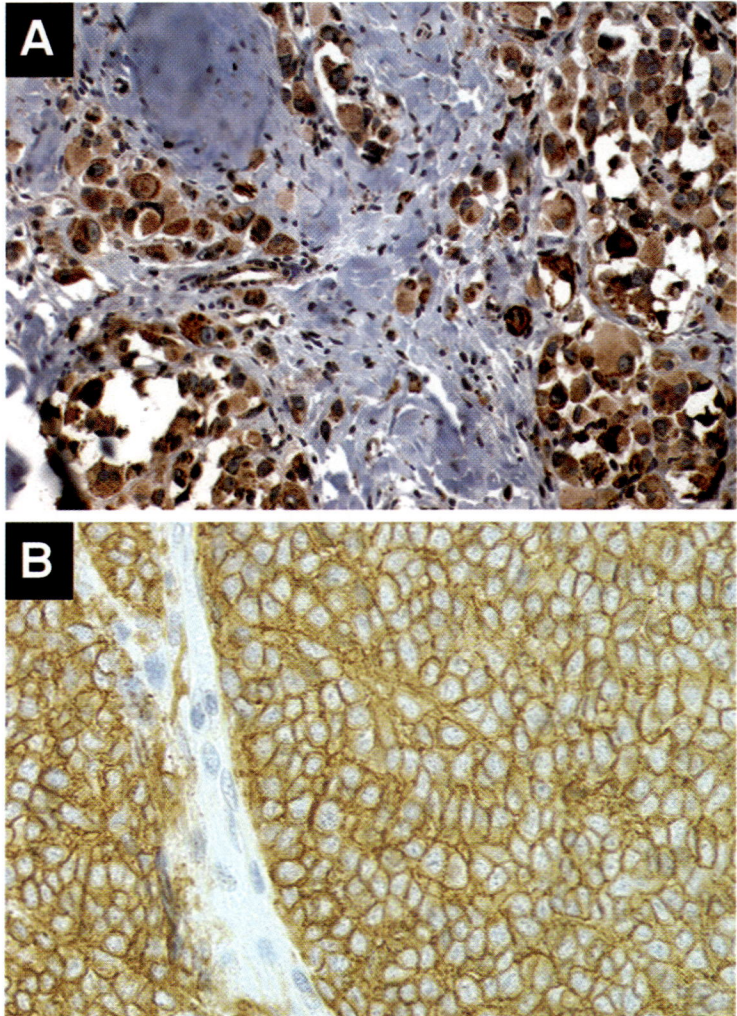

Fig. 2. Validation of cDNA microarray data by immunostaining of pediatric cancer tissue microarrays. (A) Strong cytoplasmic immunostaining for FGFR4 in a case of RMS, obtained with a polyclonal antibody to the C terminal (Santa Cruz Biotechnology, Santa Cruz, CA, USA). On a tissue microarray, all 26 RMS (17 alveolar, 9 embryonal) showed moderate to strong cytoplasmic immunostaining. Stromal elements were generally negative or only weakly positive. (B) Diffuse membranous positivity for NPYY1, a receptor for neuropeptide Y, in EWS using a commercial antibody (DiaSorin, Stillwater, MN, USA). Focal to diffuse membranous positivity was seen in most cases on a tissue microarray of EWS cases.

(Fig. 2B). EphrinB1 is a membrane-bound ligand for the EphB1 and EphB2 receptor tyrosine kinases, and this signaling is known to play a key role in neural development. NPYY1 is a receptor for neuropeptide Y in the central and peripheral nervous systems and in the gut.

In a study restricted to cell lines, Wai et al. used oligonucleotide arrays (Affymetrix, Santa Clara, CA, USA) representing 1700 cancer-associated genes to study the expression profiles of six EWS cell lines and four NB cell lines *(7)*. While most of the NB-associated and EWS-associated genes differed in the two studies, there were notable exceptions, such as high *MYC* and cyclin D1 (*CCND1*) in EWS. This is not unexpected, because the EWS-FLI1 protein of EWS may directly or indirectly regulate both *MYC* and *CCND1* transcription *(8)*. The study by Wai et al. was also notable in combining expression profiling with comparative genomic hybridization to show that some of the variability in gene expression between samples may be directly related to gains of genetic material *(7)*. This phenomenon was also highlighted in a study from the Spieker and Versteeg group, which used expression profiling data that were cross-referenced with chromosomal map position data and compared to other neuroblastoma cell lines to identify *MEIS1* as an amplified and overexpressed gene within a known amplicon derived from chromosome band 2p15 in neuroblastoma cell line IMR32 *(9)*. They also found that *MEIS1* is overexpressed in about 25% of neuroblastomas, in the absence of *MEIS1* amplification.

Monitoring the temporal and global changes in gene expression using DNA microarray profiling methods has also been effective in identifying targets of transcription factors. An example is the investigation of the molecular effects of tumor-specific chromosome translocations that encode chimeric transcription factors. These chimeric transcription factors are thought to exert their oncogenic effects through the dysregulation of gene expression, and DNA microarrays provide an opportunity to observe the broad effects of oncogenic transcription factors on gene expression and potentially elucidate their role in oncogenesis. For instance, *PAX3-FKHR* chimeric oncogene, which is found in the skeletal muscle cancer, alveolar rhabdomyosarcoma (ARMS), and results from a t(2;13). This translocation is found in the majority of ARMS, and leads to the fusion of the DNA-binding domain of PAX3, which is a gene involved in muscle differentiation, with the *trans*-activation domain of FKHR. The PAX3-FKHR gene product retains the DNA binding specificity of PAX3 and potentially acts by increasing expression of genes containing PAX3 binding sites. Investigators, utilizing murine cDNA microarrays, have found that PAX3-FKHR triggers a myogenic differentiation pathway producing a population of rhabdomyoblasts, which may eventually lead to the development of fully malignant muscle cancer upon accumulation of other genetic aberrations *(10)*. Some of these genes were also expressed in human ARMS cell lines.

An example of the application of expression profiling by serial analysis of gene expression (SAGE) (*see* Chapter 4) to transcriptional targets in pediatric cancer has been the identification of MYCN targets in neuroblastomas. As discussed above, MYCN is a transcription factor that is frequently amplified in neuroblastomas, where it is associated with poor prognosis and is used to stratify treatment. Upon analysis of 42,000 mRNA transcript tags in generated SAGE libraries of *MYCN*-transfected and control neuroblastoma cells, van Limpt et al. *(11)* found 114 up-regulated genes. The majority of these genes have a role in ribosome assembly and activity, and Northern blot analysis confirmed up-regulation of all tested transcripts. Their data suggested that *MYC* family genes function as major regulators of the protein synthesis machinery and provide a valuable resource for data mining and comparisons. Other studies with similar experimental designs are summarized in Table 1.

Table 1
Potential Direct or Indirect Targets of Oncogenic Transcription Factors in Pediatric Cancers Detected by Expression Profiling-Based Experiments

Transcription factor	Cell line transfected	Putative target genes induced	Refs.
PAX3-FKHR	NIH 3T3 mouse fibroblast	MyoD, Slug, Myl4 Myogenin, Six1.	(10)
PAX3	DAOY human medulloblastoma	MYOD, STX.	(19)
WT1	U2OS human osteosarcoma	Amphiregulin.	(20)
EWS-WT1	U2OS human osteosarcoma	Interleukin (IL)-2/15Rβ.	(21)
MYCN	SHEP-2 human neuroblastoma	nm23-H1, nm23-H2, multiple ribosomal proteins, multiple genes involved in protein synthesis and turnover.	(22,23)

Finally, bulk sequencing of cDNA libraries from cancers and normal tissues is another source of expression profiles. The National Cancer Institute (NCI) of the National Institutes of Health (NIH) has initiated the Cancer Genome Anatomy Project (CGAP), which catalogues the gene expression profiles based on sequencing of cDNA libraries derived from normal, precancer, and cancer cells (http://cgap.nci.nih.gov/). Their Web site includes tools to compare expression profiles between different cDNA libraries. An example in pediatric cancer has been the generation and sequencing of libraries derived from EWS by CGAP (http://cgap.nci.nih.gov/Tissues/LibInfo?ORG=Hs&LID=31).

BRAIN TUMORS

A key clinical question is whether one can predict the prognosis of patients based on the expression profile of their cancer at presentation. The central hypotheses being tested here are: (*i*) the genetic information that confers prognosis for a given cancer is already present in the cancer at presentation; and (*ii*) this information can be used to more accurately predict the prognosis. The algorithms used to prove these hypothesis are essentially the same. In a recent study relevant to these issues, Pomeroy et al. *(12)* performed expression analysis on a series of embryonal tumors of the central nervous system. They studied 99 tumor samples, including 60 medulloblastomas, 8 supratentorial primitive neuroectodermal tumors (PNETs), 5 atypical teratoid/rhabdoid tumors (AT/RTs), and 10 malignant gliomas, on Affymetrix microarrays representing 6817 genes *(12)*. A classification system based on DNA microarray gene expression data demonstrated that medulloblastomas are molecularly distinct from other brain tumors, including PNETs, AT/RTs, and malignant gliomas. The relationship of medulloblastomas to cerebellar granule cells and the central role of activation of the SHH pathway in this tumor type were confirmed. Thus, cerebellum-specific genes, such as *ZIC1* and *NSCL1*, and SHH downstream targets, such as *PTCH*, *GLI*, and *MYCN*, were part of the expression profile of medulloblastomas. Another interesting finding was that brain AT/RTs were more similar to their renal counterparts than to other brain tumors. The investigators also showed that the clinical outcome of children with medulloblastomas can be predicted on the basis of the tumor gene expression profiles, independent of stage and other prognostic markers such as TRKC expression (although the latter was part of the favorable expression profile).

It is interesting to compare the results of two prior smaller studies to those of the above study *(13)*. Michiels and colleagues examined genes differentially expressed in medulloblastoma and fetal brain using the SAGE technique. Medulloblastoma displayed significantly higher expression of *ZIC1* and *OTX2*, both normally expressed in the cerebellar germinal layers. These data are consistent with the more recent data of Pomeroy et al. *(12)*. In contrast, in terms of prognostically significant expression profiles, there was less overlap between the latter data *(12)* and those from an earlier study by MacDonald et al., who used expression profiles of 23 primary medulloblastomas to identify genes whose expression differed significantly between those clinically designated as metastatic and those that did not metastasize *(14)*. They found platelet-derived growth factor receptor α (*PDGFRα*) and members of the downstream *RAS*/mitogen-activated protein kinase (*MAPK*) signal transduction pathway to be overexpressed in the subset of primary medulloblastomas that later metastasized. Perhaps surprisingly, these genes were not among the 50 genes most highly associated with unfavorable outcome in the larger study of Pomeroy et al. *(12)*. This inconsistency may stem from different study groups or subtly different end points (metastatic potential vs treatment failure) or may be due to idiosyncrasies of individual data sets, highlighting the need for prospective trials on larger numbers of patients.

ACUTE LEUKEMIAS

Several clinicopathological and biological prognostic markers have also been identified allowing stratification of therapy depending on the grade of the cancers and presence or absence of specific markers. For example, patients with acute lymphoblastic leukemia (ALL) who are between 10–21 yr of age or who are between 1–9 yr of age with a white blood cell (WBC) count of $\geq 50,000/\mu L$ are been determined to be at high risk of relapse and are stratified to receive more intense chemotherapy. Also, patients with ALL, which have the t(9:22), are at very high risk, are not given standard therapy, and are instead given a bone marrow transplant in first remission if a donor is available. The Downing group performed a large-scale demonstration of the diagnostic power of expression profiling in pediatric ALLs *(15)*. Pediatric ALL is an appealing forum for the initial clinical implementation of expression profiling, because the role of morphology is already limited, and the complexity and expense of the usual ancillary testing (flow immunophenotyping, cytogenetics, molecular diagnostics) weakens the same arguments regarding clinical expression profiling. In what is one of the largest expression profiling studies to date, they used Affymetrix microarrays, representing about 12,600 genes, to analyze the pattern of genes expressed in 360 pediatric ALLs. Expression profiling accurately separated cases of each of the prognostically important ALL subtypes, including 43 T cell ALL, those conventionally defined by known genetic changes such as the specific gene fusions, *E2A-PBX1* (27 cases), *BCR-ABL* (15 cases), and *TEL-AML1* (79 cases), the mixed-lineage leukemia gene (*MLL*) rearrangement (20 cases), and 64 cases with hyperdiploidy (>50 chromosomes). These distinctive expression profiles contained many genes whose expression was not previously known to be restricted to particular genetic subtypes of ALL. These genes can now be evaluated as surrogate markers for these subgroups (e.g., *MERTK* in *E2A-PBX1*-positive ALL, *HOXA9* and *MEIS1* in *MLL* rearranged cases). Among the non-T cell ALLs, they observed that specific expression profiles more strongly reflected the specific gene fusions rather than the particular B cell differentiation stage. Moreover, they also identified a novel type of ALL, accounting for about 4% of their ALL study group, based on its unique expression profile *(15)*.

In a study focused on 59 pediatric T cell ALLs, Ferrando and colleagues made somewhat different observations. They found that several different genes (*HOX11*, *TAL1*, *LYL1*, *LMO1*, *LMO2*) classically activated by specific chromosomal translocations involving the T cell receptor genes are often aberrantly expressed in the apparent absence of chromosomal abnormalities. This phenomenon was not evident among the non-T cell ALLs studied by the Downing group *(15)*. Also, in contrast to the latter study, they found that subset-specific expression profiles clearly reflected T cell differentiation stage. Furthermore, they identified a novel T cell ALL subset characterized by *HOX11L2* expression and poor treatment response.

Finally, an earlier study by Armstrong et al. focused on acute leukemias with chromosomal translocations involving the *MLL* gene *(16)*. These are seen both in infants and postchemotherapy, but their study contained primarily samples derived from the former. By comparing gene expression profiles in 17 *MLL*-rearranged acute leukemias (of which 15 were in infants or children), 20 B cell lineage ALL, and 20 acute myeloid leukemias, using Affymetrix arrays representing 12,600 genes, they confirmed that *MLL* leukemias constitute a distinct disease, derived from an early hematopoietic progenitor expressing multilineage markers. *MLL* leukemias were also noted to overexpress certain *HOX* and *HOX*-related genes (*HOXA9*, *HOXA5*, *MEIS1*), an observation subsequently reproduced by the Downing group *(15)*. Together, these various studies suggest that pediatric acute leukemias will be a particularly fruitful and interesting area of expression profiling studies.

WILMS TUMOR

The first systematic expression profiling study of WT has recently been published. Li et al. *(17)* used Affymetrix oligonucleotide microarrays representing approx 12,000 genes to study six nonanaplastic nonsyndromic WTs and six midgestation fetal kidney specimens, as well as heterologous normal and neoplastic tissues (eight samples of BL). They thus identified 357 genes differentially expressed between WTs and fetal kidneys, of which a subset of 27 genes constituted, relative to heterologous tissues, a "WT signature" set. The latter included several genes encoding transcription factors, of which at least four (*PAX2*, *EYA1*, *HBF2*, *HOXA11*) play key roles in cell survival and proliferation in early metanephric development. Li et al. *(17)* also related their human microarray data to existing rat microarray data derived from a study of the developing rat kidney *(18)*. This comparison showed that WTs overexpress genes corresponding to the earliest stage of metanephric development and underexpress genes corresponding to subsequent stages. They concluded that the blastematous elements in WTs represent a differentiation arrest at the earliest committed stage in mesenchymal–epithelial transition.

REFERENCES

1. Linet, M. S., Ries, L. A., Smith, M. A., Tarone, R. E., and Devesa, S. S. (1999) Cancer surveillance series: recent trends in childhood cancer incidence and mortality in the United States. *J. Natl. Cancer Inst.* **91,** 1051–1058.
2. Triche, T. J., Schofield, D., and Buckley, J. (2001) DNA microarrays in pediatric cancer. *Cancer J.* **7,** 2–15.
3. Khan, J., Wei, J., Ringnér, M., et al. (2001) Classification and diagnostic prediction of cancers using gene expression profiling and artificial neural networks. *Nat. Med.* **7,** 673–679.
4. Shaoul, E., Reich-Slotky, R., Berman, B., and Ron, D. (1995) Fibroblast growth factor receptors display both common and distinct signaling pathways. *Oncogene* **10,** 1553–1561.
5. deLapeyriere, O., Ollendorff, V., Planche, J., et al. (1993) Expression of the Fgf6 gene is restricted to developing skeletal muscle in the mouse embryo. *Development* **118,** 601–611.
6. Wei, J., Khan, J., Welford, S., et al. (2001) Elucidation of molecular targets of the EWS/FLI1 fusion gene found in Ewing's sarcoma using cDNA microarrays. *Proc. Am. Assoc. Cancer Res.* **42,** 427–427.
7. Wai, D. H., Schaefer, K. L., Schramm, A., et al. (2002) Expression analysis of pediatric solid tumor cell lines using oligonucleotide microarrays. *Int. J. Oncol.* **20,** 441–451.
8. Dauphinot, L., De Oliveira, C., Melot, T., et al. (2001) Analysis of the expression of cell cycle regulators in Ewing cell lines: EWS-FLI-1 modulates p57KIP2 and c-Myc expression. *Oncogene* **20,** 3258–3265.
9. Spieker, N., van Sluis, P., Beitsma, M., et al. (2001) The MEIS1 oncogene is highly expressed in neuroblastoma and amplified in cell line IMR32. *Genomics* **71,** 214–221.
10. Khan, J., Bittner, M. L., Saal, L. H., et al. (1999) cDNA microarrays detect activation of a myogenic transcription program by the PAX3-FKHR fusion oncogene. *Proc. Natl. Acad. Sci. USA* **96,** 13,264–13,269.
11. van Limpt, V., Chan, A., Caron, H., et al. (2000) SAGE analysis of neuroblastoma reveals a high expression of the human homologue of the Drosophila Delta gene. *Med. Pediatr. Oncol.* **35,** 554–558.
12. Pomeroy, S. L., Tamayo, P., Gaasenbeek, M., et al. (2002) Prediction of central nervous system embryonal tumour outcome based on gene expression. *Nature* **415,** 436–442.
13. Michiels, E. M., Oussoren, E., Van Groenigen, M., et al. (1999) Genes differentially expressed in medulloblastoma and fetal brain. *Physiol. Genomics* **1,** 83–91.
14. MacDonald, T. J., Brown, K. M., LaFleur, B., et al. (2001) Expression profiling of medulloblastoma: PDGFRA and the RAS/MAPK pathway as therapeutic targets for metastatic disease. *Nat. Genet.* **29,** 143–152.
15. Yeoh, E. J., Ross, M. E., Shurtleff, S. A., et al. (2002) Classification, subtype discovery, and prediction of outcome in pediatric acute lymphoblastic leukemia by gene expression profiling. *Cancer Cell* **1,** 133–143.
16. Armstrong, S. A., Staunton, J. E., Silverman, L. B., et al. (2002) MLL translocations specify a distinct gene expression profile that distinguishes a unique leukemia. *Nat. Genet.* **30,** 41–47.
17. Li, C. M., Guo, M., Borczuk, A., et al. (2002) Gene expression in Wilms' tumor mimics the earliest committed stage in the metanephric mesenchymal-epithelial transition. *Am. J. Pathol.* **160,** 2181–2190.
18. Stuart, R. O., Bush, K. T., and Nigam, S. K. (2001) Changes in global gene expression patterns during development and maturation of the rat kidney. *Proc. Natl. Acad. Sci. USA* **98,** 5649–5654.
19. Mayanil, C. S., George, D., Freilich, L., et al. (2001) Microarray analysis detects novel Pax3 downstream target genes. *J. Biol. Chem.* **276,** 49,299–49,309.

20. Lee, S. B., Huang, K., Palmer, R., et al. (1999) The Wilms tumor suppressor WT1 encodes a transcriptional activator of amphiregulin. *Cell* **98,** 663–673.
21. Wong, J. C., Lee, S. B., Bell, M. D., et al. (2002) Induction of the interleukin-2/15 receptor beta-chain by the EWS-WT1 translocation product. *Oncogene* **21,** 2009–2019.
22. Boon, K., Caron, H. N., van Asperen, R., et al. (2001) N-myc enhances the expression of a large set of genes functioning in ribosome biogenesis and protein synthesis. *EMBO J.* **20,** 1383–1393.
23. Godfried, M. B., Veenstra, M., Sluis, P., et al. (2002) The N-myc and c-myc downstream pathways include the chromosome 17q genes nm23-H1 and nm23-H2. *Oncogene* **21,** 2097–2101.

17
Transcriptomes of Soft Tissue Tumors
Pathologic and Clinical Implications

Sabine C. Linn, Rob B. West, and Matt van de Rijn

INTRODUCTION

Soft tissue tumors (STTs) are rare tumors for which over 100 different diagnostic entities have been defined *(1)*. For clinical purposes, STTs can be divided into three categories: benign, intermediate, and malignant *(2)*. The incidence of the malignant category is approx 7800 cases annually in the U.S. Slightly over 50% of these new patients will ultimately die of their disease *(3)*. When diagnosed correctly at an early stage, soft tissue sarcomas are curable by surgery, often in combination with adjuvant chemotherapy and/or radiotherapy.

The development of useful classifications has been a challenge to pathologists and clinicians due to the rarity of STTs, combined with the large variability in clinical presentation, location, histopathological diagnosis, and outcome. In general, two classifications are used in parallel: a pathologic classification mainly based on phenotypic characteristics of the tumor cells, and a clinical classification largely based on clinical tumor aggressiveness *(2)*. The advent of DNA microarray technology will allow identification of previously unrecognized subsets of tumors within existing tumor categories and the discovery of new diagnostic and prognostic markers. This may significantly improve existing classifications. In addition, this new technique may facilitate the discovery of potential targets for specific therapies, like imatinib mesylate (STI571), which targets the tyrosine kinase receptor *KIT (4)* in gastrointestinal stromal tumors (GISTs) *(5)*. Validation of potential diagnostic and prognostic markers will be made possible with tissue microarray technology *(6)* using immunohistochemistry (IHC) on large numbers of paraffin-embedded samples of STTs. In this chapter, we will point out some of the issues regarding existing classifications and discuss the novel applications of gene expression profiling and tissue microarrays in the study and understanding of STTs.

Pathologic Classification

The histologic classification presented in the latest version of *Enzinger and Weiss' Soft Tissue Tumors* recognizes 140 types of soft tissue tumors *(1)*. Table 1 lists the major categories in the pathologic classification of STTs. The aim of the pathologic classification is to categorize STTs according to their differentiation characteristics.

From: *Expression Profiling of Human Tumors: Diagnostic and Research Applications*
Edited by: Marc Ladanyi and William L. Gerald © Humana Press Inc., Totowa, NJ

Table 1
Histological Classifications of STTs

Major categories in the classification of STTs[a]

1. Fibrous and myofibroblastic tumors.
2. Fibrohistiocytic tumors.
3. Lipomatous tumors.
4. Smooth muscle tumors.
5. Skeletal muscle tumors.
6. Vascular tumors.
7. Perivascular tumors.
8. Synovial tumors.
9. Neural tumors.
10. Osseous and cartilaginous tumors.
11. Miscellaneous tumors.

[a]Adapted from Kempson et al. (2).

The initial characterization of an STT usually starts with the search for specific cellular features, such as filaments arranged as cross striations (skeletal muscle tumors), cytoplasmic lipid vacuoles (lipomatous tumors), or extracellular matrix material (e.g., cartilage, bone) (2). Immunohistochemical analysis of paraffin-embedded tumors is of additional value in the distinction of various STTs from each other. Table 2 shows some of the most commonly used markers.

The diagnosis of an STT is often surrounded by uncertainty due to the lack of experience of most general pathologists with these neoplasms. In a large study conducted in the UK, 22% of a group of 449 tumors originally diagnosed as sarcomas were revised into a nonsarcoma category (7). Most misdiagnoses came from carcinomas diagnosed as sarcomas. Furthermore, of the remaining sarcomas, 39% were reclassified into a different histologic subtype. The subtypes with the poorest level of agreement were leiomyosarcoma, malignant fibrous histiocytoma, liposarcoma, fibrosarcoma, and rhabdomyosarcoma. However, this study was carried out before the availability of IHC. Nevertheless, a recent study presented similar findings (25% major discrepancies on a total of 266 cases) and concluded that these misdiagnoses were more likely due to unfamiliarity with these lesions than with the increasing use of needle biopsy or the failure to perform IHC (8). Clearly, one reason for this high discrepancy rate is the rarity of sarcomas, with only one to two new cases per year per pathologist, if these lesions are distributed equally amongst pathologists, which in practice is not the case (9). Another important reason is the intrinsic difficulty in recognizing the subtle differences in the histology of STTs (Fig. 1). Lastly, the dearth of specific diagnostic markers limits the usefulness of IHC in distinguishing these neoplasms (Table 2). It should be noted, for example, that several of these markers react with more than one tumor type. Actin reactivity can be seen in both leiomyosarcoma and rhabdomyosarcoma. S100 staining is found in benign and malignant nerve sheath tumors, but also in clear cell sarcoma; outside the field of STTs, S100 also reacts with melanoma and a variety of carcinomas. CD34, initially described on hematopoietic stem cells was rapidly found to react with vascular neoplasms. However, in subsequent studies, it was also seen to

Table 2
Frequently Used Monoclonal Antibodies in the Diagnosis of STT[a]

Monoclonal antibodies	Tumor	Percent reactivity
Smooth muscle actin[b]	Leiomyosarcoma	90%
Muscle actin	Leiomyosarcoma	90%
Desmin	Leiomyosarcoma	75%
Muscle actin	Rhabdomyosarcoma	90%
Desmin	Rhabdomyosarcoma	95%
Myoglobin	Rhabdomyosarcoma	30%
Cytokeratin[b]	Biphasic synovial sarcoma	95–100%
Epithelial membrane antigen[b]	Biphasic synovial sarcoma	95%
Cytokeratin	Monophasic synovial sarcoma	50%
Epithelial membrane antigen	Monophasic synovial sarcoma	50%
Cytokeratin	Epithelioid sarcoma	>95%
Epithelial membrane antigen	Epithelioid sarcoma	>95%
S100[b]	Malignant peripheral nerve sheath tumor	50%
S100	Clear cell sarcoma	80%
CD99[b]	Ewing's sarcoma/PNET	>95%
CD34[b]	Angiosarcoma	80%
CD31	Angiosarcoma	90%

[a]Adapted from Kempson et al. (2).
[b]These antibodies are not specific and will react with a wide variety of other lesions.

react with solitary fibrous tumors (SFTs), gastrointestinal stromal tumors, and a variety of other neoplasms *(10–12)*. The issue is further complicated by the fact that, in many tumor types, only a subset of lesions may react for certain markers. For example, only 50% of malignant peripheral nerve sheath tumors react for S100. It is clear that the field of soft tissue tumor pathology would be greatly helped by the identification of additional tumor markers.

A variety of translocations have been identified in STTs (Table 3) *(13)*. The translocations shown in Table 3 appear specific for a particular STT diagnosis. In addition, they occur in the majority of these lesions and, hence, are valid diagnostic markers. For example, many studies have reported that t(X:18) is not only a sensitive but also a specific marker for synovial sarcoma *(14–19)*. Several techniques exist for the detection of translocations, most of which require (or work better on) fresh frozen material.

Clinical information may also help to assess the likelihood of a given STT diagnosis. The evaluation of risk factors, such as prior radiation therapy or a positive family history, can be of use. For instance, are there any indications of a germline mutation in a tumor suppressor gene, such as Li-Fraumeni syndrome (*TP53*) *(20,21)*, Gardner's syndrome (adenomatous polyposis coli [*APC*]) *(22)*, or neurofibromatosis type I (*NF-1*) *(23,24)*? Furthermore, it is valuable to know the results of imaging studies, to exclude the possibility of a neoplasm arising from an organ invading surrounding soft tissue *(2)*. Information on age is important; the prevalence of MFH and liposarcoma peaks in the seventh decade, while for instance, fibrosarcoma peaks in the fourth decade *(3)*. Likewise, the anatomic site is informative; liposarcoma, malignant fibrous hystiocytoma (MFH), and synovial sarcoma are, for instance, the most common lower

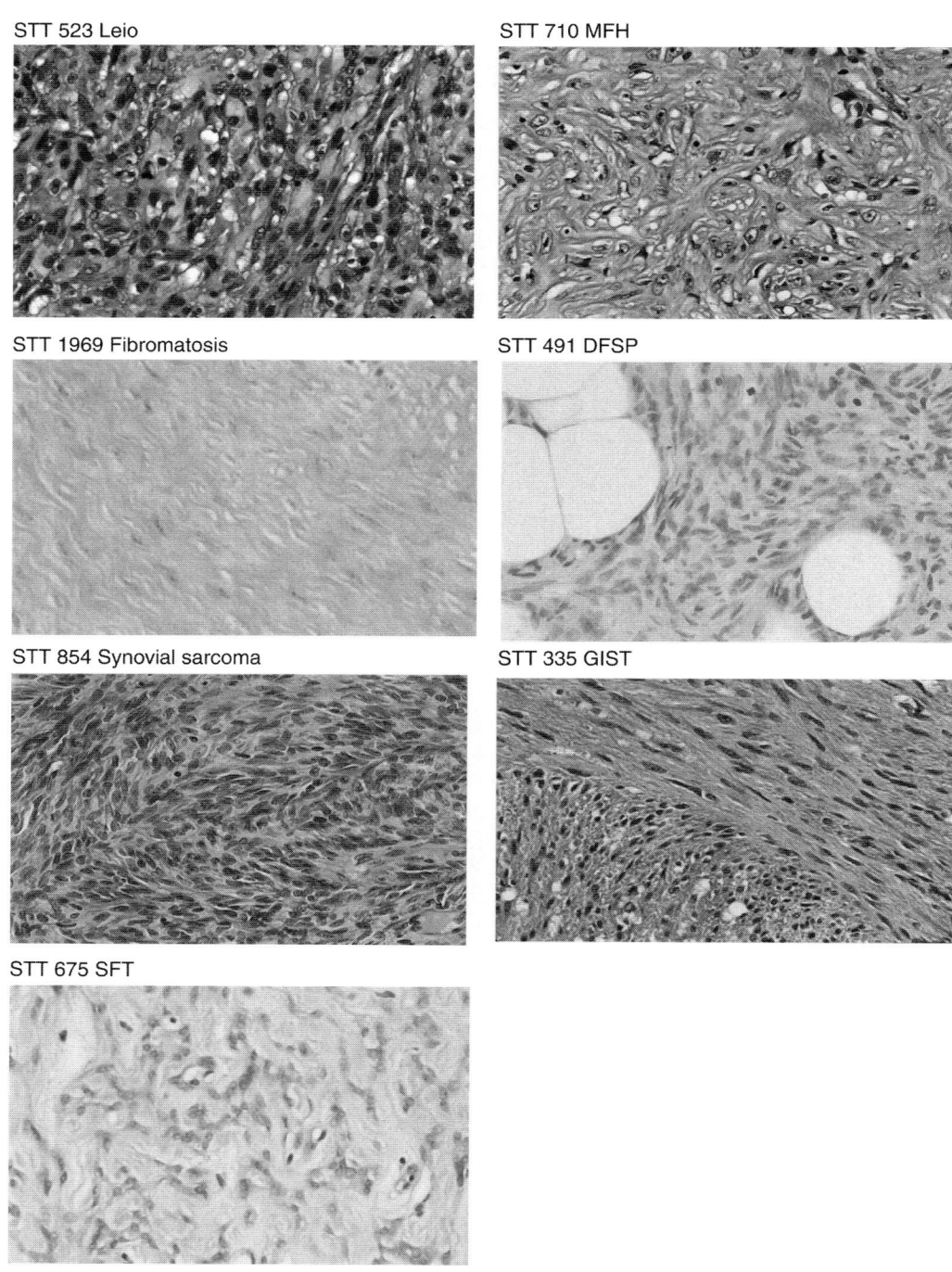

Fig. 1. Representative histology of specimens used for this study, including: leiomyosarcoma, MFH, fibromatosis–desmoid tumor, DFSP, synovial sarcoma, GIST, and SFT.

Table 3
Major Chromosomal Translocations in Sarcomas

Tumor	Translocation	Fusion product
Alveolar rhabdomyosarcoma	t(2;13)(q35;q14)	*PAX3-FKHR*
	t(1;13)(p36;q14)	*PAX7-FKHR*
Alveolar soft part sarcoma	t(X;17)(p11;q25)	*ASPL-TFE3*
Clear cell sarcoma	t(12;22)(q13;q12)	*EWS-ATF1*
Dermatofibrosarcoma protuberans	t(17;22)(q22;q13)	*COL1A1-PDGFB*
Desmoplastic small round cell tumor	t(11;22)(p13;q12)	*EWS-WT1*
Ewing's sarcoma/peripheral primitive neuroectodermal tumor	t(11;22)(q24;q12)	*EWS-FLI1*
	t(21;22)(q22;q12)	*EWS-ERG*
	t(7;22)(p22;q12)	*EWS-ETV1* (rare)
	t(17;22)(q12;q12)	*EWS-E1AF* (rare)
	t(2;22)(q33;q12)	*EWS-FEV* (rare)
Infantile fibrosarcoma	t(12;15)(p13;q25)	*ETV6-NTRK3*
Inflammatory myofibroblastic tumor	t(1;2)(q22;p23)	*TPM3-ALK*
	t(2;19)(p23;p13)	*TPM4-ALK*
Extraskeletal Myxoid chondrosarcoma	t(9;22)(q22;q12)	*EWS-CHN*
	t(9;17)(q22;q11)	*TAF2N-CHN* (rare)
Myxoid liposarcoma	t(12;16)(q13;p11)	*TLS-CHOP*
	t(12;22)(q13;q12)	*EWS-CHOP* (rare)
Synovial sarcoma	t(X;18)(p11;q11)	*SYT-SSX1*
		SYT-SSX2
		SYT-SSX4 (rare)

extremity sarcomas, while leiomyosarcomas prevail at visceral sites *(3)*. Other useful clinical information includes the size of the tumor and its depth (dermal, subcutaneous, or deep). Knowledge of all these clinical variables and marker studies is undoubtedly of great help, but with the possible exception of the translocation studies mentioned above, they in themselves are not pathognomonic in individual cases, and uncertainty about the diagnosis often remains.

Clinical Disease Categories

The clinician is less interested in the detailed phenotypic characteristics of a STT, but rather wants to know the expected biological behavior of a given lesion, which directs the choice of therapy and provides prognostic information. The World Health Organization (WHO) classification is not very helpful in this regard, as it only groups STTs in the clinically benign, intermediate, or malignant category. Kempson et al. have designed a more sophisticated "managerial classification," which further divides these three categories into seven groups, providing more information on the likelihood of a local recurrence after excision, whether a recurrence can become destructive, and the chances of a given lesion to metastasize or to be disseminated at the outset *(2)*.

Prognostic Variables and Grading of Soft Tissue Sarcomas

The prognosis of extremity sarcomas is generally better than for other sarcomas, while retroperitoneal and mediastinal locations confer a worse prognosis *(2,3)*. This might at least partly be due to the general earlier detection of extremity sarcomas and consequently smaller tumor size at the time of diagnosis (25). Besides the anatomical location, tumor depth, size, histologic grade, and the presence of nodal or distant metastases are all recognized as prognostic variables. The current version of the American Joint Committee on Cancer (AJCC) staging system for sarcomas is based on these parameters (26). It has also been shown that a microscopically positive surgical margin in soft tissue sarcomas of the extremity is an independent adverse prognostic factor for local relapse (25).

The histologic grading of adult sarcomas is a controversial issue (27). The purpose of grading is to predict the likelihood of a given sarcoma to recur or metastasize, based on morphologic criteria such as cellularity, pleomorphism, extent of tumor cell necrosis, mitotic activity, and degree of differentiation (2). Most pathologists agree that necrosis and mitotic rate are the most important determinants of aggressive tumor behavior. Grading is problematic because the histologic subtypes of sarcoma must be taken into account; e.g., several sarcomas (epithelioid sarcoma, synovial sarcoma) can have a poor prognosis despite a bland cytology and a low mitotic rate. Furthermore, there is no consensus on which of the several existing grading systems is most informative (25,28).

Many molecular aberrations have been detected in sarcomas, though few have been identified as a prognostic marker (29,30). Mutations in *TP53*, overexpression of p53, and high Ki-67 proliferation index are associated with poor prognosis (31). However, in series including multiple STT histologies, both markers have been associated with tumor grade (31) and, as such, have little independent prognostic value. Within synovial sarcoma, the *SYT-SSX2* fusion transcript has been associated with a longer disease-free and overall survival than the *SYT-SSX1* fusion transcript (32–34). In MFH, the presence of additional chromosomal material on 19p conferred a worse prognosis (35). However, it is clear that additional prognosticators would greatly help management of these lesions.

Treatment Options

The mainstay in the treatment of STTs is surgery. Adjuvant external radiation therapy or brachytherapy is an option for local control in selected patients, depending mainly on the size, grade, localization, and surgical margins of the tumor (36). Adjuvant chemotherapy is still a matter of debate (37), although preoperative radiation and/or chemotherapy are sometimes used for large high-grade extremity sarcomas that would otherwise require amputation (38–40). The most active classical drugs in soft tissue sarcomas are anthracyclines, ifosfamide, and dacarbazine, with modest overall response rates between 20–30% in advanced disease (3,41). Hyperthermia may enhance the effects of chemotherapy (42,43), but currently remains investigational in the U.S. What all these treatments have in common is that their mechanisms of action are relatively nonspecific. The recent advances in the development of more targeted therapies, such as imatinib mesylate for GIST (44,45), have raised high hopes for the future (46), with the anticipation that genome-wide analysis of STT may identify additional targets for tumor-specific therapies.

Synopsis

The field of STT pathology is plagued, perhaps more than any other group of tumors, by diagnostic uncertainty. In addition, there is insufficient knowledge regarding the prognostication of tumors within each diagnostic group. A variety of grading schemes exists, but for the wide diversity of STTs, no one grading system has been accepted or even shown to be appropriate for all lesions. The differential diagnostic and prognostic issues have lately become even more relevant, as it was demonstrated that certain sarcomas respond very well to tumor-specific therapies (44,45). What is needed in the field of STT pathology, therefore, is: (i) better diagnostic markers; (ii) better prognostic markers; and (iii) identification of specific targets for therapy. The expectation is that genome-wide screening of large numbers of STTs will not only lead to a more refined tumor classification, but will also address these three practical issues.

TISSUE PROCESSING FOR DNA MICROARRAY STUDIES
Specimen Types

For the initial diagnosis of STT, three options exist to obtain tissue: incisional biopsy, core-needle biopsy, and fine-needle biopsy (cytology). The core-needle biopsy procedure is rapidly gaining acceptance as the first approach to obtaining diagnostic material. However, the limited amount of tissue usually precludes grading in the case of a malignancy. Complete excision with a small rim of normal tissue is a good alternative for small superficial STTs (<5 cm). Large STTs frequently are diagnosed by incisional biopsy. Because tissue architecture is such an important diagnostic parameter, cytology has so far had only limited application in the diagnosis of STTs (3,47,48).

Tissue Handling, Processing, and Availability

Most STTs are of significant size, and incisional biopsies usually yield sufficient quantities of tumor material, which allows one to cut frozen sections, send part of the specimen for cytogenetic analysis, and process the remainder for paraffin sections. Once enough material is present for these studies, paraffin-embedding of the remaining frozen material used for frozen sections usually does not contribute significantly to diagnosis. If one stores remaining frozen material at –80°C, there will usually be enough tissue for DNA microarray analysis. We use spotted cDNA microarrays, which require 2 µg of mRNA for each experiment, an amount that can be isolated from approx 200 mg of fresh frozen tissue. For core biopsy material, where less tissue is available, linear amplification of mRNA prior to array analysis is an option (49,50).

Due to the rarity of STT, tissue availability is a major problem for gene expression experiments, but this can be overcome by multicenter collaborative studies. For our investigations, we have initiated a collaborative effort that uses STT samples from the Stanford University Medical Center, the University of British Columbia, in Vancouver (Torsten O. Nielsen, MD, PhD), the University of Seattle (Brian P. Rubin, MD, PhD), and the Cleveland Clinic Foundation (John R. Goldblum, MD).

Tissue Heterogeneity

Few studies have yet addressed the problem of tissue heterogeneity in other organ systems, mainly due to technical issues involving amplification procedures on microdissected tissue (50) or sorting material based on cell surface markers (51). Most STTs are quite homogeneous in their histologic features, yet it would be of interest to examine different areas from tumors for heterogeneity. Some examples where this would be of genuine interest and where the areas of different morphology would be relatively easy to dissect are: (*i*) areas of dedifferentiation in liposarcomas *(52)*; (*ii*) round cell areas in myxoid liposarcoma *(53)*; and (*iii*) areas of fibrosarcomatous transformation in dermatofibrosarcoma protuberans (DFSP) *(54,55)*.

Studies in Experimental Model Systems

The ultimate goal of studying tumor biology of STTs is to identify those molecular events that drive malignant transformation. Gene expression studies on cell lines that have been transfected with DNA containing a specific translocation could be of value (e.g., synovial sarcoma, DFSP, Ewing sarcoma–peripheral primitive neuroectodermal tumor [PNET]) (*see* also Table 3), especially when compared to expression studies on native tumors. A problem with this approach is that for many STTs, the cell of origin, which may influence the genes affected by the translocation, is not known *(56,57)*. Culturing human STTs could circumvent this drawback, but their altered *in vitro* phenotype, as a result of cell culture *(58)*, may complicate the interpretation of the gene expression patterns resulting from the translocation. It has already been shown in several studies that gene expression patterns of cell lines differ from the native human tumors from which the cell lines have been derived *(59–61)*.

Xenograft studies have been instrumental in the study of new promising drugs in STTs, such as pharmacological studies of imatinib mesylate for DFSP in mice *(62,63)*. An interesting study was recently published that used a xenograft model in combination with microarray analysis to screen for genes involved in melanoma metastasis *(64)*. To date, no xenograft studies in STTs have been published exploiting the power of microarray analysis.

Some transgenic or knock-out mouse models develop sarcoma-like tumors and may provide tools for the study of human STTs *(65–72)*. However, lack of knowledge concerning the cell of origin for most STTs complicates the development of STT-specific transgenic or conditional knock-out mouse models.

RESULTS OF GENE EXPRESSION PROFILING OF STTs

General and Specific Aspects of the Transcriptome

The genome-wide expression pattern (transcriptome) of different STTs will lead to the identification of novel markers that may be used to determine the cell of origin for these neoplasms. In this manner, a novel subclassification of connective tissue cells will emerge. Thus, the study of STTs may lead to the identification of functional differences between histologically similar cells, similar to that in the lymphoma field, where molecular analysis of different lymphomas led to the identification of functionally distinct subsets of lymphocytes. The molecular characterization of normal connective tissues could have applications beyond the field of STTs and could lead, for example, to increased insight into the pathogenesis of a variety of non-neoplastic connective tissue disorders.

Examples of STT for which the cell of origin is unknown are many, but include synovial sarcoma, DFSP, SFT, and the controversial MFH. The latter is subject of an ongoing debate concerning this tumor as a separate entity from other sarcomas, such as poorly differentiated leiomyosarcomas and liposarcomas *(73)*.

Many histologic subtypes within STTs have characteristic genetic aberrations, such as translocations, which are thought to be instrumental in the transformed phenotype (Table 3). Most events downstream of chromosomal translocations remain unknown, but it is intriguing to suppose that genes affected by these translocations may be found in clusters that contain genes specifically over- or underexpressed in these tumors. The deregulation of particular pathways could give clues to the identification of potential novel therapeutic targets.

For some STT categories, such as SFT, leiomyosarcoma, and desmoid fibromatosis, little is known about the underlying genetic events. A careful examination of their characteristic gene clusters, possibly in combination with array-based comparative genomic hybridization data, might lead to novel insights regarding their pathogenesis.

Proof of Principle Studies and Novel Insights

To date, few studies have been published on gene microarray analyses of STTs. A major contribution came from the group of Khan, Meltzer, and coworkers that tested the application of artificial neural networks in the classification of small round blue cell tumors of childhood by using gene expression profiles *(59)*. A significant number of potential new markers for this diagnostically challenging group of lesions was identified that warrant further study. The same group recently published their findings of a remarkable homogeneous gene expression profile in a collection of mainly large GISTs with an aggressive clinical behavior and proven mutations in the KIT gene *(74)*. A considerable overlap was found between their GIST-specific gene cluster and our microarray results for GISTs *(75)*.

Stanford Studies of Spindle Cell and Pleomorphic STTs

In our laboratory, we have recently performed a proof of principle study on a variety of STTs (http://genome-www.stanford.edu/sarcoma/) *(75)*. This study combined 41 different STT in a single cluster analysis. In Fig. 2 another collection of 41 STTs is shown, which partially overlaps with our initial work. In contrast to our first study, where a mixture of gene arrays containing 22,000 and arrays containing 42,000 cDNA elements were used, the current samples were all run on microarrays consisting of 42,000 cDNA elements. After a procedure that selected well-measured and informative cDNA spots, 7080 genes were chosen to partition the tumors into discrete groups using hierarchical clustering *(76)*. The tumor categories of fibromatosis, DFSP, synovial sarcoma, GIST, and SFT all clustered tightly on individual branches, indicating the ability of cDNA microarray analysis to separate histologically different entities. In contrast, a subset of the leiomyosarcomas grouped together on an ill-defined branch with the MFHs. This is an interesting observation, especially in the context of the ongoing debate surrounding the diagnostic entity of MFH *(73)*. The branching pattern indicates that the transcriptomes of these leiomyosarcomas and MFHs are quite similar. It remains possible however, that only a small group of genes distinguishes these entities and that this expression profile is overwhelmed by genes that are shared, such as proliferation genes (Figs. 2B1,C1 and 3), host cell genes, and inflammatory response genes (Fig. 2B3,C3) analogous to the initial branching pattern found for diffuse large B cell lymphomas *(61)*. Inclusion of larger numbers of MFHs, leiomyosarcomas, and also liposarcomas, in combination with the use of selective gene lists for clustering, may resolve this issue.

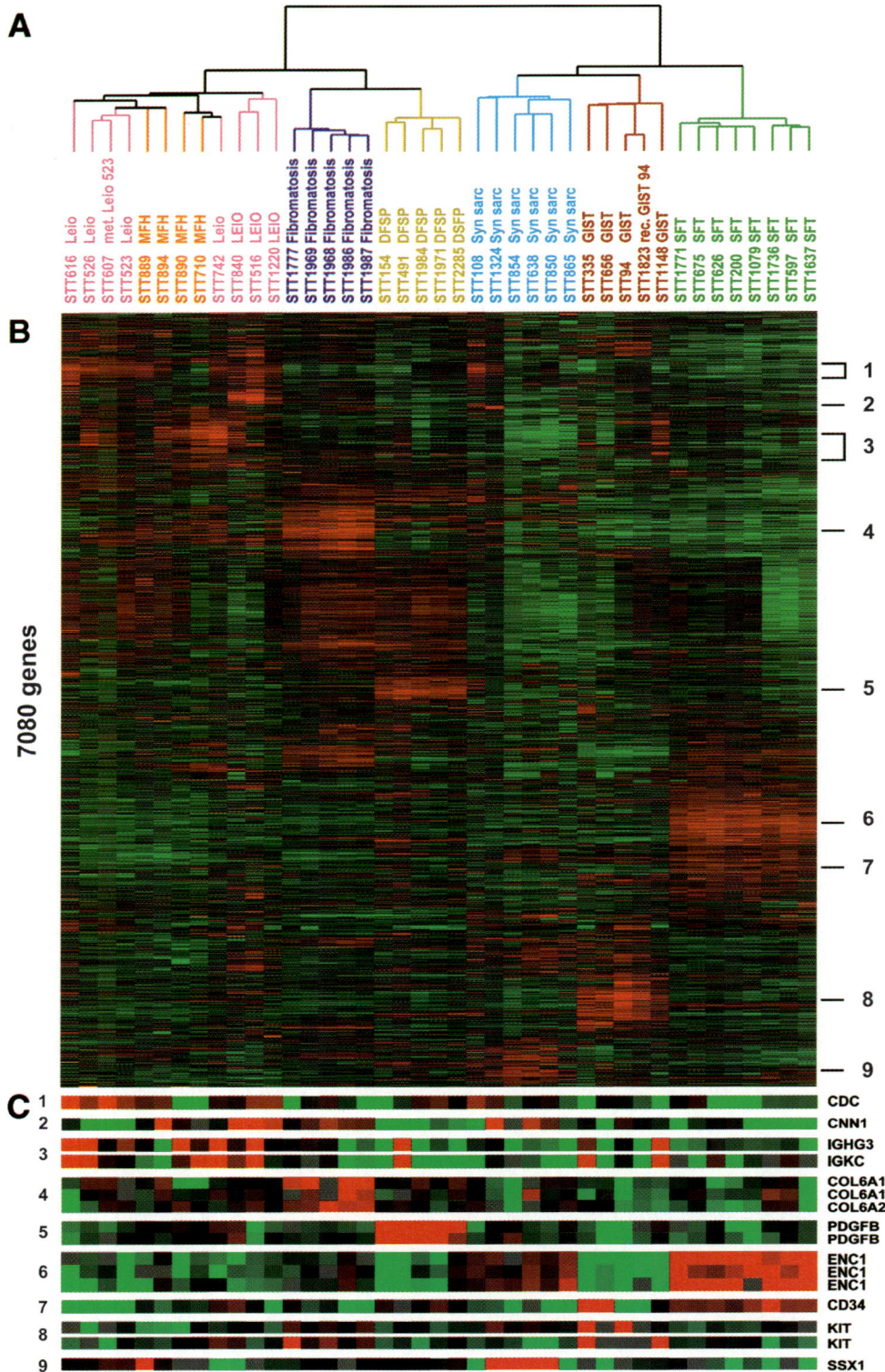

Fig. 2.

For one tumor (STT523), a leiomyosarcoma of the thigh, a pulmonary metastasis was available. The two samples appeared on closely related branches of the dendrogram. For another tumor (GIST), originally occurring in the ileum of a 27-yr-old man, his second (STT94) and fifth (STT1823) peritoneal recurrence were available. These two specimens, obtained with a time interval of 3 yr, showed the highest degree of similarity of all tumors of the cluster. The tight clustering of primary tumors and their lymph node metastases has been reported previously for lung carcinoma *(77)* and breast carcinoma *(60)*.

GENE EXPRESSION PROFILE OF GISTs

The GISTs were widely separated from the two categories of leiomyosarcoma and were characterized by a highly expressed gene cluster that contained *KIT* (Fig. 2A). This confirms the hypothesis that, unlike leiomyosarcomas, GISTs originate from the interstitial cell of Cajal, or at least from cells differentiating towards an interstitial cell of Cajal-phenotype *(78)*. These cells have been associated with pacemaker activity in the bowel wall, and *KIT* expression is essential for the normal development of their intestinal pacemaker network *(79)*. The known prevalence of activating *KIT* mutations in GISTs *(80,81)* and the central position of *KIT* in the GIST-specific gene cluster add

Fig. 2. *(previous page)* Gene expression matrix of 41 STT specimens and 7080 genes obtained with average-linkage hierarchical clustering *(76)*. Arrays comprising 42,000 cDNA elements were used. Experiments were performed as described elsewhere *(60)* (http://genome-www.stanford.edu/molecularportraits/). Genes were selected based on the following criteria: uninterpretable spots were manually flagged and excluded. Of the remaining spots only those were included with a ratio of signal over background of at least 1.6 in either Cy3 or Cy5 channels; those that had at least 80% well measured data points across the 46 arrays and a fluorescence ratio at least threefold greater than the geometric mean ratio in the specimens examined in at least two arrays. A row in the matrix represents the relative level of expression for a gene, centered at the geometric mean of its expression level across the 41 samples. Gene expression levels are displayed in red, relative high expression; black, mean expression; green, relative low expression; or grey, no well-measured information. The red or green color intensity represents the magnitude of the deviation from the mean. Each column represents the relative expression levels of all the selected genes for a single neoplasm. The tumor dendrogram **(A)** is displayed above and describes the degree of relatedness between tumor samples, with short branches denoting a high degree of similarity. Leio/LEIO, leiomyosarcoma; MFH, malignant fibrous histiocytoma; DFSP, dermatofibrosarcoma protuberans; syn sarc, Synovial sarcoma; GIST, gastrointestinal stromal tumor; SFT, solitary fibrous tumor; Met, metastasis; Rec, recurrence. **(B)** Gene expression matrix with numbers on the right indicating distinct gene expression clusters. 1. Proliferation gene cluster. 2. Muscle gene cluster of calponin-positive leiomyosarcomas. 3. Immunoglobulin gene cluster. 4. Fibromatosis-specific gene cluster. 5. DFSP-specific gene cluster. 6. SFT-specific gene cluster. 7. The commonly used immunohistochemical marker CD34. 8. GIST-specific gene cluster. 9. Synovial sarcoma-specific gene cluster. **(C)** Examples of genes present in the clusters detailed in panel B. A white line between two genes means that the genes were not located adjacent to each other in the gene cluster. 1. *CDC*, cell division cycle. 2. *CNN1*, calponin 1, basic smooth muscle. 3. *IGHG3*, immunoglobulin heavy constant γ-3; *IGKC*, immunoglobulin-κ constant. 4. *COL6A1*, collagen, type VI, α-1; *COL6A2*, collagen, type VI, α-2. 5. *PDGFB*, platelet-derived growth factor-β polypeptide. 6. *ENC1*, ectodermal-neural cortex-1. 7. *CD34*, CD34 antigen. 8. *KIT*, c-kit/CD117. 9. *SSX1*, synovial sarcoma, X breakpoint-1.

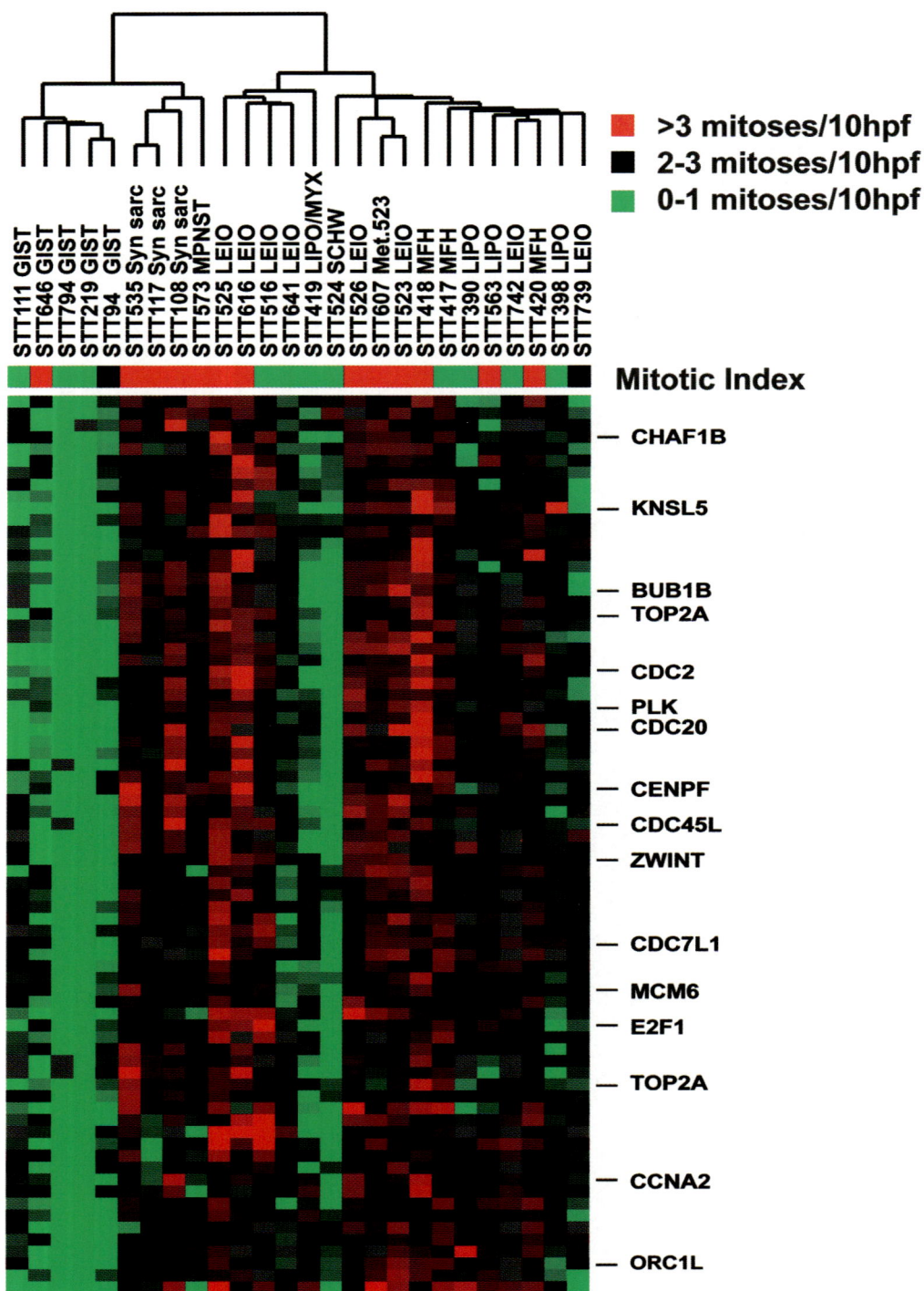

Fig. 3.

further support to the concept that aberrant c-kit activity is instrumental in the transformation of these tumors *(74,75,81)* (Fig. 2B8,C8). The GISTs also have decreased cell cycle activity, as indicated by relative underexpression of proliferation genes (Fig. 3). These findings support the thought that GISTs are tumors of low complexity, presumably initially driven by a single mutation in *KIT*, while acquiring additional genetic aberrations during tumor progression *(82)*. The impressive responses to the tyrosine kinase inhibitor imatinib mesylate are in agreement with this view *(44)*.

GENE EXPRESSION PROFILE OF LEIOMYOSARCOMAS

A subset of three leiomyosarcomas are clustered on a separate branch (Fig. 2). These cases were characterized by relative overexpression of a set of muscle-related genes, such as calponin (Fig. 2B2,C2). The other five leiomyosarcomas, which clustered with four MFHs on another branch, mostly lacked expression of these muscle-related genes. Thus, our findings discerned two subgroups of leiomyosarcoma that were both distinct from GIST *(75)*. Larger series of leiomyosarcomas are needed to correlate these results with histological and clinical features.

GENE EXPRESSION PROFILES OF STTS WITH KNOWN TRANSLOCATIONS

We studied the transcriptomes of two spindle cell STTs with known translocations: DFSP and synovial sarcoma. In both subtypes of STTs, we found relatively high expression of at least one of the translocation partners (Fig. 2B5,C5,B9,C9).

Most, if not all, DFSPs are characterized by specific chromosomal rearrangements, such as supernumerary ring chromosomes that contain sequences from chromosome 17 and 22 or the t(17;22)(q22;q13) translocation *(83)*. These rearrangements result in the expression of a *COL1A1-PDGFB* fusion protein that is cleaved within the endoplasmic reticulum into a functional PDGF-BB molecule *(83,84)*. In vitro studies with NIH3T3 mouse fibroblasts transfected with the fusion gene have shown that the production of the PDGF-BB ligand leads to autocrine PDGF receptor stimulation *(84)*. As the breakpoint for *PDGFB* always lies within intron 1 *(84)*, the sequences for *PDGFB* used on our arrays cannot distinguish between the native or fusion transcript. Nevertheless, we found a very interesting distinct gene expression pattern for DFSPs, which was centered on *PDGFB* (Linn et al., manuscript in preparation), supporting a central role for the PDGF pathway in the pathogenesis of DFSP *(84)*. Preclinical studies

---

Fig. 3. *(previous page)* Comparison of mitotic index, determined by microscopic evaluation of 10 high power fields (hpf) on paraffin-embedded sections, with the proliferation gene cluster of 26 tumor specimens run on arrays containing 22,000 cDNA elements. For explanation of gene selection procedure, data analysis, rows, columns, coloration, and dendrogram, *see* Fig. 2. MPNST, malignant peripheral nerve sheath tumor; LIPO/MYX, myxoid liposarcoma; SCHW, schwannoma; LIPO, liposarcoma. For other tumor abbreviations, *see* Fig. 2. *CHAF1B*, chromatin assembly factor 1, subunit B; *KNSL5*, kinesin-like 5. *BUB1B*, budding uninhibited by benzimidazoles 1, β; *TOP2A*, topoisomerase II α; *CDC2*, cell division cycle 2, G1 to S and G2 to M; *PLK*, polo-like kinase; *CDC20*, cell division cycle 20, *Saccharomyces cerevisiae*, homolog; *CENPF*, centromere protein F; *CDC45L*, CDC45 cell division cycle 45, *S. cerevisiae*, homolog-like; *ZWINT*, ZW10 interactor; *CDC7L1*, CDC7 cell division cycle 7, *S. cerevisiae*, homolog-like 1; MCM6, minichromosome maintenance deficient 6; *E2F1*, E2F transcription factor 1; *TOP2A*, topoisomerase DNA II α 170 kDa; *CCNA2*, cyclin A2; *ORC1L*, origin recognition complex, subunit 1 (yeast homolog)-like.

have demonstrated a growth inhibitory effect of the tyrosine kinase inhibitor imatinib mesylate in primary cultures derived from human DFSPs *(62)*, suggesting that this drug may be of use in patients with inoperable DFSP. Indeed, a pilot study with imatinib mesylate in a patient with unresectable metastatic DFSP showed very promising results *(84a)*.

The chromosomal translocation t(X;18)(p11.2;q11.2), which results in a fusion transcript between the synovial sarcoma translocation gene *SYT* on 18p11.2 and either one of the highly homologous synovial sarcoma X-breakpoint genes *SSX1* or *SSX2* on Xp11 *(85,86)*, has been found in over 90% of synovial sarcomas *(13)*. In our series, we found relative high expression of *SSX*. The *SSX* sequence present on the array contains the 3' end of the gene and, thus, cannot discriminate between native *SSX* and the *SYT-SSX* fusion transcript. However, it is likely that we measured the *SYT-SSX* transcript.

GENE EXPRESSION PROFILES OF FIBROMATOSES AND SFTs

Fibromatoses–desmoid tumors are locally infiltrative and proliferative processes with a propensity to recur, but they do not metastasize, and malignant transformations are rare *(87)*. Most desmoid tumors have a deletion and/or inactivation of the *APC* gene *(88)*. In addition, trisomy 8 and/or 20 has been found in some desmoid tumors *(89,90)*. Histologically, these neoplastic spindle cells are surrounded by a collagenous matrix *(87)*. Indeed, we found various collagen genes, such as *COL6A1* and *COL6A2*, relatively highly expressed in these tumors (Fig. 2B4,C4). In addition, a large number of other genes, many of which generate components of the extracellular matrix, were found in this cluster.

SFT is a fibrous and myofibroblastic proliferation in which the constituent cells, at least focally, are virtually always separated by strip-like bands of collagen. The tumor cells are almost always CD34 positive (Fig. 2B7,C7) *(10)*. Several cytogenetic abnormalities have been described in SFT, yet no consistent pattern has emerged from these studies *(91,92)*. In a search for markers potentially more specific than CD34 in SFT, we identified the ectodermal neural-cortex 1 (*ENC1*) gene, among others (Fig. 2B6,C6) (West et al., manuscript in preparation). *ENC1*, originally identified as a p53-induced gene (*PIG10*) *(93)*, has been implicated in the differentiation of neural and fat cells *(94,95)*. Recently, it has also been reported to be a downstream target of the β-catenin/ T cell factor complex and to be up-regulated in colorectal cancer *(96)*. Future studies will tell whether antibodies directed against ENC1 protein can discriminate between SFTs and other tumors.

CLUSTERING PATTERNS OF A PRIMARY AND ITS METASTASIS

Especially in the case of carcinomas, it has been argued that the clustering pattern of the primary tumors was not so much driven by the gene expression profiles of the tumor cells as by the presence of normal tissue. Although this argument holds less for STTs, it is a strong counterargument that a lung metastasis of a leiomyosarcoma (STT523) clustered together with its extremity primary on the same terminal branch when using the earlier arrays (22,000 cDNA spots) (Fig. 3). The same pair clustered less tightly when using the newer arrays (42,000 cDNA spots), but this was probably due to the marginal quality of the metastasis array experiment (Fig. 2A; STT607—note the amount of noninformative spots for this case). This observation, together with similar findings in other tumor types *(60,77)*, suggest that only few genes are necessary for tumor progression and may support the cancer stem cell theory *(97)*.

Clustering of Controversial Cases

For these studies we used "classical" cases with a noncontroversial diagnosis. When initial classification of well-defined STT based on their transcriptome has been completed, it will be of interest to include cases with an ambiguous diagnosis. We expect that at least a subset of diagnostically difficult lesions would "find" their diagnosis in a branching pattern determined by the expression of a thousand genes or more rather than by the analysis of histology and a handful of IHC markers. A proof of principle study testing this concept has recently been reported for carcinomas *(98)*.

Correlation of Proliferation Gene Cluster with Mitotic Index

Most of the scientific literature today is based on the assumption that there exists a tight correlation between genotype and phenotype. Figure 3 is an example of this correlation that exists between the expression levels of genes involved in proliferation and the proliferation phenotype of a tumor as expressed by its "mitotic index." It should be kept in mind that STTs generally have a much lower proliferation rate than carcinomas.

Gene Expression Results of STTs When Compared with Other Malignancies

When comparing gene expression profiles of STTs, it is important to remember that the differences in expression levels are relative to each other. In other words, gene expression levels that appear significantly different when comparing different histological subtypes of STTs may vanish when profiles of STTs are clustered with other malignancies, like carcinomas or lymphomas (Fig. 4A,B). Obviously, when searching for new diagnostic markers, the desired discriminative value of the marker should be known in advance, and comparisons with other tissues and/or malignancies made accordingly. These precautions at least reduce the chance of raising an antibody against a gene product that ultimately appears to react with numerous tissues. For example, in Fig. 4a, the markers *GAS2*-related on chromosome 22 *(GAR22)* and argininosuccinate lyase *(ASL)* appear quite specific for SFT. However, when analyzed in the context of lymphomas, normal bladder, and a variety of carcinomas, many STT appear to have at least moderate levels of mRNA for these genes (Fig. 4b). Obviously, it depends on the relative sensitivity of immunohistochemical analysis, the quality of the antibody used, and other factors, whether this moderate level of expression would result in a positive staining reaction or whether only the highly expressing SFT would stain. Other reasons for discrepancies between differences observed at the RNA expression level that cannot be reproduced at the protein level include posttranslational modifications of the protein and slow protein turnover.

Validation of cDNA Microarray Results

Our microarray results described above have shown good correlations with known immunohistochemical markers, cytogenetic markers, and histological features of STTs. A further validation of identified markers with cDNA microarrays is expected to come from tissue microarray (TMA) studies. An extensive overview of this technique is given in Chapter 5 of this volume. Briefly, a TMA is a collection of 300–1000 different

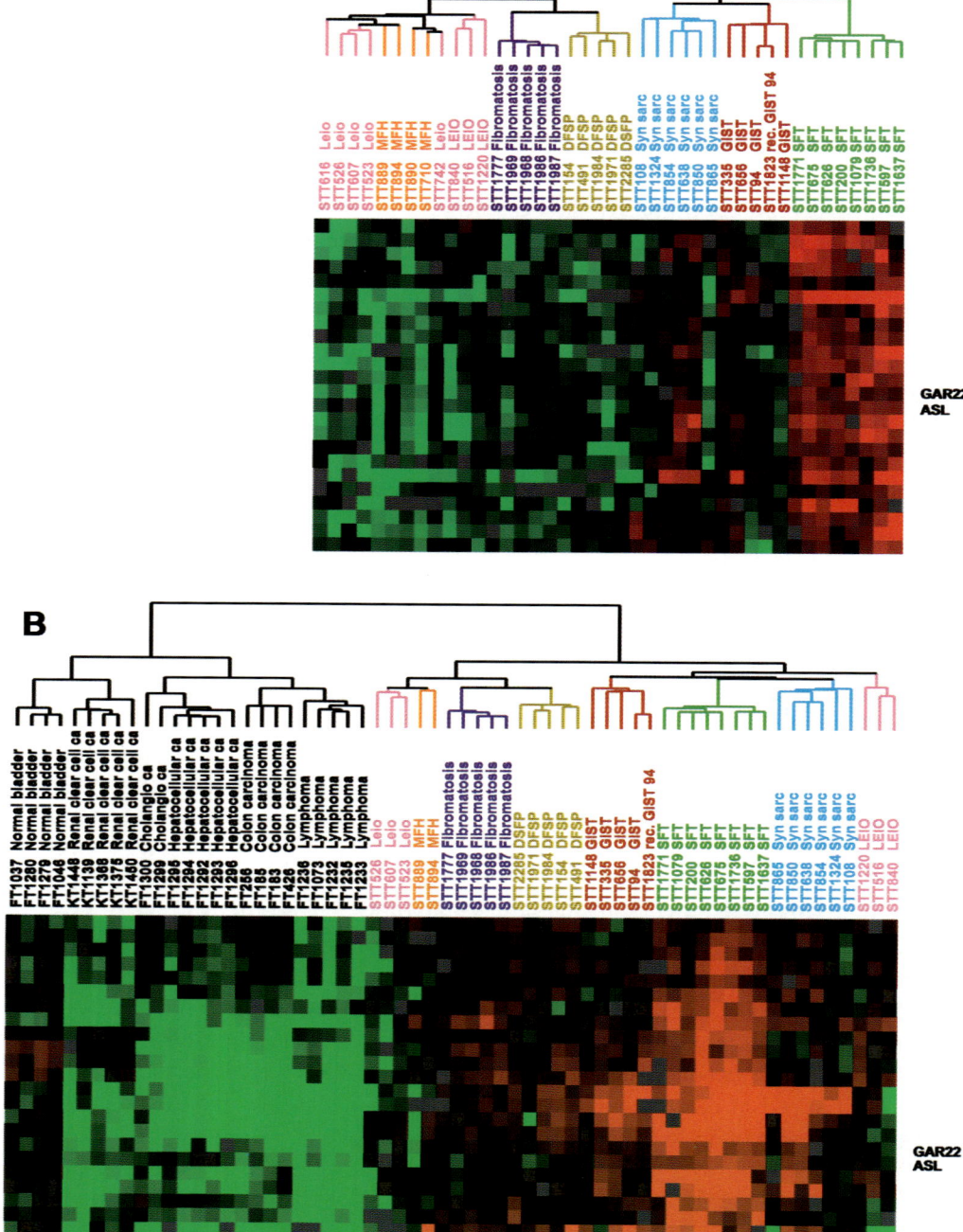

Fig. 4.

formalin-fixed paraffin-embedded tissue cores of 0.6–1.0 mm diameter, obtained from various donor blocks, and arranged in a recipient paraffin block *(6)*. Around 200 sections of 4–8 μm can be generated from this new paraffin block.

TMAs allow for the rapid examination of hundreds of specimens with new markers *(99,100)*. The advantages are obvious: (*i*) each stain is done on several hundred samples in a single experiment, hence, instead of dealing with hundreds of different glass slides, one only manages a handful; (*ii*) it is inexpensive; (*iii*) there is less interexperimental variation; and (*iv*) one uses a limited amount of tissue per case, so more tissue will be left for other experiments *(6)*. Although one tissue core stands for approx 0.3% of the currently considered representative amount of material, data obtained with two cores are 95% concordant with data generated by conventional methods *(101)*.

FUTURE PROSPECTS

In future studies, we hope to increase the number of cases analyzed by gene microarrays. We expect that this will lead not only to a better classification of STTs, but also will identify new subcategories within specific STT types. Correlations with outcome data may identify potential prognostic markers. These prognostic markers will be tested on TMAs; e.g., we have constructed a 312-core TMA containing 72 primary rhabdomyosarcoma cases and identified a new prognostic marker, besides confirming the prognostic value of Ki-67 and CD44 (Linn et al., manuscript in preparation). New potential diagnostic markers will be tested on a TMA with duplicate cores for 400 cases of STT (West et al., manuscript in preparation). We hope that our efforts, together with contributions from many other research groups in the field, will ultimately lead to better treatment options and improved survival for patients with STTs.

ACKNOWLEDGMENTS

The authors would like to thank Pat Brown and David Botstein and the members of their laboratories for providing gene arrays and for many helpful discussions. This work is also dependent on the efforts of a number of collaborators outside our institution, including John Goldblum, Torsten Nielsen, and Brian Rubin. Sabine Linn is a recipient of a Dutch Cancer Society Fellowship.

Fig. 4. *(previous page)* Example of relative aspect of gene expression measurements using cDNA microarrays. For explanation of gene selection procedure, data analysis, rows, columns, coloration, and dendrogram, *see* Fig. 2. Before hierarchical clustering, we center each gene at the geometric mean of its expression level across all samples used for that particular clustering analysis. Obviously, the geometric mean of the expression level of each gene will change when, in addition to STTs, other specimens, in this case normal bladders, lymphomas, and several carcinomas are included in the analysis (a vs b). (a) Relative gene expression levels for the genes *GAR22* and *ASL* appear high for SFT and underexpressed or average (green or black) for most other STT. *GAR22*, *GAS2*-related on chromosome 22; *ASL*, argininosuccinate lyase. (b) Relative gene expression levels for the genes *GAR22* and *ASL* are now average to high in most STT and underexpressed or low in the various non-STTs. Ca, carcinoma.

NOTE ADDED IN PROOF

Tow additional cDNA microarray studies of synovial sarcoma have recently appeared in which the profile of gene expression in this sarcoma was compared to that of several other spindle cell sarcomas *(102,103)*.

REFERENCES

 1. Weiss, S. W. and Goldblum, J. R. (2001) *Enzinger and Weiss's Soft Tissue Tumors, 4th ed.* Mosby, Philadelphia, pp. 1–19.
 2. Kempson, R. L., Fletcher, C. D. M., Evans, H. E., Hendrickson, M. R., and Sibley, R. K. (2001) Tumors of the soft tissues, in *Atlas of Tumor Pathology, 3rd series. Fascicle 30.* (Rosai, J. and Sobin, L. H., eds.), Armed Forces Institute of Pathology, Washington, D.C., pp. 1–21.
 3. Brennan, M. F., Alektiar, K. M., and Maki, R. G. (2001) Sarcomas of the soft tissue and bone, in *Cancer: Principles and Practice of Oncology.* (DeVita, V. T. J., Hellman, S., and Rosenberg, S. A., eds.), Lippincott, Williams & Wilkins, Philadelphia, pp. 1841–1891.
 4. Buchdunger, E., Cioffi, C. L., Law, N., et al. (2000) Abl protein-tyrosine kinase inhibitor STI571 inhibits in vitro signal transduction mediated by c-kit and platelet-derived growth factor receptors. *J. Pharmacol. Exp. Ther.* **295,** 139–145.
 5. van Oosterom, A. T., Judson, I., Verweij, J., et al. (2001) Safety and efficacy of imatinib (STI571) in metastatic gastrointestinal stromal tumours: a phase I study. *Lancet* **358,** 1421–1423.
 6. Kononen, J., Bubendorf, L., Kallioniemi, A., et al. (1998) Tissue microarrays for high-throughput molecular profiling of tumor specimens. *Nat. Med.* **4,** 844–847.
 7. Harris, M., Hartley, A. L., Blair, V., et al. (1991) Sarcomas in north west England: I. Histopathological peer review. *Br. J. Cancer* **64,** 315–320.
 8. Arbiser, Z. K., Folpe, A. L., and Weiss, S. W. (2001) Consultative (expert) second opinions in soft tissue pathology. Analysis of problem-prone diagnostic situations. *Am. J. Clin. Pathol.* **116,** 473–476.
 9. Harris, M. and Hartley, A. L. (1997) Value of peer review of pathology in soft tissue sarcomas. *Cancer Treat. Res.* **91,** 1–8.
10. van de Rijn, M., Lombard, C. M., and Rouse, R. V. (1994) Expression of CD34 by solitary fibrous tumors of the pleura, mediastinum, and lung. *Am. J. Surg. Pathol.* **18,** 814–820.
11. van de Rijn, M., Hendrickson, M. R., and Rouse, R. V. (1994) CD34 expression by gastrointestinal tract stromal tumors. *Hum. Pathol.* **25,** 766–771.
12. van de Rijn, M. and Rouse, R. V. (1994) CD34: a review. *Appl. Immunohistochem.* **2,** 71–80.
13. Fletcher, J. A. (1997) Cytogenetics of soft tissue tumors. *Cancer Treat. Res.* **91,** 9–29.
14. van de Rijn, M., Barr, F. G., Xiong, Q. B., Hedges, M., Shipley, J., and Fisher, C. (1999) Poorly differentiated synovial sarcoma: an analysis of clinical, pathologic, and molecular genetic features. *Am. J. Surg. Pathol.* **23,** 106–112.
15. van de Rijn, M., Barr, F. G., Collins, M. H., Xiong, Q. B., and Fisher, C. (1999) Absence of SYT-SSX fusion products in soft tissue tumors other than synovial sarcoma. *Am. J. Clin. Pathol.* **112,** 43–49.
16. Lasota, J., Jasinski, M., Debiec-Rychter, M., Szadowska, A., Limon, J., and Miettinen, M. (1998) Detection of the SYT-SSX fusion transcripts in formaldehyde-fixed, paraffin-embedded tissue: a reverse transcription polymerase chain reaction amplification assay useful in the diagnosis of synovial sarcoma. *Mod. Pathol.* **11,** 626–633.
17. Hiraga, H., Nojima, T., Abe, S., et al. (1998) Diagnosis of synovial sarcoma with the reverse transcriptase-polymerase chain reaction: analyses of 84 soft tissue and bone tumors. *Diagn. Mol. Pathol.* **7,** 102–110.
18. Guillou, L., Coindre, J., Gallagher, G., et al. (2001) Detection of the synovial sarcoma translocation t(X;18) (SYT;SSX) in paraffin-embedded tissues using reverse tran-

scriptase-polymerase chain reaction: a reliable and powerful diagnostic tool for pathologists. A molecular analysis of 221 mesenchymal tumors fixed in different fixatives. *Hum. Pathol.* **32,** 105–112.

19. Naito, N., Kawai, A., Ouchida, M., et al. (2000) A reverse transcriptase-polymerase chain reaction assay in the diagnosis of soft tissue sarcomas. *Cancer* **89,** 1992–1998.
20. Li, F. P. and Fraumeni, J. F., Jr. (1969) Soft-tissue sarcomas, breast cancer, and other neoplasms. A familial syndrome? *Ann. Intern. Med.* **71,** 747–752.
21. Malkin, D., Li, F. P., Strong, L. C., et al. (1990) Germ line p53 mutations in a familial syndrome of breast cancer, sarcomas, and other neoplasms. *Science* **250,** 1233–1238.
22. Okamoto, M., Sato, C., Kohno, Y., et al. (1990) Molecular nature of chromosome 5q loss in colorectal tumors and desmoids from patients with familial adenomatous polyposis. *Hum. Genet.* **85,** 595–599.
23. Viskochil, D., Buchberg, A. M., Xu, G., et al. (1990) Deletions and a translocation interrupt a cloned gene at the neurofibromatosis type 1 locus. *Cell* **62,** 187–192.
24. Legius, E., Marchuk, D. A., Collins, F. S., and Glover, T. W. (1993) Somatic deletion of the neurofibromatosis type 1 gene in a neurofibrosarcoma supports a tumour suppressor gene hypothesis *Nat. Genet.* **3,** 122–126.
25. Pisters, P. W., Leung, D. H., Woodruff, J., Shi, W., and Brennan, M. F. (1996) Analysis of prognostic factors in 1,041 patients with localized soft tissue sarcomas of the extremities. *J. Clin. Oncol.* **14,** 1679–1689.
26. Fleming, I. D., Cooper, J. S., Henson, D. E., et al. (1997) *AJCC Cancer Staging Manual.* 5th ed. ed. Lippincott Williams & Wilkins, Philadelphia.
27. Oliveira, A. M. and Nascimento, A. G. (2001) Grading in soft tissue tumors: principles and problems. *Skeletal Radiol.* **30,** 543–559.
28. Guillou, L., Coindre, J. M., Bonichon, F., et al. (1997) Comparative study of the National Cancer Institute and French Federation of Cancer Centers Sarcoma Group grading systems in a population of 410 adult patients with soft tissue sarcoma. *J. Clin. Oncol.* **15,** 350–362.
29. Cooper, C. S. and Cornes, P. (1997) Molecular genetics of soft tissue sarcomas. *Cancer Treat. Res.* **91,** 31–50.
30. Graadt van Roggen, J. F., Bovee, J. V., Morreau, J., and Hogendoorn, P. C. (1999) Diagnostic and prognostic implications of the unfolding molecular biology of bone and soft tissue tumours. *J. Clin. Pathol.* **52,** 481–489.
31. Drobnjak, M., Latres, E., Pollack, D., et al. (1994) Prognostic implications of p53 nuclear overexpression and high proliferation index of Ki-67 in adult soft-tissue sarcomas. *J. Natl. Cancer Inst.* **86,** 549–554.
32. Kawai, A., Woodruff, J., Healey, J. H., Brennan, M. F., Antonescu, C. R., and Ladanyi, M. (1998) SYT-SSX gene fusion as a determinant of morphology and prognosis in synovial sarcoma. *N. Engl. J. Med.* **338,** 153–160.
33. Nilsson, G., Skytting, B., Xie, Y., et al. (1999) The SYT-SSX1 variant of synovial sarcoma is associated with a high rate of tumor cell proliferation and poor clinical outcome. *Cancer Res.* **59,** 3180–3184.
34. Ladanyi, M., Antonescu, C. R., Leung, D. H., et al. (2002) Impact of SYT-SSX fusion type on the clinical behavior of synovial sarcoma: a multi-institutional retrospective study of 243 patients. *Cancer Res.* **62,** 135–140.
35. Rydholm, A., Mandahl, N., Heim, S., Kreicbergs, A., Willen, H., and Mitelman, F. (1990) Malignant fibrous histiocytomas with a 19p+ marker chromosome have increased relapse rate. *Genes Chromosom. Cancer* **2,** 296–299.
36. Wylie, J. P., O'Sullivan, B., Catton, C., and Gutierrez, E. (1999) Contemporary radiotherapy for soft tissue sarcoma. *Semin. Surg. Oncol.* **17,** 33–46.
37. Bramwell, V. H. (2001) Adjuvant chemotherapy for adult soft tissue sarcoma: is there a standard of care? *J. Clin. Oncol.* **19,** 1235–1237.
38. Casper, E. S., Gaynor, J. J., Harrison, L. B., Panicek, D. M., Hajdu, S. I., and Brennan, M. F.

(1994) Preoperative and postoperative adjuvant combination chemotherapy for adults with high grade soft tissue sarcoma. *Cancer* **73,** 1644–1651.
39. Eilber, F. C., Rosen, G., Eckardt, J., et al. (2001) Treatment-induced pathologic necrosis: a predictor of local recurrence and survival in patients receiving neoadjuvant therapy for high-grade extremity soft tissue sarcomas. *J. Clin. Oncol.* **19,** 3203–3209.
40. Lejeune, F. J., Kroon, B. B., Di Filippo, F., et al. (2001) Isolated limb perfusion: the European experience. *Surg. Oncol. Clin. N. Am.* **10,** 821–32, ix.
41. Benjamin, R. S., Rouesse, J., Bourgeois, H., and van Hoesel, Q. G. (1998) Should patients with advanced sarcomas be treated with chemotherapy? *Eur. J. Cancer* **34,** 958–965.
42. Issels, R. D., Prenninger, S. W., Nagele, A., et al. (1990) Ifosfamide plus etoposide combined with regional hyperthermia in patients with locally advanced sarcomas: a phase II study. *J. Clin. Oncol.* **8,** 1818–1829.
43. Wiedemann, G. J., d'Oleire, F., Knop, E., et al. (1994) Ifosfamide and carboplatin combined with 41.8 degrees C whole-body hyperthermia in patients with refractory sarcoma and malignant teratoma. *Cancer Res.* **54,** 5346–5350.
44. van Oosterom, A. T., Judson, I., Verweij, J., et al. (2001) Safety and efficacy of imatinib (STI571) in metastatic gastrointestinal stromal tumours: a phase I study. *Lancet* **358,** 1421–1423.
45. Joensuu, H., Roberts, P. J., Sarlomo-Rikala, M., et al. (2001) Effect of the tyrosine kinase inhibitor STI571 in a patient with a metastatic gastrointestinal stromal tumor. *N. Engl. J. Med.* **344,** 1052–1056.
46. Traxler, P., Bold, G., Buchdunger, E., et al. (2001) Tyrosine kinase inhibitors: from rational design to clinical trials. *Med. Res. Rev.* **21,** 499–512.
47. Kissin, M. W., Fisher, C., Webb, A. J., and Westbury, G. (1987) Value of fine needle aspiration cytology in the diagnosis of soft tissue tumours: a preliminary study on the excised specimen. *Br. J. Surg.* **74,** 479–480.
48. Kindblom, L. G. (1983) Light and electron microscopic examination of embedded fine-needle aspiration biopsy specimens in the preoperative diagnosis of soft tissue and bone tumors. *Cancer* **51,** 2264–2277.
49. Van Gelder, R. N., von Zastrow, M. E., Yool, A., Dement, W. C., Barchas, J. D., and Eberwine, J. H. (1990) Amplified RNA synthesized from limited quantities of heterogeneous cDNA. *Proc. Natl. Acad. Sci. USA* **87,** 1663–1667.
50. Luo, L., Salunga, R. C., Guo, H., et al. (1999) Gene expression profiles of laser-captured adjacent neuronal subtypes. *Nat. Med.* **5,** 117–122.
51. St Croix, B., Rago, C., Velculescu, V., et al. (2000) Genes expressed in human tumor endothelium. *Science* **289,** 1197–1202.
52. Sakamoto, A., Oda, Y., Adachi, T., et al. (2001) H-ras oncogene mutation in dedifferentiated liposarcoma. Polymerase chain reaction-restriction fragment length polymorphism analysis. *Am. J. Clin. Pathol.* **115,** 235–242.
53. Smith, T. A., Easley, K. A., and Goldblum, J. R. (1996) Myxoid/round cell liposarcoma of the extremities. A clinicopathologic study of 29 cases with particular attention to extent of round cell liposarcoma. *Am. J. Surg. Pathol.* **20,** 171–180.
54. Wang, J., Morimitsu, Y., Okamoto, S., et al. (2000) COL1A1-PDGFB fusion transcripts in fibrosarcomatous areas of six dermatofibrosarcomas protuberans. *J. Mol. Diagn.* **2,** 47–52.
55. Kiuru-Kuhlefelt, S., El Rifai, W., Fanburg-Smith, J., Kere, J., Miettinen, M., and Knuutila, S. (2001) Concomitant DNA copy number amplification at 17q and 22q in dermatofibrosarcoma protuberans. *Cytogenet. Cell Genet.* **92,** 192–195.
56. Greco, A., Fusetti, L., Villa, R., et al. (1998) Transforming activity of the chimeric sequence formed by the fusion of collagen gene COL1A1 and the platelet derived growth factor b-chain gene in dermatofibrosarcoma protuberans. *Oncogene* **17,** 1313–1319.

57. Nagai, M., Tanaka, S., Tsuda, M., et al. (2001) Analysis of transforming activity of human synovial sarcoma-associated chimeric protein SYT-SSX1 bound to chromatin remodeling factor hBRM/hSNF2 alpha. *Proc. Natl. Acad. Sci. USA* **98,** 3843–3848.
58. Ross, D. T., Scherf, U., Eisen, M. B., et al. (2000) Systematic variation in gene expression patterns in human cancer cell lines. *Nat. Genet.* **24,** 227–235.
59. Khan, J., Wei, J. S., Ringner, M., et al. (2001) Classification and diagnostic prediction of cancers using gene expression profiling and artificial neural networks. *Nat. Med.* **7,** 673–679.
60. Perou, C. M., Sorlie, T., Eisen, M. B., et al. (2000) Molecular portraits of human breast tumours. *Nature* **406,** 747–752.
61. Alizadeh, A. A., Eisen, M. B., Davis, R. E., et al. (2000) Distinct types of diffuse large B-cell lymphoma identified by gene expression profiling. *Nature* **403,** 503–511.
62. Sjoblom, T., Shimizu, A., O'Brien, K. P., et al. (2001) Growth inhibition of dermatofibrosarcoma protuberans tumors by the platelet-derived growth factor receptor antagonist STI571 through induction of apoptosis. *Cancer Res.* **61,** 5778–5783.
63. Greco, A., Roccato, E., Miranda, C., Cleris, L., Formelli, F., and Pierotti, M. A. (2001) Growth-inhibitory effect of STI571 on cells transformed by the COL1A1/PDGFB rearrangement. *Int. J. Cancer* **92,** 354–360.
64. Clark, E. A., Golub, T. R., Lander, E. S., and Hynes, R. O. (2000) Genomic analysis of metastasis reveals an essential role for RhoC. *Nature* **406,** 532–535.
65. Donehower, L. A., Harvey, M., Slagle, B. L., et al. (1992) Mice deficient for p53 are developmentally normal but susceptible to spontaneous tumours. *Nature* **356,** 215–221.
66. Harvey, M., McArthur, M. J., Montgomery, C. A., Jr., Butel, J. S., Bradley, A., and Donehower, L. A. (1993) Spontaneous and carcinogen-induced tumorigenesis in p53-deficient mice. *Nat. Genet.* **5,** 225–229.
67. Teitz, T., Chang, J. C., Kitamura, M., Yen, T. S., and Kan, Y. W. (1993) Rhabdomyosarcoma arising in transgenic mice harboring the beta-globin locus control region fused with simian virus 40 large T antigen gene. *Proc. Natl. Acad. Sci. USA* **90,** 2910–2914.
68. Hahn, H., Wojnowski, L., Zimmer, A. M., Hall, J., Miller, G., and Zimmer, A. (1998) Rhabdomyosarcomas and radiation hypersensitivity in a mouse model of Gorlin syndrome. *Nat. Med.* **4,** 619–622.
69. Serrano, M., Lee, H., Chin, L., Cordon-Cardo, C., Beach, D., and DePinho, R. A. (1996) Role of the INK4a locus in tumor suppression and cell mortality. *Cell* **85,** 27–37.
70. Kamijo, T., Zindy, F., Roussel, M. F., et al. (1997) Tumor suppression at the mouse INK4a locus mediated by the alternative reading frame product p19ARF. *Cell* **91,** 649–659.
71. McClatchey, A. I., Saotome, I., Mercer, K., et al. (1998) Mice heterozygous for a mutation at the Nf2 tumor suppressor locus develop a range of highly metastatic tumors. *Genes Dev.* **12,** 1121–1133.
72. Perez-Losada, J., Sanchez-Martin, M., Rodriguez-Garcia, M. A., et al. (2000) Liposarcoma initiated by FUS/TLS-CHOP: the FUS/TLS domain plays a critical role in the pathogenesis of liposarcoma. *Oncogene* **19,** 6015–6022.
73. Fletcher, C. D., Gustafson, P., Rydholm, A., Willen, H., and Akerman, M. (2001) Clinicopathologic re-evaluation of 100 malignant fibrous histiocytomas: prognostic relevance of subclassification. *J. Clin. Oncol.* **19,** 3045–3050.
74. Allander, S. V., Nupponen, N. N., Ringner, M., et al. (2001) Gastrointestinal stromal tumors with KIT mutations exhibit a remarkably homogeneous gene expression profile. *Cancer Res.* **61,** 8624–8628.
75. Nielsen, T. O., West, R. B., Linn, S. C., et al. (2002) Molecular portraits of soft tissue tumors. *Lancet* **359,** 1301–1307.
76. Eisen, M. B., Spellman, P. T., Brown, P. O., and Botstein, D. (1998) Cluster analysis and display of genome-wide expression patterns. *Proc. Natl. Acad. Sci. USA* **95,** 14,863–14,868.

77. Garber, M. E., Troyanskaya, O. G., Schluens, K., et al. (2001) Diversity of gene expression in adenocarcinoma of the lung. *Proc. Natl. Acad. Sci. USA* **98,** 13,784–13,789.
78. Kindblom, L. G., Remotti, H. E., Aldenborg, F., and Meis-Kindblom, J. M. (1998) Gastrointestinal pacemaker cell tumor (GIPACT): gastrointestinal stromal tumors show phenotypic characteristics of the interstitial cells of Cajal. *Am. J. Pathol.* **152,** 1259–1269.
79. Maeda, H., Yamagata, A., Nishikawa, S., et al. (1992) Requirement of c-kit for development of intestinal pacemaker system. *Development* **116,** 369–375.
80. Hirota, S., Isozaki, K., Moriyama, Y., et al. (1998) Gain-of-function mutations of c-kit in human gastrointestinal stromal tumors. *Science* **279,** 577–580.
81. Rubin, B. P., Singer, S., Tsao, C., et al. (2001) KIT activation is a ubiquitous feature of gastrointestinal stromal tumors. *Cancer Res.* **61,** 8118–8121.
82. El Rifai, W., Sarlomo-Rikala, M., Andersson, L. C., Knuutila, S., and Miettinen, M. (2000) DNA sequence copy number changes in gastrointestinal stromal tumors: tumor progression and prognostic significance. *Cancer Res.* **60,** 3899–3903.
83. Simon, M. P., Pedeutour, F., Sirvent, N., et al. (1997) Deregulation of the platelet-derived growth factor B-chain gene via fusion with collagen gene COL1A1 in dermatofibrosarcoma protuberans and giant-cell fibroblastoma. *Nat. Genet.* **15,** 95–98.
84. Shimizu, A., O'Brien, K. P., Sjoblom, T., et al. (1999) The dermatofibrosarcoma protuberans-associated collagen type Ialpha1/platelet-derived growth factor (PDGF) B-chain fusion gene generates a transforming protein that is processed to functional PDGF-BB. *Cancer Res.* **59,** 3719–3723.
84a. Rubin, B. P., Schuetze, S. M., Eary, J. F., et al. (2002) Molecular targeting of platelet-derived growth factor B by imatinib mesylate in a patient with metastatic dermatofibrosarcoma protuberans. *J. Clin. Oncol.* **20,** 3586–3591.
85. Clark, J., Rocques, P. J., Crew, A. J., et al. (1994) Identification of novel genes, SYT and SSX, involved in the t(X;18)(p11.2;q11.2) translocation found in human synovial sarcoma. *Nat. Genet.* **7,** 502–508.
86. Crew, A. J., Clark, J., Fisher, C., et al. (1995) Fusion of SYT to two genes, SSX1 and SSX2, encoding proteins with homology to the Kruppel-associated box in human synovial sarcoma. *EMBO J.* **14,** 2333–2340.
87. Weiss, S. W. and Goldblum, J. R. (2001) *Enzinger and Weiss's Soft Tissue Tumors, 4th ed.* Mosby, Philadelphia, pp. 309–346.
88. Miyaki, M., Konishi, M., Kikuchi-Yanoshita, R., et al. (1993) Coexistence of somatic and germ-line mutations of APC gene in desmoid tumors from patients with familial adenomatous polyposis. *Cancer Res.* **53,** 5079–5082.
89. Dal Cin, P., Sciot, R., Aly, M. S., et al. (1994) Some desmoid tumors are characterized by trisomy 8. *Genes Chromosom. Cancer* **10,** 131–135.
90. Dal Cin, P., Sciot, R., Van Damme, B., De Wever, I., and Van den Berghe, H. (1995) Trisomy 20 characterizes a second group of desmoid tumors. *Cancer Genet. Cytogenet.* **79,** 189.
91. Fletcher, C. D., Dal Cin, P., De Wever, I., et al. (1999) Correlation between clinicopathological features and karyotype in spindle cell sarcomas. A report of 130 cases from the CHAMP study group. *Am. J. Pathol.* **154,** 1841–1847.
92. Debiec-Rychter, M., De Wever, I., Hagemeijer, A., and Sciot, R. (2001) Is 4q13 a recurring breakpoint in solitary fibrous tumors? *Cancer Genet. Cytogenet.* **131,** 69–73.
93. Polyak, K., Xia, Y., Zweier, J. L., Kinzler, K. W., and Vogelstein, B. (1997) A model for p53-induced apoptosis. *Nature* **389,** 300–305.
94. Kim, T. A., Lim, J., Ota, S., et al. (1998) NRP/B, a novel nuclear matrix protein, associates with p110(RB) and is involved in neuronal differentiation. *J. Cell Biol.* **141,** 553–566.

95. Zhao, L., Gregoire, F., and Sook, S. H. (2000) Transient induction of ENC-1, a Kelch-related actin-binding protein, is required for adipocyte differentiation. *J. Biol. Chem.* **275,** 16,845–16,850.
96. Fujita, M., Furukawa, Y., Tsunoda, T., Tanaka, T., Ogawa, M., and Nakamura, Y. (2001) Up-regulation of the ectodermal-neural cortex 1 (ENC1) gene, a downstream target of the beta-catenin/T-cell factor complex, in colorectal carcinomas. *Cancer Res.* **61,** 7722–7726.
97. Reya, T., Morrison, S. J., Clarke, M. F., and Weissman, I. L. (2001) Stem cells, cancer, and cancer stem cells. *Nature* **414,** 105–111.
98. Su, A. I., Welsh, J. B., Sapinoso, L. M., et al. (2001) Molecular classification of human carcinomas by use of gene expression signatures. *Cancer Res.* **61,** 7388–7393.
99. Natkunam, Y., Warnke, R. A., Montgomery, K., Falini, B., and van de Rijn, M. (2001) Analysis of MUM1/IRF4 protein expression using tissue microarrays and immunohistochemistry. *Mod. Pathol.* **14,** 686–694.
100. Higgins, J. P., Montgomery, K., Wang, L., et al. (2003) Expression of FKBP12 in benign and malignant vascular endothelium: an immunohistochemical study on conventional sections and tissue micorarrays. *Am. J. Surg. Pathol.* **27,** 58–64.
101. Camp, R. L., Charette, L. A., and Rimm, D. L. (2000) Validation of tissue microarray technology in breast carcinoma. *Lab. Invest.* **80,** 1943–1949.
102. Allander, S. V., Illei, P. B., Chen, Y., et al. (2002) Expression profiling of synovial sarcoma by cDNA microarrays. Association of *ERBB2, IGFBP2,* and *ELF3* with epithelial differentiation. *Am. J. Pathol.* **161,** 1587–1595.
103. Nagayama, S., Katagirl, T., Tsunoda, T., et al. (2002) Genome-wide analysis of gene expression in synovial sarcomas using cDNA microarray. *Cancer Res.* **62,** 5859–5866.

18
Gene Expression Profiling in Lymphoid Malignancies

Wing C. Chan and Louis M. Staudt

INTRODUCTION

An ideal tumor classification system should be accurate, reproducible, easy to use, and above all, biologically meaningful and clinically relevant. The traditional approach has relied heavily on the morphologic features of the tumor with modifications based on correlative clinicopathologic studies. The older lymphoma classification systems discussed in the Working Formulation (1) are based on this principle but despite this simple approach, they have made significant contributions to the diagnosis and treatment of lymphoma. In the past two decades, there has been emarkable advances in our understanding of the immune system, the process of oncogenesis, and in how some key genes and genetic pathways influence the behavior of tumor cells. The more recent classification systems (2,3) attempt to incorporate our current knowledge from multiple disciplines to divide lymphomas into distinct clinicopathologic entities. However, there is clearly marked biologic heterogeneity within each of these entities, as illustrated by the significant survival differences of individuals within each type of lymphoma, when cases are segregated according to the International Prognostic Index (IPI) (4) (Fig. 1).

The biologic characteristics of a tumor are determined by the set of genetic lesions in its genome (5). These genetic lesions introduce a gene expression signature unique to the tumor cells that is distinct from their normal counterpart. The tumor gene expression signature is, therefore, a reflection of the genetic lesions present, and it can also serve as a predictor of the biologic behavior of the tumor. We may hypothesize that by studying the gene expression profile of a large series of lymphomas, it is possible to identify patterns that correlate with unique biologic and clinical behaviors. These distinctive profiles will also help us elucidate the molecular mechanisms that determine various tumor characteristics and the differences in treatment response and survival. We may ultimately be able to derive a classification system based on the molecular abnormalities present in individual tumors. This will allow optimal treatment decisions and accurate prognostication. The improved understanding of the molecular mechanisms that define the behavior of a tumor will be expected to provide new molecular targets for the development of novel therapeutic interventions.

From: *Expression Profiling of Human Tumors: Diagnostic and Research Applications*
Edited by: Marc Ladanyi and William L. Gerald © Humana Press Inc., Totowa, NJ

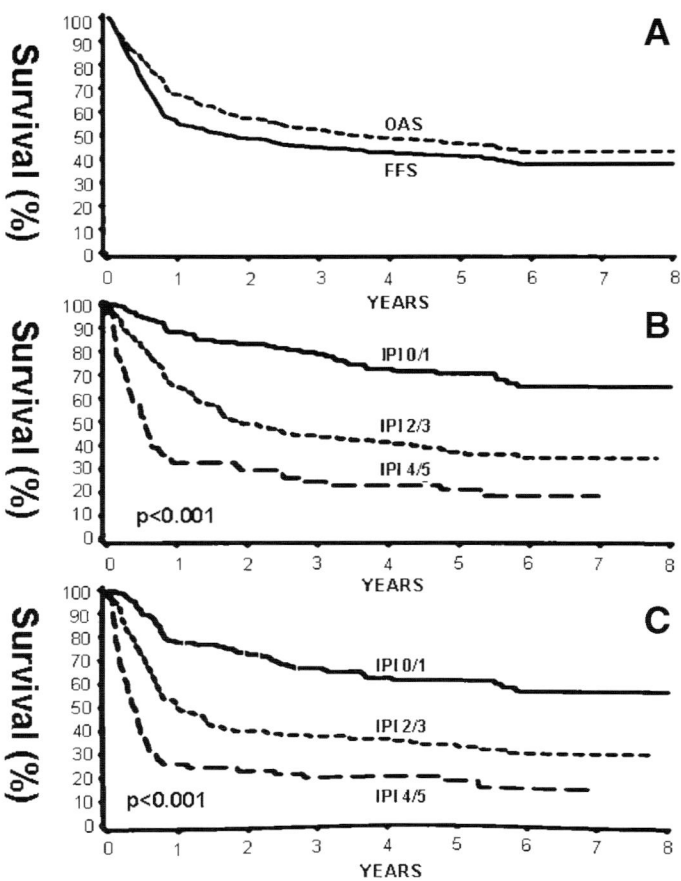

Fig. 1. Survival curves of patients with DLBCL. Panel A shows the overall (OAS) and failure-free survival (FFS) of the entire group of patients. A significant difference in survival can be seen in these patients when they are segregated according to the IPI (panels B for OAS and C for FFS) with the low clinical risk (low IPI) patients having much better survival than intermediate or high risk patients (middle and lower curves respectively). Reproduced with permission from ref. 4.

STUDY DESIGN AND OTHER TECHNICAL CONSIDERATIONS

There are a number of important practical considerations in gene expression profiling studies. These studies are frequently retrospective, and utilize tissues obtained from a number of institutions. The tissues are, therefore, unlikely to be processed in an entirely uniform fashion, with the expected introduction of some artifactual variability. Aside from the variables due to tissue processing, there is also the marked intrinsic heterogeneity in gene expression among tumors. The above considerations, together with the large number of parameters measured, make it essential to study a large number of cases in order to draw meaningful conclusions.

For solid tumors, RNA is often extracted from whole tissues, and the profile will include contributions from stromal elements and infiltrating immune cells. This will add to the complexity of data analysis, but at the same time, will also provide some important information on the host response, which is not obtained if purified tumor cells are used. The ambiguities introduced by whole tissue analysis can be mitigated to some extent by prior knowledge of the gene expression profiles of the different reactive cellular elements frequently present in malignant tumors.

Data analysis (including structure detection in the data set, model fitting, class prediction, and class discovery), is a major challenge in gene expression profiling studies *(6,7)*. It is essential to have the corresponding clinical data and, preferably, cytogenetic and molecular genetic data to help analyze the results obtained. A thorough pathologic examination of all the samples, with appropriate immunophenotypic analysis, is extremely important to ensure that the diagnosis is correct and the samples are adequate. Unique pathologic and immunologic features should be recorded in a format that can be used for future computational analysis. It is also useful to have profiles of different subpopulations of normal B cells for comparison with the profiles from the tumor populations. In addition, a large number of cell lines have been derived from a variety of lymphomas, with different cytogenetic and genetic abnormalities *(8)*. Gene expression profiles can be obtained from these cell lines, and they can serve as useful reagents to assist in the interpretation of data and the discovery of unique gene expression profiles in association with different tumor characteristics and genetic abnormalities.

Validation of the experimental data, analysis, and conclusions is an extremely important aspect of gene expression profiling studies. Validation of the expression of selected genes, which appear to be clinically and biologically significant from initial analysis, can be achieved by a number of methods. At the mRNA expression level, quantitative reverse transcription polymerase chain reaction (PCR) *(9–12)* can be applied to an aliquot of the RNA sample used for microarray experiments. To confirm expression in tumor cells, RNA extracted from microdissected frozen sections of tumor blocks from the same cases can be assayed. Alternatively, *in situ* hybridization *(13,14)* may be performed for tissue localization and a semiquantitative assessment of the expression level. As mRNA expression cannot be equated to protein expression level, it is of interest and informative to investigate the expression of the corresponding protein. If a suitable antibody is available, an immunohistochemical assay *(15,16)* on corresponding tissue sections of the tumor studied can be performed. Other techniques, such as western blot or assays for activities of specific enzymes, may be employed *(17)*. However, to determine tissue localization by the latter techniques will require some form of cell isolation procedure, and the amount of purified samples that can be obtained vs the sensitivity of the technique becomes an important consideration.

The validity of data analysis can be assessed by computational–statistical methods *(18)* and by examining the conclusions drawn using independent clinical and biologic parameters. For example, presumptive new tumor classes can be examined for distinct morphologic characteristics, clinical behavior, cytogenetic associations, and biological features. Some examples of this type of biologic validation of gene expression profiling studies will be described in later sections. Valid scientific observations should be reproducible, which means that the study of another series of cases should provide similar results, not only in the same laboratory, but also in different laboratories using the same or different platforms of analysis. To compare experimental data from across laboratories demands that primary data be accessible and readily translatable–standardized. The experimental methods and the analytical approach should be presented in sufficient details for independent analysis by other laboratories. Since gene expression profiling is still a new approach in studying cancer biology, open exchange of information and techniques among laboratories is important for validating this approach and advancing the field *(16,19)*.

MICROARRAY ANALYSIS OF B CELL LYMPHOPROLIFERATIVE DISORDERS

Gene Expression Profiling Studies on Diffuse Large B Cell Lymphoma

Since a gene expression profile is limited by the genes on the microarray, it is important to include as many relevant genes as possible. The Lymphochip, utilized by Alizedah and coworkers *(20)*, is a cDNA microarray specifically designed for the study of lymphoproliferative disorders, since it includes many genes from a normalized germinal center B cell cDNA library and cDNA libraries from a number of different lymphoid malignancies. The Lymphochip also contains a curated set of genes known to be important in lymphocyte biology and oncogenesis. About one-quarter of the genes that are present are duplicated on the Lymphochip to provide quality control for the uniformity of the spots across the array and the hybridization process. The reference standard mRNA *(19)* consists of a mixture of mRNA from nine different lymphoma cell lines and is used in all experiments. The expression level of each mRNA species in the tumor sample is expressed as a ratio to the corresponding mRNA in the reference standard, thus allowing gene expression in different tumor samples to be compared with each other.

The initial study using the Lymphochip included 42 tumor biopsies of diffuse large B cell lymphoma (DLBCL), 9 samples of follicular lymphoma (FL), 11 samples of chronic lymphocytic leukemia (CLL), as well as representative normal or activated T and B lymphocytes, and reactive lymphoid tissue *(19)*.

Unsupervised hierarchical clustering *(21)* divided the lymphoproliferative disorders into three main clusters corresponding largely to DLBCL, FL, and CLL. It should be noted that DLBCL differed in several ways from FL and CLL, which may influence the clustering results. The DLBCL mRNA was extracted from frozen tissue samples containing stromal elements and infiltrating T cells and macrophages, whereas most of the FL and all CLL samples were purified CD19 positive cells. A gene expression signature of cellular proliferation was highly expressed in DLBCL, since there was a high proliferative fraction in these tumors in contrast to FL and CLL. The differential and coordinate expression of large sets of genes, such as the proliferation signature and the lymph node stromal gene signature, can contribute significantly to the way tumors are segregated by hierarchical clustering or other pattern recognition algorithms. Predominant gene expression signatures may obscure more subtle similarities or differences among cases, either within the same tumor type or between tumor groups. Care must be taken to consider the competing effects of different gene expression signatures on the clustering of tumors by gene expression profiling.

To examine whether the DLBCL can be further divided into subgroups, the cases were examined for expression of genes defining the germinal center (GC) B cell signature. Two subgroups were apparent with one subgroup expressing genes in the normal GC-B cell signature, whereas the other subgroup express these genes at low levels (Fig. 2).

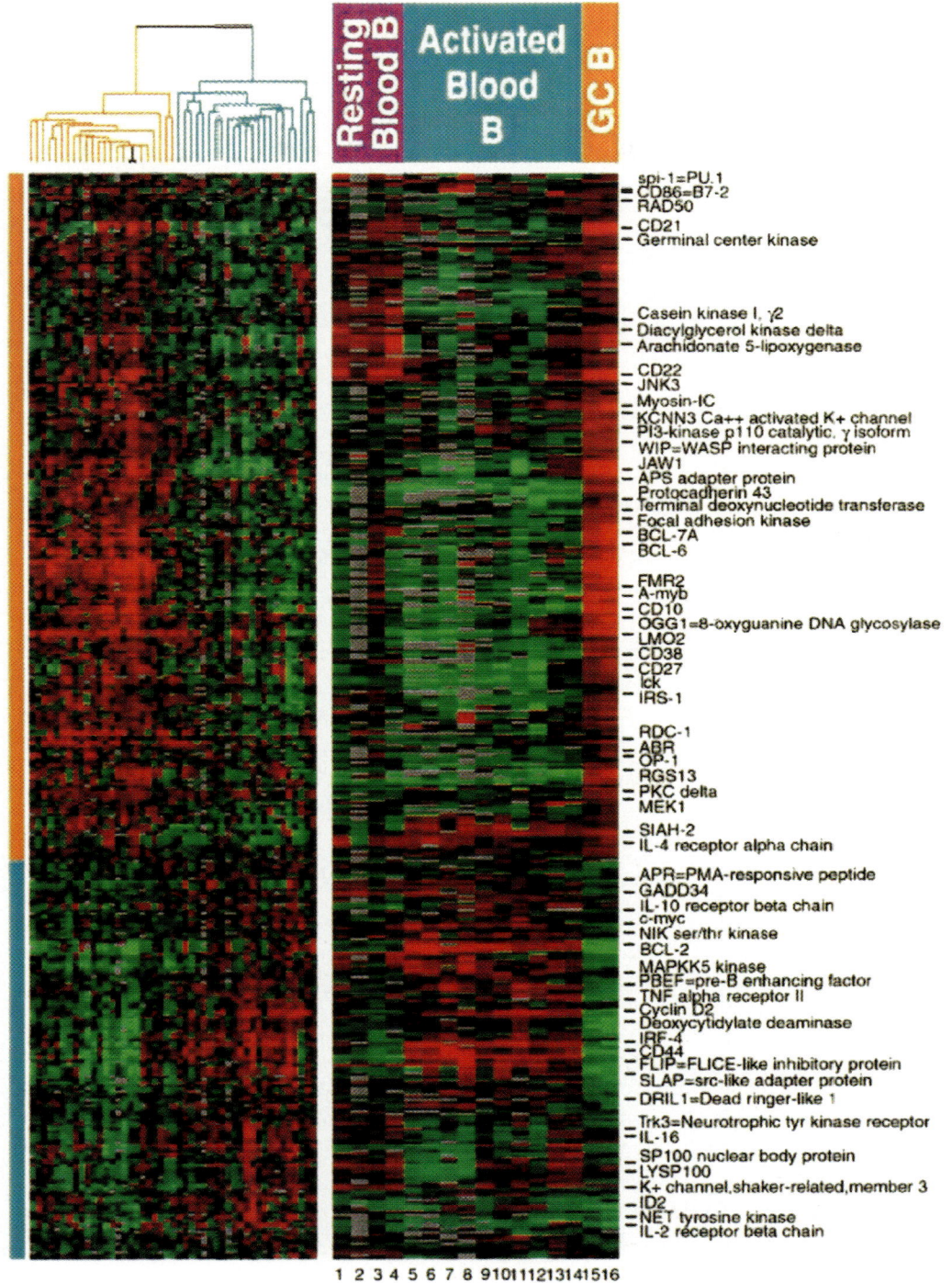

Fig. 2. DLBCL can be divided into two subgroups by gene expression profiling: The GCB group (orange dendrogram) with gene expression profile resembling normal GC-B cells, and the AB group with a profile resembling peripheral blood B cells activated by mitogenic stimuli. Reproduced with permission from ref. *19*.

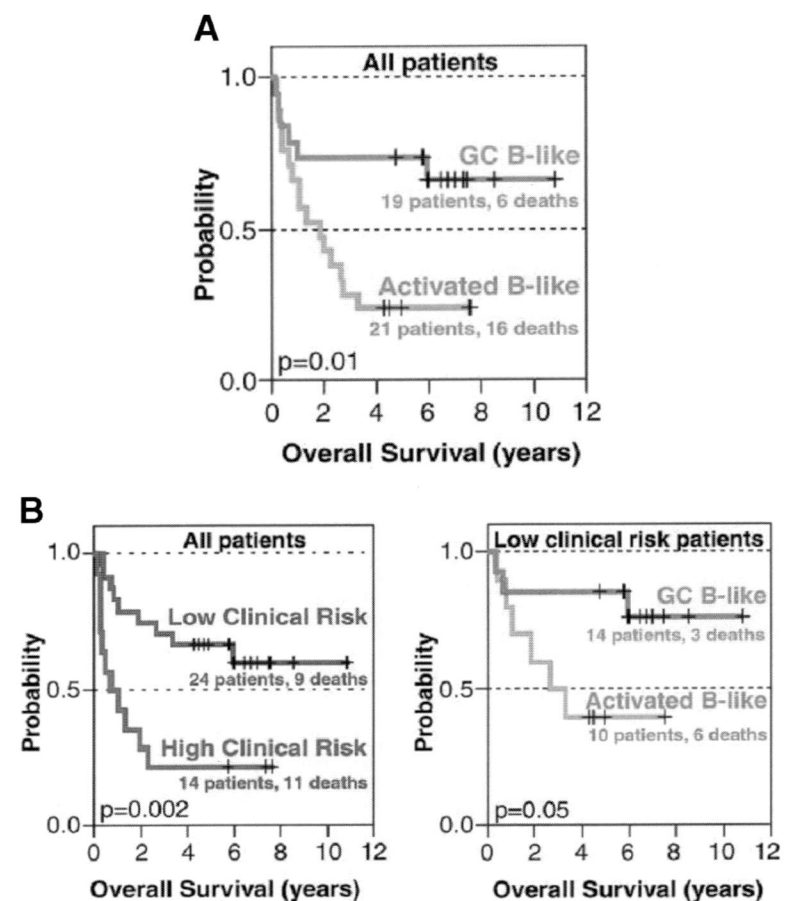

Fig. 3. Patients in the GCB group have significantly better survival than those in the ABC Group (A). This survival difference is observed even with the subset of patients in the low clinical risk category. Reproduced with permission from ref. *19*.

In contrast, many of the genes that are induced during mitogenic activation of peripheral blood (PB) B cells were selectively expressed in this latter group of DLBCL. Therefore, DLBCL seems to be divisible into at least two major subgroups: one with a gene expression profile similar to normal GC cells, termed GC-B cell-like (GCB) DLBCL, and the other with an expression profile similar to activated PB-B cells, termed activated-B cell-like (ABC) DLBCL.

Evidence in Support of the Existence of Two Subgroups of DLBCL

The overall survival of these two subgroups of DLBCL was compared, and the GCB subgroup was found to have significantly better survival. This survival advantage was seen even when patients with low IPI were analyzed separately (Fig. 3). Two biologically different entities do not necessarily have to have different clinical survival rates, but when significant survival differences are observed, it lends credence to the distinction.

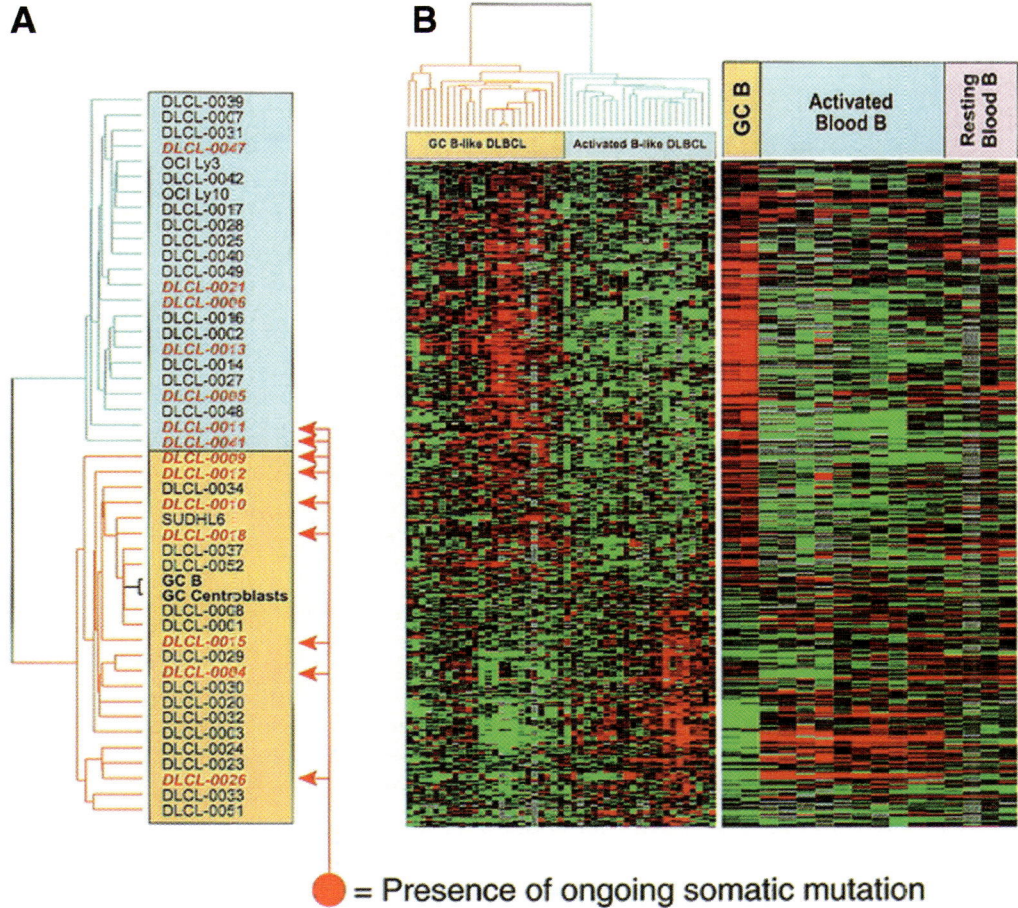

● = Presence of ongoing somatic mutation

Fig. 4. DLBCLs in the GCB group have ongoing somatic hypermutation of their V_H genes. All seven cases in the GCB group show ongoing somatic hypermutation whereas, only two out of seven in the AB group do. Reproduced with permission from ref. 22.

One of the cardinal features of GC-B cells is the occurrence of ongoing somatic hypermutations in their rearranged immunoglobulin heavy chain variable region (V_H) genes. If GCB DLBCLs have an active GC-B cell gene expression program, it is expected that they would have ongoing somatic mutation of their rearranged V_H genes. This hypothesis was explored by Lossos and coworkers (22), who examined seven cases of the GCB DLBCL and seven cases of ABC DLBCL. All seven GCB DLBCLs had ongoing V_H gene hypermutation, whereas only two out of seven ABC DLBCLs showed this characteristic (Fig. 4). The two ABC DLBCL cases with V_H gene mutations had a lower frequency of ongoing mutation than observed in GCB DLBCLs.

The t(14;18) (q32;q21) involving the *bcl-2* gene is a hallmark of FL and is believed to be the initiating event in its pathogenesis (23,24). This translocation is also present in about 20% of *de novo* DLBCL (25,26). It is possible that *bcl-2* translocation may also serve as an initiating event in this group of DLBCL, and if this is true, this group of tumors may also originate from GC-B cells and may retain the GCB expression profile

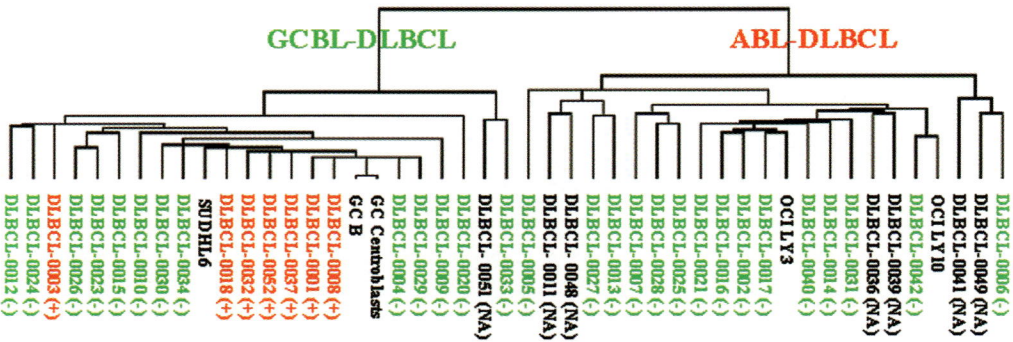

Fig. 5. Bcl-2 translocation detected only in the GCB group. *Bcl-2* translocation to the IgH locus was examined by FISH on interphase nuclei (35 out of 42 cases were studied). Cases with (–) are negative for the translocation, while cases with (+) are positive, and all positive cases are within the GCB group. Six of the seven positive cases cluster tightly together and with normal GC-B cells as well as a cell line with *bcl-2* translocation (SUDHL-6), indicating a closely related gene expression profile. Reproduced with permission from ref. *15*.

as a fully developed DLBCL. Thirty-five cases of DLBCL with known gene expression profiles were examined for t(14;18) by fluorescence *in situ* hybridization (FISH) *(15)*. Seven cases were found to have the translocation, and all seven had the GCB profile (Fig. 5). It is interesting that six out of seven of the cases clustered tightly with normal GC-B cells and a t(14;18)+ cell line (SU-DHL-6), indicating closely related gene expression profiles. The exclusive presence of t(14;18) in GCB DLBCLs strongly supports the view that the different DLBCL gene expression subgroups are pathogenetically distinct diseases.

Another pathogenetic distinction between the DLBCL subgroups involves the nuclear factor (NF)-κB signaling pathway. NF-κB refers to a family of transcription factors that participate in a variety of immune and inflammatory responses, and they have potent anti-apoptotic effects. NF-κB is present in a latent form in the cytoplasm of many cells as a complex with an inhibitory protein, IkBα. A variety of cell surface receptors can activate a kinase complex termed IkBα kinase (IKK) that phosphorylates IkB, leading to its ubiquitination and degradation in the proteosome. When IkBα is degraded, NF-κB can travel to the nucleus and transcriptionally activate a set of target genes that carry out its biological functions. ABC DLBCLs were noted to express several of these NF-κB target genes highly, and this was not a feature of GCB DLBCLs *(27)*. Cell line models of ABC DLBCL also expressed these target genes and had constitutively nuclear NF-κB secondary to constitutive activity of IKK. Importantly, interference with the NF-κB pathway in ABC DLBCL cell lines induced cell death, whereas NF-κB inhibition had no effect on GCB DLBCL cell lines. These results highlight the pathogenetic differences between ABC and GCB DLBCLs and validate the NF-κB pathway as a new molecular target for therapeutic intervention in the subgroup of DLBCL, which is relatively refractory to current anthracycline-based combination chemotherapy.

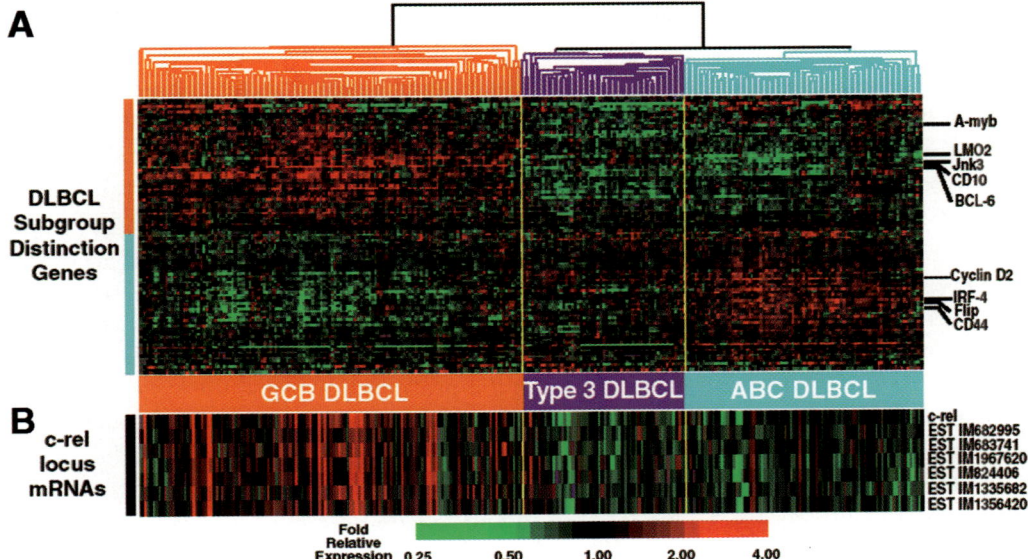

Fig. 6. Gene expression profiling of 274 cases of DLBCL confirms the presence of a GCB subgroup (orange dendrogram) and an AB subgroup (blue dendrogram). A smaller subgroup, type 3 (purple dendrogram) is also identified and characterized by the low expression of genes generally overexpressed in the GCB and ABC groups. Panel B shows the overexpression of *c-rel* mainly in cases of the GCB group, and *c-rel* gene amplification was detected in 18 cases with all of them in the GC-B group also. A group of ESTs are coordinately expressed with c-rel and correspond to a group of genes close to the c-rel locus. Reproduced with permission from ref. *28*.

Validation by Studying a New Larger Series of DLBCL Cases

A large consortium of collaborating institutions, termed the Lymphoma–Leukemia Molecular Profiling Project (LLMPP) has just completed a new study of 274 cases of DLBCL and confirmed the existence of the GCB and ABC DLBCL subgroups *(28)* (Fig. 6). Clinically, a significantly different overall survival was again observed between these two groups. A third subgroup (Type 3), which did not highly express genes characteristic of either GCB or ABC DLBCL, was also delineated, and this subgroup had an overall survival rate similar to that of ABC DLBCL.

A survey of these LLMPP cases with PCR for the t(14;18) showed that all positive cases were confined to GCB DLBCL, confirming the previous FISH analysis on a smaller number of cases *(15)*. Another interesting observation was that *c-rel* gene overexpression and genomic amplification was found exclusively in the GCB DLBCL subgroup *(28)* (Fig. 6B). This finding again shows the association of an independent biologic variable, the amplification of *c-rel*, with a distinct subgroup of DLBCL defined by gene expression profiling.

Identification of Prognostic Markers

While there is good evidence supporting the notion that the DLBCL gene expression subgroups represent pathogenetically and clinically distinct diseases, there remains substantial heterogeneity in clinical outcome within each subgroup *(19)*. Additional genes, which have significant influence on tumor biology and patient survival and are not accounted for by the DLBCL subgroup distinction, must, therefore, exist. It is, therefore, an essential next step to identify individual genes or gene expression signatures that can segregate functionally and clinically distinct groups within the already defined DLBCL subgroups. Response to treatment and survival can be used to guide this discovery process, looking for genes with expression patterns that correlate with favorable or poor survival. This type of "supervised" analysis has been conducted on the LLMPP DLBCL cases, and several recurrent biological themes were observed among the genes that predicted clinical outcome (28). As expected from previous work, genes in the GC-B cell signature were associated with good prognosis, and this entire signature could be represented by three genes after data reduction. On the other hand, the overexpression of many of the genes in the proliferation signature was associated with poor prognosis, and this signature could also be represented by three genes. Furthermore, many other genes that are predictive of favorable outcome could be classified into two gene expression signatures involving the host immune response, the major histocompatibility complex (MHC) class II and reactive lymph node signatures. Only one other gene (bmp-6) outside of these four signatures can significantly improve the predictor when added to the model. These genes could be combined into a multivariate gene expression outcome predictor that could be useful clinically in the management of DLBCL patients (28) (Table 1).

Shipp and colleagues (16), performed an analysis on 58 cases of DLBCL using an oligonucleotide microarray containing 6817 genes. They used the gene expression data and the available clinical data to develop an outcome predictor through a supervised learning approach. After analysis with multiple cross-validation loops, they decided on a 13-gene predictor of clinical outcome. Three of the genes in the predictors were present on the Lymphochip microarray, and the ability of these genes to predict outcome was assessed using the published data set of Alizadeh et al. (19). The expression of these genes correlated with outcome in this previous series of patients (neuron-derived orphan receptor [NOR] 1, $p = 0.05$; phosphodiesterase [PDE]4B, $p = 0.07$; and protein kinase C [PKC]-β2 isoform, $p = 0.04$), demonstrating the reproducibility of gene expression profiling in identifying clinical prognostic markers. Interestingly, two of these genes that predicted poor outcome, PKC-β2 and PDE4B, are highly expressed in ABC DLBCL relative to GCB DLBCL (19), and thus, their predictive power may reflect the different clinical outcomes of these two DLBCL subgroups.

The known functions of some of the genes in these predictors suggest the molecular mechanisms that are responsible for their influence on patient survival. Some of these genes are involved in the regulation of cellular proliferation, or the determination of B cell responses and cell fate on receptor signaling, and hence, may determine tumor cell growth, survival, and the susceptibility to apoptosis after the administration of chemotherapy. Other genes may be an indicator of the immune interaction between the host and the tumor (16,28). The mechanisms of action of the some of the remaining genes are not obvious. Further studies should aim at precisely defining or confirming the mechanisms by which each of the molecular markers influences survival to gain a better understanding on how to modify these processes to improve the treatment and survival of patients.

Table 1
Predictors for Survival in DLBCL

Signature	Representative genes in outcome predictor	Outcome prediction
GC-B cell	BCL-6	Good
	CCAT-1/Centerin	
	GCAT2/HGAL	
MHC class II	DP-α	Good
	DQ-α	
	DR-α	
	DR-β	
Lymph node	α-Actinin	Good
	Collagen type III α 1	
	Connective tissue growth factor	
	Fibronectin	
	KIAA0233	
	Plasminogen activator, urokinase	
Proliferation	c-myc	Bad
	E21G3	
	Nucleophosmin/nucleoplasmin 3	
Other	BMP6	Bad

STUDIES ON CHRONIC LYMPHOCYTIC LEUKEMIA

Though CLL is the most prevalent leukemia, very little is known about the molecular pathogenesis of this disorder. Recent examination of immunoglobulin (Ig) genes in CLL cells revealed two forms of the disease, one with somatic mutations of Ig genes (Ig-mutated CLL) and one with Ig genes that are germ line in sequence (Ig-nonmutated CLL). Ig-nonmutated CLL patients have a more aggressive clinical course, whereas Ig-mutated CLL patients often require late or no treatment *(29,30)*. These observations suggested that CLL might encompass two different diseases.

This possibility was tested by gene expression profiling experiments, which revealed that all CLL cases shared expression of a common set of CLL "signature" genes, which distinguished CLL from other normal and malignant cells *(31,32)*. Notable among these signature genes are the high expression of *Wnt-3* and *Ror-1*, which may participate in cellular proliferation. The overexpression of *EPAC* and *CDC25* may activate the *Raf/ERK* pathway, while up-regulation of *bcl-2* and down-regulation of a number of proapoptotic genes may contribute to the apoptosis-resistant phenotype. This observation suggests that Ig-unmutated and Ig-mutated CLL share a common oncogenic mechanism and/or have a common cell of origin. However, a directed search revealed a relatively small set of genes (<160 genes) that distinguished the two forms of CLL *(31)*. A selected subset of these genes could be combined to create a predictor of the CLL subtype distinction that correctly assigned 25 of 27 *(31)*. The most differentially expressed gene between the CLL subtypes was *ZAP70*, and this single gene alone or in combination with one or two additional genes could be used as the basis of a simple diagnostic test for this clinically important distinction between Ig-nonmutated and

Ig-mutated CLL *(31)*. A 23-gene classifier has also been proposed by Klein and colleagues *(32)*, and using this classifier, they were able to correctly predict 12 of 14 cases regarding their mutational status.

GENE DISCOVERY

Many of the elements on the Lymphochip microarray represent genes with unknown function. These genes were identified as expressed sequence tags (ESTs) during high-throughput sequencing of lymphoid cDNA libraries. When the expression pattern of a gene of unknown function correlates closely with the expression profile of genes of known function, it is reasonable to hypothesize that the novel gene may share functional properties with the known genes. In this way, gene expression profiling may provide insight into the function of some of these novel genes.

A number of genes are differentially expressed between GCB and ABC DLBCLs with high statistical significance (Table 2). Many of the differentially expressed genes that are highly expressed in GCB DLBCL are novel genes from the GC-B cell library and may well be important in the physiology of normal GC-B cells. For example, one of these novel genes corresponds to a recently cloned gene, termed centerin, from a differential display experiment *(33)*. Centerin is a member of the SERPIN gene family located on chromosome 14q32. Several novel genes are expressed in very similar patterns to centerin and may be serving similar or related functions in GC-B cells. It will be very interesting to characterize this group of genes further, as they may provide insights into GC differentiation and function, and into the biology of lymphomas that arise from GC-B cells.

Gene discovery can also be guided by genetic data on the assumption that certain genetic abnormalities may be associated with a unique and identifiable gene expression profile. Hence, a correlation between a genetic abnormality and a gene expression profile may reveal the functional consequence of the genetic lesion, and conversely, in some instances, the unique profile may help to identify the candidate gene involved in the abnormal genetic locus. Cytogenetic data are, therefore, a very important source of information. Cytogenetic data may be enriched by one of the multicolor karyotyping techniques such as M-FISH or spectral karyotyping (SKY) *(34)*. Unfortunately, generally only a relatively minor fraction of cases has the requisite materials for these studies. Comparative genomic hybridization (CGH) is a useful method to obtain amplification and deletion data over the entire genome *(34,35)*. Since this technique does not require tumor metaphase preparation, it can be performed on all cases with adequately represented tumor tissue and sufficiently preserved DNA. Newer CGH techniques use cDNA *(36)* or bacterial artificial chromosomes (BAC) *(37,38)* microarrays as indicators of hybridization instead of a normal metaphase spread. The new platforms afford much higher resolution and promise to enrich our knowledge of genomic changes in tumors that can be correlated with tumor gene expression profiles. There are specific genetic abnormalities that are known to have a significant influence on patient survival. The status of these genes may be assessed by appropriate assays, preferably in a high-throughput format. Examples include: loss of heterozygosity (LOH). Mutation of p53; LOH methylation of p16 and p15 and *c-myc* translocation. This additional information may be added to correlative studies aimed at further elucidating the functional changes associated with the genetic abnormalities and at predicting clinical outcome.

Table 2
Examples of Differentially Expressed Genes in Subtypes of DLBCL

Up-regulated in GCB subtype

Unknown UG Hs. 169565; Clone=825217
Unknown UG Hs. 120716; Clone=1334260
Unknown UG Hs. 224323; Clone=1338448
Unknown UG Hs. 136345; Clone=746300
JAW1; Clone=815539
A-myb; Clone=1367994
Unknown UG Hs. 208410; Clone=135036
Unknown UG Hs. 49614; Clone=814622

Up-regulated in ABC subtype

Unknown UG Hs. 169081; Clone=1355435
Deoxycytidylate deaminase; Clone1302032
T-cell protein-tyropsine phosphatase; Clone=665903
Potassium voltage-gated channel; Clone=1337856
Zinc finger protein 42 MZF-1; Clone=490387
T-cell protein-tyrosine phosphatase; Clone=740402
Cyclin D2; Clone=1357360
MCL1; Clone=711870

HOST INTERACTION WITH LYMPHOMA

Some of the gene expression signatures that predict clinical outcome, and some of the clinical parameters in the IPI suggest an important interplay between the host and the tumor. Tumor invasion requires the breakdown of the surrounding tissue and the generation of new tissue stroma and blood vessels. Matrix metalloproteinases (MMP) have been implicated in tumor invasion and metastasis (39,40–42). In DLBCL, MMP-2, MMP-9, and tissue inhibitor of metalloproteinase (TIMP)-3 are generally expressed at a high level, whereas MMP-11, MMP-12, and TIMP-2 are present at a low level. The expression of fibronectin and osteonectin are highly correlated with the expression MMP-2 and MMP-9 suggestive that these MMPs are likely to be important in tissue remodeling in DLBCL. Immunohistochemical studies demonstrated that MMP-2 was present in macrophages and vascular endothelial cells, but not DLBCL cells, suggesting that macrophages recruited to the tumor are the source of MMP-2 and they may play an important role in tissue remodeling. The cellular origin of the TIMPs and MMPs and their interactions and roles in tumor invasion and metastasis of DLBCL merits further investigations.

A T cell infiltrate is always present in DLBCL, but the extent of infiltration is very variable and is correlated with the CD3 delta mRNA level. The expression of certain groups of genes tend to be correlated with the levels of CD3-δ transcript. Genes that are overexpressed in T cell-rich cases include many cytokines, cytokine receptors, MHC class II molecules, adhesion molecules, and molecules associated with T and NK cell activation. These results are suggestive of the presence of an activated T cell population in the DLBCL. This activated population is quantitatively and likely qualitatively variable in different cases. How the T cell response alters the biology of the tumor and survival of the patient and what determines the quantity and quality of this T cell infiltrate are interesting topics for future studies.

PERSPECTIVES

Gene expression profiling studies of malignant diseases were started only a few years ago, but the results obtained in this short period of time have clearly demonstrated the potentials of this approach to identify clinically relevant subgroups of lymphoproliferative disorders. The molecular mechanisms responsible for these differences are beginning to be unraveled, and it is highly likely that some of the genes and/or genetic pathways involved will be suitable targets for future therapeutic intervention. Currently, the options for treating malignant lymphomas have expanded beyond the application of multi-agent chemotherapy, and have included monoclonal antibodies, antisense nucleic acids, and antagonists to several molecular targets. The options are still quite limited, but hopefully, this situation will begin to improve as additional important molecular targets are identified by gene expression profiling and other novel approaches. Studies on DLBCL have shown that a relatively small panel of genes can serve as a useful predictor of survival. It is a good possibility that this will be true for other types of lymphomas, and a diagnostic panel containing all the relevant molecular markers will be defined in the near future. This panel may become available in a microarray or other formats. Every tumor can be profiled at the time of the diagnostic biopsy for prognostication and treatment decisions. As the molecular mechanisms that determine the biology of the tumor and clinical survival of the patients are defined, novel molecular targets for therapeutic intervention will be identified. Comprehensive molecular diagnostics of lymphoma and individualized therapy based on molecular lesions may become the standard of care with marked improvement in survival and reduction in treatment-associated toxicity and complications.

REFERENCES

1. Non-Hodgkin's lymphoma pathologic classification project: National Cancer Institute sponsored study of classifications of non-Hodgkin's lymphomas: summary and description of a Working Formulation for clinical usage. (1982) *Cancer* **49,** 2112–2135.
2. Harris, N. L., Jaffe, E. S., Stein, H., et al. (1994) A revised European-American classification of lymphoid neoplasms: a proposal from the International Lymphoma Study Group. *Blood* **84,** 1361–1392.
3. Jaffe, E. S., Harris, N. L., Stein, H., and Vardiman, J. W. (2001) *WHO Classification of Tumours; Pathology and Genetics of Tumours of Haematopoietic and Lymphoid Tissues.* IRCA Press, Lyon.
4. Armitage, J. O. and Weisenburger, D. D. (1998) New approach to classifying non-Hodgkin's lymphomas: clinical features of the major histologic subtypes. Non-Hodgkin's Lymphoma Classification Project. *J. Clin. Oncol.* **16,** 2780–2795.
5. Pearson, P. L. and Van der Luijt, R. B. (1998) The genetic analysis of cancer. *J. Intern. Med.* **243,** 413–417.
6. Ermolaeva, O., Rastogi, M., Pruitt, K. D., et al. (1998) Data management and analysis for gene expression arrays. *Nat. Genet.* **20,** 19–23.
7. Sherlock, G. (2000) Analysis of large-scale gene expression data. *Curr. Opin. Immunol.* **12,** 201–205.

8. Drexler, H. D. (2001) *The Leukemia and Lymphoma Cell Line-Facts Book*. Academic Press, London.
9. Cross, N. C. (1995) Quantitative PCR techniques and applications. *Br. J. Haematol.* **89**, 693–697.
10. Orlando, C., Pinzani, P., and Pazzagli, M. (1998) Developments in quantitative PCR. *Clin. Chem. Lab. Med.* **36**, 255–269.
11. Gerard, C. J., Olsson, K., Ramanathan, R., Reading, C., and Hanania, E. G. (1998) Improved quantitation of minimal residual disease in multiple myeloma using real-time polymerase chain reaction and plasmid-DNA complementarity determining region III standards. *Cancer Res.* **58**, 3957–3964.
12. Luthra, R., McBride, J. A., Cabanillas, F., and Sarris, A. (1998) Novel 5' exonuclease-based real-time PCR assay for the detection of t(14;18)(q32;q21) in patients with follicular lymphoma. *Am. J. Pathol.* **153**, 63–68.
13. Zaidi, A. U., Enomoto, H., Milbrandt, J., and Roth, K. A. (2000) Dual fluorescent in situ hybridization and immunohistochemical detection with tyramide signal amplification. *J. Histochem. Cytochem.* **48**, 1369–1375.
14. Nuovo, G. J. (1998) In situ localization of PCR-amplified DNA and cDNA. *Mol. Biotechnol.* **10**, 49–62.
15. Huang, J. Z., Sanger, W. G., Greiner, T. C., et al. (2002) The t(14;18) defines a unique subset of diffuse large B-cell lymphoma with a germinal center B-cell gene expression profile. *Blood* **99**, 2285–2290.
16. Shipp, M. A., Ross, K. N., Tamayo, P., et al. (2002) Diffuse large B-cell lymphoma outcome prediction by gene-expression profiling and supervised machine learning. *Nat. Med.* **8**, 68–74.
17. Emmert-Buck, M. R., Roth, M. J., Zhuang, Z., et al. (1994) Increased gelatinase A (MMP-2) and cathepsin B activity in invasive tumor regions of human colon cancer samples. *Am. J. Pathol.* **145**, 1285–1290.
18. Siedow, J. N. (2001) Making sense of microarrays. *Genome Biol.* **2**, 4003.1–4003.2.
19. Alizadeh, A. A., Eisen, M. B., Davis, R. E., et al. (2000) Distinct types of diffuse large B-cell lymphoma identified by gene expression profiling. *Nature* **403**, 503–511.
20. Alizadeh, A., Eisen, M., Davis, R. E., et al. (1999) The lymphochip: a specialized cDNA microarray for the genomic-scale analysis of gene expression in normal and malignant lymphocytes. *Cold Spring Harb. Symp. Quant. Biol.* **64**, 71–78.
21. Eisen, M. B., Spellman, P. T., Brown, P. O., and Botstein, D. (1998) Cluster analysis and display of genome-wide expression patterns. *Proc. Natl. Acad. Sci. USA* **95**, 14,863–14,868.
22. Lossos, I. S., Alizadeh, A. A., Eisen, M. B., et al. (2000) Ongoing immunoglobulin somatic mutation in germinal center B cell-like but not in activated B cell-like diffuse large cell lymphomas. *Proc. Natl. Acad. Sci. USA* **97**, 10,209–10,213.
23. Horsman, D. E., Gascoyne, R. D., Coupland, R. W., Coldman, A. J., and Adomat, S. A. (1995) Comparison of cytogenetic analysis, southern analysis, and polymerase chain reaction for the detection of t(14; 18) in follicular lymphoma. *Am. J. Clin. Pathol.* **103**, 472–478.
24. Yunis, J. J., Frizzera, G., Oken, M. M., McKenna, J., Theologides, A., and Arnesen, M. (1987) Multiple recurrent genomic defects in follicular lymphoma. A possible model for cancer. *N. Engl. J. Med.* **316**, 79–84.
25. Weiss, L. M., Warnke, R. A., Sklar, J., and Cleary, M. L. (1987) Molecular analysis of the t(14;18) chromosomal translocation in malignant lymphomas. *N. Engl. J. Med.* **317**, 1185–1189.
26. Cornillet, P., Rimokh, R., Berger, F., et al. (1991) Involvement of the BCL2 gene in 131 cases of non-Hodgkin's B lymphomas: analysis of correlations with immunological findings and cell cycle. *Leuk. Lymphoma* **4**, 355–362.

27. Davis, R. E., Brown, K. D., Siebenlist, U., and Staudt, L. M. (2001) Constitutive nuclear factor kappaB activity is required for survival of activated B cell-like diffuse large B cell lymphoma cells. *J. Exp. Med.* **194,** 1861–1874.
28. Rosenwald, A., Wright, G., Chan, W. C., et al. (2002) The use of molecular profiling to predict survival after chemotherapy for diffuse large-B-cell lymphoma. *N. Engl. J. Med.* **346,** 1937–1947.
29. Damle, R. N., Wasil, T., Fais, F., et al. (1999) Ig V gene mutation status and CD38 expression as novel prognostic indicators in chronic lymphocytic leukemia. *Blood* **94,** 1840–1847.
30. Hamblin, T. J., Davis, Z., Gardiner, A., Oscier, D. G., and Stevenson, F. K. (1999) Unmutated Ig V(H) genes are associated with a more aggressive form of chronic lymphocytic leukemia. *Blood* **94,** 1848–1854.
31. Rosenwald, A., Alizadeh, A. A., Widhopf, G., et al. (2001) Relation of gene expression phenotype to immunoglobulin mutation genotype in B cell chronic lymphocytic leukemia. *J. Exp. Med.* **194,** 1639–1647.
32. Klein, U., Tu, Y., Stolovitzky, G. A., et al. (2001) Gene expression profiling of B cell chronic lymphocytic leukemia reveals a homogeneous phenotype related to memory B cells. *J. Exp. Med.* **194,** 1625–1638.
33. Frazer, J. K., Jackson, D. G., Gaillard, J. P., et al. (2000) Identification of centerin: a novel human germinal center B cell- restricted serpin. *Eur. J. Immunol.* **30,** 3039–3048.
34. Ried, T., Liyanage, M., du Manoir, S., et al. (1997) Tumor cytogenetics revisited: comparative genomic hybridization and spectral karyotyping. *J. Mol. Med.* **75,** 801–814.
35. Kallioniemi, A., Kallioniemi, O. P., Sudar, D., et al. (1992) Comparative genomic hybridization for molecular cytogenetic analysis of solid tumors. *Science* **258,** 818–821.
36. Pollack, J. R., Perou, C. M., Alizadeh, A. A., et al. (1999) Genome-wide analysis of DNA copy-number changes using cDNA microarrays. *Nat. Genet.* **23,** 41–46.
37. Lichter, P., Joos, S., Bentz, M., and Lampel, S. (2000) Comparative genomic hybridization: uses and limitations. *Semin. Hematol.* **37,** 348–357.
38. Wessendorf, S., Fritz, B., Wrobel, G., et al. (2002) Automated screening for genomic imbalances using matrix-based comparative genomic hybridization. *Lab. Invest.* **82,** 47–60.
39. Aoudjit, F., Masure, S., Opdenakker, G., Potworowski, E. F., and St-Pierre, Y. (1999) Gelatinase B (MMP-9), but not its inhibitor (TIMP-1), dictates the growth rate of experimental thymic lymphoma. *Int. J. Cancer* **82,** 743–747.
40. Kossakowska, A. E., Huchcroft, S. A., Urbanski, S. J., and Edwards, D. R. (1996) Comparative analysis of the expression patterns of metalloproteinases and their inhibitors in breast neoplasia, sporadic colorectal neoplasia, pulmonary carcinomas and malignant non-Hodgkin's lymphomas in humans. *Br. J. Cancer* **73,** 1401–1408.
41. Vacca, A., Moretti, S., Ribatti, D., et al. (1997) Progression of mycosis fungoides is associated with changes in angiogenesis and expression of the matrix metalloproteinases 2 and 9. *Eur. J. Cancer* **33,** 1685–1692.
42. Kossakowska, A. E., Hinek, A., Edwards, D. R., et al. (1998) Proteolytic activity of human non-Hodgkin's lymphomas. *Am. J. Pathol.* **152,** 565–576.

19
Gene Expression Profiling of Brain Tumors

Meena K. Tanwar and Eric C. Holland

INTRODUCTION

Microarray analysis is a practical and efficient method for gene expression profiling of human tumors. However, it can become complicated depending on the characteristics of the tumor system under study. In the case of brain tumors, the two major issues that must be taken into consideration are: the functional, regional, and cellular heterogeneity of the normal brain, and the diversity of intracranial neoplasms. These tumors differ in their histology, epidemiology, genetic alterations, cells of origin, and prognoses. Through large-scale analyses, such as microarrays, it is possible to gain a better understanding of brain tumor biology and to address important clinical issues.

Development and Normal Brain

The normal adult brain is composed of a heterogeneous population of cells that arise from multipotent stem cells. The developing mammalian embryo consists of three main layers of cells: the endoderm, the ectoderm, and the mesoderm. The ectoderm is the outermost layer and gives rise to all major tissues of the central nervous system (CNS) and peripheral nervous system. The neural plate, which is the dorsal region of the ectoderm, folds to form the neural tube and the neural epithelial cells that line the wall of the neural tube. Multipotent neuronal stem cells from this region give rise to neurons and glia, including oligodendrocytes and astrocytes *(1–3)*. The differentiation of neuronal and glial cells into mature CNS cell types is regulated by signaling from growth factors and other environmental influences. For example, glial progenitor cells can differentiate into type 1 astrocytes, type 2 astrocytes, or oligodendrocytes in the presence or absence of growth factors such as platelet-derived growth factor (PDGF), basic fibroblast growth factor (bFGF), or ciliary neuronotrophic factor (CNTF).

Within the brain, the appearance, function, and presumably gene expression patterns of neurons, astrocytes, and oligodendrocytes are also topologically complex. For example, neurons are found primarily in the gray matter, oligodendrocytes in the white matter, and astrocytes in both. Neurons are highly variable in appearance and function and can be regionally clustered in cortical layers or as nuclei, which are functional units of adjacent neurons. Taken as a whole, the normal brain is composed of a vastly heterogeneous population of cells, organized regionally. Similarly, microarray analyses may reflect these topological differences, depending on what part of the brain

From: *Expression Profiling of Human Tumors: Diagnostic and Research Applications*
Edited by: Marc Ladanyi and William L. Gerald © Humana Press Inc., Totowa, NJ

the samples are obtained from. This is an important consideration when interpreting the gene expression patterns of different regions of normal brain.

Brain Tumors

Brain tumors account for approximately 2% of cancer-related deaths overall and 20% of malignancies under the age of 15 *(4)*. In the U.S., there are approx 17,000 newly diagnosed primary brain tumors per year and approx 11,500 deaths *(5)*. According to the World Health Organization (WHO), there are six major categories and up to 130 subcategories of CNS tumors *(6)*. The major divisions of CNS tumors are: neuroepithelial tumors, tumors of the meninges, lymphomas and hematopoietic tumors, germ cell tumors, tumors of the sellar region, and metastatic tumors. Primary CNS tumors are those that originate in or adjacent to the brain parenchyma, whereas secondary tumors are metastases that originate from a distant site, but reestablish and grow in the CNS.

Primary Intraparenchymal Tumors

Primary brain tumors are those that arise either within (intra) or adjacent (extra) to the brain parenchyma, which consists of neurons and their supporting cells. The most common primary intraparenchymal brain tumors are gliomas and medulloblastomas, which account for approx 65% of primary CNS tumors and 20% of pediatric CNS tumors, respectively *(6,7)*. Gliomas are classified as neuroepithelial tumors and are thought to arise from a common glial-precursor cell population. Gliomas are subdivided into astrocytomas, oligodendrogliomas, and ependymomas. Astrocytomas are the largest subgroup of gliomas and are classified into four grades according to the WHO. Grade I tumors are pilocytic astrocytomas. They occur most often in children and young adults and are histologically distinct from diffuse astrocytomas of grades II–IV, in that they are well circumscribed tumors with little infiltration into surrounding brain and limited malignant potential *(8)*. Grade II diffuse low-grade astrocytomas tend to infiltrate surrounding normal brain tissue, have an increased number and size of astrocytes, rare mitoses, and no nuclear atypia, microvascular proliferation, or necrosis. The median age of occurrence of grade II gliomas is 35 yr old, and survival after diagnosis is about 7 yr, however the strongest prognostic factor is age. These low-grade gliomas (LGGs) have some tendency for progression to grade III or IV gliomas after 4 to 5 yr.

Grade III gliomas, also referred to as anaplastic astrocytomas (AA) occur, in large part, as a result of tumor progression from low-grade astrocytoma. Histologically, AAs typically have increased cellularity, nuclear atypia, mitotic activity, and no microvascular proliferation or necrosis. The average age of onset is 41 yr, and they commonly progress to grade IV glioblastoma multiforme (GBM) within 2 yr of diagnosis.

GBMs are the highest grade of astrocytic tumors and have the worst prognosis. They account for approx 50% of gliomas and fall into two categories. Primary GBMs refer to tumors that arise *de novo*, with no previous clinical history of glioma and are more common in older adults (median age of 55 yr). Secondary GBMs are the result of tumor progression from a preexisting lower grade lesion, and the median age of occurrence is 45 yr old. These two classes of GBMs can also be distinguished on the basis of chromosomal alterations, such as *EGFR* amplification in primary GBMs and mutations in

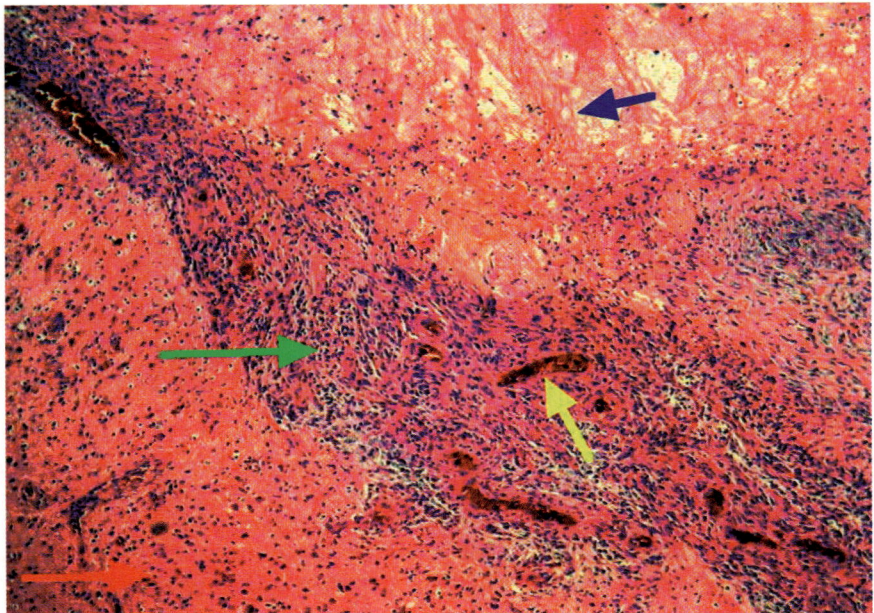

Fig. 1. Histological section of a GBM, demonstrating tumor heterogeneity. This section illustrates areas of dense tumor (green arrow), tumors cells invading adjacent normal brain (red arrow), areas of vascular proliferation (yellow arrow), and areas of necrosis (blue arrow).

TP53 in secondary GBMs. Histologically, both types of tumor are diffuse with heterogeneous, anaplastic, and poorly differentiated cells, high mitotic indices, and prominent neovascularization and/or necrosis (Fig. 1). More often than not, the blood brain barrier is disrupted due to aberrant blood vessel generation. This type of vascular abnormality, which is not a characteristic of normal brain and lower grade astrocytomas, results in leakage of intravascular molecules into the tumor mass and allows for detection of high-grade gliomas by intravenous contrast enhancement on magnetic resonance imaging (MRI) scans. Frequently, there are areas of enhancing tumor surrounded by nonenhancing regions consisting of tumor cells invading adjacent normal brain (Fig. 2). Patients diagnosed with GBM have very poor prognoses, with a median survival time of about 50 wk even after surgical resection, radiation, and chemotherapy.

Oligodendrogliomas are another subset of gliomas, which account for approx 5% of all intracranial neoplasms and 14% of gliomas *(9–11)*. They differ from astrocytic tumors in that the tumor cells have small rounded nuclei and clear scant cytoplasm, and they resemble oligodendrocyte morphology. These tumors frequently have deletions on chromosomes 1p and 19q, however the tumor suppressor genes on these chromosome arms are unknown *(12)*. Anaplastic oligodendrogliomas display features of malignancy, such as high mitotic activity, microvascular proliferation, and necrosis. Reports on patient prognosis have varied considerably from 3–10 yr for grade II tumors *(13–15)* and <1 yr to 4 yr for grade III anaplastic oligodendrogliomas *(13,15)*. Combined yet distinct astrocytic and oligodendrocyte cell morphology is observed in oligoastrocytomas or anaplastic oligoastrocytomas. Genetically, alterations character-

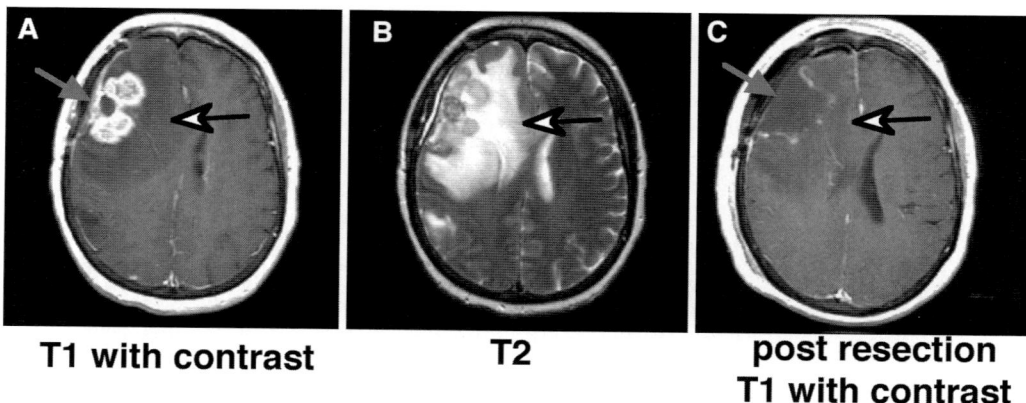

Fig. 2. MRI scans illustrating heterogeneity within CNS tumors. (**A**) T1-weighted scan demonstrating a contrast enhancing mass in the right frontal region (closed arrow). Within the enhancing portion of the tumor there are regions that do not enhance correlating with necrosis. Surrounding the enhancing mass, a region of nonenhancing tumor and edema is seen, corresponding to tumor cells invading into the adjacent brain tissue (open arrow). (**B**) T2-weighted scan highlighting the invading tumor and edema (open arrow). (**C**) Postresection T1-weighted scan with contrast, illustrating the removal of the enhancing mass (closed arrow). The surrounding edema and invading tumor cells are still present (open arrow) and will eventually give rise to tumor recurrence.

istic of both tumor types are seen, but are detected uniformly across the tumor cell populations, indicating clonal expansion from a common precursor cell *(16)*. Survival for patients with oligoastrocytomas has been reported as 3–6.3 yr *(17–19)* and <1 yr to 3 yr for anaplastic oligoastrocytomas *(19,20)*.

Ependymomas are the smallest subgroup of gliomas and are more common in children and adolescents than in adults. They consist of neoplastic ependymal cells and are presumed to originate from the lining of the ventricle walls. These tumors are well demarcated, have low cellularity and mitoses, and have perivascular pseudorosettes and ependymal rosettes. Ependymomas do not exhibit the genetic mutations, deletions, and amplifications seen in astrocytomas and oligodendrogliomas. Allelic loss of DNA sequences on chromosome 17p has been found in 9 of 18 pediatric ependymomas. By chromosomal location, this appears to correspond to *TP53*, but a candidate tumor suppressor gene has not yet been identified *(21)*. Unlike astrocytic tumors, children with ependymal tumors have worse prognoses than adults *(22)*.

The most common intraparenchymal tumors and the leading cause of cancer-related deaths in the pediatric age group are the medulloblastomas. These tumors appear to arise from the external granule cell layer (EGL) during cerebellar development. The molecular abnormalities in these tumors indicate that a blockade in differentiation of EGL cells may contribute to the formation of these lesions *(23)*. Although many of these tumors can be successfully treated with combinations of surgery, radiation, and chemotherapy, many children with medulloblastomas are not cured, and those that are cured experience significant side effects from treatment. Clearly, improvement in current treatment strategies is needed.

Metastatic Intraparenchymal Tumors

Metastatic brain tumors from systemic cancers are the most common intracranial neoplasm, with a slightly higher incidence than gliomas. Due to their high prevalence, lung cancer and breast cancer are the most common metastatic brain tumors *(24)*. Other tumors that have a high incidence of metastasis to the brain, but are less common, include melanoma, renal cell carcinoma, choriocarcinoma, and colorectal cancer, however rare metastases from other tumor types are also seen. Unlike diffuse gliomas, brain metastases are spheroid and well demarcated from the normal brain parenchyma. Furthermore, they are often surrounded by an extensive zone of edema, due to the disruption of the blood brain barrier caused by neovascularization, and large tumors also exhibit a central zone of necrosis. Disruption of the blood brain barrier allows the detection of metastatic tumors through contrast enhancement on MRIs. In most cases, the histological characteristics of metastatic lesions in the brain are similar to those of the primary tumor of origin. Metastases are seen more frequently in older patients, and important prognostic factors include age, location and number of metastases, and progression and status of the primary lesion. In the case of a single metastatic legion, surgery is often an option depending on the tumor location. However, when multiple metastases are present, surgery is frequently not possible. The median survival time for metastatic brain cancer patients is highly variable. On average, survival of patients with single metastases treated with stereotactic radiosurgery and whole-brain radiation therapy (WBRT) is 10 mo *(25–30)*, 4 mo in the case of multiple metastases treated with WBRT *(31)*, and 1 to 2 mo with no treatment *(32,33)*.

Extraparenchymal Tumors

While the majority of intracranial neoplasms arise within the brain parenchyma, there are a number of tumors that develop outside of this region. The most common of the extraparenchymal tumors are the meningiomas. These tumors comprise approx 20% of brain tumors *(34)*, are primarily benign, and arise from the cap cells of the acrachnoid. Other examples of extraparenchymal tumors include acoustic neuromas, craniopharyngiomas, and pituitary adenomas and carcinomas. Pituitary adenomas account for 10% of all intracranial neoplasms. They are functionally classified according to secretion of growth hormone (GH), prolactin (PRL), adrenocorticotrophic hormone (ACTH), or thyroid stimulating hormone (TSH). These tumors do not arise in the brain parenchyma and are not invasive; therefore, histologically they consist primarily of tumor cells and blood vessels. Although most of these tumors are benign in character, malignant forms (pituitary carcinoma) that grow rapidly, spread, and recur, do exist. The molecular basis for the aggressive behavior of these tumors has yet to be identified *(35)*. In general, all extraparenchymal tumor specimens, unlike diffuse gliomas, consist of solid tumor cells without contamination of normal brain.

MATERIALS AND METHODS
Tissue Collection and Processing

Obtaining high quality brain tumor tissue requires the surgeon to properly resect and preserve the tissue for research purposes. Frequently, brain tumors are resected using a suction method in which most of the tumor is removed in small pieces, through a vacuum. However, in order to collect sufficient tumor specimen for experimental analysis, it is preferable to excise the tumor *en bloc*, resecting around the tumor mass to maximize the amount of tumor tissue collected. The macroscopic appearance of brain tumors, especially of high-grade gliomas, is variable. The surgeon can visually distinguish highly cellular, diffusely infiltrating, and necrotic regions. These regions will clearly have different expression profiles. Therefore, close and productive communications between the surgeon and molecular biologist is critical if useful data are to be derived from these experiments.

In most cases, the tissue specimen must be preserved by flash freezing in liquid nitrogen and subsequent storage at −80°C or in a liquid nitrogen chamber. This requires that the liquid nitrogen be available in the operating room, allowing for immediate tissue preservation and minimal RNA degradation. Once the tissue is collected, it can be ground to powder form in the presence of liquid nitrogen or dry ice, using a sterile mortar and pestle or mechanical homogenizer. Subsequent RNA, DNA, or protein extraction can then be performed.

Control Tissue

Normal brain controls are even more difficult to obtain than brain tumor tissue because of the obvious ethical problem in removing normal brain tissue from patients. This problem has forced many groups to take the alternate route of using surgically removed, and otherwise discarded, nontumorigenic tissue. Although these tissue specimens are not cancerous, they should not be considered completely normal either. Frequently, tissue is used from brain trauma or epileptic patients. Many groups have published microarray results using these types of tissue as their normal controls *(36–39)*. Another source of "normal" brain tissue that has been used is from postmortem specimens *(36)*, and it has been shown that good quality mRNA can be obtained even from postmortem brain specimens *(40)*.

One alternative to these methods is the use of normal tissue removed during resection of metastatic lesions. As discussed above, metastatic brain tumors are generally spheroid, noninvasive, and easily distinguishable from surrounding normal brain. During resection of deep metastatic lesions, it is frequently necessary to remove overlying normal cortex, which can be collected, preserved, and used as normal brain controls. However, gene expression analysis of any proposed control brain tissue, including the above mentioned examples, compared to actual normal brain has not been done. Such comparisons will be necessary in order to determine the validity of the control samples.

A further consideration in obtaining normal brain tissue controls for microarray analysis is the heterogeneous mixture of component cells. Each region of the brain is composed of many cell types. This could influence the results of a microarray experiment, depending on what region of normal brain was used in the analysis. It may be necessary to collect tumor samples and normal brain tissue from corresponding brain regions for more accurate expression analysis.

Tumor Heterogeneity

A major problem area in microarray analyses of various tumors is that the expression profiles often reflect changes seen in a mixed population of tumor and nontumor cells, including stromal cells, endothelial cells, and inflammatory cells. The substantial regional heterogeneity seen within tumor samples (Fig. 1) raises concern as to whether gene expression changes are due to actual effects of tumor progression or to the varied contents of the sample. Furthermore, even among adjacent tumor cells, there is substantial heterogeneity of structure and cell morphology, such as pseudopalisading around regions of necrosis and giant tumor cells adjacent to clusters of small cells.

One approach to addressing intratumoral heterogeneity is the use of laser capture microdissection (LCM). This technique allows the selection of the exact cells or cell type desired for microarray analysis. For example, in the case of gliomas with a heterogeneous population of normal cells and tumor cells in a given specimen, it is possible to selectively sort cancer cells and similarly, normal cells for gene expression analysis, by using immunohistochemical markers *(41)*. Furthermore, cells adjacent to necrotic areas could be compared to cells infiltrating normal brain near the tumor periphery, using LCM technology on a given tumor specimen.

RESULTS

Validation of Expression Profiles

Once the results from gene expression profiling experiments are obtained, it is critical that the data be verified, in order to draw sound conclusions. Standard verification techniques include Northern blot or real-time reverse transcription polymerase chain reaction (RT-PCR) to validate transcript levels. Most array experiments have shown that independent methods of verification correlate well with the array data *(36–39,42–44)*.

Not all increases in mRNA expression yield a corresponding increase in protein expression. Transcript levels seen in tumor samples do not always translate to increased protein expression because of issues such as transcript stability and translation efficiency. Western blot analysis is a quantitative way of verifying protein expression, while immunohistochemistry is required for identifying the cell-specific expression of a particular protein. Because of the heterogeneity of tumor samples, it is important to determine whether the tumor cells are expressing the gene of interest, or if it is actually expressed by intermittent normal, inflammatory, stromal, or endothelial cells. Many groups have verified subsets of array genes to be overexpressed at the protein level as well *(36,37,39,43,45,46)*.

Clinical and Pathologic Issues and Microarray Utility

With proper experimental design, many important clinical and pathologic issues can be addressed and potentially resolved through microarray analysis studies. One clinical issue of concern is detection of gliomas. Microarray analysis can be used to elucidate genes, encoding secreted proteins, which could prove to be accurate diagnostic markers for glioma presence and grade. Gliomas are only detected and diagnosed when patients present with symptoms such as seizures, headache, dizziness, or altered mental status. Often, in the case of low-grade diffuse glioma, the tumor may be present for years prior to detection. If left undetected, these low-grade tumors are allowed to continue proliferating and can progress to grade III glioma or GBM. Glioma detection and diagnosis often occur after the tumor has progressed to a more malignant phenotype, and at these later stages of tumor progression, patient prognosis is much worse.

Another issue in treatment of gliomas is the variability in patient survival and response to therapy. Using microarray analysis, it may be possible to identify a group of genes with expression patterns that correlate with prognosis. This group of genes may indicate the aggressiveness of the tumor, whether the tumor will respond better to a particular type of treatment, or if the tumor is likely to recur.

Microarray analysis could also be utilized in identifying new molecular targets for drug therapy. By discovering novel genes overexpressed in gliomas, it is possible to further understand the molecular changes involved in gliomagenesis. These changes could reveal new pathways involved in tumorigenesis, which may prove to be better and/or more potent targets for drug therapy.

Microarray analysis of metastatic brain tumors and their primary tumors of origin may elucidate important gene expression changes involved in the process of metastasis. This type of technology could help identify collections of genes that dictate the brain as the site of metastasis instead of other tissues. Furthermore, as with the gliomas, gene targets can be identified for more effective drug development and therapy.

Yet another utility of gene expression profiling may be in aiding pathologic diagnosis and classification of CNS tumors. Currently, diagnosis of CNS tumors is made by histologic analysis, however significant discrepancies commonly occur between pathologists in their diagnoses of given tumors. Furthermore, survival of patients with defined subgroups of gliomas defined by histologic criteria alone is significantly overlapping. The added information provided by gene expression profiling analysis could greatly improve the number of accurate brain tumor diagnoses.

Analysis of the Literature

Since microarrays were first demonstrated to be highly effective tools for gene expression profiling, many array experiments have been performed to reveal novel gene expression changes specific to various types and grades of brain tumors *(36–39,42–48)*. In these analyses of glioma gene expression, using either oligonucleotide *(38,39,48)* or cDNA arrays *(36,44–46,48)*, genes such as epidermal growth factor receptor (*EGFR*), *CDK4*, *MDM-2*, *CD44*, *vimentin*, *fibronectin*, insulin-like growth factor binding protein (*IGFBP-2*), *IGFBP-5*, and secreted protein acidic and rich in cysteine precursor (*SPARC*), that have been shown to be overexpressed in gliomas or glioblastoma cell lines, were similarly found to be differentially expressed on the arrays. Furthermore, similar expression patterns, such as increased or decreased expression of novel genes not previously associated with gliomas, were identified as differentially expressed on both of the oligonucleotide and cDNA arrays. For example, chitinase-3-like-1 (*CHI3L1*) and macrophage migration inhibitory factor (MIF) were found to be differentially expressed on arrays of GBM vs normal brain *(38,46)*. Additionally, tyrosine protein kinase (*TYR03*), glutamate receptor (*AMPA-2*) and *apolipoprotein D* were found to be down-regulated in GBMs as compared to controls *(36,38,39)*.

Although many gene expression profiling analyses have been performed, very few studies have actually demonstrated that these molecular changes can be utilized to further characterize brain tumor type, grade, prognosis, treatment response, recurrence, or the propensity for metastasis to the brain. One study showed that overexpression of *Laminin-8* in GBMs was an indication of shorter time to tumor recurrence (4.3 mo), whereas overexpression of *Laminin-9* by GBMs had a mean recurrence time of 9.7 mo

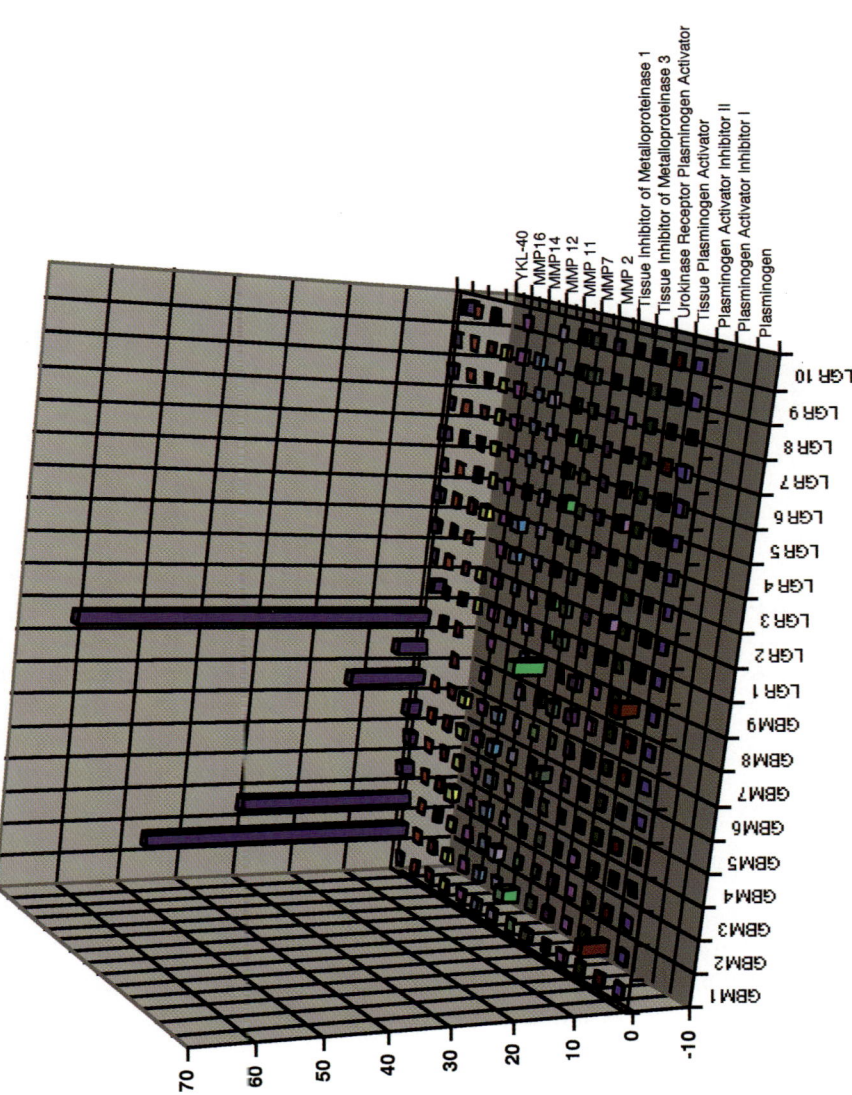

Fig. 3. Comparison of differential mRNA levels encoding *YKL-40* with transcripts encoding proteins involved in degradation of the extracellular matrix. *YKL-40* mRNA is significantly overexpressed compared to all endoproteases on the array. Order of genes graphed beginning from the back: *YKL-40*, matrix metalloproteinase-16 (*MMP-16*), *MMP-14*, *MMP-12*, *MMP-11*, *MMP-7*, *MMP-2*, tissue inhibitor of metalloproteinase 1 (*TIMP-1*), *TIMP-3*, urokinase receptor plasminogen activator, tissue plasminogen activator, plasminogen activator inhibitor-II (*PAI-II*), *PAI-I*, plasminogen.

(37,42). Another study showed that *CHI3L1* (also referred to as *YKL-40*) was overexpressed in GBMs, but not in AAs or LGGs at the mRNA level, and compared to genes encoding other extracellular matrix-related proteins, *YKL-40* was significantly elevated (Fig. 3). *YKL-40* was found to encode a secreted glycoprotein, which could be detected in the serum of glioma patients. This protein was found to be significantly elevated in serum of patients with GBM compared to normal controls or patients with lower grade gliomas, as well as in patients with lower grade tumors compared to normal controls *(46)*.

In one microarray study, it was shown that based on an eight-gene clustering model, medulloblastoma patients were categorized as "survivors" or "poor outcomes." Eighty percent of patients in the "survivor" group lived at least 5 yr, whereas 83% of patients in the "poor outcomes" group lived less than 5 yr *(47)*. In this study, the medulloblastoma patients with "poor outcomes" tended to show elevated expression of genes leading to multidrug resistance, ribosome biogenesis, and cell proliferation, while medulloblastomas with good outcomes showed elevation in genes involved in cerebellar development. The identification of general functional differences between tumor populations, such as this, is only possible by the simultaneous analysis of large number of genes by technologies such as microarray analysis.

Technical Limitations

RNA to Protein

Microarray analysis is a large-scale and productive technique that allows for the detection of gene expression changes at the transcript level. However, one major drawback to this approach is that it largely ignores the consequences of differential stability and turnover of mRNA transcripts. As a result, microarray data are not necessarily predictive of processes downstream of transcription. Increased RNA levels do not always lead to increased protein production, and similarly, lack of differential RNA expression implies, but ultimately is not indicative of, lack of differential protein translation. This is increasingly important with the realization that activated signaling pathways in high-grade gliomas regulate the translational efficiencies of existing mRNAs.

FUTURE DIRECTIONS

To date, the majority of gene expression studies on CNS tumors have been done on the more prevalent and better characterized tumors such as gliomas and medulloblastomas. However, major advances from microarray technology may come instead, from their use on rare tumors, for which sample availability is limited. Therefore, in studying these tumors, we need to derive as much information from each specimen as possible. In that sense, the information output from microarray-based studies in CNS tumors is just beginning.

REFERENCES

1. Goldman, J. E. (2000) Glial differentiation and lineages. *J. Neurosci. Res.* **59,** 410–412.
2. Noble, M. and Mayer-Proschel, M. (1997) Growth factors, glia and gliomas. *J. Neurooncol.* **35,** 193–209.
3. Rajan, P. and McKay, R. D. (1998) Multiple routes to astrocytic differentiation in the CNS. *J. Neurosci.* **18,** 3620–3629.

4. Kaye, A. H. and Laws, E. R., Jr. (2001) *Brain Tumors: An Encyclopedic Approach.* Churchill Livingstone, New York, p. 3.
5. Boring, C. C., Squires, T. S., and Tong, T. (1992) Cancer statistics, 1992. *CA Cancer J. Clin.* **42,** 19–38.
6. Kleihues, P. and Cavenee, W. K. (2000) *World Health Organization Classification of Tumors Pathology and Genetics: Tumours of the Nervous System.* IARC Press, Lyon.
7. Russell, D. S. and Rubinstein, L. J. (1989) *Pathology of Tumors of the Nervous System.* Williams and Wilkins, Baltimore.
8. Berger, M. S. and Wilson, C. B. (1999) *The Gliomas.* W.B. Saunders Company, Philadelphia, p. 172.
9. Mork, S. J., Lindegaard, K. F., Halvorsen, T. B., et al. (1985) Oligodendroglioma: incidence and biological behavior in a defined population. *J. Neurosurg.* **63,** 881–889.
10. Tola, M. R., Casetta, I., Granieri, E., et al. (1994) Intracranial gliomas in Ferrara, Italy, 1976 to 1991. *Acta Neurol. Scand.* **90,** 312–317.
11. Zulch, K. J. (1986) *Brain Tumors: Their Biology and Pathology.* Springer-Verlag, Berlin.
12. Ueki, K., Nishikawa, R., Nakazato, Y., et al. (2002) Correlation of histology and molecular genetic analysis of 1p, 19q, 10q, TP53, EGFR, CDK4, and CDKN2A in 91 astrocytic and oligodendroglial tumors. *Clin. Cancer Res.* **8,** 196–201.
13. Dehghani, F., Schachenmayr, W., Laun, A., and Korf, H. W. (1998) Prognostic implication of histopathological, immunohistochemical and clinical features of oligodendrogliomas: a study of 89 cases. *Acta Neuropathol. (Berl)* **95,** 493–504.
14. Heegaard, S., Sommer, H. M., Broholm, H., and Broendstrup, O. (1995) Proliferating cell nuclear antigen and Ki-67 immunohistochemistry of oligodendrogliomas with special reference to prognosis. *Cancer* **76,** 1809–1813.
15. Shaw, E. G., Scheithauer, B. W., O'Fallon, J. R., Tazelaar, H. D., and Davis, D. H. (1992) Oligodendrogliomas: the Mayo Clinic experience. *J. Neurosurg.* **76,** 428–434.
16. Kraus, J. A., Bolln, C., Wolf, H. K., et al. (1994) TP53 alterations and clinical outcome in low grade astrocytomas. *Genes Chromosom. Cancer* **10,** 143–149.
17. Hart, M. N., Petito, C. K., and Earle, K. M. (1974) Mixed gliomas. *Cancer* **33,** 134–140.
18. Jaskolsky, D., Zawirski, M., Papierz, W., and Kotwica, Z. (1987) Mixed gliomas. Their clinical course and results of surgery. *Zentralbl. Neurochir.* **48,** 120–123.
19. Shaw, E. G., Scheithauer, B. W., O'Fallon, J. R., and Davis, D. H. (1994) Mixed oligoastrocytomas: a survival and prognostic factor analysis. *Neurosurgery* **34,** 577–582.
20. Kim, L., Hochberg, F. H., Thornton, A. F., et al. (1996) Procarbazine, lomustine, and vincristine (PCV) chemotherapy for grade III and grade IV oligoastrocytomas. *J. Neurosurg.* **85,** 602–607.
21. von Haken, M. S., White, E. C., Daneshvar-Shyesther, L., et al. (1996) Molecular genetic analysis of chromosome arm 17p and chromosome arm 22q DNA sequences in sporadic pediatric ependymomas. *Genes Chromosom. Cancer* **17,** 37–44.
22. Horn, B., Heideman, R., Geyer, R., et al. (1999) A multi-institutional retrospective study of intracranial ependymoma in children: identification of risk factors. *J. Pediatr. Hematol. Oncol.* **21,** 203–211.
23. Wechsler-Reya, R. J. and Scott, M. P. (1999) Control of neuronal precursor proliferation in the cerebellum by Sonic Hedgehog. *Neuron* **22,** 103–114.
24. Delattre, J. Y., Krol, G., Thaler, H. T., and Posner, J. B. (1988) Distribution of brain metastases. *Arch Neurol.* **45,** 741–744.
25. Adler, J. R., Cox, R. S., Kaplan, I., and Martin, D. P. (1992) Stereotactic radiosurgical treatment of brain metastases. *J. Neurosurg.* **76,** 444–449.
26. Alexander, E., 3rd, Moriarty, T. M., Davis, R. B., et al. (1995) Stereotactic radiosurgery for the definitive, noninvasive treatment of brain metastases. *J. Natl. Cancer Inst.* **87,** 34–40.
27. Coffey, R. J., Flickinger, J. C., Bissonette, D. J., and Lunsford, L. D. (1991) Radiosurgery for solitary brain metastases using the cobalt-60 gamma unit: methods and results in 24 patients. *Int. J. Radiat. Oncol. Biol. Phys.* **20,** 1287–1295.

28. Engenhart, R., Kimmig, B. N., Hover, K. H., et al. (1993) Long-term follow-up for brain metastases treated by percutaneous stereotactic single high-dose irradiation. *Cancer* **71,** 1353–1361.
29. Loeffler, J. S., Alexander, E., 3rd, Kooy, H. M., et al. (1991) Principles and practice of oncology, in *PPO Updates: Radiosurgery for Brain Metastases. Vol. 5(2)* (Devita, V. T., Hellman, S., and Rosenberg, S. A., eds.), J.B. Lippincott, Philadelphia, pp. 1–12.
30. Somaza, S., Kondziolka, D., Lunsford, L. D., Kirkwood, J. M., and Flickinger, J. C. (1993) Stereotactic radiosurgery for cerebral metastatic melanoma. *J. Neurosurg.* **79,** 661–666.
31. Zimm, S., Wampler, G. L., Stablein, D., Hazra, T., and Young, H. F. (1981) Intracerebral metastases in solid-tumor patients: natural history and results of treatment. *Cancer* **48,** 384–394.
32. Cairncross, J. G. and Posner, J. B. (1983) Oncology of the nervous system, in *The Management of Brain Metastases* (Walker, M. D., ed.), Martinus Nijhoff Publishers, Boston, pp. 341–377.
33. Markesbery, W. R., Brooks, W. H., Gupta, G. D., and Young, A. B. (1978) Treatment for patients with cerebral metastases. *Arch. Neurol.* **35,** 754–756.
34. Walker, A. E., Robins, M., and Weinfeld, F. D. (1985) Epidemiology of brain tumors: the national survey of intracranial neoplasms. *Neurology* **35,** 219–226.
35. Kaye, A. H. and Laws, E. R., Jr. (2001) *Brain Tumors: An Encyclopedic Approach.* Churchill Livingstone, New York, pp. 803–849.
36. Huang, H., Colella, S., Kurrer, M., Yonekawa, Y., Kleihues, P., and Ohgaki, H. (2000) Gene expression profiling of low-grade diffuse astrocytomas by cDNA arrays. *Cancer Res.* **60,** 6868–6874.
37. Ljubimova, J. Y., Lakhter, A. J., Loksh, A., et al. (2001) Overexpression of alpha4 chain-containing laminins in human glial tumors identified by gene microarray analysis. *Cancer Res.* **61,** 5601–5610.
38. Markert, J. M., Fuller, C. M., Gillespie, G. Y., et al. (2001) Differential gene expression profiling in human brain tumors. *Physiol. Genomics* **5,** 21–33.
39. Rickman, D. S., Bobek, M. P., Misek, D. E., et al. (2001) Distinctive molecular profiles of high-grade and low-grade gliomas based on oligonucleotide microarray analysis. *Cancer Res.* **61,** 6885–6891.
40. Bahn, S., Augood, S. J., Ryan, M., Standaert, D. G., Starkey, M., and Emson, P. C. (2001) Gene expression profiling in the post-mortem human brain—no cause for dismay. *J. Chem. Neuroanat.* **22,** 79–94.
41. Sugiyama, Y., Sugiyama, K., Hirai, Y., Akiyama, F., and Hasumi, K. (2002) Microdissection is essential for gene expression profiling of clinically resected cancer tissues. *Am. J. Clin. Pathol.* **117,** 109–116.
42. Evans, C. O., Young, A. N., Brown, M. R., et al. (2001) Novel patterns of gene expression in pituitary adenomas identified by complementary deoxyribonucleic acid microarrays and quantitative reverse transcription-polymerase chain reaction. *J. Clin. Endocrinol. Metab.* **86,** 3097–3107.
43. Fuller, G. N., Rhee, C. H., Hess, K. R., et al. (1999) Reactivation of insulin-like growth factor binding protein 2 expression in glioblastoma multiforme: a revelation by parallel gene expression profiling. *Cancer Res.* **59,** 4228–4232.
44. Sehgal, A., Boynton, A. L., Young, R. F., et al. (1998) Application of the differential hybridization of Atlas Human expression arrays technique in the identification of differentially expressed genes in human glioblastoma multiforme tumor tissue. *J. Surg. Oncol.* **67,** 234–241.
45. Sallinen, S. L., Sallinen, P. K., Haapasalo, H. K., et al. (2000) Identification of differentially expressed genes in human gliomas by DNA microarray and tissue chip techniques. *Cancer Res.* **60,** 6617–6622.

46. Tanwar, M. K., Gilbert, M. R., and Holland, E. C. (2002) Gene expression microarray analysis reveals YKL-40 to be a potential serum marker for malignant character in human glioma. *Cancer Res.* **62,** 4364–4368.
47. Pomeroy, S. L., Tamayo, P., Gaasenbeek, M., et al. (2002) Prediction of central nervous system embryonal tumour outcome based on gene expression. *Nature* **415,** 436–442.
48. Watson, M. A., Perry, A., Budhjara, V., Hicks, C., Shannon, W. D., and Rich, K. M. (2001) Gene expression profiling with oligonucleotide microarrays distinguishes World Health Organization grade of oligodendrogliomas. *Cancer Res.* **61,** 1825–1829.

20
Expression Profiling of Bone Tumors

Deborah Schofield, Daniel Wai, and Timothy J. Triche

INTRODUCTION

A variety of neoplasms arise in bone. Many of these, such as osteogenic sarcoma, are relatively unique to bone, while others, such as hemangiomas and lymphomas, are more ubiquitous in their origins. This chapter will focus on the malignant tumors whose primary site of origin is bone and, in particular, those tumors whose cell of origin is mesenchymal rather than hematopoietic. The most common mesenchymal tumors of bone are osteosarcomas, chondrosarcomas, and Ewing's sarcomas. As Ewing's sarcomas are discussed in more detail in Chapter 16, they will be addressed here in a comparative fashion with osteosarcoma and other common sarcomas of childhood.

BACKGROUND

Osteosarcomas

These account for approx 20% of all primary tumors of bone, with the majority arising in the second decade, the period of maximal skeletal growth (1). While most of these neoplasms arise sporadically, familial cases have been reported, and there is an increased incidence of osteogenic sarcoma in patients with Li-Fraumeni, Bloom, or Rothmund-Thomson Syndromes, and in patients diagnosed with retinoblastoma (2–11). There is also an increased incidence of osteosarcoma in patients who have undergone radiation therapy (possibly potentiated by chemotherapy) (12,13) and patients with a variety of otherwise benign conditions and/or lesions including Paget's disease, osteochondromas, enchondromas, fibrous dysplasia, Mazabraud's disease, bone infarcts, metallic implants, and chronic osteomyelitis. Although the osteogenic sarcomas arising in these patients are indistinguishable from the sporadic cases, they occasionally carry a worse prognosis.

Corresponding to the peak age of incidence, most osteosarcomas arise within the regions of bone where most growth occurs, i.e., the metaphyseal region of the distal femur, proximal tibia, and proximal humerus. Although there is both radiographic and histologic heterogeneity, the diagnosis of conventional osteosarcoma is usually relatively straightforward—they typically arise within the medullary cavity of the bone and have a destructive growth pattern, with ill-defined borders and a lytic and/or sclerotic radiographic appearance.

From: *Expression Profiling of Human Tumors: Diagnostic and Research Applications*
Edited by: Marc Ladanyi and William L. Gerald © Humana Press Inc., Totowa, NJ

Microscopically, production of osteoid matrix by cytologically malignant stromal cells is the diagnostic hallmark of osteosarcoma. Other than the presence of osteoid, which varies from minimal to abundant, the histologic appearance of intramedullary osteosarcomas is amazingly diverse. A wide variety of different types have been described, including osteoblastic, chondroblastic, fibroblastic, malignant fibrous histiocytoma-like, osteoblastoma-like, giant-cell-rich, small-cell, epithelioid, and telangiectatic *(14)*. The small-cell osteosarcoma *(15)*, while uncommon, is one of the osteosarcoma variants that may present an interesting diagnostic dilemma, as some have a histologic appearance similar to the Ewing's family of tumors and this diagnosis may be entertained in the differential, particularly when dealing with small biopsies with inconspicuous osteoid formation *(15–17)*. Different grading systems have been applied to conventional osteosarcomas; however, there is no significant correlation with clinical outcome, and all are considered to be high grade tumors *(18)*. The poorer prognosis historically associated with telangiectatic osteosarcomas has seemingly been ameliorated with the advent of combined-modality treatment *(19,20)*. In addition to the conventional osteosarcomas, there is a select group of rare, but distinctive tumors that are characterized by a more indolent clinical course. These include the well-differentiated intramedullary osteosarcoma (which may require appropriate radiographic and clinical correlations to establish the diagnosis), the fibroblastic parosteal osteosarcoma, and the cartilage-rich periosteal osteosarcoma *(21–25)*. The first two of these tumors behave in a low-grade fashion, while the latter is considered to be of low to intermediate grade malignancy.

Following a diagnostic biopsy, the therapy of most cases of osteogenic sarcoma involves the administration of adjuvant chemotherapy followed by definitive surgery. When resected, the tumor is extensively sampled and examined microscopically to assess the responsiveness of the tumor to therapy, as determined by the extent of tumor necrosis. More aggressive postoperative therapy may be administered to patients with tumors classified as "nonresponders," although the efficacy of this approach has not been well-documented. Although it is generally accepted that the degree of necrosis correlates with prognosis, there is debate surrounding the amount of tumor necrosis required for a lesion to be considered a "responder"—this cut-off is probably around 98%, meaning that less than 2% viable tumor cells remain in the resected specimen *(26–30)*. Some of the controversy surrounding this issue has related to differences in the intensity and timing of presurgical chemotherapy. Regardless, it is obvious that it would be most valuable to identify responsive and nonresponsive tumors prior to the administration of therapy (rather than after), such that trials of novel or more aggressive therapies could be undertaken while the micrometastasis burden is lowest. A specific example of this approach currently under investigation in the use of the monoclonal antibody, herceptin, in patients whose tumors express the *Her2/neu* gene *(31)*. However, debate surrounds both the incidence of gene expression, the association of its native expression or amplification with prognosis, and the most accurate method of detection *(32–35)*.

Chondrosarcomas

These are the second most common primary malignant tumor of bone *(36)*. In contrast to osteosarcomas, however, the incidence of these neoplasms gradually increases with age, as the majority occur in patients over 50 yr *(37–40)*. Conditions or lesions predisposing to the development of chondrosarcoma include some of those associated with a higher incidence of osteogenic sarcoma, such as Paget's disease, radiation therapy, and osteochondromas. In addition, patients with Ollier's disease and Maffucci's syndrome, along with solitary enchondromas are at increased risk for the development of a secondary chondrosarcoma *(41–49)*.

Most patients with chondrosarcoma present with pain of months to years duration and the bones most commonly involved are the ilium, proximal/mid femur and humerus. The tumors, particularly those arising *de novo*, are intramedullary and radiographically appear as a radiolucent lesion with variable numbers of punctate opacities.

The diagnosis of chondrosarcoma, similar to osteosarcoma, covers a clinically and histologically heterogeneous group of tumors. However, >90% of chondrosarcomas are characterized by the uniform and isolated formation of chondroid matrix by malignant cells and are referred to as conventional chondrosarcomas. Although abundant neoplastic cartilage formation may be seen in an osteogenic sarcoma, the opposite is not true. The formation of neoplastic osteoid generally rules out a diagnosis of chondrosarcoma. The microscopic appearance of conventional chondrosarcomas is relatively consistent and, unlike osteosarcomas, grading based upon standard histologic criteria, such as invasiveness, degree of cellularity, nuclear pleomorphism, and mitoses, correlates well with clinical behavior *(18,50–53)*. Ninety percent of these tumors are low to intermediate in grade (grade I–II), with limited metastatic potential, and follow an indolent clinical course. Some of the low-grade lesions may be histologically indistinguishable from their benign enchondroma counterparts and other features, such as location, skeletal maturity, radiographic appearance, and a presenting symptom of pain are used to establish a diagnosis. The remaining 10% of conventional chondrosarcomas are high-grade (grade III) and carry a metastatic potential similar to osteogenic sarcomas. The "nonconventional" types of chondrosarcoma, which are clear cell, dedifferentiated, and mesenchymal, are each characterized by distinct clinical and morphologic features, and have metastatic potentials similar to, or greater than, the high-grade conventional chondrosarcoma.

The primary therapy for chondrosarcoma is complete surgical removal. In cases where this cannot be achieved, there is a role for radiation therapy, and adjuvant chemotherapy may be considered in cases of the high-grade lesions *(54,55)*.

CLINICAL AND PATHOLOGIC ISSUES

The preceding discussion highlights a wide spectrum of different issues that may potentially be addressed by analysis of gene expression patterns of bone tumors—issues related to both diagnosis and underlying pathobiology that affect prognosis and therapeutic responsiveness. In general, these can be considered in three broad categories: (*i*) diagnostic gene expression profiles; (*ii*) prognostic gene expression profiles; and (*iii*) gene targets for potential therapeutic development and intervention.

The diagnosis of some cases of occasional osteosarcoma or chondrosarcoma is a challenge. Can gene expression patterns be used to confirm, or rule out, a diagnosis of osteosarcoma (i.e., small-cell osteosarcoma vs Ewing's family tumor) or chondrosarcoma (i.e., well-differentiated chondrosarcoma vs enchondroma) in these atypical cases? Are different gene expression patterns associated with different histologic appearances, particularly among osteosarcomas? Once a diagnosis of osteosarcoma or chondrosarcoma is established, can gene expression patterns identify those tumors that are biologically aggressive or indolent and, thus, be used to subclassify them accordingly? In addition, can gene expression patterns be correlated with therapeutic responsiveness and thereby used to confirm or identify new mechanisms of drug resistance and sensitivity? Moreover, can uniquely overexpressed genes be identified that may serve as potential targets for either directed or innovative therapies?

Finally, once a diagnosis is established, risk assigned, and therapeutic markers assessed, can gene expression patterns provide insight into mechanisms of tumorigenesis?—thereby further expanding our understanding of and enhancing our ability to successfully treat these malignant neoplasms. For example, can expression patterns associated with differing grades of chondrosarcoma be used to identify specific genes, such as *p53 (56)*, involved in tumor progression? Can expression patterns associated with primary and metastatic tumors be used to identify genes integral to the evolution or acquisition of metastatic potential? Are there unique gene expression patterns associated with tumors arising in patients that have predisposing syndromes or conditions, despite phenotypic homogeneity?

And, in addition to the more generic questions raised above, how do expression array data correlate with specific studies and corresponding hypotheses that have been individually reported over time? For example, what is the incidence and/or prognostic effect of *Her2/neu*, *CDK4*, *p53*, or *pRB* overexpression in osteosarcomas? How do expression levels of these specific and other genes correlate with the expression level of additional genes in both related and unrelated cellular pathways? That is, what is the evidence for networking among genes that determine a biologic or clinical behavior?

These issues are now being addressed in depth with the recent advent of technology that, for the first time, allows global gene expression profiling and the identification of any expressed gene, as well as its association with parameters of interest such as class, biologic aggressiveness, and therapeutic responsiveness. Much of the following text will discuss comparative data from historical single or small gene group analyses with comparable data derived from gene expression microarrays. In particular, the correlation, or lack thereof, between these methods and even from one microarray study to another will be considered in detail.

CLINICAL APPLICATIONS OF MICROARRAY TECHNOLOGY

Technical details of cDNA and oligonucleotide gene expression microarrays are discussed in detail elsewhere in this book. It suffices to note that the most important issues when handling clinical material are not the underlying technology (e.g., spotted vs *in situ* synthesized oligonucleotides), but quality assurance and data normalization *(57)*. The single most common mistake we encounter is the widespread belief that a single microarray generates useful data. This is not the case *(58)*. Rather, comparison of a new, presumably "unknown" sample with an archive of similar and dissimilar samples is imperative, and reveals subtle differences relevant to class distinction (e.g., diagnosis) as well as prognosis and likely therapeutic responsiveness *(59,60)*. Failure to realize this, coupled with superficial understanding of the need for rigorous statistical analysis of data after global normalization and rejection of flawed chips and datasets, has led to widespread skepticism. Many doubt the relevance of microarray data to actual gene expression values, let alone actual expressed protein levels or their activation state. Despite this, properly processed datasets subjected to proper statistical analysis can yield highly relevant information sufficient to diagnose and prognosticate, as well as identify associated genes *(57,61,62)*.

In our studies, we have found commercially prepared *in situ* synthesized oligonucleotide microarrays of 12,000–40,000 sequences (e.g., GeneChips™, Affymetrix, Santa Clara, CA, USA) to be of particular value, as they are of generally uniform quality, report comparable expression values from sample to sample, are highly reproducible (e.g., cc > 0.93), and allow comparison of data over time, between laboratories, and between institutions. This has proved to be of critical importance, as "homemade" arrays are considerably less amenable to interlaboratory comparison, and specific gene expression values are often a function of the probe or probe sets chosen by the laboratory that synthesizes the arrays, rendering comparisons less facile. For these reasons, we have undertaken our studies with commercially produced arrays, processed in a College of American Pathologists-Clinical Laboratory Improvement Amendment (CAP/CLIA) certified laboratory, with the future intent of using this information for potential patient management (e.g., diagnosis and assessment of therapeutic responsiveness related to specific patterns of gene expression).

BIOINFORMATICS

Microarrays generate enormous amounts of data. We have dealt with this challenge in several ways, including the purchase and use of commercially available software (such as GeneSpring (Silicon Genetics) and Affymetrix Microarray Suite), as well as shareware available from the Web, such as Cluster and TreeView, developed by Michael Eisen. In the past 2 yr, however, we have developed our own proprietary software, Genetrix™, which combines several useful features in a single package: rigorous data quality assessment and correction–normalization, followed by a suite of analytic tools that allow virtually unlimited data and gene subsets to be analyzed by all common methods, in context with clinical covariate data. The latter point is relevant here, as we have been unable to find software that allows for clinically applicable analysis of microarray data. Genetrix was developed specifically to do so. However, a detailed description is beyond the scope of this article. All the analyses found in this chapter were performed using Genetrix.

TISSUE PROCESSING

As with all organ systems and tumor types, the nature of the tissue samples obtained ranges from biopsy to extensive resection, from relatively pure tumor to a mixture of tumor and normal tissue elements, and from hypo- to densely cellular. Generically, all samples are snap-frozen as quickly as possible, usually in isopentane immersed in liquid nitrogen. Most samples are frozen in optimal cutting temperature (OCT) compound so that an hematoxylin and eosin (H&E)-stained frozen section slide can be prepared on all samples, and so that either macro- or microdissection can be performed if desired. Rarely, primarily in those cases in which the sample is known to be pure tumor, but associated with abundant osteoid formation, such that it is difficult to obtain high-quality frozen sections, tissue is directly snap-frozen in liquid nitrogen and processed.

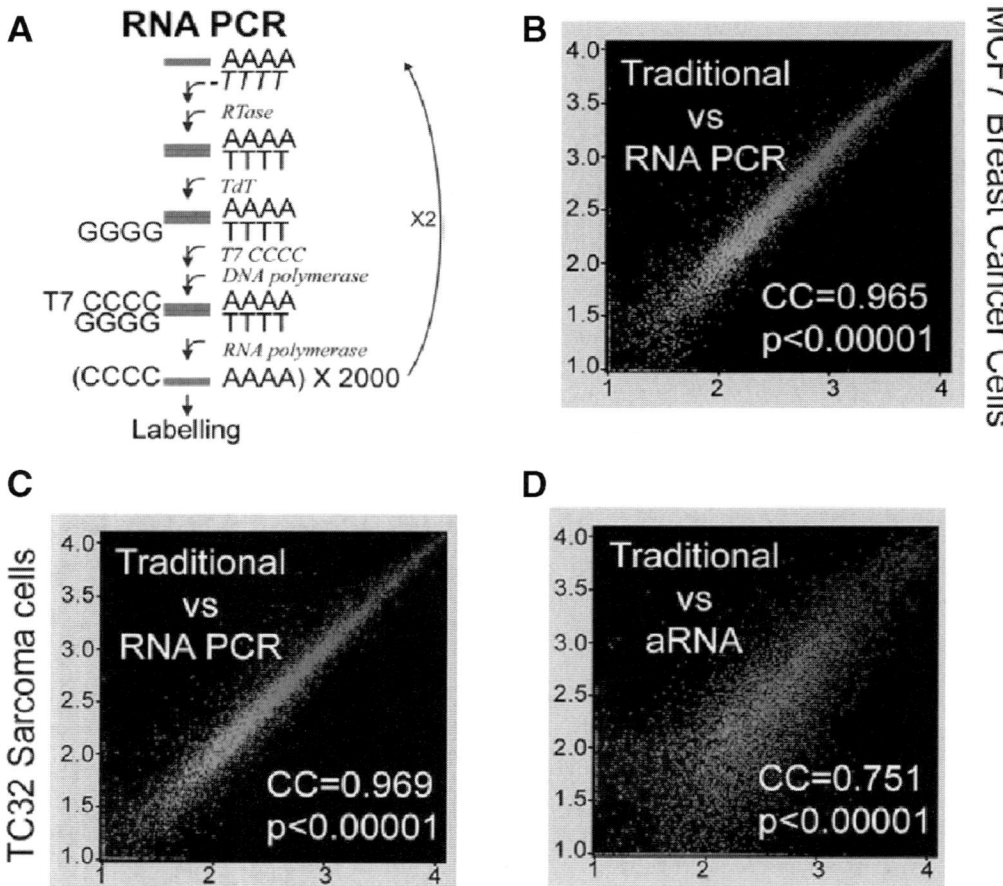

Fig. 1. RNA-PCR amplification of cancerous gene expression. **(A)** RNA-PCR utilizes poly(dT)$_{24}$- and oligo(dC)-promoter primers in first- and second-strand cDNA synthesis, respectively. Subsequent promoter-driven transcription amplifies mRNAs up to 250-fold per cycle (adapted from ref. 65). mRNA populations from **(B)** breast cancer cells and **(C)** Ewing's sarcoma cells are generated with high-fidelity (CC approx 0.97) by RNA-PCR (x-axes) when compared to nonamplified tumor–tissue mRNA (y-axes) and especially when compared to **(D)** a comparison of nonamplified (y-axis) vs antisense (aRNA or Eberwine) amplification (x-axis). Examples provided courtesy of Dr. Cheng Ming Chuong, Pathology Dept., Keck School of Medicine, USC.

In addition to the presence of osteoid, there are two additional issues involved in the processing of some bone tumor specimens: cellular heterogeneity, and the production of abundant extracellular matrix material rich in collagens, proteoglycans, and/or minerals. With regards to the latter issue, it has been shown that high quality RNA can be easily isolated from samples such as normal cartilage, suggesting that even hypocellular, matrix-rich tumors are amenable to expression analysis *(63)*. Cellular heterogeneity is by no means unique to bone tumors. It is the hallmark of most clinical material; other bone lesions such as Langerhans cell histiocytosis and nonossifying fibromas that classically contain a mixed population of both neoplastic and non-neoplastic cells are specific examples. Obtaining meaningful gene expression profiles on these tumors involves isolation of the specific cell population of interest by either cell sorting or microdissection, both of which are technologies that require additional labor and yield decreased amounts of RNA. Fortunately, methods are becoming available that reliably and linearly amplify scant populations of mRNA, such that even single-cell gene expression profiles may become feasible (though likely undesirable, given likely fluctuations in patterns of gene expression associated with cell cycle and other cellular perturbations that are likely "averaged" in whole tissue extracts) *(64–66)*. An example of such a profile, obtained from fewer than 100 Ewing's tumor cells, as well as breast cancer cells, is illustrated in Fig. 1. Note that the amplification procedure generates mRNA populations (x-axis) virtually indistinguishable from nonamplified mRNA extracted directly from tissue (y-axis), with a correlation coefficient of approx 0.97. This matches the reproducibility found between arrays using separate aliquots of the same mRNA without amplification in our experience. This method, when combined with laser capture microscopy (LCM) promises to make virtually any clinical sample, even cytology specimens, amenable to gene expression profiling.

Limitations of Clinical Samples

While the optimal gene expression studies are performed on human tissue samples, there are inherent limitations, including restricted quantities and types of samples, RNA degradation associated with tumor necrosis and delays in handling, and cellular heterogeneity *(66)*. Therefore, alternative sources of tumor and model systems, such as cell culture, xenografts and animal models, are frequently used to perform investigative studies, especially confirmatory studies of genes first identified by exploratory profiling of clinical samples. By definition, these samples cannot be controlled or repeated, nor can they generally be subjected to experimental manipulations designed to provoke or suppress a cellular response of interest (e.g., apoptosis, necrosis, etc.). Some of these surrogate sources have been used to study gene expression patterns in both osteosarcoma and chondrosarcoma. While results obtained from these types of studies need to be validated on larger numbers of human tissue samples prior to extrapolation in the clinic, interesting and promising data have already been generated, as discussed below.

RESULTS

Model Systems

Investigators have used cDNA microarrays to identify a consistent group of genes that are both up- and down-regulated when cell lines derived for osteosarcomas are compared to normal human osteoblasts *(67)*. The results were confirmed using reverse transcription polymerase chain reaction (RT-PCR) on both osteosarcoma cell lines and human tissue samples. The most significantly up-regulated genes included heat-shock protein and polyaderaylate-binding protein-like 1, while fibronectin 1 and thrombospondin 1 were among the group of genes that were down-regulated.

Others have performed preferential amplification–acquisition of coding sequences on multiple samples to identify coding sequences differentially expressed between human osteosarcoma cell lines and an osteoblast cell line *(68)*. In this study, differential expression of a subset of genes including a group of cyclins (D and E) and cyclin-dependent kinases, transcription factors, *E2F4* and *E2F5*, that interact with the retinoblastoma (RB) protein, and chondrocyte-derived ezrin-like protein was documented. Uniform down-regulation of mitogen-activated protein kinase 5 in the cell lines led to the postulate that this gene may function as a tumor suppressor in osteosarcoma. Results were confirmed with RT-PCR.

Based upon studies performed on a murine model of metastatic osteosarcoma, investigators have used cDNA microarrays to identify genes potentially involved in the variable metastatic potential of individual osteosarcomas *(69)*. One of the genes up-regulated in tumors with a greater propensity to metastasize was ezrin, a protein that plays a role in motility, invasion, and cellular adherence. Its potential relevance was confirmed by confirmatory Northern analysis and immunostaining, along with its documented expression in human osteosarcoma cell lines. However, as will be discussed later in "Metastasis Associated Genes", these results correlate only loosely with our studies of human osteosarcoma metastases.

Human Tumor Studies

While meaningful and invaluable information can be derived from both cell line and animal model studies, ultimate validation on human tumor samples is required *(70)*. Published studies describing microarray analysis of gene expression patterns on human bone tumor samples are limited. Using oligonucleotide arrays (U95A; Affymetrix), we have analyzed the gene expression patterns of a group of 43 osteosarcomas to address and illustrate some of the issues highlighted in previous sections. We have combined these data with similar U95A data from 63 other bone and soft tissue sarcomas (e.g., small cell osteosarcoma, Ewing's sarcoma, embryonal and alveolar rhabdomyosarcoma, and some complex phenotype sarcomas). The raw expression data from our 106 samples were imported into Genetrix, and the software performed a global normalization, which modeled the gene-probe intensities across all the samples simultaneously. Based upon estimates of standard error, identification of outliers, and adjustments for scanner saturation, we identified 18 samples for which the hybridization data were deemed unsuitable for further analysis. Our remaining cohort of 88 samples consisted of 32 osteosarcomas and 56 cases of other tumor types. The results allow a striking demonstration of the power of this technology to molecularly classify tumors, reliably identify genes associated with outcome, and potentially identify networks of genes responsible for clinical behaviors such as development of metastases and drug resistance. Candidate target genes for future therapeutic drug development are also readily identified.

MOLECULAR CLASSIFICATION OF BONE AND SOFT TISSUE SARCOMAS

Figure 2 demonstrates a molecular classification of osteosarcoma in context with multiple other bone and soft tissue sarcomas, using a three-dimensional, three-axis principal component analysis (PCA). This method is well described in the literature *(71–75)*. In essence, the multidimensionality of these complex datasets (e.g., approx 12,000 genes by 88 samples) is reduced to a visualizable two- or three-dimensional representation by first defining the axis through this space that maximally separates either samples or genes. This process is then repeated at right angles to the first, and (if a three-dimensional representation) yet again. The result maps each sample (or gene)

Expression Profiling of Bone Tumors 367

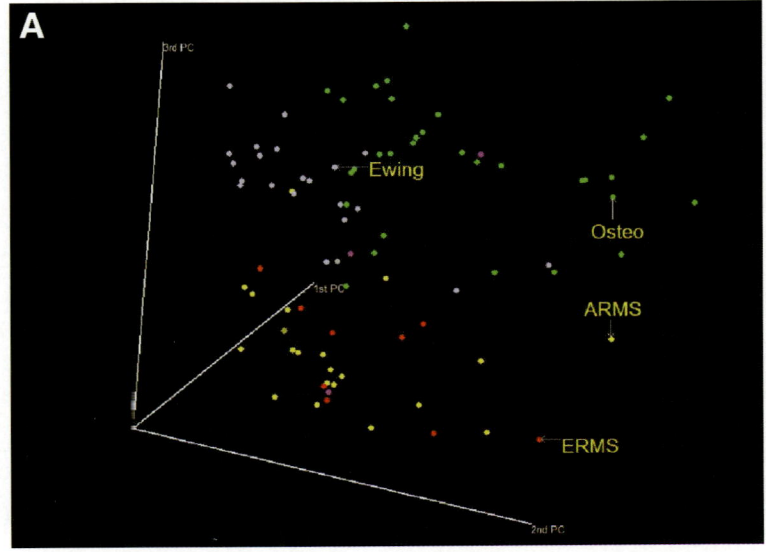

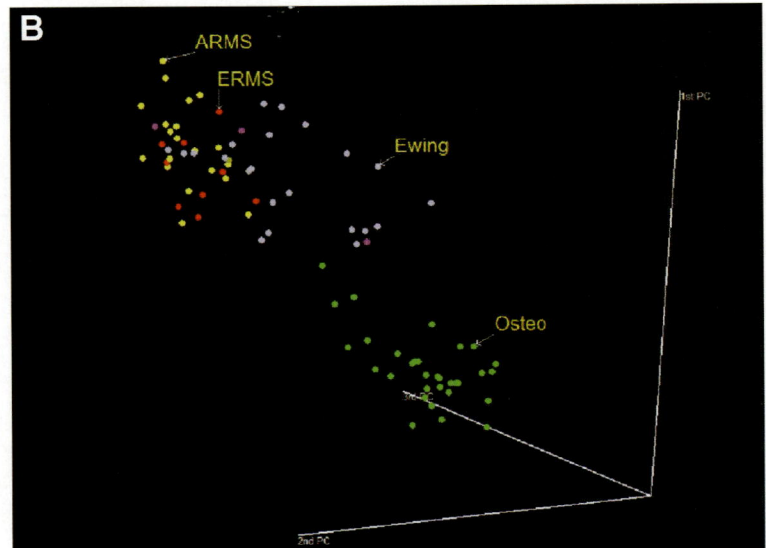

Fig. 2. Three-dimensional PCA plots. The raw expression data was normalized and used to generate PCA plots for 88 tumors based upon **(A)** all 12,600 genes on U95A GeneChips, and **(B)** a subset of only 30 genes identified to be statistically preferentially associated with osteosarcomas. Note that classification of known discrete tumor groups, especially between the two common bone sarcomas, Ewing's and osteosarcoma, is not uniformly possible (A). In contrast, when using the defined subset of osteosarcoma-associated genes, osteosarcomas are readily distinguished from all other tumors. This illustrates that subsets of genes can be more powerful classifiers than global gene transcript profiling.

into this quasi-three-dimensional space, wherein each sample or group can be identified in spatial relationship to the other. Figure 2A illustrates the result when all 88 cases are analyzed using all 12,600 genes on the Affymetrix U95A GeneChip. Note that, although the osteosarcomas (green dots) are reasonably separate from the other tumors, there is,

Table 1
Genes Associated with Osteosarcoma

Rank[a]	Gene
1	Distal-less homeobox 5.
2	Matrix metalloproteinase 13 (collagenase 3).
3	Gap junction protein, α 1, 43 kDa (connexin 43).
4	Ribosome binding protein 1 homolog 180 kDa (dog).
5	Chondroitin sulfate proteoglycan 4 (melanoma-associated).
6	T-complex-associated-testis-expressed 1-like.
7	KIAA0869 protein.
8	Procollagen-lysine, 2-oxoglutarate 5-dioxygenase (lysine hydroxylase) 2.
9	S100 calcium binding protein A10 (annexin II ligand, calpactin I, light polypeptide [p11]).
10	Ectonucleotide pyrophosphatase/phosphodiesterase 2 (autotaxin).
11	75 kDa Infertility-related sperm protein.
12	Matrix metalloproteinase 9 (gelatinase B, 92 kDa gelatinase, 92 kDa type IV collagenase).
13	Acid phosphatase 5, tartrate resistant.
14	Old astrocyte specifically induced substance.
15	Twist homolog (acrocephalosyndactyly 3; Saethre-Chotzen syndrome) (*Drosophila*).
16	Tropomyosin 1 (α).
17	Sialyltransferase 7D.
18	ATPase, H+ transporting, lysosomal membrane sector associated protein M8-9.
19	S100 calcium binding protein A4.
20	Cartilage linking protein 1.
21	Vitamin D (1,25- dihydroxyvitamin D3) receptor.
22	PTPRF interacting protein, binding protein 2 (liprin β 2).
23	Cathepsin K (pycnodysostosis).
24	SEC24 related gene family, member D (*S. cerevisiae*).
25	Moesin.
26	Metallothionein 1E (functional).
27	Integrin-binding sialoprotein (bone sialoprotein, bone sialoprotein II).
28	Procollagen C-endopeptidase enhancer.
29	Transforming growth factor, β-induced, 68 kDa.
30	Procollagen-proline, 2-oxoglutarate 4-dioxygenase, α polypeptide II.
31	Short stature homeobox 2.
32	Tyrosylprotein sulfotransferase 1.
33	Tumor necrosis factor (ligand) superfamily, member 11.
34	Serine (or cysteine) proteinase inhibitor, clade H (HSP 47), member 2.
35	Calumenin.
36	KIAA1199 protein.
37	DKFZp586E2023.
38	Chromosome 21 open reading frame 80.
39	DKFZP564F0522 protein.
40	Ras homolog gene family, member C.
41	Inhibitor of DNA binding 1, dominant negative helix-loop-helix protein.
42	Cathepsin Z.
43	Tissue inhibitor of metalloproteinase 2.
44	E74-like factor 4 (ets domain transcription factor).
45	Nucleobindin 2.
46	Myosin VI.
47	Peptidylprolyl isomerase C (cyclophilin C).
48	Tyrosylprotein sulfotransferase 2.
49	Procollagen-proline, 2-oxoglutarate 4-dioxygenase, beta polypeptide.
50	Interleukin 10 receptor, β.

[a]Genes are ranked by statistical significance and magnitude of association with osteosarcoma (see text for details).

nonetheless, some degree of co-mingling, such that a clean separation of osteosarcoma from other tumors is not evident; Ewing's sarcomas, in particular, are closely intermingled with osteosarcomas. In contrast, when the tumors are analyzed using a reduced gene subset of only 30 genes identified as being statistically preferentially associated with osteosarcomas as opposed to other sarcomas, based on multiple t-tests, the osteosarcomas are clearly clustered distinct from any other childhood sarcoma (Fig. 2B). This phenomenon has been reported by several other authors, who have chosen their subset genes by many different methods (75–77).

As might be expected, osteosarcomas can be segregated from other sarcomas of childhood and adolescence by their gene expression pattern. Some genes preferentially up-regulated in osteosarcomas compared to other tumor groups (Table 1) include numerous obvious bone-associated genes. Seven collagen-associated genes (four procollagen genes, two cartilage genes, and bone sialoprotein) speak to the osteogenic phenotype. Four extracellular matrix-associated genes (matrix metalloproteinases 9 and 13, cathepsin K, *TIMP2*) are evidence of the potent extracellular matrix degrading capability of osteosarcoma. Noteworthy is cathepsin K, an osteoclastic protease with potent collagenolytic activity. Although it theoretically may also play a role in the metastatic process, it is interesting to note that its expression has been linked to that of the metalloproteinases 9 and 13 (noted in the same group) in the process of fracture healing (78), a process not unlike the growth of osteosarcoma and sometimes histologically confused with it. Many other genes are less conspicuously associated with bone: a PTPRF interacting protein (liprin β 2) a signaling protein tyrosine phosphatase which has been associated with axonal guidance and breast development but not previously associated with bone; *TGF-β2*, which though widely expressed in tissues has been especially associated with developing bone and cartilage; and cathepsin Z and E74-like factor 4 (an ETS domain transcription factor) with no obvious association with osteosarcoma as opposed to other sarcomas. Interestingly, S100 proteins A4 and A10 are members of a family of calcium-binding proteins involved in signal transduction reported to induce invasiveness of primary tumors and promote metastasis.

SMALL CELL OSTEOSARCOMA

Although there were not enough osteosarcomas in our dataset to comment on differential gene expression patterns associated with specific histologic subtypes, it is nonetheless interesting to note that two cases of small cell osteosarcomas not only definitively cluster within the overall group of osteosarcomas, but also tend to cluster together when using either PCA analysis as described above, or conventional two-axis (e.g., sample by gene) hierarchical clustering (Fig. 3). Note that in the latter case, two samples from the same patient cluster immediately adjacent to one another (left arrow), and close to a second case (right arrow), well within the family of osteosarcomas (purple dots) depicted here, and quite separate from all other tumors, including Ewing's tumors, which cluster as a group immediately adjacent to the osteosarcomas. The two small cell osteosarcoma cases do not seem to share overlapping gene expression patterns with the Ewing's family of tumors, which is an observation supported by lack of a documentable *EWS/FLI-1* transcript in at least one of the two tumors. Thus, the gene expression profile results would support the concept that these tumors, often confused histologically with Ewing's sarcoma of bone, are,

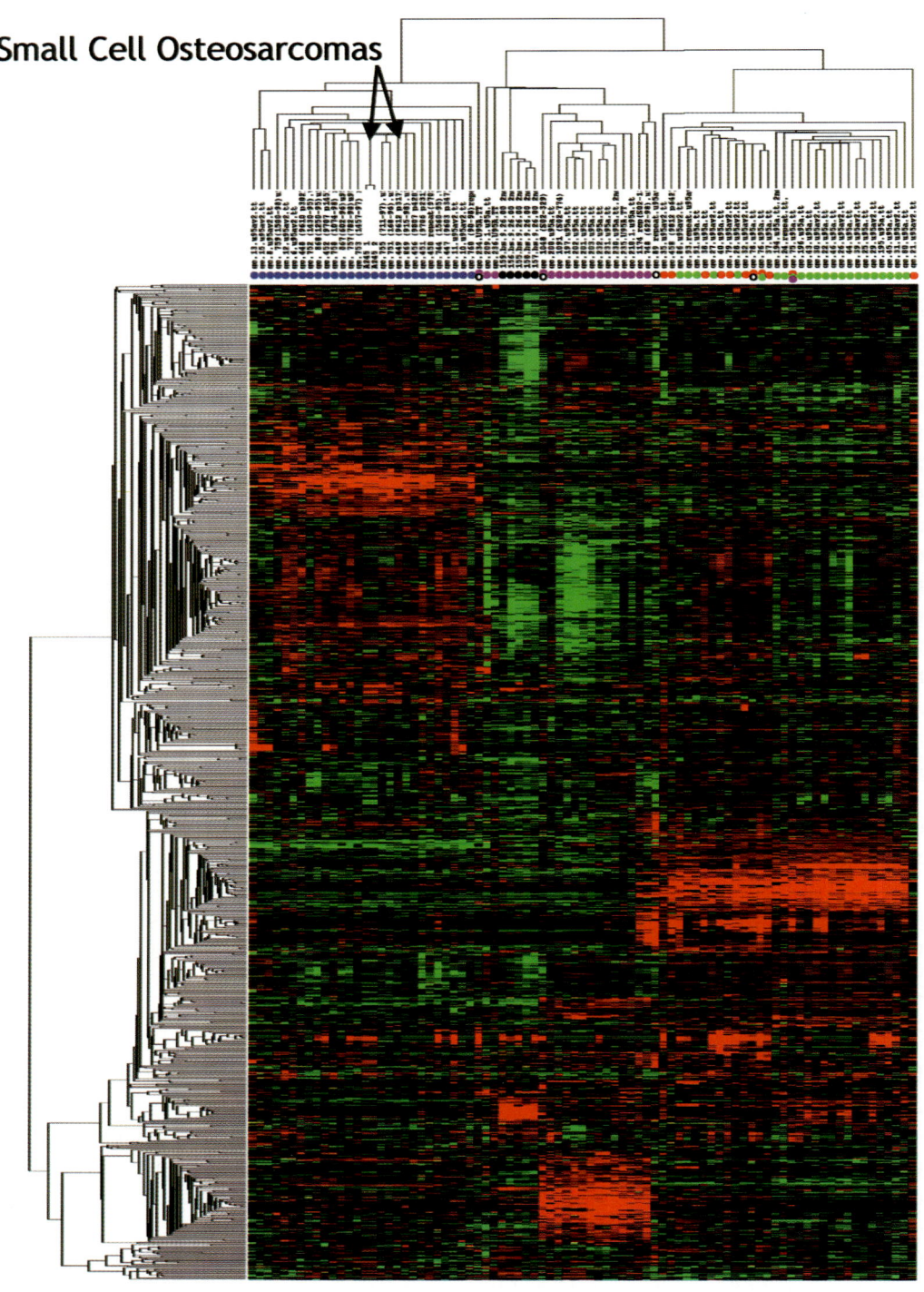

Fig. 3. Hierarchical clustering of expression microarray data. In this cluster diagram, columns represent tumor samples, whereas rows indicate individual genes. Based solely on their gene–expression profiles, cluster analysis divided the tumors into three major groups with osteosarcomas (blue) being grouped on the left, neuroblastomas (black) and Ewing's sarcomas

in reality, bona fide members of the osteosarcoma family of bone tumors and are not related to Ewing's tumors. A larger series of cases will be needed to confirm this impression.

PROGNOSIS-ASSOCIATED GENES

A second major goal of gene expression profiling is to determine outcome from primary biopsy material as a potential guide to choice of therapy and overall patient management. Thus, realization at the time of diagnosis that a given therapy will likely prove to be inadequate or ineffective would dictate alternate therapy, at a time (e.g., prior to development of therapeutic resistance) when the effect of these agents might not be limited by overall drug resistance. There is increasing evidence that this is possible *(60,76,80–88)*. Thus, it is of some potential importance to identify those genes that might connote differential outcome, specifically those genes that, alone or (more likely) as a group, appear to be strongly associated with clinical outcome. To do this, of course, one must have access to appropriate clinical material, specifically tumors with linked clinical data on survival status and other variables of potential interest. In the case of osteosarcomas, clinical cooperative group cases offer a particularly compelling opportunity, as all patients are treated on relatively few therapeutic protocols, and are thus readily comparable. In addition, institutional cases (from one or more institutions) offer another possibility that is generally not possible with clinical trials: access to both primary and metastatic tumor material, often from the same patient. In our studies, we have collected multiple metastases from some patients, and numerous single (pulmonary) metastases from many patients. As a result, we can begin to assess those genes most associated with the two most important predictors of outcome: metastases and death.

We looked first at primary tumor material from patients with a known outcome. Figure 4 is a typical survival curve generated by comparing two groups of patients. In this case, expression of 12,600 genes was analyzed in 26 tumors in which outcome was known and documented. Multiple *t*-tests were performed to identify those genes most associated with favorable or unfavorable prognosis (e.g., alive or dead). The 10 genes most strongly associated with either outcome on the basis of this statistical analysis are listed in Table 2 in rank order from most to least significantly associated. Note that the top two genes associated with death are important membrane receptor genes. The first, a Patched (*PTCH*)-related gene, *TRC8*, whose product, like PTCH, control Sonic Hedgehog (SHH) signaling through SMO (Smoothed) to GLI, an important transcription factor controlling the expression of several genes. *ABC3*, the second, encodes member of the ATP-binding cassette family of membrane transporters important in modulating drug resistance (e.g., multiple drug resistance protein [MRP]). Interestingly, these two genes are also statistically strongly associated with poor outcome along with Tomosyn, or "friend of syntaxin," which is a gene important in neurotransmission in the brain. Its

---

(Figure 3 caption continued) (magenta) in the middle, and alveolar (green) and embryonal rhabdomyosarcomas (red) on the right. Relative gene expression is indicated by color and ranges from very high (intense red) to very low (intense green). The arrows indicate three small cell osteosarcoma cases, which are grouped relatively close to one another within the osteosarcoma cluster and whose expression profiles are quite dissimilar to the Ewing's sarcomas.

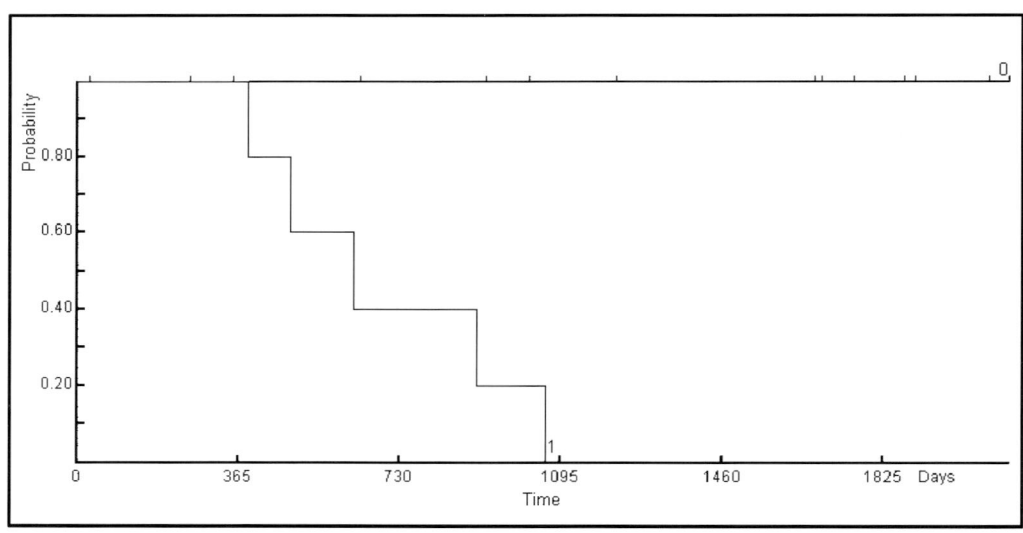

Fig. 4. Probability of survival. The expression of *TRC8*, *ABC3*, and *tomosyn* was examined in osteosarcoma patients with known outcome ($n = 26$ primary tumors). High expression of these three genes is strongly associated with a poor prognosis with death likely occurring in <3 yr. (Cox regression: Chi-square = 16.29; $p = 0.00005$. Comparison statistics: Chi-square = 17.48; $p = 0.00003$.)

Table 2
Ten Best and Ten Worst Prognostic Genes

	High expression associated with death
Rank[a]	Gene
1	Patched related protein translocated in renal cancer.
2	ATP-binding cassette, subfamily C (CFTR/MRP), member 3.
3	Tomosyn.
4	Histone deacetylase 5.
5	Survival motor neuron pseudogene.
6	Protein tyrosine phosphatase, receptor type, R.
7	Peripherin.
8	Nuclear factor of activated T-cells, cytoplasmic, calcineurin-dependent 4.
9	Dedicator of cyto-kinesis 3.
10	Timeless homolog (*Drosophila*).
	High expression associated with survival (and low expression with death)
Rank[b]	Gene
1	I factor (complement).
2	Guanine nucleotide binding protein (G protein), α-inhibiting activity polypeptide 2.
3	Lysosomal-associated multispanning membrane protein-5.
4	Zinc finger protein-like 1.
5	Proteoglycan 1, secretory granule.
6	Major histocompatibility complex, class I, E.
7	Protein tyrosine phosphatase, receptor type, C.
8	Zx53d03.r1.
9	Wingless-type MMTV integration site family, member 5A.
10	Hemopoietic cell kinase.

[a]Genes are ranked by statistical significance and magnitude of association with patient death.
[b]Genes are ranked by statistical significance and magnitude of association with patient survival.
MMTV, mouse mammary tumor virus (see text for deatils).

relevance here is not clear, but the association is strong. Accordingly, the survival analysis illustrated in Fig. 4 was performed using just these three genes. As is evident, all the high expressers of these three genes were dead within 3 yr, which is a strong indication that these genes are associated with an adverse outcome. The more interesting issue, of course, is why? One must suspect that both unregulated cell signaling through PTCH and drug resistance through cystic fibrosis transmembrane conductance regulation (CFTR)/MRP are relevant here. The role of tomosyn is completely obscure at this point. In fact, in the absence of biologic validation, the role of this gene in outcome must remain speculative at best. Nonetheless, these three genes, when expressed at high levels, reliably predict death in a cohort where prediction of this outcome is otherwise not possible. These genes, or others like them, may yet prove useful as prognostic indicators, even at the time of diagnosis.

METASTASIS-ASSOCIATED GENES

Metastasis is a complicated process, and it is perhaps naïve to expect to identify metastasis-specific genes in a comparison of metastatic and nonmetastatic osteosarcomas *(89)*. However, such a comparison might shed light on the relative expression of known metastasis-associated genes or metastasis-suppressive genes in clinically derived human osteosarcomas. Diverse factors such as tumor type vs metastatic type (the "seed" and "soil" question) are simplified in this case: osteosarcoma patients ultimately die, or not, depending on whether they develop pulmonary metastases. Those who do not are unlikely to die. Those who do have a greater than 80% risk of dying. Clinically, the problem of pulmonary metastases is central to any hoped for improvement in outcome in this disease. It is, thus, of more than academic interest to explore whether there is a genetic pattern to development of pulmonary metastases in osteosarcoma patients. Fortunately, the cohort of patients reviewed here includes 32 osteosarcoma patients of known stage (e.g., metastatic *[6]* or not *[26]*), which permits at least a cursory study of metastasis-associated genes.

We first considered genes reported previously in the literature on osteosarcoma. Although few papers specifically consider human osteosarcoma and its metastases, consideration of mouse models might offer some insight. Khanna et al. *(69)* used a mouse orthotopic model to identify 53 genes out of 3166 present on their arrays that were associated with pulmonary metastasis. We have compared their results with our own and find a relatively poor association. Specifically, the 27 identifiable of 31 genes noted to be up-regulated in pulmonary metastases could be compared to the equivalent genes present on U95A arrays. Of these 27 genes, 23 were represented by 58 probe sets (due to either replicate clones present on the arrays, or inclusion of gene subsets not further characterized by Khanna et al.). These are listed in Table 3. As is seen in Fig. 5A, there was little association between metastatic status and overexpression in our series. This is even more striking when both the magnitude and significance of association between these 23 genes (represented by 58 probe sets) are compared to their expression in metastatic (upper right quadrant) vs primary tumors (lower left quadrant) (Fig. 5B). Interestingly, 27 of the probe sets are relatively overexpressed in the metastases, but 32 are more associated with the primary tumors! Clearly, there is little to suggest a compelling pattern of expression in metastases, at least for these 27 genes in these 6 metastases, among 32 human osteosarcomas.

Table 3
Analysis of Metastasis-Associated Genes[a]

	Up-regulated in metastases[b]
Rank[c]	Gene
1	Proprotein convertase subtilisin/kexin type 5.
2	V-myb myeloblastosis viral oncogene homolog (avian).
3	Proteolipid protein 2 (colonic epithelium-enriched).
4	CCAAT/enhancer binding protein (C/EBP), α.
5	Integrin, α V (vitronectin receptor, α polypeptide, antigen CD51).
6	Crystallin, α A.
7	Tubulin, β, 5.
8	Proprotein convertase subtilisin/kexin type 2.
9	V-myb myeloblastosis viral oncogene homolog (avian).
10	Protein kinase, interferon-inducible double-stranded RNA dependent.
11	Crystallin, α A.
12	Hepatocyte nuclear factor 3, α.
13	V-myb myeloblastosis viral oncogene homolog (avian).
14	Integrin, β 2 antigen CD18 (p95).
15	V-myb myeloblastosis viral oncogene homolog (avian).
16	Cyclin D1 (PRAD1: parathyroid adenomatosis 1).
17	Connective tissue growth factor.
18	V-myb myeloblastosis viral oncogene homolog (avian).
19	Nuclear factor (erythroid-derived 2)-like 1.
20	Proteolipid protein1 (Pelizaeus-Merzbacher disease, spastic paraplegia 2, uncomplicated).
21	E2F transcription factor 5, p130-binding.
22	Lectin, galactoside-binding, soluble, 3 (galectin 3).
23	Crystallin, α A.
24	Nuclear factor (erythroid-derived 2), 45 kDa.
25	Integrin, β 4.
26	A disintegrin and metalloproteinase domain 8.

	Down-regulated in metastases[b]
Rank[d]	Gene
1	Tubulin, β.
2	Tubulin, β polypeptide.
3	Nuclear factor (erythroid-derived 2)-like 2.
4	Tubulin, β.
5	Integrin, α V (vitronectin receptor, α polypeptide, antigen CD51).
6	Clusterin.
7	V-myb myeloblastosis viral oncogene homolog (avian).
8	Tubulin, β 2.
9	Tubulin, β, 4.
10	Nuclear factor (erythroid-derived 2)-like 3.
11	Integrin, α V (vitronectin receptor, α polypeptide, antigen CD51).
12	Protein kinase, interferon-inducible dsRNA-dependent activator.
13	Tubulin, β, 2.
14	Cyclin D1 (PRAD1: parathyroid adenomatosis 1).
15	Proprotein convertase subtilisin/kexin type 1.

Table 3 *(continued)*

	Down-regulated in metastases[b]
Rank[d]	Gene
16	Villin 2 (ezrin).
17	Cyclin D1 (PRAD1: parathyroid adenomatosis 1).
18	Metallothionein 2A.
19	Asparagine synthetase.
20	Tubulin, β, 5.
21	Hepatocyte nuclear factor 3, β.
22	Tubulin, β polypeptide.
23	Integrin, β 4.
24	Hepatocyte nuclear factor 3, β.
25	E2F transcription factor 5, p130-binding.
26	E2F transcription factor 5, p130-binding.
27	Caudal type homeo box transcription factor 2.
28	V-myb myeloblastosis viral oncogene homolog (avian).
29	V-myb myeloblastosis viral oncogene homolog (avian).
30	Tubulin, β polypeptide 4, member Q.
31	Caudal type homeo box transcription factor 2.
32	Tubulin, β polypeptide.

[a]As identified by Khanna et al. *(69)*.
[b]As observed in our 32 institutional osteosarcoma cases (26 primaries and 6 metastases).
[c]Genes are ranked by statistical significance and magnitude of association with metastases.
[d]Genes are ranked by statistical significance and magnitude of association with primary tumors (see text for details).

It is important to note that the relative fold changes reported on spotted arrays, as employed by Khanna et al. *(69)*, cannot be compared to the expression levels reported on Affymetrix arrays. However, it is relevant to note that all of these genes were overexpressed in metastases compared to primary tumors in that series. There is very little evidence of a uniform pattern of overexpression in this series. Note, for example, that connective tissue growth factor, reported to be more than threefold overexpressed in metastases, is virtually equally expressed in metastatic and primary tumors, albeit at the highest level of any of the genes analyzed here. Tubulin-β subunit shows an even more confounding pattern, with overexpression in primary tumors of tubulin-β2, but underexpression of tubulin-β5 in these same tumors. These data cannot be compared to Khanna et al.'s data *(69)*, as only the tubulin β-chain was analyzed in that study, further highlighting the difficulty of comparing data across array platforms (with differing probe sets) and across species (e.g., mouse to human).

We then considered the genes that were over- and underexpressed in our cases. As seen in Table 4, column A, overexpressed genes are listed in rank order based on a statistical model that computes multiple *t*-tests for each gene in each sample. On this basis, virtually none of the genes found by Khanna et al. *(69)* are found among the top 100 genes in our dataset. Likewise, Table 4, column B presents similarly derived data for the top 100 underexpressed genes in metastases, again in rank order from most to least significant. Perhaps the most striking feature of this list is the lack of genes obviously associated with metastases *(89)*. Notable in their absence are genes encoding

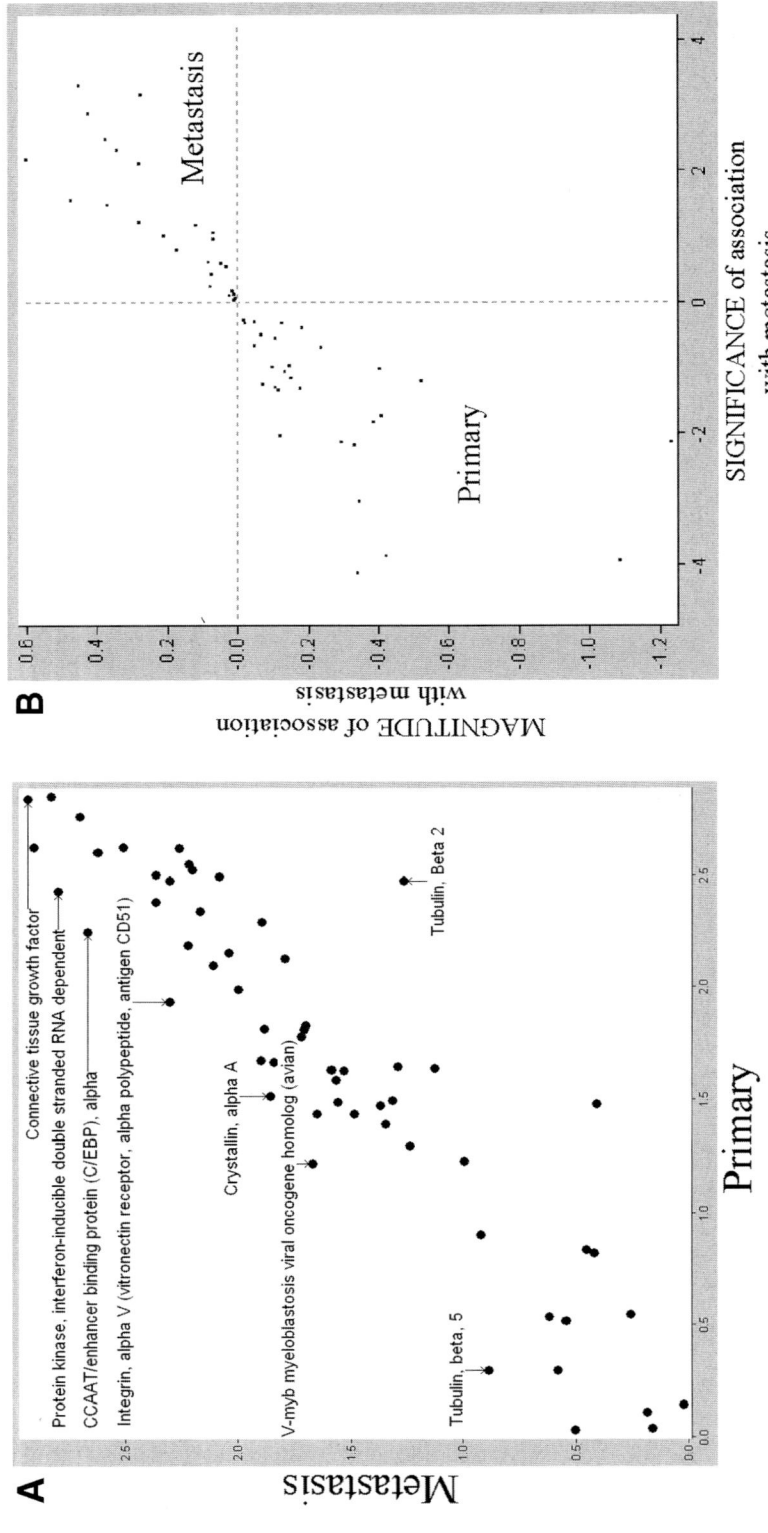

Fig. 5. Scatter plots of metastasis-associated genes as identified by Khanna et al. (69). Twenty-seven out of 31 genes reported to be up-regulated in metastases are represented by 58 probe sets on U95A GeneChips. (A) A log-expression scatter plot did not reveal any correlation with metastasis in our 32 osteosarcomas (26 primary tumors and 6 metastases). (B) Plotting their significance and magnitude of association with metastasis actually grouped 32 probe sets with primary tumors and only 26 with metastases. Thus, there was little association of these murine osteosarcoma metastasis-associated genes with these human osteosarcoma metastases.

Table 4
Metastasis-Associated Genes[a]

	Metastasis overexpressed genes
Rank[b]	Gene
1	Immunoglobulin λ-like polypeptide 2.
2	G antigen 3.
3	G antigen 4.
4	Troponin I, cardiac.
5	T-box 1.
6	G antigen 5.
7	High-mobility group (nonhistone chromosomal) protein isoform I-C.
8	Tumor necrosis factor receptor superfamily, member 5.
9	Wk24e08.x1.
10	Zinc finger protein 220.
11	Defensin, β 1
12	Carbonic anhydrase IX.
13	G antigen 7.
14	G antigen 2.
15	DNA from chr. 19-cosmid f24590 containing CAPNS and POL2RI, genomic sequence.
16	CD5 antigen (p56-62).
17	G antigen 3.
18	Casein, β.
19	Ribosomal protein S11.
20	Slug homolog, zinc finger protein (chicken).
21	Chemokine (C-C motif) receptor 4.
22	Complement component 3a receptor 1.
23	(λ) DNA for immunoglobin light chain.
24	Hypothetical protein MGC4293.
25	Ligase III, DNA, ATP-dependent.
26	GDNF family receptor α 2.
27	Trinucleotide repeat containing 4.
28	G antigen 6.
29	Ubiquitin-conjugating enzyme E2L 3.
30	Gap junction protein, β 1, 32 kDa (connexin 32, CMT neuropathy, X-linked).
31	Excision repair cross-complementing rodent repair deficiency, complementation gr. 4.
32	MHC class II transactivator.
33	E2F transcription factor 2.
34	Envoplakin.
35	Mitogen-activated protein kinase 8 interacting protein 1.
36	U3 snoRNP-associated 55-kDa protein.
37	39a1.
38	ZFM1 protein alternatively spliced product.
39	Phosphatidylinositol glycan, class A (paroxysmal nocturnal hemoglobinuria).
40	Clone IMAGE-35527 unknown protein.
41	Protein phosphatase 3 (formerly 2B), catalytic subunit, γ isoform (calcineurin A γ).
42	Guanylate cyclase activator 1A (retina).
43	Macrophage stimulating, pseudogene 9.
44	Protein phosphatase 3 (formerly 2B), catalytic subunit, β isoform (calcineurin A β).

(continued)

Metastasis overexpressed genes

Rank[b]	Gene
45	Double homeobox, 1.
46	Topoisomerase (DNA) III β.
47	Kell blood group.
48	Phosphatidylinositol glycan, class H.
49	Potassium inwardly-rectifying channel, subfamily J, member 12.
50	Hairy and enhancer of split (*Drosophila*) homolog 2.
51	Burkitt lymphoma receptor 1, GTP-binding protein.
52	Transglutaminase 3 (E polypeptide, protein-glutamine-γ-glutamyltransferase).
53	Receptor tyrosine kinase-like orphan receptor 1.
54	Collagen, type I, α 1.
55	BTB (POZ) domain containing 2.
56	Histamine receptor H1.
57	U2 small nuclear ribonucleoprotein auxiliary factor (65 kDa).
58	SWI/SNF-related, matrix-associated, actin-dependent regulator of chromatin, c1.
59	Tumor necrosis factor (ligand) superfamily, member 6.
60	Exchanger (human, liver, mRNA, 2665 nt).
61	Netrin 2-like (chicken).
62	CDC-like kinase 3.
63	Putative methyltransferase.
64	RAGE binding protein (P12).
65	Clone 23826.
66	HIV-1 rev binding protein 2.
67	Paired box gene 8.
68	Calcium channel, voltage-dependent, L type, α 1D subunit.
69	Ribosomal protein L3-like.
70	Natriuretic peptide precursor A.
71	Kinase.
72	DKFZp566C093.
73	Purine-rich element binding protein A.
74	G protein-coupled receptor 18.
75	Matrix Gla protein.
76	Zinc finger protein 169.
77	Adducin 2 (β).
78	Serine (or cysteine) proteinase inhibitor, clade D (heparin cofactor), member 1.
79	Kynurenine 3-monooxygenase (kynurenine 3-hydroxylase).
80	Potassium voltage-gated channel, KQT-like subfamily, member 1.
81	Yc92c11.s1.
82	Deoxyribonuclease I-like 2.
83	Hypothetical gene DKFZp570I0164.
84	Calcitonin gene-related peptide-receptor component protein.
85	Peripheral benzodiazepine receptor-associated protein 1.
86	Breast cancer suppressor element Ishmael Upper CP1.
87	Myotubular myopathy 1.
88	Regulator of G-protein signaling 14.
89	KIAA0763 gene product.
90	32f11.
91	Poly(A) binding protein, cytoplasmic 1.

Table 4 *(continued)*

Rank[b]	Metastasis overexpressed genes
	Gene
92	Fibroblast growth factor 13.
93	KIAA0514 gene product.
94	Hypothetical protein FLJ22624.
95	Putative GR6 protein.
96	Mannosidase, α, class 2A, member 1.
97	Keratin, hair, acidic, 3A.
98	Epithelial V-like antigen 1.
99	Fibroblast growth factor 8 (androgen-induced).
100	Interleukin 9 receptor.

Rank[c]	Metastasis underexpressed genes
	Gene
1	KIAA0310 gene product.
2	Ring finger protein 3.
3	Thymosin, β, identified in neuroblastoma cells.
4	TNF-α-inducible cellular protein containing leucine zipper domains.
5	KIAA0451 gene product.
6	Macrophage erythroblast attacher.
7	Neural precursor cell expressed, developmentally down-regulated 4.
8	KIAA1750 protein.
9	Apoptosis inhibitor 5.
10	CGI-150 protein.
11	Sema domain, Ig domain, TM domain, short cytoplasmic domain, semaphorin 4F.
12	Adaptor-related protein complex 2, mu 1 subunit.
13	PFTAIRE protein kinase 1.
14	Isocitrate dehydrogenase 3 (NAD+) β.
15	Proteasome (prosome, macropain) 26S subunit, non-ATPase, 1.
16	ATP synthase, H+ transporting, mitochondrial F0 complex, subunit e.
17	Translin-associated factor X.
18	KIAA0308 protein.
19	Casein kinase 1, γ 2.
20	Hypothetical protein TCBAP0758.
21	Smoothelin.
22	ATP-binding cassette, subfamily C (CFTR/MRP), member 5.
23	TNF α-inducible cellular protein containing leucine zipper domains.
24	Small nuclear ribonucleoprotein 70 kDa polypeptide (RNP antigen).
25	Rho-specific guanine nucleotide exchange factor p114.
26	RYK receptor-like tyrosine kinase.
27	Proteasome (prosome, macropain) 26S subunit, non-ATPase, 8.
28	ATPase, Class VI, type 11B.
29	Hypothetical protein FLJ11021 similar to splicing factor, arginine/serine-rich 4.
30	KIAA0728 protein.
31	Desmuslin.
32	ATP synthase, H+ transporting, mitochondrial F0 complex, subunit e.

(continued)

Rank[c]	Metastasis underexpressed genes Gene
33	Ligase I, DNA, ATP-dependent.
34	Protein phosphatase 2 (formerly 2A), catalytic subunit, β isoform.
35	Translocase of outer mitochondrial membrane 70 homolog A (yeast).
36	Ubiquitously transcribed tetratricopeptide repeat gene, Y chromosome.
37	API5-like 1.
38	Protein-L-isoaspartate (D-aspartate) O-methyltransferase.
39	KIAA0537 gene product.
40	Mitogen inducible 2.
41	Proteasome (prosome, macropain) subunit, β type, 1.
42	ADP-ribosylation factor-like 3.
43	Nucleobindin 2.
44	CAAX box 1.
45	Hypothetical protein FLJ21007.
46	Ubiquitin carboxyl-terminal esterase L1 (ubiquitin thiolesterase).
47	Adenosine kinase.
48	Multiple inositol polyphosphate histidine phosphatase, 1.
49	ADP-ribosylation factor 4-like.
50	DKFZp564A026.
51	Hypothetical protein FLJ11191.
52	ADP-ribosyltransferase (NAD+; poly [ADP-ribose] polymerase).
53	Zinc finger protein 133 (clone pHZ-13).
54	Transglutaminase 1.
55	Integrin, β 5.
56	Tuberous sclerosis 1.
57	Damage-specific DNA binding protein 2 (48 kDa).
58	Mitogen-activated protein kinase kinase kinase 4.
59	KIAA0431 protein.
60	Pre-B-cell leukemia transcription factor 3.
61	Bone γ-carboxyglutamate (gla) protein (osteocalcin).
62	Trophinin.
63	Integrin, β 5.
64	DKFZp564L0822.
65	E74-like factor 2 (ets domain transcription factor).
66	High-mobility group (nonhistone chromosomal) protein 17-like 1.
67	Succinate-CoA ligase, ADP-forming, β subunit.
68	Excision repair cross-complementing repair deficiency, complementation group 1.
69	Carbonic anhydrase II.
70	Vacuolar protein sorting 45A (yeast).
71	M-phase phosphoprotein 9.
72	Golgi phosphoprotein 1.
73	Beclin 1 (coiled-coil, myosin-like BCL2 interacting protein).
74	DJ1033B10.12 (collagen, type XI, α 2 [COL11A2]).
75	Tubulin-specific chaperone E.
76	Homeo box D3.
77	CGI-60 protein.
78	COX17 homolog, cytochrome c oxidase assembly protein (yeast).
79	Hypothetical protein FLJ10618.

Table 4 *(continued)*

	Metastasis underexpressed genes
Rank[c]	Gene
80	Tubulin, β.
81	Defender against cell death 1.
82	Myotubularin-related protein 1.
83	Hypothetical protein FLJ11193.
84	Cyclin-dependent kinase inhibitor 2D (p19, inhibits CDK4).
85	Protein kinase (cAMP-dependent, catalytic) inhibitor γ.
86	Zinc finger protein 288.
87	Transforming growth factor β-stimulated protein TSC-22.
88	Polymerase (RNA) II (DNA directed) polypeptide I (14.5 kDa).
89	Suppressor of *S. cerevisiae* gcr2.
90	NADH dehydrogenase (ubiquinone) Fe-S protein 4 (18 kDa).
91	Nicotinamide nucleotide transhydrogenase.
92	Excision repair cross-complementing repair deficiency, complementation group 1.
93	Sorting nexin 4.
94	Leptin receptor overlapping transcript-like 1.
95	Cisplatin resistance-associated overexpressed protein.
96	Thyroid hormone receptor interactor 7.
97	Similar to KIAA0010 gene product (*H. sapiens*).
98	Heat-shock transcription factor 2.
99	Protein tyrosine phosphatase, receptor type, M.
100	Zinc finger protein 195.

[a]As identified in our 32 institutional osteosarcoma cases (26 primaries and 6 metastases).

[b]Genes are ranked by statistical significance and magnitude of association with metastases.

[c]Genes are ranked by statistical significance and magnitude of association with primary tumors (see text for details).

extracellular matrix remodeling proteins like matrix metalloproteases, tissue inhibitors thereof, TGF-β, insulin-like growth factor (IGF)1, TGF-α, epidermal growth factor receptor (EGFR)/HER2, CXCR4 or CCR7, or even generic evidence of activation of the RAS/mitogen-activated protein kinase (MAPK) cell signaling pathway, or evidence of vascular proliferation (vascular endothelial growth factor [VEGF], etc.) However, expression of many of these genes in osteosarcomas, as opposed to metastases only, is evident from the literature. What is lacking at this point is any real understanding of the functional significance of these genes. Most studies have documented one or more members and hypothesized a specific role; in reality, these genes function within networks of interactive genes. The nature of these functional interactions is not intuitively obvious from simple inspection of gene lists, a point which is increasingly evident to those performing any form of gene expression profiling. There is, consequently, a real need for interpretative software that will help to dynamically link these gene networks.

We did note one very striking aspect of our metastatic tumors when compared to nonmetastatic tumors. When metastatic tumors were contrasted with nonmetastatic tumors, it became immediately evident that there is a group of genes that are relatively overexpressed in metastatic tumors (Fig. 6). Of interest, this group includes six genes in the G antigen (*GAGE*) family, in particular *GAGE-2, -3, -4, -5, -6,* and *-7*, which lie

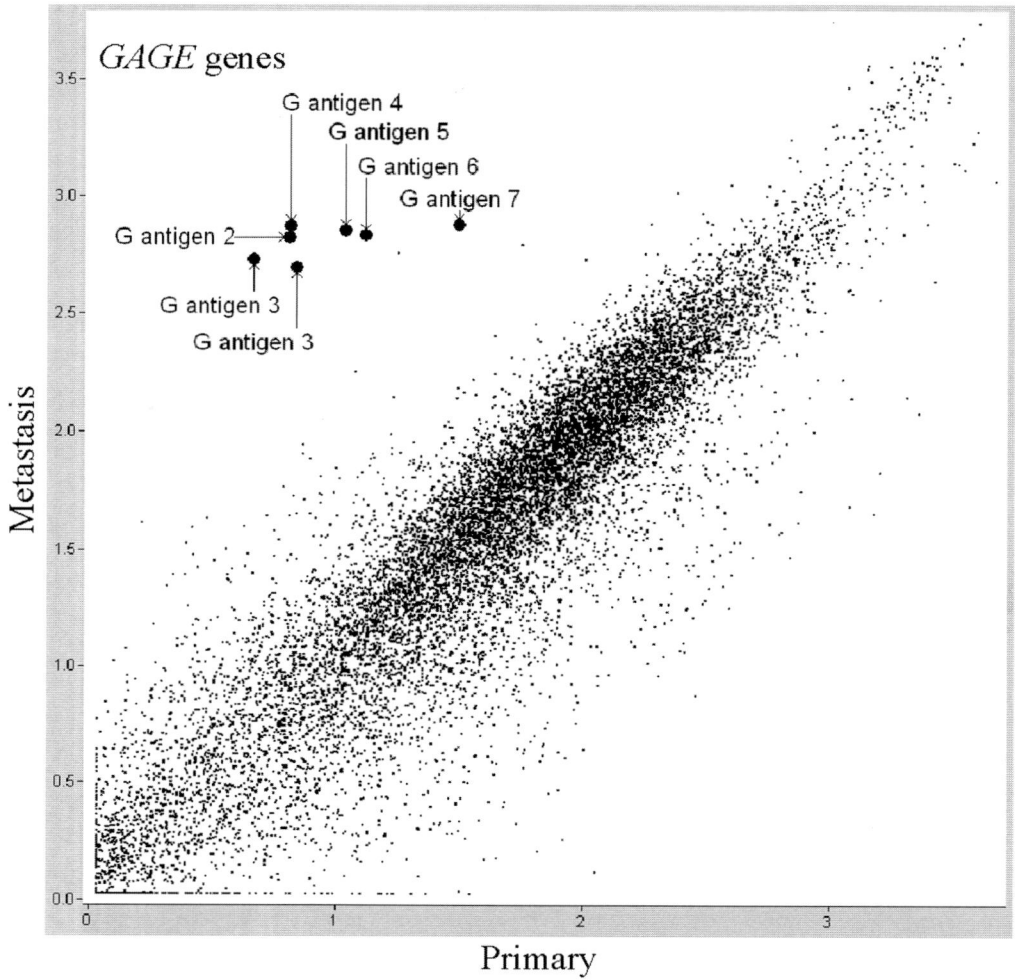

Fig. 6. Scatter plot of 12,600 genes for 32 osteosarcomas. For each gene, the log expression across 26 primary osteosarcomas was averaged and compared to the average log expression in 6 pulmonary metastases. Seven probe sets representing six members of the G-antigen (*GAGE*) family, including *GAGE-2, -3, -4, -5, -6,* and *-7*, are highly expressed in the metastatic lesions.

within a tight cluster in close proximity to one another (see labels, Fig. 6). Surprisingly, no other genes lie within this cluster. This observation provoked a more detailed consideration of these findings. Due to significant sequence homology between *GAGE* family members *(90)*, it seems likely that there may be significant cross-hybridization between members that might artifactually cause this cluster, and that separation of expression levels for individual *GAGE* genes could be difficult. However, plotting of averaged *GAGE* expression levels on a line graph for individual metastatic and primary tumors (Fig. 7A) summarizes and confirms that several *GAGE* family genes (e.g., 2–7) are expressed at high levels in most metastases, but not in the majority of primary tumors. *GAGE-1*, despite marked sequence homology with *GAGE-2*, showed no variation between tumors (data not shown), indicating that the individual probe sets do

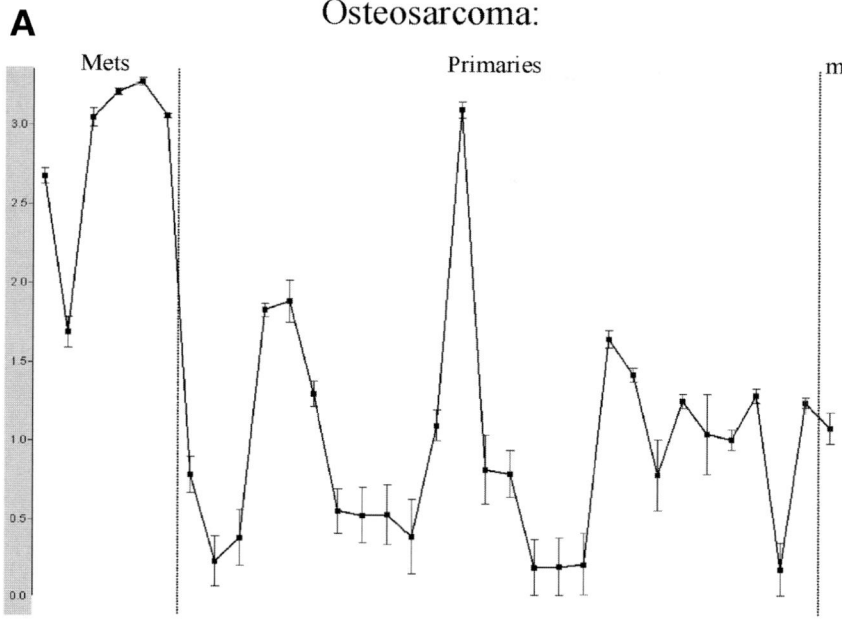

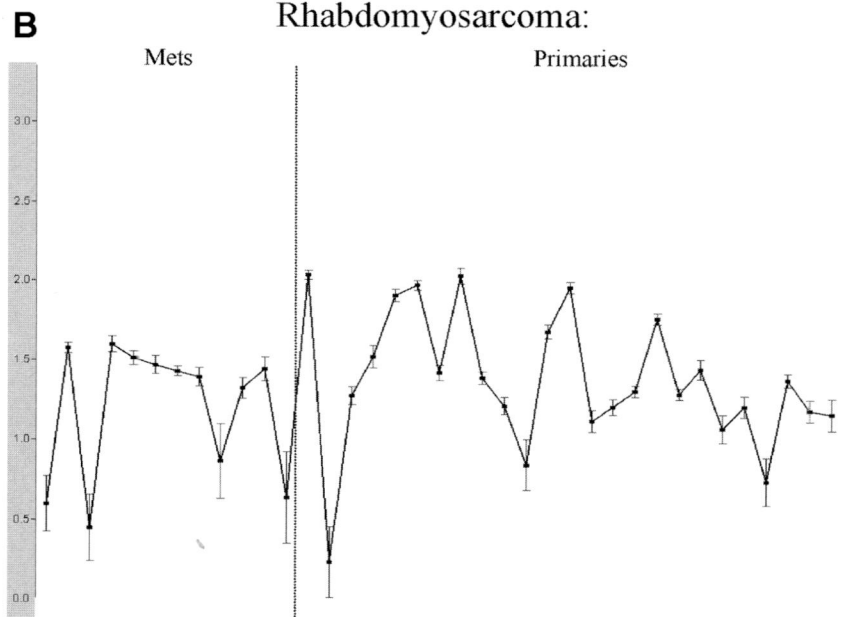

Fig. 7. Line graphs of *GAGE* expression in sarcomas. The average log expression of six members of the G-antigen (*GAGE*) family, in particular *GAGE-2, -3, -4, -5, -6*, and *-7*, are shown for individual primary and metastatic tumors. **(A)** *GAGE* expression is dramatically increased in five of six metastatic osteosarcomas (Mets), vs 25 of 26 primary lesions, and is barely detectable in a normal muscle sample (m). **(B)** *GAGE* family genes are not differentially expressed between primary and metastatic rhabdomyosarcomas, even though detectable levels are present across the tumor system.

distinguish individual *GAGE* family genes, despite a single base difference in the coding region in question. Note that for comparison, normal muscle expresses this same group of antigens at barely detectable levels (e.g., log approx 1.2), consistent with published data *(90,91)*. Notably, this is also comparable to the expression level for these same genes in primary osteosarcomas and markedly less than virtually all metastases (note error bars). Interestingly, this pattern of elevated expression in metastases is true of osteosarcoma but not rhabdomyosarcoma. Figure 7B shows no significant difference in expression levels between metastatic and primary rhabdomyosarcoma, despite the readily detectable levels of *GAGE* expression overall in this tumor system, as previously reported *(92)*.

The *GAGE* genes are part of a *GAGE*-like superfamily of cancer–testis genes, including, but not limited to, *MAGE, BAGE, PAGE, SSX, ESO,* and *XAGE*. Peptides derived from these proteins bind to membrane-bound class I major histocompatibility complex (MHC) molecules and are thus recognizable by cytotoxic T cells. The *GAGE* genes have been localized to chromosome X, band p11.2-11.3, and their expression is largely restricted to tumors and gametogenic tissue. Of interest, expression of *GAGE* and/or *GAGE*-like family genes have been reported in a variety of human cancers, including mesothelioma, Ewing's sarcoma, melanoma, lymphoma, and neuroblastoma *(92,93)*. Their presence has been used to document the presence of minimal residual disease, and they have been put forth as potential targets for antigen-based immunotherapy *(93)*. In addition, it has been shown that DNA hypomethylating agents can up-regulate the expression level of some of these antigens and may, therefore, be entertained as ancillary agents when devising these innovative strategies *(94)*. The real interest in the preferential expression of these genes is that they may represent ideal immunotherapy targets, as cytolytic T cells (CTLs) have been documented in tumors expressing *GAGE* or other testis-related antigens. Thus, the striking expression of *GAGE* genes seen in the metastases studied here suggests an immunotherapy approach to osteosarcoma metastases might be useful. This is potentially of great clinical interest, given the paucity of therapeutic options for patients with pulmonary metastases, the most common cause of death in this disease.

VALIDATION STUDIES

The intricacies of microarray technologies, combined with the tremendous spectrum and inherent complexity of the data, provide ample opportunity to arrive at misleading conclusions. While methodologic and technical modifications have improved data quality, validation of final results remains a requisite part of any study. The primary methods used to validate gene expression studies are immunohistochemistry (IHC) and real-time quantitative RT-PCR (Q-RT-PCR), with the two technologies providing unique, but overlapping and complementary information. Q-RT-PCR assesses the level of transcript expression at the RNA level (similar to oligonucleotide and cDNA expression arrays), while IHC assesses the level of expression at the protein level. IHC, in contrast to Q-RT-PCR, provides information regarding the cellular localization of this expressed protein within a given tumor sample and often allows important distinctions not possible by extractive methods such as Q-RT-PCR, notably tumor cell vs stromal cell contributions to overall gene expression profiles. *In situ* hybridization on tissue sections can localize mRNA, similarly to IHC, but is less often employed due to its perceived greater complexity and the relative instability of mRNA as opposed to protein in tissue sections. Q-RT-PCR is the most common validation method employed and can be performed by a limited number of different methods, all yielding relatively comparable and reliable results if understood and handled appropriately *(95)*.

Two of the expression studies performed on model systems described earlier used Q-RT-PCR as a confirmatory assay, while the third used a combination of Northern analysis and immunohistochemical staining. While only limited studies have been performed to date on the genes identified in our studies, we have observed a consistent pattern (albeit semiquantitative at best) of gene expression when array-generated values are compared with Q-RT-PCR using SYBR®-green. In the case of the *GAGE* family genes preferentially identified in metastatic osteosarcomas as described above, two sets of consensus primers were used that amplified *GAGE-3, -4, -5, -6,* and *-7* and *GAGE-1, -2,* and *-8* collectively. *GAGE* expression threshold levels were normalized to an average of β-actin and glyceraldehyde-3-phosphate dehydrogenase (*GAPDH*) threshold values. For the array data, a mean value for the two groups corresponding to the real-time PCR sets was derived by simple averaging of the normalized values reported by the arrays. Although actual quantitation was not possible, there was a close correlation between the ratios of *GAGE* expression to GAPDH/β-actin values in each data set, such that the rank order (e.g., high to low) between the four tumors was recapitulated by both the array and real-time PCR data.

CONCLUSION

The studies reported here largely confirm an emerging consensus among those studying patterns of gene expression in cancer and other diseases. First, molecular classification of disease appears to be the simplest and most reliable use of this technology. Tumors, in particular, are readily classified in many useful ways that relate to histogenesis, behavior, and even prognosis, often using fewer than 100 genes chosen from 12,000–40,000. Thus, the promise of a new ontology of cancer based on molecular genetic characteristics is already within reach and is widely pursued across most forms of human cancer. Numerous reports have appeared and will continue to appear documenting the potential utility of gene expression profiles in cancer diagnosis *(77,96)*.

The use of this technology to predict outcomes, even from pretreatment biopsies, is potentially of great value in developing an effective treatment strategy. Clearly, if a given expression profile dictates that response, or lack thereof, to a given chemotherapeutic regimen, is pre-ordained, there is little to recommend a therapeutic approach that is doomed to failure (if nonresponsive) or excessively toxic (if responsive). There is already good evidence that this, too, is possible. Several authors have documented their ability to identify high and low risk patient cohorts, and these preliminary observations may dictate alternative therapeutic approaches as they are confirmed on larger prospective patient cohorts. Confirmation of this application of the technology will require several years and close cooperation with, for example, large cancer clinical cooperative group trials. Here, too, there is a burgeoning sense of optimism that this use of the technology is both feasible and may soon be within grasp *(97,98)*.

There is also good evidence to date that expression profiling will both identify potential therapeutic gene targets and document their response to newly developed small molecule therapies. The nearly universal awareness of the marked efficacy of

Gleevec (Imantinib) in chronic myelogenous leukemia and gastrointestinal stromal tumors (GIST) is but a prelude to a potential onslaught of new agents that target specific genes. Many such agents are in early stage clinical trials as this is written, and many more are under development. It is not unreasonable to imagine a time when therapies will be chosen on the basis of suitable expression of target genes. This is already becoming evident, for example, with GISTs, where only those patients with activating mutations of exon 11 in the c-KIT receptor have been found to respond to Gleevec therapy. Thus, screening of newly diagnosed patients for such mutations will likely determine their eligibility in the future. This paradigm will only expand in scope as more agents and more gene targets are identified *(99,100)*.

By far the greatest frustration associated with gene expression profiling is an increasing appreciation that simple tallies of genes in a given cohort often means little and may not even be reproducible, for instance, between laboratories and across technology platforms, as discussed earlier in this chapter. There is a considerable risk that the optimism attendant to the early success of classification and outcome prediction applications of gene expression profiling will be replaced by excessive pessimism in the face of seemingly unfathomable biologic complexity revealed by this same technology. In reality, this technology is still in its infancy, and appropriate tools to unravel this complexity have yet to emerge. There is little doubt that increasingly sophisticated bioinformatic tools, capable of mining vast datasets and ferreting out valid gene–gene interactions, will emerge. Even more exciting is the prospect that these same methods will ultimately discern reproducible patterns of gene function within gene networks that ultimately dictate cellular behavior and response to stimuli, such as host immune defense, response to chemotherapy, radiation sensitivity, and even ability to metastasize. Clearly, multiple sources of information will be required, likely leading to the integration of information from multiple technologies, including gene expression, gene activation (e.g., phosphorylation and related events), protein expression and interaction, and genomic polymorphisms that dictate gene function. This, of course, will exponentially increase the sophistication required of analytic tools to interpret these data, yet that is precisely the promise of this approach. If successful, gene expression profiling will be an integral part of a whole-genome analysis applied to many biological problems and disease settings that promises to truly unravel the biologic underpinnings of diseases like cancer. Gene expression profiling is a good beginning, but not an end in and of itself. The combination of technology, analytic methods, and most importantly, biomedical researchers and physicians receptive to this fundamentally different approach to biology will ultimately determine the success of this effort. Hopefully, the examples cited above will encourage others to explore this approach for their own purposes.

REFERENCES

1. Arndt, C. A. and Crist, W. M. (1999) Common musculoskeletal tumors of childhood and adolescence. *N. Engl. J. Med.* **341,** 342–352.
2. Taghian, A., de Vathaire, F., Terrier, P., et al. (1991) Long-term risk of sarcoma following radiation treatment for breast cancer. *Int. J. Radiat. Oncol. Biol. Phys.* **21,** 361–367.
3. Porter, D. E., Holden, S. T., Steel, C. M., Cohen, B. B., Wallace, M. R., and Reid, R. (1992) A significant proportion of patients with osteosarcoma may belong to Li-Fraumeni cancer families. *J. Bone Joint Surg. Br.* **74,** 883–886.

4. Varughese, M., Leavey, P., Smith, P., Sneath, R., Breatnach, F., and O'Meara, A. (1992) Osteogenic sarcoma and Rothmund Thomson syndrome. *J. Cancer Res. Clin. Oncol.* **118,** 389–390.
5. Drouin, C. A., Mongrain, E., Sasseville, D., Bouchard, H. L., and Drouin, M. (1993) Rothmund-Thomson syndrome with osteosarcoma. *J. Am. Acad. Dermatol.* **28,** 301–305.
6. Vennos, E. M. and James, W. D. (1995) Rothmund-Thomson syndrome. *Dermatol. Clin.* **13,** 143–150.
7. Pratt, C. B., Meyer, W. H., Luo, X., et al. (1997) Second malignant neoplasms occurring in survivors of osteosarcoma. *Cancer* **80,** 960–965.
8. Lopez-Ben, R., Pitt, M. J., Jaffe, K. A., and Siegal, G. P. (1999) Osteosarcoma in a patient with McCune-Albright syndrome and Mazabraud's syndrome. *Skeletal Radiol.* **28,** 522–526.
9. Anbari, K. K., Ierardi-Curto, L. A., Silber, J. S., et al. (2000) Two primary osteosarcomas in a patient with Rothmund-Thomson syndrome. *Clin. Orthop.* 213–223.
10. Ishikawa, Y., Miller, R. W., Machinami, R., Sugano, H., and Goto, M. (2000) Atypical osteosarcomas in Werner Syndrome (adult progeria). *Jpn. J. Cancer Res.* **91,** 1345–1349.
11. Lipton, J. M., Federman, N., Khabbaze, Y., et al. (2001) Osteogenic sarcoma associated with Diamond-Blackfan anemia: a report from the Diamond-Blackfan Anemia Registry. *J. Pediatr. Hematol. Oncol.* **23,** 39–44.
12. Tucker, M. A., D'Angio, G. J., Boice, J. D., Jr., et al. (1987) Bone sarcomas linked to radiotherapy and chemotherapy in children. *N. Engl. J. Med.* **317,** 588–593.
13. Newton, W. A., Jr., Meadows, A. T., Shimada, H., Bunin, G. R., and Vawter, G. F. (1991) Bone sarcomas as second malignant neoplasms following childhood cancer. *Cancer* **67,** 193–201.
14. Dahlin, D. C. and Coventry, M. B. (1967) Osteogenic sarcoma. A study of six hundred cases. *J. Bone Joint Surg. Am.* **49,** 101–110.
15. Sim, F. H., Unni, K. K., Beabout, J. W., and Dahlin, D. C. (1979) Osteosarcoma with small cells simulating Ewing's tumor. *J. Bone Joint Surg. Am.* **61,** 207–215.
16. Devaney, K., Vinh, T. N., and Sweet, D. E. (1993) Small cell osteosarcoma of bone: an immunohistochemical study with differential diagnostic considerations [see comments]. *Hum. Pathol.* **24,** 1211–1225.
17. Nakajima, H., Sim, F. H., Bond, J. R., and Unni, K. K. (1997) Small cell osteosarcoma of bone. Review of 72 cases. *Cancer* **79,** 2095–2106.
18. Unni, K. K. and Dahlin, D. C. (1984) Grading of bone tumors. *Semin. Diagn. Pathol.* **1,** 165–172.
19. Huvos, A. G., Rosen, G., Bretsky, S. S., and Butler, A. (1982) Telangiectatic osteogenic sarcoma: a clinicopathologic study of 124 patients. *Cancer* **49,** 1679–1689.
20. Rosen, G., Huvos, A. G., Marcove, R., and Nirenberg, A. (1986) Telangiectatic osteogenic sarcoma. Improved survival with combination chemotherapy. *Clin. Orthop.* 164–173.
21. Campanacci, M., Picci, P., Gherlinzoni, F., Guerra, A., Bertoni, F., and Neff, J. R. (1984) Parosteal osteosarcoma. *J. Bone Joint. Surg. Br.* **66,** 313–321.
22. Unni, K. K., Dahlin, D. C., Beabout, J. W., and Ivins, J. C. (1976) Parosteal osteogenic sarcoma. *Cancer* **37,** 2466–2475.
23. Campanacci, M. and Giunti, A. (1976) Periosteal osteosarcoma. Review of 41 cases, 22 with long-term follow-up. *Ital. J. Orthop. Traumatol.* **2,** 23–35.
24. Schajowicz, F., McGuire, M. H., Santini Araujo, E., Muscolo, D. L., and Gitelis, S. (1988) Osteosarcomas arising on the surfaces of long bones. *J. Bone Joint Surg. Am.* **70,** 555–564.
25. Ritschl, P., Wurnig, C., Lechner, G., and Roessner, A. (1991) Parosteal osteosarcoma. 2-23-year follow-up of 33 patients. *Acta Orthop. Scand.* **62,** 195–200.
26. Rosen, G., Caparros, B., Huvos, A. G., et al. (1982) Preoperative chemotherapy for osteogenic sarcoma: selection of postoperative adjuvant chemotherapy based on the response of the primary tumor to preoperative chemotherapy. *Cancer* **49,** 1221–1230.

27. Glasser, D. B., Lane, J. M., Huvos, A. G., Marcove, R. C., and Rosen, G. (1992) Survival, prognosis, and therapeutic response in osteogenic sarcoma. The Memorial Hospital experience. *Cancer* **69,** 698–708.
28. Winkler, K., Bielack, S. S., Delling, G., Jurgens, H., Kotz, R., and Salzer-Kuntschik, M. (1993) Treatment of osteosarcoma: experience of the Cooperative Osteosarcoma Study Group (COSS). *Cancer Treat. Res.* **62,** 269–277.
29. Provisor, A. J., Ettinger, L. J., Nachman, J. B., et al. (1997) Treatment of nonmetastatic osteosarcoma of the extremity with preoperative and postoperative chemotherapy: a report from the Children's Cancer Group. *J. Clin. Oncol.* **15,** 76–84.
30. Bacci, G., Ferrari, S., Delepine, N., et al. (1998) Predictive factors of histologic response to primary chemotherapy in osteosarcoma of the extremity: study of 272 patients preoperatively treated with high-dose methotrexate, doxorubicin, and cisplatin. *J. Clin. Oncol.* **16,** 658–663.
31. Gorlick, R., Huvos, A. G., Heller, G., et al. (1999) Expression of HER2/erbB-2 correlates with survival in osteosarcoma. *J. Clin. Oncol.* **17,** 2781–2788.
32. Morris, C. D., Gorlick, R., Huvos, G., Heller, G., Meyers, P. A., and Healey, J. H. (2001) Human epidermal growth factor receptor 2 as a prognostic indicator in osteogenic sarcoma. *Clin. Orthop.* 59–65.
33. Thomas, D. G., Giordano, T. J., Sanders, D., Biermann, J. S., and Baker, L. (2002) Absence of HER2/neu gene expression in osteosarcoma and skeletal Ewing's sarcoma. *Clin. Cancer Res.* **8,** 788–793.
34. Maitra, A., Wanzer, D., Weinberg, A. G., and Ashfaq, R. (2001) Amplification of the HER-2/neu oncogene is uncommon in pediatric osteosarcomas. *Cancer* **92,** 677–683.
35. Akatsuka, T., Wada, T., Kokai, Y., et al. (2002) ErbB2 expression is correlated with increased survival of patients with osteosarcoma. *Cancer* **94,** 1397–1404.
36. Gurney, J. G., Davis, S., Severson, R. K., Fang, J.-Y., Ross, J. A., and Robison, L. L. (1996) Trends in cancer incidence among children in the U. S. *Cancer* **78,** 532–541.
37. Shah, S. H., Muzaffar, S., Soomro, I. N., Pervez, S., and Hasan, S. H. (1999) Clinico-morphological pattern and frequency of bone cancer. *J. Pak. Med. Assoc.* **49,** 110–112.
38. Bjornsson, J., McLeod, R. A., Unni, K. K., Ilstrup, D. M., and Pritchard, D. J. (1998) Primary chondrosarcoma of long bones and limb girdles. *Cancer* **83,** 2105–2119.
39. Uchida, Y., Kawai, A., Taguchi, K., Yokoi, T., Pu, J., and Inoue, H. (1996) Clinicopathology of chondrosarcoma. *Acta Med. Okayama* **50,** 191–196.
40. Dorfman, H. D. and Czerniak, B. (1995) Bone cancers. *Cancer* **75,** 203–210.
41. Balcer, L. J., Galetta, S. L., Cornblath, W. T., and Liu, G. T. (1999) Neuro-ophthalmologic manifestations of Maffucci's syndrome and Ollier's disease. *J. Neuroophthalmol.* **19,** 62–66.
42. Ramina, R., Coelho Neto, M., Meneses, M. S., and Pedrozo, A. A. (1997) Maffucci's syndrome associated with a cranial base chondrosarcoma: case report and literature review. *Neurosurgery* **41,** 269–272.
43. Damron, T. A., Sim, F. H., and Unni, K. K. (1996) Multicentric chondrosarcomas. *Clin. Orthop.* 211–219.
44. Brazier, D. J., Roberts-Harry, J., and Crockard, A. (1993) Intracavernous chondrosarcoma associated with Ollier's disease. *Br. J. Ophthalmol.* **77,** 599–600.
45. Asirvatham, R., Rooney, R. J., and Watts, H. G. (1991) Ollier's disease with secondary chondrosarcoma associated with ovarian tumour. A case report. *Int. Orthop.* **15,** 393–395.
46. Lucas, D., Tupler, R., and Enneking, W. F. (1990) Multicentric chondrosarcomas associated with Ollier's disease. Review and case report. *J. Fla. Med. Assoc.* **77,** 24–28.
47. Bushe, K. A., Naumann, M., Warmuth-Metz, M., Meixensberger, J., and Muller, J. (1990) Maffucci's syndrome with bilateral cartilaginous tumors of the cerebellopontine angle. *Neurosurgery* **27,** 625–628.

48. Schwartz, H. S., Zimmerman, N. B., Simon, M. A., Wroble, R. R., Millar, E. A., and Bonfiglio, M. (1987) The malignant potential of enchondromatosis. *J. Bone Joint Surg. Am.* **69**, 269–274.
49. Cannon, S. R. and Sweetnam, D. R. (1985) Multiple chondrosarcomas in dyschondroplasia (Ollier's disease). *Cancer* **55**, 836–840.
50. Rizzo, M., Ghert, M. A., Harrelson, J. M., and Scully, S. P. (2001) Chondrosarcoma of bone: analysis of 108 cases and evaluation for predictors of outcome. *Clin. Orthop.* 224–233.
51. Kreicbergs, A., Slezak, E., and Soderberg, G. (1981) The prognostic significance of different histomorphologic features in chondrosarcoma. *Virchows Arch. A. Pathol. Anat. Histol.* **390**, 1–10.
52. Sanerkin, N. G (1980) The diagnosis and grading of chondrosarcoma of bone: a combined cytologic and histologic approach. *Cancer* **45**, 582–594.
53. Evans, H. L., Ayala, A. G., and Romsdahl, M. M. (1977) Prognostic factors in chondrosarcoma of bone: a clinicopathologic analysis with emphasis on histologic grading. *Cancer* **40**, 818–831.
54. Lee, F. Y., Mankin, H. J., Fondren, G., et al. (1999) Chondrosarcoma of bone: an assessment of outcome. *J. Bone Joint Surg. Am.* **81**, 326–338.
55. Scanlon, P. W. (1972) Split-dose radiotherapy for radioresistant bone and soft tissue sarcoma: ten years' experience. *Am. J. Roentgenol. Radium Ther. Nucl. Med.* **114**, 544–552.
56. Oshiro, Y., Chaturvedi, V., Hayden, D., et al. (1998) Altered p53 is associated with aggressive behavior of chondrosarcoma: a long term follow-up study. *Cancer* **83**, 2324–2334.
57. Bustin, S. A. and Dorudi, S. (2002) The value of microarray techniques for quantitative gene profiling in molecular diagnostics. *Trends Mol. Med.* **8**, 269–272.
58. Schadt, E. E., Li, C., Ellis, B., and Wong, W. H. (2001) Feature extraction and normalization algorithms for high-density oligonucleotide gene expression array data. *J. Cell Biochem. Suppl.* **37**, 120–125.
59. Schofield, D. and Triche, T. J. (2002) cDNA microarray analysis of global gene expression in sarcomas. *Curr. Opin. Oncol.* **14**, 406–411.
60. Triche, T. J., Schofield, D., and Buckley, J. (2001) DNA microarrays in pediatric cancer. *Cancer J.* **7**, 2–15.
61. Nielsen, T. O., West, R. B., Linn, S. C., et al. (2002) Molecular characterisation of soft tissue tumours: a gene expression study. *Lancet* **359**, 1301–1307.
62. Macgregor, P. F. and Squire, J. A. (2002) Application of microarrays to the analysis of gene expression in cancer. *Clin. Chem.* **48**, 1170–1177.
63. Baelde, H. J., Cleton-Jansen, A. M., van Beerendonk, H., Namba, M., Bovee, J. V., and Hogendoorn, PC. (2001) High quality RNA isolation from tumours with low cellularity and high extracellular matrix component for cDNA microarrays: application to chondrosarcoma. *J. Clin. Pathol.* **54**, 778–782.
64. Ying, S. Y., Lui, H. M., Lin, S. L., and Chuong, C. M. (1999) Generation of full-length cDNA library from single human prostate cancer cells. *BioTechniques* **27**, 410–412, 414.
65. Lin, S. L., Chuong, C. M., Widelitz, R. B., and Ying, S. Y. (1999) In vivo analysis of cancerous gene expression by RNA-polymerase chain reaction. *Nucleic Acids Res.* **27**, 4585–4589.
66. Sotiriou, C., Khanna, C., Jazaeri, A. A., Petersen, D., and Liu, E. T. (2002) Core biopsies can be used to distinguish differences in expression profiling by cDNA microarrays. *J. Mol. Diagn.* **4**, 30–36.
67. Wolf, M., El-Rifai, W., Tarkkanen, M., et al. (2000) Novel findings in gene expression detected in human osteosarcoma by cDNA microarray. *Cancer Genet. Cytogenet.* **123**, 128–132.
68. Fuchs, B., Zhang, K., Schabel, A., Bolander, M. E., and Sarkar, G. (2001) Identification of twenty-two candidate markers for human osteogenic sarcoma. *Gene* **278**, 245–252.

69. Khanna, C., Khan, J., Nguyen, P., et al. (2001) Metastasis-associated differences in gene expression in a murine model of osteosarcoma. *Cancer Res.* **61,** 3750–3759.
70. Ross, D. T., Scherf, U., Eisen, M. B., et al. (2000) Systematic variation in gene expression patterns in human cancer cell lines. *Nat. Genet.* **24,** 227–235.
71. Quackenbush, J. (2001) Computational analysis of microarray data. *Nat. Rev. Genet.* **2,** 418–427.
72. Sturn, A., Quackenbush, J., and Trajanoski, Z. (2002) Genesis: cluster analysis of microarray data. *Bioinformatics* **18,** 207–208.
73. Brazma, A., Hingamp, P., Quackenbush, J., et al. (2001) Minimum information about a microarray experiment (MIAME)-toward standards for microarray data. *Nat. Genet.* **29,** 365–371.
74. Hegde, P., Qi, R., Abernathy, K., et al. (2000) A concise guide to cDNA microarray analysis. *BioTechniques* **29,** 548–556.
75. Golub, T. R., Slonim, D. K., Tamayo, P., et al. (1999) Molecular classification of cancer: class discovery and class prediction by gene expression monitoring. *Science* **286,** 531–537.
76. Sorlie, T., Perou, C. M., Tibshirani, R., et al. (2001) Gene expression patterns of breast carcinomas distinguish tumor subclasses with clinical implications. *Proc. Natl. Acad. Sci. USA* **98,** 10,869–10,874.
77. Ramaswamy, S., Tamayo, P., Rifkin, R., et al. (2001) Multiclass cancer diagnosis using tumor gene expression signatures. *Proc. Natl. Acad. Sci. USA* **98,** 15,149–15,154..
78. Uusitalo, H., Hiltunen, A., Soderstrom, M., Aro, H. T., and Vuorio, E. (2000) Expression of cathepsins B, H, K, L, and S and matrix metalloproteinases 9 and 13 during chondrocyte hypertrophy and endochondral ossification in mouse fracture callus. *Calcif. Tissue Int.* **67,** 382–390.
79. Rudland, P. S., Platt-Higgins, A., Renshaw, C., et al. (2000) Prognostic significance of the metastasis-inducing protein S100A4 (p9Ka) in human breast cancer. *Cancer Res.* **60,** 1595–1603
80. Otte, M., Zafrakas, M., Riethdorf, L., et al. (2001) MAGE-A gene expression pattern in primary breast cancer. *Cancer Res.* **61,** 6682–6687.
81. Fuller, G. N., Hess, K. R., Rhee, C. H., et al. (2002) Molecular classification of human diffuse gliomas by multidimensional scaling analysis of gene expression profiles parallels morphology-based classification, correlates with survival, and reveals clinically-relevant novel glioma subsets. *Brain Pathol.* **12,** 108–116.
82. Iwao, K., Matoba, R., Ueno, N., et al. (2002) Molecular classification of primary breast tumors possessing distinct prognostic properties. *Hum. Mol. Genet.* **11,** 199–206.
83. van 't Veer, L. J., Dai, H., van de Vijver, M. J., et al. (2002) Gene expression profiling predicts clinical outcome of breast cancer. *Nature* **415,** 530–536.
84. Volm, M., Koomagi, R., Mattern, J., and Efferth, T. (2002) Expression profile of genes in non-small cell lung carcinomas from long-term surviving patients. *Clin. Cancer Res.* **8,** 1843–1848.
85. Stratowa, C., Loffler, G., Lichter, P., et al. (2001) CDNA microarray gene expression analysis of B-cell chronic lymphocytic leukemia proposes potential new prognostic markers involved in lymphocyte trafficking. *Int. J. Cancer* **91,** 474–480.
86. Lakhani, S. R. and Ashworth, A. (2001) Microarray and histopathological analysis of tumours: the future and the past? *Nat. Rev. Cancer* **1,** 151–157.
87. Pomeroy, S. L., Tamayo, P., Gaasenbeek, M., et al. (2002) Prediction of central nervous system embryonal tumour outcome based on gene expression. *Nature* **415,** 436–442.
88. Singh, D., Febbo, P. G., Ross, K., et al. (2002) Gene expression correlates of clinical prostate cancer behavior. *Cancer Cell* **1,** 203–209.
89. Chambers, A. F., Groom, A. C., and MacDonald, I. C. (2002) Metastasis: Dissemination and growth of cancer cells in metastatic sites. *Nat. Rev. Cancer* **2,** 563–572.

90. De Backer, O., Arden, K. C., Boretti, M., et al. (1999) Characterization of the GAGE genes that are expressed in various human cancers and in normal testis. *Cancer Res.* **59,** 3157–3165.
91. Van den Eynde, B., Peeters, O., De Backer, O., Gaugler, B., Lucas, S., and Boon, T. (1995) A new family of genes coding for an antigen recognized by autologous cytolytic T lymphocytes on a human melanoma. *J. Exp. Med.* **182,** 689–698.
92. Dalerba, P., Frascella, E., Macino, B., et al. (2001) MAGE, BAGE and GAGE gene expression in human rhabdomyosarcomas. *Int. J. Cancer* **93,** 85–90.
93. Cheung, I. Y. and Cheung, N. K. (2001) Detection of microscopic disease: comparing histology, immunocytology, and RT-PCR of tyrosine hydroxylase, GAGE, and MAGE. *Med. Pediatr. Oncol.* **36,** 210–212.
94. Sigalotti, L., Coral, S., Altomonte, M., et al. (2002) Cancer testis antigens expression in mesothelioma: role of DNA methylation and bioimmunotherapeutic implications. *Br. J. Cancer* **86,** 979–982.
95. Ginzinger, D. G. (2002) Gene quantification using real-time quantitative PCR. An emerging technology hits the mainstream. *Exp. Hematol.* **30,** 503–512.
96. Yeang, C. H., Ramaswamy, S., Tamayo, P., et al. (2001) Molecular classification of multiple tumor types. *Bioinformatics* **17(Suppl 1),** S316–S322.
97. Staunton, J. E., Slonim, D. K., Coller, H. A., et al. (2001) Chemosensitivity prediction by transcriptional profiling. *Proc. Natl. Acad. Sci. USA* **98,** 10,787–10,792.
98. Butte, A. J., Tamayo, P., Slonim, D., Golub, T. R., and Kohane, I. S. (2000) Discovering functional relationships between RNA expression and chemotherapeutic susceptibility using relevance networks. *Proc. Natl. Acad. Sci. USA* **97,** 12,182–12,186.
99. Marton, M. J., DeRisi, J. L., Bennett, H. A., et al. (1998) Drug target validation and identification of secondary drug target effects using DNA microarrays. *Nat. Med.* **4,** 1293–1301.
100. Huang, S. (2001) Genomics, complexity and drug discovery: insights from Boolean network models of cellular regulation. *Pharmacogenomics* **2,** 203–222.

Index

A

Acute leukemias, *see* Leukemia
Anaplastic astrocytoma, *see* Brain tumor
Angiosarcoma, *see* Soft tissue tumor
APC, colorectal cancer mutations, 151

B

BCL2,
    diffuse large B cell lymphoma translocation, 335, 336
    prostate cancer profiling, 176
Biopathology Centers (BPC),
    cooperative groups, 103
    data management, 109
    facilities,
        histology laboratory, 107
        processing and distribution laboratories, 106
        space, 106
        storage facility, 106, 107
    funding sources, 104
    informatics,
        centralized inventory system, 111, 112
        common data analysis, 110
        design objectives, 109
        informed consent and confidentiality, 110
        linking and delinking of data, 110, 111
        minimum data requirement for banking, 110
        programming standards, 109
        virtual private networks, 112
    performance metrics,
        investigator profiles and publications, 114
        molecular assessment, 113, 114
        morphologic assessment, 112, 113
    specimens,
        procurement kits, 107
        storage, 107
        transport, 108, 109
    tissue sources, 104
Bladder cancer profiling,
    approaches, 219, 220, 222
    cell line studies, 226, 227
    classification of tumors, 223–226
    cDNA microarray applications, 224, 225
    GeneChip, 228, 229
    hierarchical clustering, 226
    lesion types, 220
    molecular pathology, 220, 222
    prospects, 229, 230
    samples,
        pooling, 226
        preparation, 222, 223
    single nucleotide polymorphism genotyping, 228, 229
    tissue microarrays, 227, 228
Bone tumors, *see* Chondrosarcoma; Ewing's sarcoma profiling; Osteosarcoma
BPC, *see* Biopathology Centers
Brain tumor,
    anaplastic astrocytoma, 346
    developmental aspects of normal brain, 345, 346
    ependymoma, 348
    epidemiology, 346
    extraparenchymal tumors, 349
    gene expression profiling,
        classification of tumors, 352
        control tissue, 350
        detection and diagnosis, 351
        heterogeneity of tumor, 350, 351
        limitations, 354
        prospects, 354
        therapeutic response, 352
        tissue collection and processing, 349, 350
        validation, 351
    glioblastoma multiforme,
        classes, 346, 347
        gene expression profiling, 352, 354
        histology, 347
        magnetic resonance imaging, 347
        serial analysis of gene expression, 55
    gliomas, 346–348
    medulloblastomas, 55, 348, 354
    metastatic tumors, 349, 352
    oligodendroglioma, 347
    pediatric brain tumor profiling,
        clinical applications, 295–297
        prognostic value, 300, 301

Breast cancer profiling,
  axillary lymph node involvement in prognosis, 121, 137, 138
  biospecimens and patient collections, 122, 123
  chemotherapy effects on expression patterns, 136, 137
  classification based on expression,
    BRCA1 mutation status, 136
    BRCA2 mutation status, 136
    ERBB2, 132, 133
    estrogen receptor status, 121, 122, 135, 136, 139
    expression cluster characteristics, 132
    p53 mutation status, 133
  clinical outcomes, 137
  clustering of subsets, 5
  cultured cells and cancer cell lines, 130–132
  prospects, 139
  serial analysis of gene expression, 56, 57
  statistical analysis, 123, 124
  summary of microarray analyses, 123
  tissue microarray studies, 69, 71
  validity assessment,
    molecular validation,
      immunohistochemistry, 138
      Northern blot, 1238
      tissue arrays, 138
    reproducibility, 124, 125
    statistical classification, 125–130
Burkitt's lymphoma profiling,
  clinical applications, 295–297
  transcription factor target identification, 299, 300
  upregulated genes, 297–299
  validation, 297–299

C

Cancer Genome Anatomy Project (CGAP), resources, 300
cDNA microarrays, see Complementary DNA microarrays
Chondrosarcoma,
  clinical presentation, 361
  epidemiology, 360, 361
  gene expression profiling,
    animal model studies, 366
    bioinformatics, 363
    cell line studies, 365, 366
    clinical sample limitations, 365
    data analysis, 362, 363
    diagnosis, 361, 362
    GeneChip, 363, 366
    molecular classification, 366, 367, 369, 385
    prognosis, 362
    prospects, 385
    tissue processing, 363–365
  grades, 361
  treatment, 361
CGAP, see Cancer Genome Anatomy Project
Chronic lymphocytic leukemia (CLL),
  forms, 339
  signature genes, 339, 340
CIT, see Cluster Identification Tool
Class prediction, see Microarray data analysis
Clinical implementation, tumor profiling data, 5, 6
CLL, see Chronic lymphocytic leukemia
Cluster Identification Tool (CIT), clear cell renal cell carcinoma aggressive versus nonaggressive class clustering, 244, 246, 247
Clustering, see Microarray data analysis
Colorectal cancer profiling,
  adenocarcinoma,
    overexpressed genes, 154, 155, 157, 164
    underexpressed genes, 155, 158, 159, 164
  adenoma analysis, 156, 160, 161, 164
  clinical manifestations, 149, 150
  colon layers and microscopic anatomy, 148, 149
  coupled two-way hierarchical clustering, 163, 164
  dendrogram, 156, 162
  DNA versus RNA levels, 147, 148, 166
  dysregulated gene abundance, 154, 155
  epidemiology, 147
  GeneChip, 153, 154, 156
  heterogeneity of samples, 165, 166
  metabolic genes, 154, 155
  molecular pathogenesis, 150–153
  overview, 147, 148
  prospects, 166
  serial analysis of gene expression, 53–55, 153, 163, 164
  staging, 150
Complementary DNA (cDNA) microarrays,
  cell-type effects, 16
  cDNA libraries,
    gene sampling, 12
    sequence authenticity, 12, 13
    size, 13
  data, see also Microarray data analysis,
    analysis, 16–18
    dissemination, 19
    flow, 15
    validation, 18
  fluorescence detection, 14, 16
  gene list interpretation, 18, 19
  hybridization, 14, 16

Index

Lymphochip, 332
microarray construction, 13, 14
principles, 11
tumor expression profiling, *see specific tumors*
Coupled two-way hierarchical clustering (CTWC), colorectal cancer profiling, 163, 164
CTWC, *see* Coupled two-way hierarchical clustering

D

Data analysis, *see* Microarray data analysis
Dermatofibrosarcoma protuberans, *see* Soft tissue tumor
Desmoid tumors, *see* Soft tissue tumor
Diffuse large B cell lymphoma (DLBCL),
  prognostic subsets, 5
  complementary DNA microarray analysis,
    bcl-2 translocation, 335, 336
    comparison with other lymphomas and leukemias, 332
    germinal center B cell signature, 332, 334
    Lymphochip, 332
    nuclear factor-κB signaling, 336
    prognostic marker identification, 337, 338
    subgroup comparison and validation, 334–337
    survival, 334, 342
  gene discovery, 340, 341
  host interaction with lymphoma, 341, 342
DLBCL, *see* Diffuse large B cell lymphoma
DNA microarrays, *see* Complementary DNA microarrays; Oligonucleotide microarrays

E

Ependymoma, *see* Brain tumor
ER, *see* Estrogen receptor
ERBB2, breast cancer profiling and classification, 132, 133
Estrogen receptor (ER), breast cancer profiling, 121, 122, 135, 136, 139
ESTs, *see* Expressed sequence tags
Ewing's sarcoma profiling,
  clinical applications, 295–297
  transcription factor target identification, 299, 300
  upregulated genes, 297–299
  validation, 297–299
Expressed sequence tags (ESTs), generation, 47

G

GAGE, expression in osteosarcoma metastasis, 382, 384
Gastrointestinal stromal tumor, *see* Soft tissue tumor

GeneChip, *see also* Oligonucleotide microarrays,
  bladder cancer profiling, 228, 229
  bone tumor profiling, 363, 366
  colorectal cancer profiling, 153, 154, 156
  data analysis, 39–41
  density of features, 32
  design of arrays, 29–31
  lung cancer profiling, 203, 204
  pancreatic cancer profiling, 266–268
  performance characteristics, 26, 27, 31
  prostate cancer profiling, 183, 184
  synthesis of arrays, 30–33
  Wilms tumor profiling, 302
Glioblastoma multiforme, *see* Brain tumor
Gliomas, *see* Brain tumor

I

IHC, *see* Immunohistochemistry
Immunohistochemistry (IHC),
  expression profiling, 4
  tissue arrays, *see* Tissue arrays
  validation of gene expression, 270, 286
In situ hybridization,
  tissue arrays, *see* Tissue arrays
  validation of gene expression, 270, 271
ISH, *see In situ* hybridization

K

k-mean clustering, microarray data analysis, 40
Laser capture microdissection (LCM),
  breast tumors, 131
  colorectal cancer, 164
  limitations, 37
  ovarian cancer, 279
  tissue heterogeneity solution in oligonucleotide microarray profiling, 36, 37
LCM, *see* Laser capture microdissection
Leiomyosarcoma, *see* Soft tissue tumor
Leukemia, *see also* Chronic lymphocytic leukemia,
  pediatric acute leukemia profiling,
    acute lymphoblastic leukemia, 301, 302
    clinical applications, 295–297
    subclassification with supervised learning, 88, 89
Lung cancer profiling,
  animal models, 201, 202
  carcinoma classification, 199, 203–206, 212
  cell lines, 202
  class discovery, 206–208
  epidemiology, 199, 200
  GeneChip, 203, 204
  heterogeneity of tumors, 201
  histopathology correlation, 210–212
  metastasis, 206, 211, 212

molecular signature of lung adenocarcinoma
subclasses, 208–210
prognostic value, 211–214
prospects, 214, 215
serial analysis of gene expression, 55, 56, 203
specimens,
handling, 201
sources, 200, 201
staging, 200
therapeutic target discovery, 212, 213
tumor suppressor gene mutations, 200
validation, 211
Lymphochip, complementary DNA microarray
analysis, 332, 340
Lymphoma, *see also* Burkitt's lymphoma profiling; Diffuse large B cell lymphoma,
classification systems, 329
gene expression profiling study design,
data analysis, 331
tissue heterogeneity, 330, 331
validation, 331, 332
host interaction with lymphoma, 341, 342
treatment outcome prediction with supervised
learning, 89, 92

M

Malignant fibrous hystiocytoma, *see* Soft tissue
tumor
MDX11, prostate cancer profiling, 176
Medulloblastomas, *see* Brain tumor
Metastasis,
brain tumors, 349, 352
lung cancer profiling, 206, 211, 212
osteosarcoma genes,
clinical significance, 373
downregulated genes, 374, 375, 379–381
upregulated genes, 373–375, 377–379,
381, 382, 384
pancreatic cancer invasion cluster genes,
263–266
prostate cancer profiling, 185, 187–190
Microarray data analysis, *see also specific tumors*,
basic data analysis overview, 74, 75
clustering,
algorithms, 76, 77
breast cancer profiling, 125–130
hierarchical clustering, 78, 79
limitations, 78, 79
overview, 76, 77
self-organizing maps, 40, 78
time series data, 77, 78
filtering, 75
goals, 73

molecular classification of cancer, 96, 97
normalization, 75
oligonucleotide microarrays, 39–41, 75
principal components analysis, 95
prospects, 96, 97
raw data quality control, 75
scaling, 75
supervised learning,
applications,
leukemia subclassification, 88, 89
lymphoma treatment outcome prediction,
89, 92
multiple tumor type classification, 92–95
breast cancer profiling, 125–130
class prediction, 3, 4, 85, 86
classification algorithms, 86
gene marker selection and validation, 80
overview, 76, 79, 80
pattern discovery, 85
permutation tests, 80, 82, 83, 85
statistical significance of supervised
classifier, 86–88
thresholding, 75

N

Neuroblastoma profiling,
clinical applications, 295–297
serial analysis of gene expression, 299
transcription factor target identification,
299, 300
upregulated genes, 297–299
validation, 297–299
NF-κB, *see* Nuclear factor-κB
Nuclear factor-κB (NF-κB), diffuse large B cell
lymphoma subgroup signaling, 336

O

Oligodendroglioma, *see* Brain tumor
Oligonucleotide microarrays, *see also* GeneChip,
applications in tumor profiling, 23, 27, 28
components, 26
costs, 41
data analysis, 39–41, *see also* Microarray data
analysis
density of features, 32
design, 28–31
expression assay format, 37, 38
high-throughput, 37, 39
hybridization, 37
principles, 25, 26
prospects for clinical use, 41, 42
quantitative analysis, 26, 31
reproducibility, 26

sample preparation, 33, 34
signal detection, 37
specificity, 26
synthesis of arrays, 30–33
target preparation, 37
tissue heterogeneity,
  complicating factors, 34, 35
  solutions, 35–37
tumor expression profiling, *see specific tumors*
Osteosarcoma,
  etiology, 359
  familial syndromes, 359
  gene expression profiling,
    animal model studies, 366
    bioinformatics, 363
    cell line studies, 365, 366
    clinical sample limitations, 365
    data analysis, 362, 363
    diagnosis, 361, 362
    GeneChip, 363, 366
    metastasis-associated genes,
      clinical significance, 373
      downregulated genes, 374, 375, 379–381
      upregulated genes, 373–375, 377–379, 381, 382, 384
    molecular classification, 366, 367, 369
    prognosis, 362, 371–373, 385
    prospects, 385
    small cell osteosarcoma, 369, 371
    table of genes, 358
    therapeutic target identification, 385, 386
    tissue processing, 363–365
    validation, 384, 385
  histology, 360
  origins, 359
  treatment, 360
Ovarian cancer profiling,
  cell lines, 284–286
  class discovery, 289, 290
  class prediction, 287–289
  diagnostic markers, 291
  downregulated genes, 280, 283
  epidemiology, 277
  histology and molecular correlates, 290, 291
  molecular pathogenesis, 278, 279
  overview of studies, 280, 281
  prognosis, 277, 278
  prospects, 291, 292
  serial analysis of gene expression, 56, 280
    specimen processing, 279, 280
    specimen procurement, 279
  upregulated genes, 280, 282, 284, 287–289
  validation, 280, 284, 286

P

p53
  bladder cancer profiling, 228, 229
  breast cancer profiling using mutation status, 133
  prostate cancer profiling, 176
Pancreatic cancer profiling,
  applications,
    diagnosis, 271
    prognosis, 271, 272
    treatment, 271
  cell lines, 260
  complementary DNA microarrays, 266
  epidemiology, 257
  GeneChip, 266–268
  heterogeneity of samples, 260, 262
  immunohistologic markers, 258, 259
  invasion cluster genes, 263–266
  molecular pathogenesis, 257, 258
  prognosis, 257
  prospects, 272
  ProteinChip, 268, 270
  serial analysis of gene expression, 53–55, 262–266, 271
  validation of gene expression, 270, 271
PCA, *see* Principal components analysis
Pediatric cancer, *see specific tumors*
Permutation test, microarray data analysis, 80, 82, 83, 85
Photolithography, oligonucleotide microarray synthesis, 32, 33
Principal components analysis (PCA), microarray data analysis, 95
Prostate cancer profiling,
  anatomy, 174
  androgen-independent cancer, 176, 177, 187, 191, 192
  androgens,
    ablation therapy response, 192
    development and progression role, 173, 174, 176, 177
    expression analysis, 187, 191
  animal models, 177
  BCL2, 176
  complementary DNA microarrays, 182, 183
  data analysis, 181
  epidemiology, 173, 175
  GeneChip, 183, 184
  heterogeneity of tissue, 178, 179
  histogenesis, 174
  histologic grade correlations, 184
  loss of heterozygosity, 175
  MDX11, 176

metastasis, 185, 187–190
normal gene expression, 182
p53, 176
prospects, 192, 194
PTEN mutations, 175, 176
serial analysis of gene expression, 57, 179, 182
transcript profiling methods, 179, 181
ProteinChip,
    pancreatic cancer profiling, 268, 270
    principles, 268
PTEN, prostate cancer mutations, 175, 176

R

RCC, *see* Renal cell carcinoma
Renal cell carcinoma (RCC),
    cDNA microarray analysis,
        applications, 237
        cell lines, 237
        clear cell tumors,
            aggressive versus nonaggressive class clustering, 244, 246, 247
            downregulated genes, 240–242
            gene ontology classification, 242–244
            overview, 238, 239
            upregulated genes, 239, 240
        comparison of histological subtypes, 238
        prospects, 250, 251
    epidemiology, 235
    histological subtypes, 235
    management, 236
    molecular pathogenesis, 235, 236
    post-gene expression profiling studies, 251
    prognostic factors, 236, 237
    prognostic set of genes,
        clinical simulation test, 247–249
        gene types, 249
        survival analysis, 249, 250
Reverse transcription-polymerase chain reaction (RT-PCR), validation of gene expression, 270, 286, 384, 385
Rhabdomyosarcoma profiling, *see also* Soft tissue tumor,
    clinical applications, 295–297
    serial analysis of gene expression, 55
    transcription factor target identification, 299, 300
    upregulated genes, 297–299
    validation, 297–299
RT-PCR, *see* Reverse transcription-polymerase chain reaction

S

SAGE, *see* Serial analysis of gene expression
Sarcomas, *see* Soft tissue tumor

Self-organizing map (SOM), microarray data analysis, 40, 78
Serial analysis of gene expression (SAGE),
    advantages, 47, 48
    data analysis, 51
    disadvantages, 50
    human genome mining and annotation, 53
    pathway dissection in cancer, 57, 58
    principles, 48–50, 263
    resources, 51, 53
    technical advances, 53
    tumor expression profiling, *see specific tumors*
Soft tissue tumor, *see also* Rhabdomyosarcoma profiling,
    animal models and cell lines, 312
    chromosomal translocations, 307, 309, 313
    clinical classification, 305
    complementary DNA microarray analysis,
        clustering patterns,
            controversial cases, 319
            metastasis, 318, 319
            mitotic index correlation, 319
            primary tumors, 318
        dermatofibrosarcoma protuberans, 317, 318
        desmoid tumors, 318
        gastrointestinal stromal tumors, 313, 315, 317
        leiomyosarcoma, 313, 315, 317
        malignant fibrous hystiocytoma, 313
        prospects, 321
        solitary fibrous tumors, 318
        specimens,
            availability, 311
            biopsy types, 311
            heterogeneity of tissue, 311, 312
            processing, 311
        synovial sarcoma, 317, 318
        therapeutic target discovery, 305, 312
        validation with tissue microarrays, 319, 321
    diagnosis, 306, 307, 309
    grading, 310, 311
    histologic classification, 305, 306, 312, 313
    monoclonal antibodies, 306, 307
    prognostic variables, 309–311
    treatment, 310
Solitary fibrous tumors, *see* Soft tissue tumor
SOM, *see* Self-organizing map
Supervised learning, *see* Microarray data analysis
Synovial sarcoma, *see* Soft tissue tumor

T

TGFβRII, renal cell carcinoma expression and prognosis, 249

TIMP-3, renal cell carcinoma expression and
    prognosis, 249
Tissue arrays,
    breast cancer profiling validation, 138
    construction of tissue microarrays,
        block construction, 62, 63, 65, 66
        instrumentation, 65, 66
        overview, 62, 64
        punch diameter, 66
        sectioning, 66
        starting materials, 63
    data analysis, 71
    detection techniques, 61
    development, 62
    frozen arrays, 68, 69
    prospects, 71, 72
    soft tissue tumor complementary DNA
        microarray validation, 319, 321
    types and applications, 67, 68
    validation studies, 69, 71
TP53, *see p53*

Tumor bank,
    challenges,
        biohazards, 115
        informed consent and confidentiality,
            115, 116
        more research with less tissue, 114, 115
    design, *see* Biopathology Centers

U

Unsupervised learning, *see* Microarray data
    analysis

V

Virtual private network (VPN), tumor banking,
    112
VPN, *see* Virtual private network

W

Wilms tumor profiling,
    clinical applications, 295–297
    GeneChip, 302